Foundations of Electromagnetic Theory

THIRD EDITION

JOHN R. REITZ
Scientific Laboratory
Ford Motor Company

FREDERICK J. MILFORD
Battelle Memorial Institute

ROBERT W. CHRISTY
Dartmouth College

ADDISON-WESLEY PUBLISHING COMPANY

Reading, Massachusetts • Menlo Park, California
London • Amsterdam • Don Mills, Ontario • Sydney

This book is in the
Addison-Wesley Series in Physics

Second printing, June 1980

ISBN 0-201-06332-8
ABCDEFGHIJ-MA-89876543210

Preface

Although Maxwell's equations were formulated about one hundred years ago, the subject of electromagnetism has not remained static. Advanced undergraduate students in science, to whom we are directing our attention, today approach the subject with a qualitative understanding of atomic phenomena. At the same time, they have acquired a good background in mathematics and for the first time are in a position to solve some of the important problems of classical physics. The present volume evolved from the teaching of courses in electricity and magnetism to physics majors at Case Institute of Technology and at Dartmouth College. A course in electromagnetism is ideally suited to a development of the concepts of vector analysis, partial differential equations, and boundary-value problems. The sections involving these techniques are written in such a way that little previous knowledge of the subject is required.

We believe that building up electricity and magnetism from the basic experimental laws is the correct approach at the intermediate level, and we have followed this approach through a rigorous exposition of the fundamentals. We have also been careful to include a number of appropriate examples to bridge the gap between the formal development of the subject and the problems. A full understanding of the electric and magnetic fields inside matter can be obtained only after the atomic nature of materials is appreciated. Hence we have used elementary atomic concepts freely in the development of macroscopic theory.

We prefer to discuss the static electric field in a material medium immediately after the vacuum electric field, and we discuss the magnetostatic field similarly. The reader may, however, study both vacuum cases together before considering either electric or magnetic fields in matter, by postponing Chapters 4, 5, 6, 7 (except Sections 7.1 and 7.2), 9, and 10 until after reading Chapter 8 or even Chapter 11. The macroscopic electromagnetic behavior of dielectrics, conductors, magnetic materials, plasmas, and superconductors is treated in separate chapters (Chapters 4, 7, 9, 14, and 15, respectively). A simple discussion of the microscopic theory of these classes of matter (except superconductors) is also provided (in Chapters 5, 7, 10, and 14).

The third edition of the book is changed principally by the addition of more material on *electromagnetic waves*. The two old chapters on Maxwell's equations have been expanded into five chapters. The book is thus adaptable either to a one-semester course or to a two-semester course in which the second semester emphasizes propagation and generation of radiation.

Much of modern physics (and engineering) involves time-dependent electromagnetic fields in which Maxwell's displacement current plays a crucial role. Chapters 16 through 20 develop the application to waves—especially the connection with optics, which is the frequency range that is now succeeding microwaves in technological interest. Chapters 16 and 17 extend the old treatment of the wave equations, introducing the idea of gauge transformations. The notions of complex dielectric function and refractive index are emphasized, with resulting conceptual clarity and simplification of formulas. Chapter 18 expands the treatment of boundary-value problems, to include examples of interest in optical filters and waveguides. Chapter 19 gives the classical microscopic theory of transverse wave propagation in matter (dielectrics, metals, plasmas); it is an extension of Chapters 5 and 7 to time-dependent fields. It also includes a simple discussion of the Kramers-Kronig dispersion relations for a linear response function. Chapter 20, on the generation of radiation by antennas and accelerated charges, includes new material on induction fields, radiation damping, and Thomson scattering.

The material in the rest of the book has been slightly rearranged, so that the discussion of static fields and steady currents is completed before the introduction of Faraday's law of induction, in Chapter 11, followed by its application to slowly varying currents in a-c circuits, plasmas, and superconductors in Chapters 13, 14, and 15. The relativistic formulation of electromagnetism has been put at the end, although it could be read at any point after Chapter 16. Some relativistic aspects are anticipated in new treatments of the magnetic force (Chapter 8) and Faraday's law (Chapter 11).

Other changes from earlier editions include the introduction of the Dirac delta function in Chapter 2 and its use to simplify several later derivations. Orthogonal transformations are moved to an Appendix, which can be read in conjunction with Chapter 1 if desired. The del-operator notation is used for vector differentiation. All tables of data and references to other books have been updated, and SI units and notation are used systematically throughout. (Reference is also made, however, to the Gaussian units, since they are widely used in the current physics literature.) A summary section at the end of each chapter identifies key ideas and formulas, and about one hundred and thirty additional problems extend and apply the concepts.

As an aid to the reader, the more difficult problems are labeled with an asterisk. Sections and chapters of the text that are starred are not essential to its further development and may be omitted in an abbreviated study.

Dearborn, Michigan   J. R. R.
Columbus, Ohio   F. J. M.
Hanover, New Hampshire   R. W. C.
January 1979

Contents

Starred sections and chapters may be omitted without loss of continuity

CHAPTER 1 Vector Analysis

In the study of electricity and magnetism a great saving in complexity of notation may be accomplished by using the notation of vector analysis. In providing this valuable shorthand, vector analysis also brings to the forefront the physical ideas involved in equations. It is the purpose of this chapter to give a brief but self-contained exposition of basic vector analysis and to provide the rather utilitarian knowledge of the field which is required for a treatment of electricity and magnetism. Those already familiar with vector analysis will find it a useful review and an introduction to the notation of the text.

1–1 DEFINITIONS

In the study of elementary physics several kinds of quantities have been encountered; in particular, the division into vectors and scalars has been made. For our purposes it will be sufficient to define a scalar as follows:

A scalar is a quantity that is completely characterized by its magnitude.

Examples of scalars are numerous: mass, time, volume, etc. A simple extension of the idea of a scalar is a *scalar field*, i.e., a function of position that is completely specified by its magnitude at all points in space.

A vector may be defined as follows:

A vector is a quantity that is completely characterized by its magnitude and direction.

As examples of vectors we cite position from a fixed origin, velocity, acceleration, force, etc. The generalization to a vector field gives a function of position that is completely specified by its magnitude and direction at all points in space.

These definitions may be refined and extended; in fact, in Appendix I they are replaced by more subtle definitions in terms of transformation properties. In addition, more complicated kinds of quantities, such as tensors, are sometimes encountered. Scalars and vectors will, however, largely suffice for our purposes until Chapter 22.

1

1–2 VECTOR ALGEBRA

Since the algebra of scalars is familiar to the reader, this algebra will be used to develop vector algebra. In order to proceed with this development it is convenient to have a representation of vectors, for which purpose we introduce a three-dimensional Cartesian coordinate system. This three-dimensional system will be denoted by the three variables x, y, z or, when it is more convenient, x_1, x_2, x_3. With respect to this coordinate system a vector is specified by its x-, y-, and z-components. Thus a vector* $\mathbf{V}$ is specified by its components V_x, V_y, V_z, where $V_x = |\mathbf{V}| \cos \alpha_1$, $V_y = |\mathbf{V}| \cos \alpha_2$, $V_z = |\mathbf{V}| \cos \alpha_3$, the α's being the angles between $\mathbf{V}$ and appropriate coordinate axes. The scalar $|\mathbf{V}| = \sqrt{V_x^2 + V_y^2 + V_z^2}$ is the *magnitude* of the vector $\mathbf{V}$, or its *length*. In the case of vector fields, each of the components is to be regarded as a function of x, y, and z. It should be emphasized at this point that we introduce a representation of the vectors with respect to a Cartesian coordinate system only for simplicity and ease of understanding; all of the definitions and operations are, in fact, independent of any special choice of coordinates.

The sum of two vectors is defined as the vector whose components are the sums of the corresponding components of the original vectors. Thus if $\mathbf{C}$ is the sum of $\mathbf{A}$ and $\mathbf{B}$, we write

$$\mathbf{C} = \mathbf{A} + \mathbf{B} \tag{1–1}$$

and

$$C_x = A_x + B_x, \qquad C_y = A_y + B_y, \qquad C_z = A_z + B_z. \tag{1–2}$$

This definition of the vector sum is completely equivalent to the familiar parallelogram rule for vector addition.

Vector subtraction is defined in terms of the negative of a vector, which is the vector whose components are the negatives of the corresponding components of the original vector. Thus if $\mathbf{A}$ is a vector, $-\mathbf{A}$ is defined by

$$(-\mathbf{A})_x = -A_x, \qquad (-\mathbf{A})_y = -A_y, \qquad (-\mathbf{A})_z = -A_z. \tag{1–3}$$

The operation of subtraction is then defined as the addition of the negative. This is written

$$\mathbf{A} - \mathbf{B} = \mathbf{A} + (-\mathbf{B}). \tag{1–4}$$

Since the addition of real numbers is associative and commutative, it follows that vector addition (and subtraction) is also associative and commutative. In vector notation this appears as

$$\mathbf{A} + (\mathbf{B} + \mathbf{C}) = (\mathbf{A} + \mathbf{B}) + \mathbf{C} = (\mathbf{A} + \mathbf{C}) + \mathbf{B} = \mathbf{A} + \mathbf{B} + \mathbf{C}. \tag{1–5}$$

In other words, the parentheses are not needed, as indicated by the last form.

Proceeding now to the process of multiplication, we note that the simplest product is a scalar times a vector. This operation results in a vector each component of which is the scalar times the corresponding component of the original

---

* Vector quantities will be denoted by boldface symbols.

vector. If c is a scalar and $\mathbf{A}$ a vector, the product $c\mathbf{A}$ is a vector, $\mathbf{B} = c\mathbf{A}$, defined by

$$B_x = cA_x, \qquad B_y = cA_y, \qquad B_z = cA_z. \tag{1-6}$$

It is clear that if $\mathbf{A}$ is a *vector field* and c a *scalar field* then $\mathbf{B}$ is a new vector field which is *not* necessarily a constant multiple of the original field.

If, now, two vectors are to be multiplied, there are two possibilities, known as the vector and scalar products. Considering first the scalar product, we note that this name derives from the scalar nature of the product, although the alternative names, inner product and dot product, are sometimes used. The definition of the scalar product, written $\mathbf{A} \cdot \mathbf{B}$, is

$$\mathbf{A} \cdot \mathbf{B} = A_x B_x + A_y B_y + A_z B_z. \tag{1-7}$$

This definition is equivalent to another, and perhaps more familiar, definition, i.e., as the product of the magnitudes of the original vectors times the cosine of the angle between these vectors. If $\mathbf{A}$ and $\mathbf{B}$ are perpendicular to each other,

$$\mathbf{A} \cdot \mathbf{B} = 0.$$

The scalar product is commutative. The length of $\mathbf{A}$ is

$$|\mathbf{A}| = \sqrt{\mathbf{A} \cdot \mathbf{A}}.$$

The vector product of two vectors is a vector, which accounts for the name. Alternative names are outer and cross product. The vector product is written $\mathbf{A} \times \mathbf{B}$; if $\mathbf{C}$ is the vector product of $\mathbf{A}$ and $\mathbf{B}$, then $\mathbf{C} = \mathbf{A} \times \mathbf{B}$, or

$$C_x = A_y B_z - A_z B_y, \qquad C_y = A_z B_x - A_x B_z, \qquad C_z = A_x B_y - A_y B_x. \tag{1-8}$$

It is important to note that the cross product depends on the order of the factors; interchanging the order introduces a minus sign:

$$\mathbf{B} \times \mathbf{A} = -\mathbf{A} \times \mathbf{B}.$$

Consequently,

$$\mathbf{A} \times \mathbf{A} = 0.$$

This definition is equivalent to the following: the vector product is the product of the magnitudes times the sine of the angle between the original vectors, with the direction given by a right-hand screw rule.*

The vector product may be easily remembered in terms of a determinant. If $\mathbf{i}$, $\mathbf{j}$, and $\mathbf{k}$ are unit vectors, i.e., vectors of unit magnitude, in the x-, y-, and z-directions, respectively, then

$$\mathbf{A} \times \mathbf{B} = \begin{vmatrix} \mathbf{i} & \mathbf{j} & \mathbf{k} \\ A_x & A_y & A_z \\ B_x & B_y & B_z \end{vmatrix}. \tag{1-9}$$

---

* Let $\mathbf{A}$ be rotated into $\mathbf{B}$ through the smallest possible angle. A right-hand screw rotated in this manner will advance in a direction perpendicular to both $\mathbf{A}$ and $\mathbf{B}$; this direction is the direction of $\mathbf{A} \times \mathbf{B}$.

If this determinant is evaluated by the usual rules, the result is precisely our definition of the cross product.

The algebraic operations discussed above may be combined in many ways. Most of the results so obtained are obvious; however, there are two triple products of sufficient importance to merit explicit mention. The triple scalar product $D = \mathbf{A} \cdot \mathbf{B} \times \mathbf{C}$ is easily found to be given by the determinant

$$D = \mathbf{A} \cdot \mathbf{B} \times \mathbf{C} = \begin{vmatrix} A_x & A_y & A_z \\ B_x & B_y & B_z \\ C_x & C_y & C_z \end{vmatrix} = -\mathbf{B} \cdot \mathbf{A} \times \mathbf{C}. \tag{1-10}$$

This product is unchanged by an exchange of dot and cross or by a cyclic permutation of the three vectors; parentheses are not needed, since the cross product of a scalar and a vector is undefined. The other interesting triple product is the triple vector product $\mathbf{D} = \mathbf{A} \times (\mathbf{B} \times \mathbf{C})$. By a repeated application of the definition of the cross product, Eq. (1-8), we find

$$\mathbf{D} = \mathbf{A} \times (\mathbf{B} \times \mathbf{C}) = \mathbf{B}(\mathbf{A} \cdot \mathbf{C}) - \mathbf{C}(\mathbf{A} \cdot \mathbf{B}), \tag{1-11}$$

which is frequently known as the *back cab rule*. It should be noted that in the cross product the parentheses are vital; without them the product is not well defined.

At this point one might well inquire as to the possibility of vector division. Division of a vector by a scalar can, of course, be defined as multiplication by the reciprocal of the scalar. Division of a vector by another vector, however, is possible only if the two vectors are parallel. On the other hand, it is possible to write general solutions to vector equations and so accomplish something closely akin to division. Consider the equation

$$c = \mathbf{A} \cdot \mathbf{X}, \tag{1-12}$$

where c is a known scalar, $\mathbf{A}$ a known vector, and $\mathbf{X}$ an unknown vector. A general solution to this equation is

$$\mathbf{X} = \frac{c\mathbf{A}}{\mathbf{A} \cdot \mathbf{A}} + \mathbf{B}, \tag{1-13}$$

where $\mathbf{B}$ is a vector of arbitrary magnitude that is perpendicular to $\mathbf{A}$, that is, $\mathbf{A} \cdot \mathbf{B} = 0$. What we have done is very nearly to divide c by $\mathbf{A}$; more correctly, we have found the general form of the vector $\mathbf{X}$ that satisfies Eq. (1-12). There is no unique solution, and this fact accounts for the vector $\mathbf{B}$. In the same fashion we may consider the vector equation

$$\mathbf{C} = \mathbf{A} \times \mathbf{X}, \tag{1-14}$$

where $\mathbf{A}$ and $\mathbf{C}$ are known vectors and $\mathbf{X}$ is an unknown vector. The general solution of this equation is

$$\mathbf{X} = \frac{\mathbf{C} \times \mathbf{A}}{\mathbf{A} \cdot \mathbf{A}} + k\mathbf{A} \tag{1-15}$$

if $\mathbf{C} \cdot \mathbf{A} = 0$, where k is an arbitrary scalar. If $\mathbf{C} \cdot \mathbf{A} \neq 0$, no solution exists. This again is very nearly the quotient of $\mathbf{C}$ by $\mathbf{A}$; the scalar k takes account of the nonuniqueness of the process. If $\mathbf{X}$ is required to satisfy both (1–12) and (1–14), then the result is unique (if it exists) and given by

$$\mathbf{X} = \frac{\mathbf{C} \times \mathbf{A}}{\mathbf{A} \cdot \mathbf{A}} + \frac{c\mathbf{A}}{\mathbf{A} \cdot \mathbf{A}}. \tag{1–16}$$

1–3 GRADIENT

The extensions of the ideas introduced above to differentiation and integration, i.e., vector calculus, will now be considered. The simplest of these is the relation of a particular vector field to the derivatives of a scalar field. It is convenient first to introduce the idea of the *directional derivative* of a function of several variables. This is just the rate of change of the function in a specified direction. The directional derivative of a scalar function φ is usually denoted by $d\varphi/ds$; it must be understood that ds represents an infinitesimal displacement in the direction being considered, and that ds is the scalar magnitude of $d\mathbf{s}$. If $d\mathbf{s}$ has the components dx, dy, dz, then

$$\frac{d\varphi}{ds} = \lim_{\Delta s \to 0} \frac{\varphi(x + \Delta x, y + \Delta y, z + \Delta z) - \varphi(x, y, z)}{\Delta s}$$

$$= \frac{\partial \varphi}{\partial x} \frac{dx}{ds} + \frac{\partial \varphi}{\partial y} \frac{dy}{ds} + \frac{\partial \varphi}{\partial z} \frac{dz}{ds}.$$

In order to clarify the idea of a directional derivative, consider a scalar function of two variables. Thus, $\varphi(x, y)$ represents a two-dimensional scalar field. We may plot φ as a function of x and y as is done in Fig. 1–1 for the function $\varphi(x, y) = x^2 + y^2$. The directional derivative at the point x_0, y_0 depends on the direction. If we choose the direction corresponding to $dy/dx = -x_0/y_0$, then we find

$$\left.\frac{d\varphi}{ds}\right|_{x_0,y_0} = \frac{\partial \varphi}{\partial x} \frac{dx}{ds} + \frac{\partial \varphi}{\partial y} \frac{dy}{ds} = \left[2x_0 - 2y_0 \frac{x_0}{y_0}\right] \frac{dx}{ds} = 0. \tag{1–17a}$$

Alternatively, if we choose $dy/dx = y_0/x_0$, we find

$$\left.\frac{d\varphi}{ds}\right|_{x_0,y_0} = \left(2x_0 + 2\frac{y_0^2}{x_0}\right)\sqrt{\frac{x_0^2}{x_0^2 + y_0^2}} = 2\sqrt{x_0^2 + y_0^2}, \tag{1–17b}$$

since $ds = \sqrt{(dx)^2 + (dy)^2}$. As a third possibility choose $dy/dx = \alpha$; then

$$\left.\frac{d\varphi}{ds}\right|_{x_0,y_0} = (2x_0 + 2\alpha y_0)(1 + \alpha^2)^{-1/2}. \tag{1–17c}$$

If this result is differentiated with respect to α and the derivative set equal to zero, then the value of α for which the derivative is a maximum or minimum is found. When we perform these operations, we obtain $\alpha = y_0/x_0$, which simply means

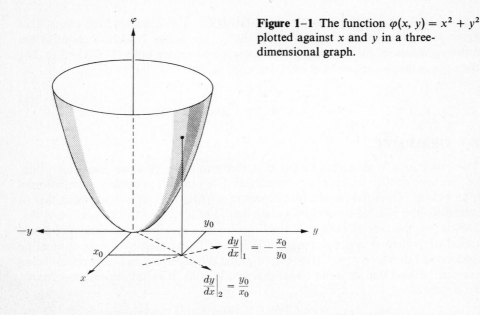

Figure 1–1 The function $\varphi(x, y) = x^2 + y^2$ plotted against x and y in a three-dimensional graph.

$$\frac{dy}{dx}\Big|_1 = -\frac{x_0}{y_0}$$

$$\frac{dy}{dx}\Big|_2 = \frac{y_0}{x_0}$$

that the direction of maximum rate of change of the function $\varphi = x^2 + y^2$ is the radial direction. If the direction is radially outward then the maximum is the maximum rate of increase; if it is radially inward it is a maximum rate of decrease or minimum rate of increase. In the direction specified by $dy/dx = -x_0/y_0$ the rate of change of $x^2 + y^2$ is zero. This direction is tangent to the circle $x^2 + y^2 = x_0^2 + y_0^2$. Clearly, on this curve, $\varphi = x^2 + y^2$ does not change. The direction in which $d\varphi/ds$ vanishes gives the direction of the curve $\varphi = constant$ through the point being considered. These lines, which are circles for the function $x^2 + y^2$, are completely analogous to the familiar contour lines or lines of constant altitude which appear on topographic maps. Figure 1–2 shows the function $\varphi = x^2 + y^2$ replotted as a contour map.

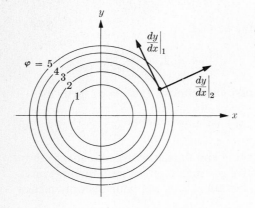

Figure 1–2 The function $\varphi(x, y)$ of Fig. 1–1 expressed as a contour map in two dimensions.

The idea of contour lines may be generalized to a function of three variables, in which case the surfaces, $\varphi(x, y, z) = constant$, are called *level surfaces* or *equipotential surfaces*. The three-dimensional analog to Fig. 1–2 is the only practical way of graphing a scalar field for a three-dimensional space.

The gradient of a scalar function may now be defined as follows:

The gradient of a scalar function φ is a vector whose magnitude is the maximum directional derivative at the point being considered and whose direction is the direction of the maximum directional derivative at the point.

It is evident that the gradient has the direction normal to the level surface of φ through the point being considered. The most common symbols for the gradient are $\mathbf{V}$ and **grad**; of these we will use the latter at first. In terms of the gradient the directional derivative is given by

$$\frac{d\varphi}{ds} = |\mathbf{grad}\ \varphi|\ \cos\theta, \tag{1-18}$$

where θ is the angle between the direction of $d\mathbf{s}$ and the direction of the gradient. This is immediately evident from the geometry of Fig. 1–3. If we write $d\mathbf{s}$ for the vector displacement of magnitude ds, then (1–18) can be written

$$\frac{d\varphi}{ds} = \mathbf{grad}\ \varphi \cdot \frac{d\mathbf{s}}{ds}. \tag{1-19}$$

This equation enables us to find the explicit form of the gradient in any coordinate system in which we know the form of $d\mathbf{s}$. In rectangular coordinates we know that $d\mathbf{s} = \mathbf{i}\ dx + \mathbf{j}\ dy + \mathbf{k}\ dz$. We also know that

$$d\varphi = \frac{\partial\varphi}{\partial x}\ dx + \frac{\partial\varphi}{\partial y}\ dy + \frac{\partial\varphi}{\partial z}\ dz.$$

From this and Eq. (1–19) it follows that

$$\frac{\partial\varphi}{\partial x}\ dx + \frac{\partial\varphi}{\partial y}\ dy + \frac{\partial\varphi}{\partial z}\ dz = (\mathbf{grad}\ \varphi)_x\ dx + (\mathbf{grad}\ \varphi)_y\ dy + (\mathbf{grad}\ \varphi)_z\ dz.$$

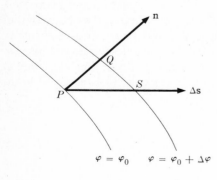

Figure 1–3 Parts of two level surfaces of the function $\varphi(x, y, z)$. $|\mathbf{grad}\ \varphi|$ at P equals the limit as $\overline{PQ} \to 0$ of $\Delta\varphi/\overline{PQ}$, and $d\varphi/ds$ is the corresponding limit of $\Delta\varphi/\overline{PS}$.

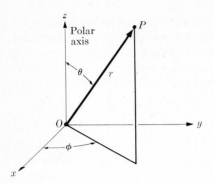

Figure 1–4 Definition of the polar coordinates r, θ, ϕ.

Equating coefficients of differentials of independent variables on both sides of the equation gives

$$\textbf{grad } \varphi = \mathbf{i}\,\frac{\partial \varphi}{\partial x} + \mathbf{j}\,\frac{\partial \varphi}{\partial y} + \mathbf{k}\,\frac{\partial \varphi}{\partial z} \tag{1–20}$$

in rectangular coordinates. In a more complicated case the procedure is the same. In spherical polar coordinates, with r, θ, ϕ as defined in Fig. 1–4, we have

$$d\varphi = \frac{\partial \varphi}{\partial r}\,dr + \frac{\partial \varphi}{\partial \theta}\,d\theta + \frac{\partial \varphi}{\partial \phi}\,d\phi, \tag{1–21}$$

and

$$d\mathbf{s} = \mathbf{a}_r\,dr + \mathbf{a}_\theta r\,d\theta + \mathbf{a}_\phi r \sin\theta\,d\phi, \tag{1–22}$$

where $\mathbf{a}_r$, $\mathbf{a}_\theta$, and $\mathbf{a}_\phi$ are unit vectors in the r, θ, and ϕ directions respectively. Applying (1–19) and equating coefficients of independent variables yields

$$\textbf{grad } \varphi = \mathbf{a}_r\,\frac{\partial \varphi}{\partial r} + \mathbf{a}_\theta\,\frac{1}{r}\frac{\partial \varphi}{\partial \theta} + \mathbf{a}_\phi\,\frac{1}{r \sin\theta}\frac{\partial \varphi}{\partial \phi} \tag{1–23}$$

in spherical coordinates.

1–4 VECTOR INTEGRATION

There are, of course, other aspects to differentiation involving vectors; however, it is convenient to discuss vector integration first. For our purposes we may consider three kinds of integrals: line, surface, and volume, according to the nature of the differential appearing in the integral. The integrand may be either a vector or a scalar; however, certain combinations of integrands and differentials give rise to uninteresting integrals. Those of most interest here are the scalar line integral of a vector, the scalar surface integral of a vector, and the volume integrals of both vectors and scalars.

If **F** is a vector, a line integral of **F** is written

$$\int_{a_C}^{b} \mathbf{F} \cdot d\mathbf{l}, \tag{1–24}$$

where C is the curve along which the integration is performed, a and b the initial and final points on the curve, and $d\mathbf{l}$ an infinitesimal vector displacement along the curve C. Since $\mathbf{F} \cdot d\mathbf{l}$ is a scalar it is clear that the line integral is a scalar. The definition of the line integral follows closely the Riemann definition of the definite integral. The segment of C between a and b is divided into a large number of small increments $\Delta\mathbf{l}_i$; for each increment an interior point is chosen and the value of $\mathbf{F}$ at that point found. The scalar product of each increment with the corresponding value of $\mathbf{F}$ is found and the sum of these computed. The line integral is then defined as the limit of this sum as the number of increments becomes infinite in such a way that each increment goes to zero. This definition may be compactly written as

$$\int_{a_C}^{b} \mathbf{F} \cdot d\mathbf{l} = \lim_{N \to \infty} \sum_{i=1}^{N} \mathbf{F}_i \cdot \Delta\mathbf{l}_i.$$

It is important to note that the line integral usually depends not only on the endpoints a and b but also on the curve C along which the integration is to be done. The line integral around a closed curve is of sufficient importance that a special notation is used for it, namely,

$$\oint_{C} \mathbf{F} \cdot d\mathbf{l}. \tag{1–25}$$

The integral around a closed curve may or may not be zero; the class of vectors for which the line integral around any closed curve is zero is of considerable importance. For this reason one often encounters line integrals around undesignated closed paths, for example,

$$\oint \mathbf{F} \cdot d\mathbf{l}. \tag{1–26}$$

This notation is useful only in those cases where the integral is independent of the contour C within rather wide limits. If any ambiguity is possible, it is wise to specify the contour. The basic approach to the evaluation of line integrals is to obtain a one-parameter description of the curve and then use this description to express the line integral as the sum of three ordinary one-dimensional integrals. In all but the simplest cases this is a long and tedious procedure; fortunately, however, it is seldom necessary to evaluate the integrals in this fashion. As will be seen later, it is often possible to show that the line integral does not depend on the path between the endpoints. In the latter case a simple path may be chosen to simplify the integration.

If $\mathbf{F}$ is again a vector, a surface integral of $\mathbf{F}$ is written

$$\int_{S} \mathbf{F} \cdot \mathbf{n} \, da, \tag{1–27}$$

where S is the surface over which the integration is to be performed, da is an infinitesimal area on S and $\mathbf{n}$ is a unit normal to da. There is a twofold ambiguity in the choice of $\mathbf{n}$, which is resolved by taking $\mathbf{n}$ to be the outward drawn normal if

S is a closed surface. If S is not closed and is finite then it has a boundary, and the sense of the normal is important only with respect to the arbitrary positive sense of traversing the boundary. The positive sense of the normal is the direction in which a right-hand screw would advance if rotated in the direction of the positive sense on the bounding curve. This is illustrated in Fig. 1–5. The surface integral of $\mathbf{F}$ over a closed surface S is sometimes denoted by

$$\oint_S \mathbf{F} \cdot \mathbf{n} \, da.$$

Comments exactly parallel to those made for the line integral can be made for the surface integral. This surface integral is clearly a scalar; it usually depends on the surface S, and cases where it does not are particularly important. The definition of the surface integral is made in a way exactly comparable to that of the line integral. The detailed formulation is left as an exercise.

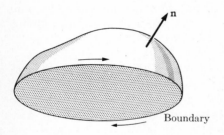

Figure 1–5 Relation of normal $\mathbf{n}$ to a surface, and the direction of traversal of the boundary.

Boundary

If $\mathbf{F}$ is a vector and φ a scalar then the two volume integrals in which we are interested are

$$J = \int_V \varphi \, dv, \qquad \mathbf{K} = \int_V \mathbf{F} \, dv. \qquad (1\text{–}28)$$

Clearly J is a scalar and $\mathbf{K}$ a vector. The definitions of these integrals reduce quickly to just the Riemann integral in three dimensions except that in $\mathbf{K}$ one must note that there is one integral for each component of $\mathbf{F}$. These integrals are sufficiently familiar to require no further comment.

1–5 DIVERGENCE

Another important operator, which is essentially a derivative, is the divergence operator. The divergence of vector $\mathbf{F}$, written div $\mathbf{F}$, is defined as follows:

The divergence of a vector is the limit of its surface integral per unit volume as the volume enclosed by the surface goes to zero. That is,

$$\text{div } \mathbf{F} = \lim_{V \to 0} \frac{1}{V} \oint_S \mathbf{F} \cdot \mathbf{n} \, da.$$

The divergence is clearly a scalar point function (scalar field), and it is defined at the limit point of the surface of integration. The above definition has several virtues: it is independent of any special choice of coordinate system, and it may be used to find the explicit form of the divergence operator in any particular coordinate system.

In rectangular coordinates the volume element $\Delta x\, \Delta y\, \Delta z$ provides a convenient basis for finding the explicit form of the divergence. If one corner of the rectangular parallelepiped is at the point $x_0,\ y_0,\ z_0$, then

$$F_x(x_0 + \Delta x, y, z) = F_x(x_0, y, z) + \Delta x \left.\frac{\partial F_x}{\partial x}\right|_{x_0,y,z},$$

$$F_y(x, y_0 + \Delta y, z) = F_y(x, y_0, z) + \Delta y \left.\frac{\partial F_y}{\partial y}\right|_{x,y_0,z}, \tag{1-29}$$

$$F_z(x, y, z_0 + \Delta z) = F_z(x, y, z_0) + \Delta z \left.\frac{\partial F_z}{\partial z}\right|_{x,y,z_0},$$

where higher-order terms in Δx, Δy, and Δz have been omitted. Since the area element $\Delta y\, \Delta z$ is perpendicular to the x-axis, $\Delta z\, \Delta x$ perpendicular to the y-axis, and $\Delta x\, \Delta y$ perpendicular to the z-axis, the definition of the divergence becomes

$$\begin{aligned}
\operatorname{div} \mathbf{F} = \lim_{V \to 0} \frac{1}{\Delta x\, \Delta y\, \Delta z} \bigg\{ &\int F_x(x_0, y, z)\, dy\, dz \\
&+ \Delta x\, \Delta y\, \Delta z\, \frac{\partial F_x}{\partial x} + \int F_y(x, y_0, z)\, dx\, dz \\
&+ \Delta x\, \Delta y\, \Delta z\, \frac{\partial F_y}{\partial y} + \int F_z(x, y, z_0)\, dx\, dy \\
&+ \Delta x\, \Delta y\, \Delta z\, \frac{\partial F_z}{\partial z} - \int F_x(x_0, y, z)\, dy\, dz \\
&- \int F_y(x, y_0, z)\, dx\, dz - \int F_z(x, y, z_0)\, dx\, dy \bigg\}.
\end{aligned} \tag{1-30}$$

The minus signs associated with the last three terms account for the fact that the outward drawn normal is in the direction of the negative axes in these cases. The limit is easily taken, and the divergence in rectangular coordinates is found to be

$$\operatorname{div} \mathbf{F} = \frac{\partial F_x}{\partial x} + \frac{\partial F_y}{\partial y} + \frac{\partial F_z}{\partial z}. \tag{1-31}$$

In spherical coordinates the procedure is similar. The volume enclosed by the coordinate intervals Δr, $\Delta\theta$, $\Delta\phi$ is chosen as the volume of integration. This volume is $r^2 \sin\theta\, \Delta r\, \Delta\theta\, \Delta\phi$. Because the area enclosed by the coordinate intervals depends on the values of the coordinates (note that this is not the case with rectangular coordinates), it is best to write $\mathbf{F} \cdot \mathbf{n} \, \Delta a$ in its explicit form:

$$\mathbf{F} \cdot \mathbf{n} \, \Delta a = F_r r^2 \sin \theta \, \Delta \theta \, \Delta \phi + F_\theta r \sin \theta \, \Delta \phi \, \Delta r + F_\phi r \, \Delta r \, \Delta \theta. \quad (1\text{-}32)$$

It is clear from this expression that $r^2 F_r \sin \theta$, rather than just F_r, must be expanded in Taylor series. Similarly, it is the coefficient of the products of coordinate intervals that must be expanded in the other terms. Making these expansions and using them to evaluate the surface integral in the definition of the divergence gives

$$\text{div } \mathbf{F} = \lim_{V \to 0} \frac{1}{r^2 \sin \theta \, \Delta r \, \Delta \theta \, \Delta \phi} \left\{ \frac{\partial}{\partial r} (F_r r^2 \sin \theta) \, \Delta r \, \Delta \theta \, \Delta \phi \right.$$

$$\left. + \frac{\partial}{\partial \theta} (F_\theta r \sin \theta) \, \Delta \theta \, \Delta r \, \Delta \phi + \frac{\partial}{\partial \phi} (F_\phi r) \, \Delta \phi \, \Delta r \, \Delta \theta \right\}. \quad (1\text{-}33)$$

Taking the limit, the explicit form of the divergence in spherical coordinates is found to be

$$\text{div } \mathbf{F} = \frac{1}{r^2} \frac{\partial}{\partial r} (r^2 F_r) + \frac{1}{r \sin \theta} \frac{\partial}{\partial \theta} (\sin \theta F_\theta) + \frac{1}{r \sin \theta} \frac{\partial F_\phi}{\partial \phi}. \quad (1\text{-}34)$$

This method of finding the explicit form of the divergence is applicable to any coordinate system provided the forms of the volume and surface elements or, alternatively, the elements of length are known.

The physical significance of the divergence is readily seen in terms of an example taken from fluid mechanics. If $\mathbf{V}$ is the velocity of a fluid, given as a function of position, and ρ is its density, then $\oint_S \rho \mathbf{V} \cdot \mathbf{n} \, da$ is clearly the net amount of fluid per unit time that leaves the volume enclosed by S. If the fluid is incompressible, the surface integral measures the total source of fluid enclosed by the surface. The above definition of the divergence then indicates that it may be interpreted as the limit of the source strength per unit volume, or the source density of an incompressible fluid.

An extremely important theorem involving the divergence may now be stated and proved.

Divergence theorem *The integral of the divergence of a vector over a volume V is equal to the surface integral of the normal component of the vector over the surface bounding V. That is,*

$$\int_V \text{div } \mathbf{F} \, dv = \oint_S \mathbf{F} \cdot \mathbf{n} \, da.$$

Consider the volume to be subdivided into a large number of small cells. Let the ith cell have volume ΔV_i and be bounded by the surface S_i. It is clear that

$$\sum_i \oint_{S_i} \mathbf{F} \cdot \mathbf{n} \, da = \oint_S \mathbf{F} \cdot \mathbf{n} \, da, \quad (1\text{-}35)$$

where in each integral on the left the normal is directed outward from the volume being considered. Since outward to one cell is inward to the appropriate adjacent cell, all contributions to the left side of (1–35) cancel except those which arise from the surface of S, and Eq. (1–35) is essentially proved. The divergence theorem is now obtained by letting the number of cells go to infinity in such a way that the volume of each cell goes to zero.

$$\oint_S \mathbf{F} \cdot \mathbf{n} \, da = \lim_{\Delta V_i \to 0} \sum_i \left\{ \frac{1}{\Delta V_i} \oint_{S_i} \mathbf{F} \cdot \mathbf{n} \, da \right\} \Delta V_i. \tag{1–36}$$

In the limit, the sum on i becomes an integral over V and the ratio of the integral over S_i to ΔV_i becomes the divergence of $\mathbf{F}$. Thus,

$$\oint_S \mathbf{F} \cdot \mathbf{n} \, da = \int_V \operatorname{div} \mathbf{F} \, dv, \tag{1–37}$$

which is the divergence theorem. We shall have frequent occasion to exploit this theorem, both in the development of the theoretical aspects of electricity and magnetism and for the very practical purpose of evaluating integrals.

1–6 CURL

The third interesting vector differential operator is the curl. The curl of a vector, written **curl F**, is defined as follows:

The curl of a vector is the limit of the ratio of the integral of its cross product with the outward drawn normal, over a closed surface, to the volume enclosed by the surface as the volume goes to zero. That is,

$$\mathbf{curl}\ \mathbf{F} = \lim_{V \to 0} \frac{1}{V} \oint_S \mathbf{n} \times \mathbf{F} \, da. \tag{1–38}$$

The parallelism between this definition and the definition of the divergence is quite apparent; instead of the scalar product of the vector with the outward drawn normal, one has the vector product. Otherwise the definitions are the same. A different but equivalent definition is more useful. This alternative definition is

The component of **curl F** *in the direction of the unit vector* **a** *is the limit of a line integral per unit area, as the enclosed area goes to zero, this area being perpendicular to* **a**. *That is,*

$$\mathbf{a} \cdot \mathbf{curl}\ \mathbf{F} = \lim_{S \to 0} \frac{1}{S} \oint_C \mathbf{F} \cdot d\mathbf{l}, \tag{1–39}$$

where the curve C, which bounds the surface S, is in a plane normal to **a**. It is easy to see the equivalence of the two definitions by considering a plane curve C and the volume swept out by this curve when it is displaced a distance ξ in the direction of the normal to its plane, as shown in Fig. 1–6. If **a** is normal to this

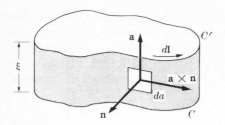

Figure 1–6 Volume swept out by displacing the plane curve C in the direction of its normal, **a**.

plane, then taking the dot product of **a** with the first definition of the curl, (1–38), gives

$$\mathbf{a} \cdot \mathbf{curl\ F} = \lim_{V \to 0} \frac{1}{V} \oint_S \mathbf{a} \cdot \mathbf{n} \times \mathbf{F}\ da. \tag{1–40}$$

Since **a** is parallel to the normal for all of the bounding surface except the narrow strip bounded by C and C', only the integral over this surface need be considered. For this surface we note that $\mathbf{a} \times \mathbf{n}\ da$ is just $\xi\ d\mathbf{l}$, where $d\mathbf{l}$ is an infinitesimal displacement along C. Since, in addition, $V = \xi S$, the limit of the volume integral is just

$$\mathbf{a} \cdot \mathbf{curl\ F} = \lim_{V \to 0} \frac{1}{\xi S} \oint \xi \mathbf{F} \cdot d\mathbf{l},$$

which reduces to the second form of our definition upon canceling the ξ's. This equivalence can be shown without the use of the special volume used here; however, so doing sacrifices much of the simplicity of the proof given above.

The form of the curl in various coordinate systems can be calculated in much the same way as was done with the divergence. In rectangular coordinates the volume $\Delta x\ \Delta y\ \Delta z$ is convenient. For the x-component of the curl only the faces perpendicular to the y- and z-axes contribute. Recalling that $\mathbf{j} \times \mathbf{k} = -\mathbf{k} \times \mathbf{j} = \mathbf{i}$, the nonvanishing contributions from the faces of the parallelepiped to the x-component of the curl give

$$(\mathbf{curl\ F})_x = \lim_{V \to 0} \frac{1}{V} \{ [-F_y(x, y, z + \Delta z) + F_y(x, y, z)]\ \Delta x\ \Delta y$$
$$+ [F_z(x, y + \Delta y, z) - F_z(x, y, z)]\ \Delta x\ \Delta z \}. \tag{1–41}$$

Making a Taylor series expansion and taking the limit gives

$$(\mathbf{curl\ F})_x = \frac{\partial F_z}{\partial y} - \frac{\partial F_y}{\partial z} \tag{1–42}$$

for the x-component of the curl. The y- and z-components may be found in exactly the same way. They are

$$(\mathbf{curl\ F})_y = \frac{\partial F_x}{\partial z} - \frac{\partial F_z}{\partial x}, \qquad (\mathbf{curl\ F})_z = \frac{\partial F_y}{\partial x} - \frac{\partial F_x}{\partial y}. \tag{1–43}$$

The form of the curl in rectangular coordinates can be easily remembered if it is noted that it is just the expansion of a three-by-three determinant, namely,

$$\text{curl } \mathbf{F} = \begin{vmatrix} \mathbf{i} & \mathbf{j} & \mathbf{k} \\ \dfrac{\partial}{\partial x} & \dfrac{\partial}{\partial y} & \dfrac{\partial}{\partial z} \\ F_x & F_y & F_z \end{vmatrix}. \qquad (1\text{--}44)$$

The problem of finding the form of the curl in other coordinate systems is only slightly more complicated and is left to the exercises.

As with the divergence, we encounter an important and useful theorem involving the curl, known as Stokes's theorem.

Stokes's theorem *The line integral of a vector around a closed curve is equal to the integral of the normal component of its curl over any surface bounded by the curve.* That is,

$$\oint_C \mathbf{F} \cdot d\mathbf{l} = \int_S \text{curl } \mathbf{F} \cdot \mathbf{n} \, da, \qquad (1\text{--}45)$$

where C is a closed curve which bounds the surface S. The proof of this theorem is quite analogous to the proof of the divergence theorem. The surface S is divided into a large number of cells. The surface of the ith cell is called ΔS_i and the curve bounding it is C_i. Since each of these cells must be traversed in the same sense, it is clear that the sum of the line integrals around the C_i's is just the line integral around the bounding curve; all of the other contributions cancel. Thus

$$\oint_C \mathbf{F} \cdot d\mathbf{l} = \sum_i \oint_{C_i} \mathbf{F} \cdot d\mathbf{l}.$$

It remains only to take the limit as the number of cells becomes infinite in such a way that the area of each goes to zero. The result of this limiting process is

$$\oint_C \mathbf{F} \cdot d\mathbf{l} = \lim_{\Delta S_i \to 0} \sum_i \frac{1}{\Delta S_i} \oint_{C_i} \mathbf{F} \cdot d\mathbf{l} \, \Delta S_i$$

$$= \int_S \text{curl } \mathbf{F} \cdot \mathbf{n} \, da,$$

which is Stokes's theorem. This theorem, like the divergence theorem, is useful both in the development of electromagnetic theory and in the evaluation of integrals. It is perhaps worth noting that both the divergence theorem and Stokes's theorem are essentially partial integrations.

1–7 THE VECTOR DIFFERENTIAL OPERATOR ∇

We now introduce an alternative notation for the three types of vector differentiation that have been discussed—namely, gradient, divergence, and curl. This is expressed by the vector differential operator *del*, defined in Cartesian coordinates as

$$\mathbf{V} = \mathbf{i}\,\frac{\partial}{\partial x} + \mathbf{j}\,\frac{\partial}{\partial y} + \mathbf{k}\,\frac{\partial}{\partial z}. \qquad (1\text{-}46)$$

Del is a differential operator in that it is used only in front of a function of (x, y, z), which it differentiates; it is a vector in that it obeys the laws of vector algebra.* In terms of del, Eqs. (1–20), (1–31), and (1–44) are expressed as follows:

grad $= \mathbf{V}$,

$$\mathbf{V}\varphi = \mathbf{i}\,\frac{\partial\varphi}{\partial x} + \mathbf{j}\,\frac{\partial\varphi}{\partial y} + \mathbf{k}\,\frac{\partial\varphi}{\partial z}. \qquad (1\text{-}20)$$

div $= \mathbf{V}\,\cdot$,

$$\mathbf{V}\cdot\mathbf{F} = \frac{\partial F_x}{\partial x} + \frac{\partial F_y}{\partial y} + \frac{\partial F_z}{\partial z}. \qquad (1\text{-}31)$$

curl $= \mathbf{V}\times$,

$$\mathbf{V}\times\mathbf{F} = \begin{vmatrix} \mathbf{i} & \mathbf{j} & \mathbf{k} \\ \dfrac{\partial}{\partial x} & \dfrac{\partial}{\partial y} & \dfrac{\partial}{\partial z} \\ F_x & F_y & F_z \end{vmatrix}. \qquad (1\text{-}44)$$

The operations expressed with del are themselves independent of any special choice of coordinate system. Any identities that can be proved using the Cartesian representation hold independently of the coordinate system. Del can be expressed in a non-Cartesian (curvilinear) orthonormal coordinate system in a form analogous to Eq. (1–46) with the appropriate distance elements, but it must be remembered in applying it that the unit vectors in such coordinate systems are themselves functions of position and have to be differentiated.† The important integral theorems, according to Eqs. (1–19), (1–45), and (1–37), are

$$\int_{a_C}^{b} \mathbf{V}\varphi\cdot d\mathbf{l} = \int_{a}^{b} d\varphi = \varphi\Big|_{a}^{b} = \varphi_b - \varphi_a, \qquad (1\text{-}47)$$

$$\int_{S} \mathbf{V}\times\mathbf{F}\cdot\mathbf{n}\,da = \oint_{C} \mathbf{F}\cdot d\mathbf{l}, \qquad (1\text{-}45)$$

$$\int_{V} \mathbf{V}\cdot\mathbf{F}\,dv = \oint_{S} \mathbf{F}\cdot\mathbf{n}\,da. \qquad (1\text{-}37)$$

---

* It is also a vector in terms of its transformation properties, as shown in Appendix I.
† An elementary discussion is given by H. T. Yang, *American Journal of Physics*, vol. 40, p. 109 (1972).

These give the integral of a derivative of a function over a region of n dimensions, in terms of the values of the function itself on the $(n-1)$-dimensional boundary of the region, for $n = 1, 2, 3$. Because the del operator obeys the rules of vector algebra, it is convenient to use it in calculations involving vector analysis, and henceforth we shall express the gradient, divergence, and curl in terms of ∇. It should be noted that ∇ is a *linear* operator:

$$\nabla(a\varphi + b\psi) = a\nabla\varphi + b\nabla\psi,$$

$$\nabla \cdot (a\mathbf{F} + b\mathbf{G}) = a\nabla \cdot \mathbf{F} + b\nabla \cdot \mathbf{G},$$

$$\nabla \times (a\mathbf{F} + b\mathbf{G}) = a\nabla \times \mathbf{F} + b\nabla \times \mathbf{G},$$

if a and b are constant scalars.

1–8 FURTHER DEVELOPMENTS

The operations of taking the gradient, divergence, or curl of appropriate kinds of fields may be repeated. For example, it makes sense to take the divergence of the gradient of a scalar field. Some of these repeated operations give zero for any well-behaved field. One is of sufficient importance to have a special name; the others can be expressed in terms of simpler operations. An important double operation is the divergence of the gradient of a scalar field. This combined operator is known as the *Laplacian operator* and is usually written ∇^2,

$$\nabla \cdot \nabla = \nabla^2.$$

In rectangular coordinates,

$$\nabla^2\varphi = \frac{\partial^2\varphi}{\partial x^2} + \frac{\partial^2\varphi}{\partial y^2} + \frac{\partial^2\varphi}{\partial z^2}. \tag{1–48}$$

This operator is of great importance in electrostatics and will be considered at length in Chapter 3.

The curl of the gradient of any scalar field is zero. This statement is most easily verified by writing it out in rectangular coordinates. If the scalar field is φ, then

$$\nabla \times (\nabla\varphi) = \begin{vmatrix} \mathbf{i} & \mathbf{j} & \mathbf{k} \\ \dfrac{\partial}{\partial x} & \dfrac{\partial}{\partial y} & \dfrac{\partial}{\partial z} \\ \dfrac{\partial\varphi}{\partial x} & \dfrac{\partial\varphi}{\partial y} & \dfrac{\partial\varphi}{\partial z} \end{vmatrix} = \mathbf{i}\left(\frac{\partial^2\varphi}{\partial y\,\partial z} - \frac{\partial^2\varphi}{\partial z\,\partial y}\right) + \cdots = 0, \tag{1–49}$$

which verifies the original statement. In operator notation,

$$\nabla \times \nabla = 0.$$

The divergence of any curl is also zero. This is verified directly in rectangular coordinates by writing

$$\mathbf{V} \cdot (\mathbf{V} \times \mathbf{F}) = \frac{\partial}{\partial x}\left(\frac{\partial F_z}{\partial y} - \frac{\partial F_y}{\partial z}\right) + \frac{\partial}{\partial y}\left(\frac{\partial F_x}{\partial z} - \frac{\partial F_z}{\partial x}\right) + \cdots = 0, \quad (1\text{--}50)$$

or

$$\mathbf{V} \cdot \mathbf{V} \times \mathbf{F} = \mathbf{V} \times \mathbf{V} \cdot \mathbf{F} = 0.$$

The other possible second-order operation is taking the curl of the curl of a vector field. It is left as an exercise to show that in rectangular coordinates

$$\mathbf{V} \times (\mathbf{V} \times \mathbf{F}) = \mathbf{V}(\mathbf{V} \cdot \mathbf{F}) - \mathbf{V}^2\mathbf{F}, \quad (1\text{--}51)$$

where the Laplacian of a vector is the vector whose rectangular components are the Laplacians of the rectangular components of the original vector. In any coordinate system other than rectangular the Laplacian of a vector is *defined* by Eq. (1–51).

Another way in which the application of the vector differential operators may be extended is to apply them to various products of two vectors and scalars. There are six possible combinations of differential operators and products; these are tabulated in Table 1–1. These identities may be readily verified in rectangular coordinates, which is sufficient to assure their validity in any coordinate system. A derivative of a product of more than two functions, or a higher than second-order derivative of a function, can be calculated by repeated applications of the identities in Table 1–1, which is therefore exhaustive. The formulas can be easily remembered from the rules of vector algebra and ordinary differentiation; the only ambiguity could be in (1–1–6) where $\mathbf{F} \cdot \mathbf{V}$ occurs (not $\mathbf{V} \cdot \mathbf{F}$).

Some particular types of functions come up often enough in electromagnetic theory that it is worth noting their various derivatives now. For the function $\mathbf{F} = \mathbf{r}$,

$$\mathbf{V} \cdot \mathbf{r} = 3,$$
$$\mathbf{V} \times \mathbf{r} = 0,$$
$$\mathbf{G} \cdot \mathbf{V}\mathbf{r} = \mathbf{G},$$
$$\mathbf{V}^2\mathbf{r} = 0.$$
$$(1\text{--}52)$$

For a function that depends only on the distance $r = |\mathbf{r}| = \sqrt{x^2 + y^2 + z^2}$,

$$\varphi(r) \quad \text{or} \quad \mathbf{F}(r): \; \mathbf{V} = \frac{\mathbf{r}}{r}\frac{d}{dr}. \quad (1\text{--}53)$$

For a function that depends on the argument $\mathbf{A} \cdot \mathbf{r}$, where $\mathbf{A}$ is a constant vector,

$$\varphi(\mathbf{A} \cdot \mathbf{r}) \quad \text{or} \quad \mathbf{F}(\mathbf{A} \cdot \mathbf{r}): \; \mathbf{V} = \mathbf{A}\frac{d}{d(\mathbf{A} \cdot \mathbf{r})}. \quad (1\text{--}54)$$

Table 1–1 Differential Vector Identities

$\mathbf{V} \cdot \mathbf{V}\varphi = \nabla^2\varphi$	(1-1-1)
$\mathbf{V} \cdot \mathbf{V} \times \mathbf{F} = 0$	(1-1-2)
$\mathbf{V} \times \mathbf{V}\varphi = 0$	(1-1-3)
$\mathbf{V} \times (\mathbf{V} \times \mathbf{F}) = \mathbf{V}(\mathbf{V} \cdot \mathbf{F}) - \nabla^2\mathbf{F}$	(1-1-4)
$\mathbf{V}(\varphi\psi) = (\mathbf{V}\varphi)\psi + \varphi\mathbf{V}\psi$	(1-1-5)
$\mathbf{V}(\mathbf{F} \cdot \mathbf{G}) = (\mathbf{F} \cdot \mathbf{V})\mathbf{G} + \mathbf{F} \times (\mathbf{V} \times \mathbf{G}) + (\mathbf{G} \cdot \mathbf{V})\mathbf{F} + \mathbf{G} \times (\mathbf{V} \times \mathbf{F})$	(1-1-6)
$\mathbf{V} \cdot (\varphi\mathbf{F}) = (\mathbf{V}\varphi) \cdot \mathbf{F} + \varphi\mathbf{V} \cdot \mathbf{F}$	(1-1-7)
$\mathbf{V} \cdot (\mathbf{F} \times \mathbf{G}) = (\mathbf{V} \times \mathbf{F}) \cdot \mathbf{G} - (\mathbf{V} \times \mathbf{G}) \cdot \mathbf{F}$	(1-1-8)
$\mathbf{V} \times (\varphi\mathbf{F}) = (\mathbf{V}\varphi) \times \mathbf{F} + \varphi\mathbf{V} \times \mathbf{F}$	(1-1-9)
$\mathbf{V} \times (\mathbf{F} \times \mathbf{G}) = (\mathbf{V} \cdot \mathbf{G})\mathbf{F} - (\mathbf{V} \cdot \mathbf{F})\mathbf{G} + (\mathbf{G} \cdot \mathbf{V})\mathbf{F} - (\mathbf{F} \cdot \mathbf{V})\mathbf{G}$	(1-1-10)

For a function that depends on the argument $\mathbf{R} = \mathbf{r} - \mathbf{r}'$, where $\mathbf{r}'$ is treated as a constant origin,

$$\mathbf{V} = \mathbf{V}_R; \tag{1-55}$$

$$\mathbf{V}_R = \mathbf{i}\,\frac{\partial}{\partial X} + \mathbf{j}\,\frac{\partial}{\partial Y} + \mathbf{k}\,\frac{\partial}{\partial Z},$$

where $\mathbf{R} = X\mathbf{i} + Y\mathbf{j} + Z\mathbf{k}$. If $\mathbf{r}$ is treated as constant instead,

$$\mathbf{V} = -\mathbf{V}' \tag{1-56}$$

where

$$\mathbf{V}' = \mathbf{i}\,\frac{\partial}{\partial x'} + \mathbf{j}\,\frac{\partial}{\partial y'} + \mathbf{k}\,\frac{\partial}{\partial z'}.$$

There are several possibilities for the extension of the divergence theorem and of Stokes's theorem. The most interesting of these is Green's theorem, which is

$$\int_V (\psi\nabla^2\varphi - \varphi\nabla^2\psi)\,dv = \oint_S (\psi\mathbf{V}\varphi - \varphi\mathbf{V}\psi) \cdot \mathbf{n}\,da. \tag{1-57}$$

This theorem follows from the application of the divergence theorem to the vector

$$\mathbf{F} = \psi\mathbf{V}\varphi - \varphi\mathbf{V}\psi.$$

Using this $\mathbf{F}$ in the divergence theorem, we obtain

$$\int_V \mathbf{V} \cdot [\psi\mathbf{V}\varphi - \varphi\mathbf{V}\psi]\,dv = \oint_S (\psi\mathbf{V}\varphi - \varphi\mathbf{V}\psi) \cdot \mathbf{n}\,da. \tag{1-58}$$

Using the identity (Table 1–1) for the divergence of a scalar times a vector gives

$$\mathbf{V} \cdot (\psi\mathbf{V}\varphi) - \mathbf{V} \cdot (\varphi\mathbf{V}\psi) = \psi\nabla^2\varphi - \varphi\nabla^2\psi. \tag{1-59}$$

Combining (1–58) and (1–59) yields Green's theorem. Some other integral theorems are listed in Table 1–2.

Table 1–2   Vector Integral Theorems

$$\int_S \mathbf{n} \times \nabla\varphi \; da = \oint_C \varphi \; d\mathbf{l} \qquad (1\text{–}2\text{–}1)$$

$$\int_V \nabla\varphi \; dv = \oint_S \varphi\mathbf{n} \; da \qquad (1\text{–}2\text{–}2)$$

$$\int_V \nabla \times \mathbf{F} \; dv = \oint_S \mathbf{n} \times \mathbf{F} \; da \qquad (1\text{–}2\text{–}3)$$

$$\int_V (\nabla \cdot \mathbf{G} + \mathbf{G} \cdot \nabla)\mathbf{F} \; dv = \oint_S \mathbf{F}(\mathbf{G} \cdot \mathbf{n}) \; da \qquad (1\text{–}2\text{–}4)$$

This concludes our brief discussion of vector analysis. In the interests of brevity, the proofs of many results have been relegated to the exercises. No attempt has been made to achieve a high degree of rigor; the approach has been utilitarian. What we will need we have developed; everything else has been omitted.

1–9 SUMMARY

Three different types of vector differentiation can be expressed by the vector differential operator del, ∇, namely, gradient, divergence, and curl:

$$\nabla\varphi = \mathbf{i}\,\frac{\partial\varphi}{\partial x} + \mathbf{j}\,\frac{\partial\varphi}{\partial y} + \mathbf{k}\,\frac{\partial\varphi}{\partial z},$$

$$\nabla \cdot \mathbf{F} = \frac{\partial F_x}{\partial x} + \frac{\partial F_y}{\partial y} + \frac{\partial F_z}{\partial z},$$

$$\nabla \times \mathbf{F} = \begin{vmatrix} \mathbf{i} & \mathbf{j} & \mathbf{k} \\ \dfrac{\partial}{\partial x} & \dfrac{\partial}{\partial y} & \dfrac{\partial}{\partial z} \\ F_x & F_y & F_z \end{vmatrix}.$$

Del is a linear operator. Repeated applications of it, or its application to products of functions, produce formulas that can be derived in rectangular coordinates, but are independent of the coordinate system. They can be remembered by the rules of vector algebra and ordinary differentiation. The derivatives of a few special functions are worth committing to memory. The most important integral theorems about the derivatives are:

$$\int_{a_C}^{b} \nabla\varphi \cdot d\mathbf{l} = \varphi \Big|_a^b,$$

$$\int_S \nabla \times \mathbf{F} \cdot \mathbf{n} \; da = \oint_C \mathbf{F} \cdot d\mathbf{l}, \qquad \text{(Stokes's theorem)}$$

$$\int_V \nabla \cdot \mathbf{F} \; dv = \oint_S \mathbf{F} \cdot \mathbf{n} \; da. \qquad \text{(Divergence theorem)}$$

These are generalizations of the fundamental theorem of calculus.

PROBLEMS

1–1 The vectors from the origin to the points A, B, C, D are

$$\mathbf{A} = \mathbf{i} + \mathbf{j} + \mathbf{k},$$
$$\mathbf{B} = 2\mathbf{i} + 3\mathbf{j},$$
$$\mathbf{C} = 3\mathbf{i} + 5\mathbf{j} - 2\mathbf{k},$$
$$\mathbf{D} = \mathbf{k} - \mathbf{j}.$$

Show that the lines $\overline{AB}$ and $\overline{CD}$ are parallel and find the ratio of their lengths.

1–2 Show that the following vectors are perpendicular:

$$\mathbf{A} = \mathbf{i} + 4\mathbf{j} + 3\mathbf{k},$$
$$\mathbf{B} = 4\mathbf{i} + 2\mathbf{j} - 4\mathbf{k}.$$

1–3 Show that the vectors

$$\mathbf{A} = 2\mathbf{i} - \mathbf{j} + \mathbf{k},$$
$$\mathbf{B} = \mathbf{i} - 3\mathbf{j} - 5\mathbf{k},$$
$$\mathbf{C} = 3\mathbf{i} - 4\mathbf{j} - 4\mathbf{k}$$

form the sides of a right triangle.

1–4 By squaring both sides of the equation

$$\mathbf{A} = \mathbf{B} - \mathbf{C}$$

and interpreting the result geometrically, prove the "law of cosines."

1–5 Show that

$$\mathbf{A} = \mathbf{i} \cos \alpha + \mathbf{j} \sin \alpha,$$
$$\mathbf{B} = \mathbf{i} \cos \beta + \mathbf{j} \sin \beta$$

are unit vectors in the xy-plane making angles α, β with the x-axis. By means of a scalar product, obtain the formula for $\cos (\alpha - \beta)$.

1–6 If $\mathbf{A}$ is a constant vector and $\mathbf{r}$ is the vector from the origin to the point (x, y, z), show that

$$(\mathbf{r} - \mathbf{A}) \cdot \mathbf{A} = 0$$

is the equation of a plane.

1–7 With $\mathbf{A}$ and $\mathbf{r}$ defined as in Problem 1–6, show that

$$(\mathbf{r} - \mathbf{A}) \cdot \mathbf{r} = 0$$

is the equation of a sphere.

1–8 Using the dot product, find the cosine of the angle between the body diagonal of a cube and one of the cube edges.

1–9 Prove the law of sines for a triangle by using the vector cross product with $\mathbf{A} + \mathbf{C} = \mathbf{B}$.

1–10 If $\mathbf{A}, \mathbf{B}, \mathbf{C}$, are vectors from the origin to the points A, B, C, show that

$$(\mathbf{A} \times \mathbf{B}) + (\mathbf{B} \times \mathbf{C}) + (\mathbf{C} \times \mathbf{A})$$

is perpendicular to the plane ABC.

1–11 Verify that Eq. (1–15) is a solution to Eq. (1–14) by direct substitution. [Note that Eq. (1–14) implies that **C** is perpendicular to **A**.]

1–12 Show that **A**, **B**, and **C** are not linearly independent if

$$\mathbf{A} \cdot \mathbf{B} \times \mathbf{C} = 0.$$

Are the vectors

$$\mathbf{A} = \mathbf{j} + 3\mathbf{k},$$

$$\mathbf{B} = \mathbf{i} - 2\mathbf{k},$$

$$\mathbf{C} = \mathbf{i} + \mathbf{j} + \mathbf{k}$$

linearly independent?

1–13 Show that the unit vector normal to the surface $\varphi(\mathbf{r}) = constant$ is

$$\mathbf{n} = \nabla\varphi / |\nabla\varphi|.$$

Find **n** for the ellipsoid

$$\varphi = ax^2 + by^2 + cz^2.$$

1–14 Find the gradient of φ in cylindrical coordinates, given that $ds = dr\mathbf{a_r} + r\,d\theta\mathbf{a_\theta} + dz\mathbf{k}$. It should be noted that r and θ have different meanings here than in Eqs. (1–21) and (1–22). In spherical coordinates r is the magnitude of the radius vector from the origin and θ is the polar angle. In cylindrical coordinates, r is the perpendicular distance from the cylinder axis and θ is the azimuthal angle about this axis.

1–15 From the definition of the divergence, obtain an expression for $\nabla \cdot \mathbf{F}$ in cylindrical coordinates.

1–16 Find the divergence of the vector

$$\mathbf{i}(x^2 + yz) + \mathbf{j}(y^2 + zx) + \mathbf{k}(z^2 + xy).$$

Also find the curl.

1–17 Is $\nabla \times \mathbf{F}$ necessarily perpendicular to **F** for every vector function **F**? Justify your answer.

1–18 For any two scalar functions φ and ψ, prove that

$$\nabla^2(\varphi\psi) = \varphi\nabla^2\psi + \psi\nabla^2\varphi + 2\nabla\varphi \cdot \nabla\psi.$$

1–19 If **r** is the vector from the origin to the point (x, y, z), prove the formulas

$$\nabla \cdot \mathbf{r} = 3; \qquad \nabla \times \mathbf{r} = 0; \qquad (\mathbf{u} \cdot \nabla)\mathbf{r} = \mathbf{u}.$$

[*Note:* **u** is any vector.]

1–20 If **A** is a constant vector, show that

$$\nabla(\mathbf{A} \cdot \mathbf{r}) = \mathbf{A}.$$

1–21 Prove identities (1–1–7) and (1–1–9) in Table 1–1.

1–22 If r is the magnitude of the vector from the origin to the point (x, y, z), and $f(r)$ is an arbitrary function of r, prove that

$$\nabla f(r) = \frac{\mathbf{r}}{r} \frac{df}{dr}.$$

1–23 Prove that

$$\mathbf{\nabla} \cdot \mathbf{F}(r) = \frac{\mathbf{r}}{r} \cdot \frac{d\mathbf{F}}{dr}.$$

1–24 Prove that

$$\mathbf{\nabla}\varphi(\xi) = \mathbf{A}\,\frac{d\varphi}{d\xi}$$

if $\xi = \mathbf{A} \cdot \mathbf{r}$.

1–25 Verify Eq. (1–51) in rectangular coordinates, where $\nabla^2\mathbf{F}$ in these coordinates is as defined in the text.

1–26 Prove identities (1–2–2) and (1–2–4) in Table 1–2. (*Hint:* Use the divergence theorem and one or more identities from Table 1–1.)

CHAPTER 2 Electrostatics

2–1 ELECTRIC CHARGE

The first observation of the electrification of objects by rubbing is lost in antiquity; however, it is common experience that rubbing a hard rubber comb on a piece of wool endows the rubber with the ability to pick up small pieces of paper. As a result of rubbing the two objects together (strictly speaking, as a result of bringing them into close contact), both the rubber and the wool acquire a new property; they are *charged*. This experiment serves to introduce the concept of *charge*. But charge, itself, is not created during this process; the total charge, or the sum of the charges on the two bodies, is still the same as before electrification. In the light of modern physics we know that microscopic charged particles, specifically electrons, are transferred from the wool to the rubber, leaving the wool positively charged and the rubber comb negatively charged.

Charge is a fundamental and characteristic property of the elementary particles which make up matter. In fact, all matter is composed of protons, neutrons, and electrons, and two of these particles bear charges. But even though on a microscopic scale matter is composed of a large number of charged particles, the powerful electrical forces associated with these particles are fairly well hidden in a macroscopic observation. The reason is that there are two kinds of charge, positive and negative, and an ordinary piece of matter contains approximately equal amounts of each kind. From the macroscopic viewpoint, then, charge refers to net charge, or excess charge. When we say that an object is charged, we mean that it has an excess charge, either an excess of electrons (negative) or an excess of protons (positive). In this and the following chapters, charge will usually be denoted by the symbol q.

It is an experimental observation that charge can be neither created nor destroyed. The total charge of a closed system cannot change. From the macroscopic point of view charges may be regrouped and combined in different ways; nevertheless, we may state that *net charge is conserved in a closed system*.

2–2 COULOMB'S LAW

Toward the end of the eighteenth century techniques in experimental science achieved sufficient sophistication to make possible refined observations of the forces between electric charges. The results of these observations, which were extremely controversial at the time, can be summarized in three statements. (a) There are two and only two kinds of electric charge, now known as positive and negative. (b) Two point charges exert on each other forces that act along the line joining them and are inversely proportional to the square of the distance between them. (c) These forces are also proportional to the product of the charges, are repulsive for like charges, and attractive for unlike charges. The last two statements, with the first as preamble, are known as *Coulomb's law* in honor of Charles Augustin de Coulomb (1736–1806), who was one of the leading eighteenth century students of electricity. Coulomb's law for point charges may be concisely formulated in the vector notation of Chapter 1 as

$$\mathbf{F}_1 = C\, \frac{q_1 q_2}{r_{12}^2}\, \frac{\mathbf{r}_{12}}{r_{12}}, \tag{2-1}$$

$$\mathbf{r}_{12} = \mathbf{r}_1 - \mathbf{r}_2,$$

where $\mathbf{F}_1$ is the force on charge q_1, $\mathbf{r}_{12}$ is the vector to q_1 from q_2, r_{12} is the magnitude of $\mathbf{r}_{12}$, and C is a constant of proportionality about which more will be said later. In Eq. (2–1) a unit vector in the direction of $\mathbf{r}_{12}$ has been formed by dividing $\mathbf{r}_{12}$ by its magnitude, a device of which frequent use will be made. If the force on q_2 is to be found, it is only necessary to change every subscript 1 to 2 and every 2 to 1. Understanding this notation is important, since in future work it will provide a technique for keeping track of field and source variables.

Coulomb's law applies to point charges. In the macroscopic sense a "point charge" is one whose spatial dimensions are very small compared with any other length pertinent to the problem under consideration, and we shall use the term "point charge" in this sense. To the best of our knowledge, Coulomb's law also applies to the interactions of elementary particles such as protons and electrons. Equation (2–1) is found to hold for the electrostatic repulsion between nuclei at distances greater than about 10^{-14} meter; at smaller distances, the powerful, but short-ranged, nuclear forces dominate the picture.

Equation (2–1) is an experimental law; nevertheless, there is both theoretical and experimental evidence to indicate that the inverse square law is exact, i.e., that the exponent of r_{12} is exactly 2. By an indirect experiment* it has been shown that the exponent of r_{12} can differ from 2 by no more than one part in 10^{15}.

---

* E. R. Williams, J. E. Faller, and H. A. Hill, *Phys. Rev. Letters*, vol. 26, p. 721 (1971). Similar experiments were performed earlier. Maxwell established the exponent of 2 to within one part in 20,000.

The constant C in Eq. (2–1) requires some comment, since it determines the system of units. The units of force and distance are presumably those belonging to one of the systems used in mechanics; the most direct procedure here would be to set $C = 1$, and choose the unit of charge such that Eq. (2–1) agrees with experiment. This is the procedure adopted in the gaussian system of units. Other procedures are also possible and may have certain advantages; e.g., the unit of charge may be specified in advance. It was shown by Giorgi in 1901 that all of the common electrical units, such as the ampere, volt, ohm, henry, etc., can be combined with one of the mechanical systems (namely, the mks or meter-kilogram-second system) to form a system of units for all electric and magnetic problems. It is an advantage of this system that the results of electric circuit calculations come out in the same electrical units as those which are used in the laboratory; we shall use the rationalized *mks* or *Giorgi system* of units in the present volume, in the form known as SI (*Système International*). Since in this system q is measured in coulombs (C), $\mathbf{r}$ in meters, and $\mathbf{F}$ in newtons (N), it is clear that C must have the dimensions of newton meters^2 per coulomb^2. The size of the unit of charge, the coulomb, is established from magnetic experiments; this requires that $C = 8.9874 \times 10^9 \; \mathrm{N \cdot m^2/C^2}$. We make the apparently complicated substitution, $C = 1/4\pi\epsilon_0$, in the interest of simplifying some of the other equations. The constant ϵ_0 will occur repeatedly; it is known as the *permittivity of free space*, and is numerically equal to $8.854 \times 10^{-12} \; \mathrm{C^2/N \cdot m^2}$. In Appendix I the definitions of the coulomb, the ampere, the permeability, and permittivity of free space are related to one another and to the velocity of light in a logical way; since a logical formulation of these definitions requires a knowledge of magnetic phenomena and of electromagnetic wave propagation, it is not appropriate to pursue them now. In Appendix II the gaussian system of units is discussed. Until Chapter 4, every formula can be changed to gaussian units simply by replacing ϵ_0 with $1/4\pi$.

If more than two point charges are present, the mutual forces are determined by the repeated application of Eq. (2–1). In particular, if a system of N charges is considered, the force on the ith charge is given by

$$\mathbf{F}_i = q_i \sum_{j \neq i}^{N} \frac{q_j}{4\pi\epsilon_0} \frac{\mathbf{r}_{ij}}{r_{ij}^3}, \tag{2–2}$$

$$\mathbf{r}_{ij} = \mathbf{r}_i - \mathbf{r}_j,$$

where the summation on the right is extended over all of the charges except the ith. This is, of course, just the superposition principle for forces, which says that the total force acting on a body is the vector sum of the individual forces which act on it.

A simple extension of the ideas of N interacting point charges is the interaction of a point charge with a continuous charge distribution. We deliberately choose this configuration to avoid certain subtleties that may be encountered when the interaction of two continuous charge distributions is considered. Before proceeding further the meaning of a continuous distribution of charge should be examined. It is well known that electric charge is found in multiples of a basic charge, that of the electron. In other words, if any charge were examined in great detail, its magnitude would be found to be an integral multiple of the magnitude of the electronic charge. For the purposes of macroscopic physics this discreteness of charge causes no difficulties simply because the electronic charge has a magnitude of 1.6019×10^{-19} C, which is extremely small. The smallness of the basic unit means that macroscopic charges are invariably composed of a very large number of electronic charges; this in turn means that in a macroscopic charge distribution any small element of volume contains a large number of electrons. One may then describe a charge distribution in terms of a charge density function defined as the limit of the charge per unit volume as the volume becomes infinitesimal. Care must be used, however, in applying this kind of description to atomic problems, since in these cases only a small number of electrons is involved, and the process of taking the limit is meaningless. Leaving aside these atomic cases, we may proceed as if a segment of charge might be subdivided indefinitely, and describe the charge distribution by means of point functions:

a *volume charge density* defined by

$$\rho = \lim_{\Delta V \to 0} \frac{\Delta q}{\Delta V},$$ (2-3)

and a *surface charge density* defined by

$$\sigma = \lim_{\Delta S \to 0} \frac{\Delta q}{\Delta S}.$$ (2-4)

From what has been said about q, it is evident that ρ and σ are net charge, or excess charge, densities. It is worth while mentioning that in typical solid materials even a very large charge density ρ will involve a change in the local electron density of only about one part in 10^9.

If charge is distributed through a volume V with a density ρ, and on the surface S that bounds V with a density σ, then the force exerted by this charge distribution on a point charge q located at $\mathbf{r}$ is obtained from (2-2) by replacing q_j with $\rho_j \, dv'_j$ (or with $\sigma_j \, da'_j$) and proceeding to the limit:

$$\mathbf{F}_q = \frac{q}{4\pi\epsilon_0} \int_V \frac{\mathbf{r} - \mathbf{r}'}{|\mathbf{r} - \mathbf{r}'|^3} \rho(\mathbf{r}') \, dv' + \frac{q}{4\pi\epsilon_0} \int_S \frac{\mathbf{r} - \mathbf{r}'}{|\mathbf{r} - \mathbf{r}'|^3} \sigma(\mathbf{r}') \, da'.$$ (2-5)

The variable $\mathbf{r}'$ is used to locate a point within the charge distribution, that is, it plays the role of the source point $\mathbf{r}_j$ in Eq. (2-2). It may appear at first sight that if point $\mathbf{r}$ falls inside the charge distribution, the first integral of Eq. (2-5) should diverge. This is not the case; the region of integration in the vicinity of $\mathbf{r}$ contributes a negligible amount, and the integral is well behaved (see Problem 2-5).

It is clear that the force on q as given by Eq. (2–5) is proportional to q; the same is true in Eq. (2–2). This observation leads us to introduce a vector field which is independent of q, namely, the force per unit charge. This vector field, known as the *electric field*, is considered in detail in the following section.

2–3 THE ELECTRIC FIELD

The electric field at a point is defined as the limit of the following ratio: the force on a test charge placed at the point, to the charge of the test charge, the limit being taken as the magnitude of the test charge goes to zero. The customary symbol for the electric field is **E**. In vector notation the definition of **E** becomes

$$\mathbf{E} = \lim_{q \to 0} \frac{\mathbf{F}_q}{q}. \tag{2–6}$$

The limiting process is included in the definition of **E** to ensure that the test charge does not affect the charge distribution that produces **E**. If, for example, positive charge is distributed on the surface of a conductor (a conductor is a material in which charge is free to move), then bringing a test charge into the vicinity of the conductor will cause the charge on the conductor to redistribute itself. If the electric field were calculated using the ratio of force to charge for a finite test charge, the field obtained would be that due to the redistributed charge rather than that due to the original charge distribution. In the special case where one of the charges of the charge distribution can be used as a test charge the limiting process is unnecessary. In this case the electric field at the location of the test charge will be that produced by all of the rest of the charge distribution; there will, of course, be no redistribution of charge, since the proper charge distribution obtains under the influence of the entire charge distribution, including the charge being used as test charge. In certain other cases, notably those in which the charge distribution is *specified*, the force will be proportional to the size of the test charge. In these cases, too, the limit is unnecessary; however, if any doubt exists, it is always safe to use the limiting process.

Equations (2–2) and (2–5) provide a ready means for obtaining an expression for the electric field due to a given distribution of charge. Let the charge distribution consist of N point charges $q_1, q_2, \ldots, q_N$ located at the points $\mathbf{r}_1, \mathbf{r}_2, \ldots, \mathbf{r}_N$ respectively, and a volume distribution of charge specified by the charge density $\rho(\mathbf{r}')$ in the volume V and a surface distribution characterized by the surface charge density $\sigma(\mathbf{r}')$ on the surface S. If a test charge q is located at the point $\mathbf{r}$, it experiences a force **F** given by

$$\mathbf{F} = \frac{q}{4\pi\epsilon_0} \sum_{i=1}^{N} q_i \frac{\mathbf{r} - \mathbf{r}_i}{|\mathbf{r} - \mathbf{r}_i|^3} + \frac{q}{4\pi\epsilon_0} \int_V \frac{\mathbf{r} - \mathbf{r}'}{|\mathbf{r} - \mathbf{r}'|^3} \rho(\mathbf{r}') \, dv'$$

$$+ \frac{q}{4\pi\epsilon_0} \int_S \frac{\mathbf{r} - \mathbf{r}'}{|\mathbf{r} - \mathbf{r}'|^3} \sigma(\mathbf{r}') \, da', \tag{2–7}$$

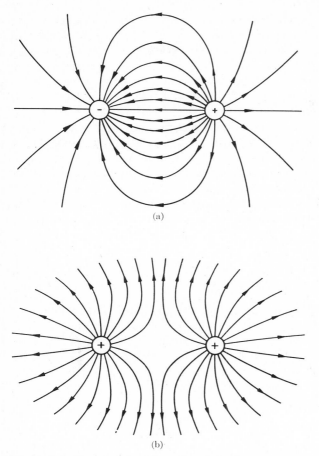

(a)

(b)

Figure 2–1 The mapping of an electric field with the aid of lines of force.

due to the given charge distribution. The electric field at **r** is the limit of the ratio of this force to the test charge q. Since the ratio is independent of q, the electric field at **r** is just

$$\mathbf{E}(\mathbf{r}) = \frac{1}{4\pi\epsilon_0} \sum_{i=1}^{N} q_i \frac{\mathbf{r} - \mathbf{r}_i}{|\mathbf{r} - \mathbf{r}_i|^3} + \frac{1}{4\pi\epsilon_0} \int_V \frac{\mathbf{r} - \mathbf{r}'}{|\mathbf{r} - \mathbf{r}'|^3} \rho(\mathbf{r}') \, dv'$$

$$+ \frac{1}{4\pi\epsilon_0} \int_S \frac{\mathbf{r} - \mathbf{r}'}{|\mathbf{r} - \mathbf{r}'|^3} \sigma(\mathbf{r}') \, da'. \tag{2–8}$$

Equation (2–8) is very general; in most cases one or more of the terms will not be needed.

The quantity we have just defined, the electric field, may be calculated at each point in space in the vicinity of a system of charges or of a charge distribution.

Thus $\mathbf{E} = \mathbf{E}(\mathbf{r})$ is a vector point function, or a vector field. This field has a number of interesting mathematical properties, which we shall proceed to develop in the following sections and in the next chapter. As an aid to visualizing the electric field structure associated with a particular distribution of charge, Michael Faraday (1791–1867) introduced the concept of *lines of force*. A line of force is an imaginary line (or curve) drawn in such a way that its direction at any point is the direction of the electric field at that point.

Consider, for example, the electric field structure associated with a single positive point charge q_1. The lines of force are radial lines radiating outward from q_1. Similarly, the lines of force associated with an isolated negative point charge are also radial lines, but this time the direction is inward (i.e., toward the negative charge). These two examples are extremely simple, but they nevertheless illustrate an important property of the field lines: the lines of force terminate on the sources of the electric field, i.e., upon the charges which produce the electric field.

Figure 2–1 shows two simple electric fields that have been mapped with the aid of lines of force.

2–4 THE ELECTROSTATIC POTENTIAL

It has been noted in Chapter 1 that if the curl of a vector vanishes, then the vector may be expressed as the gradient of a scalar. The electric field given by Eq. (2–8) satisfies this criterion. To verify this, we note that taking the curl of Eq. (2–8) involves differentiating with respect to r. This variable appears in the equation only in functions of the form $(\mathbf{r} - \mathbf{r}')/|\mathbf{r} - \mathbf{r}'|^3$, and hence it will suffice to show that functions of this form have zero curl. Using the formula from Table 1–1 for the curl of the product (vector times scalar) gives

$$\nabla \times \frac{\mathbf{r} - \mathbf{r}'}{|\mathbf{r} - \mathbf{r}'|^3} = \frac{1}{|\mathbf{r} - \mathbf{r}'|^3} \nabla \times (\mathbf{r} - \mathbf{r}') + \left[\nabla \frac{1}{|\mathbf{r} - \mathbf{r}'|^3} \right] \times [\mathbf{r} - \mathbf{r}']. \qquad (2-9)$$

A direct calculation (see Problem 1–19) shows that

$$\nabla \times (\mathbf{r} - \mathbf{r}') = 0, \qquad (2-10)$$

and (see Problem 1–22) that

$$\nabla \frac{1}{|\mathbf{r} - \mathbf{r}'|^3} = -3 \frac{\mathbf{r} - \mathbf{r}'}{|\mathbf{r} - \mathbf{r}'|^5}. \qquad (2-11)$$

These results, together with the observation that the vector product of a vector with a parallel vector is zero, suffice to prove that

$$\nabla \times \frac{\mathbf{r} - \mathbf{r}'}{|\mathbf{r} - \mathbf{r}'|^3} = 0. \qquad (2-12)$$

Since each contribution of Eq. (2–8) to the electric field is of this form, we have demonstrated that the curl of the electric field is zero. Equation (2–12) indicates that a scalar function exists whose gradient is the electric field; it remains to find this function. That is, we now know that a function exists which satisfies

$$\mathbf{E}(\mathbf{r}) = -\nabla\varphi(\mathbf{r}), \tag{2–13}$$

but we have yet to find the form of the function φ. It should be noted that it is conventional to include the minus sign in Eq. (2–13) and to call φ the electrostatic potential.

It is easy to find the electrostatic potential due to a point charge q_1; it is just

$$\varphi(\mathbf{r}) = \frac{1}{4\pi\epsilon_0} \frac{q_1}{|\mathbf{r} - \mathbf{r}_1|}, \tag{2–14}$$

as is readily verified by direct differentiation. With this as a clue it is easy to guess that the potential which gives the electric field of Eq. (2–8) is

$$\varphi(\mathbf{r}) = \frac{1}{4\pi\epsilon_0} \sum_{i=1}^{N} \frac{q_i}{|\mathbf{r} - \mathbf{r}_i|} + \frac{1}{4\pi\epsilon_0} \int_V \frac{\rho(\mathbf{r}')}{|\mathbf{r} - \mathbf{r}'|} \, dv'$$
$$+ \frac{1}{4\pi\epsilon_0} \int_S \frac{\sigma(\mathbf{r}')}{|\mathbf{r} - \mathbf{r}'|} \, da', \tag{2–15}$$

which is also easily verified by direct differentiation. It may seem that Eqs. (2–14) and (2–15) were obtained in a rather arbitrary fashion; however, since all that is required of φ is that it satisfy Eq. (2–13), and since this has been verified directly, the means by which φ was obtained is immaterial.

The electrostatic potential φ can be obtained directly as soon as its existence is established. Since φ is known to exist, we may write

$$\int_{\text{ref}}^{\mathbf{r}} \mathbf{E}(\mathbf{r}') \cdot d\mathbf{r}' = -\int_{\text{ref}}^{\mathbf{r}} \nabla\varphi \cdot d\mathbf{r}', \tag{2–16}$$

where ref stands for a reference point at which φ is zero. From the definition of the gradient,

$$\nabla\varphi \cdot d\mathbf{r}' = d\varphi. \tag{2–17}$$

Using Eq. (2–17) in Eq. (2–16) converts it into the integral of a perfect differential, which is easily done. The result is

$$-\int_{\text{ref}}^{\mathbf{r}} \nabla\varphi \cdot d\mathbf{r}' = -\varphi(\mathbf{r}) = \int_{\text{ref}}^{\mathbf{r}} \mathbf{E}(\mathbf{r}') \cdot d\mathbf{r}', \tag{2–18}$$

which is really the inverse of Eq. (2–13). If the electric field due to a point charge is used in Eq. (2–18), and the reference point or lower limit in the integral is taken at infinity, with the potential there zero, the result is

$$\varphi(\mathbf{r}) = \frac{q}{4\pi\epsilon_0 r}. \tag{2–19}$$

This, of course, is just a special case of Eq. (2–14), namely, the case where r_1 is zero. This derivation can be extended to obtain Eq. (2–15); however, the procedure is too cumbersome to include here.

Another interesting and useful aspect of the electrostatic potential is its close relation to the potential energy associated with the conservative electrostatic force. The potential energy associated with an arbitrary conservative force is

$$U(\mathbf{r}) = -\int_{ref}^{\mathbf{r}} \mathbf{F}(\mathbf{r}') \cdot d\mathbf{r}', \tag{2–20}$$

where $U(\mathbf{r})$ is the potential energy at $\mathbf{r}$ relative to the reference point at which the potential energy is arbitrarily taken to be zero. Since in the electrostatic case $\mathbf{F} = q\mathbf{E}$, it follows that if the same reference point is chosen for the electrostatic potential and for the potential energy, then the electrostatic potential is just the potential energy per unit charge. This idea is sometimes used to introduce the electrostatic potential; we feel, however, that the introduction by means of Eq. (2–13) emphasizes the importance of the electrostatic potential in determining the electrostatic field. There is, of course, no question about the equivalence of the two approaches.

The utility of the electrostatic potential in calculating electric fields can be seen by contrasting Eqs. (2–8) and (2–15). Equation (2–8) is a vector equation; to obtain the electric field from it, it is necessary to evaluate three sums or three integrals for each term. At best this is a tedious procedure; in some cases it is almost impossible to do the integrals. Equation (2–15), on the other hand, is a scalar equation and involves only one sum or integral per term. Furthermore, the denominators appearing in this equation are all of the form $|\mathbf{r} - \mathbf{r}'|$, which simplifies the integrals compared with those of Eq. (2–8). This simplification is sometimes sufficient to make the difference between doing the integrals and not doing them. It may be objected that after doing the integrals of Eq. (2–15) it is still necessary to differentiate the result; this objection is readily answered by observing that differentiation can always be accomplished if the derivatives exist, and is in fact usually much easier than integration. In Chapter 3 it will be seen that the electrostatic potential is even more important in those problems where the charge distribution is not specified, but must rather be determined in the process of solving the problem.

In the mks system the unit of energy is the newton-meter or joule. The unit of potential is joule/coulomb, but this unit occurs so frequently that it is given a special name, the volt (V). The unit of the electric field is the newton/coulomb or the volt/meter.

2–5 CONDUCTORS AND INSULATORS

So far as their electrostatic behavior is concerned, materials may be divided into two categories: *conductors* of electricity and *insulators* (*dielectrics*). Conductors are substances, like the metals, which contain large numbers of essentially free charge carriers. These charge carriers (electrons in most cases) are free to wander throughout the conducting material; they respond to almost infinitesimal electric fields, and they continue to move as long as they experience a field. These free carriers carry the electric current when a steady electric field is maintained in the conductor by an external source of energy.

Dielectrics are substances in which all charged particles are bound rather strongly to constituent molecules. The charged particles may shift their positions slightly in response to an electric field, but they do not leave the vicinity of their molecules. Strictly speaking, this definition applies to an ideal dielectric, one which shows no conductivity in the presence of an externally maintained electric field. Real physical dielectrics show a feeble conductivity, but in a typical dielectric the conductivity is 10^{20} times smaller than that of a good conductor. Since 10^{20} is a tremendous factor, it is usually sufficient to say that dielectrics are nonconductors.

Certain materials (semiconductors, electrolytes) have electrical properties intermediate between conductors and dielectrics. So far as their behavior in a static electric field is concerned, these materials behave like conductors. However, their transient response is somewhat slower; i.e., it takes longer for these materials to reach equilibrium in a static field.

In this and the following four chapters we shall be concerned with materials in *electrostatic* fields. Dielectric polarization, although a basically simple phenomenon, produces some rather complicated effects; hence we shall delay its study until Chapter 4. Conductors, on the other hand, may be treated quite easily in terms of concepts that have already been developed.

Since charge is free to move in a conductor, even under the influence of very small electric fields, the charge carriers (electrons or ions) move until they find positions in which they experience no net force. When they come to rest, the interior of the conductor must be a region devoid of an electric field; this must be so because the charge carrier population in the interior is by no means depleted, and if a field persisted, the carriers would continue to move. *Thus, under static conditions, the electric field in a conductor vanishes.* Furthermore, since $\mathbf{E} = 0$ in a conductor, the potential is the same at all points in the conducting material. In other words, *under static conditions, each conductor forms an equipotential region of space.*

2–6 GAUSS'S LAW

An important relationship exists between the integral of the normal component of the electric field over a closed surface and the total charge enclosed by the surface. This relationship, known as Gauss's law, will now be investigated in more detail. The electric field at point $\mathbf{r}$ due to a point charge q located at the origin is

$$\mathbf{E}(\mathbf{r}) = \frac{q}{4\pi\epsilon_0} \frac{\mathbf{r}}{r^3}. \qquad (2\text{–}21)$$

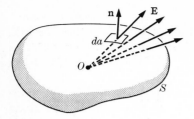

Figure 2–2 An imaginary closed surface S that encloses a point charge at the origin.

Consider the surface integral of the normal component of this electric field over a closed surface (such as that shown in Fig. 2–2) that encloses the origin and, consequently, the charge q; this integral is just

$$\oint_S \mathbf{E} \cdot \mathbf{n} \, da = \frac{q}{4\pi\epsilon_0} \oint_S \frac{\mathbf{r} \cdot \mathbf{n}}{r^3} \, da. \tag{2-22}$$

The quantity $(\mathbf{r}/r) \cdot \mathbf{n} \, da$ is the projection of da on a plane perpendicular to $\mathbf{r}$. This projected area divided by r^2 is the solid angle subtended by da, which is written $d\Omega$. It is clear from Fig. 2–3 that the solid angle subtended by da is the same as the solid angle subtended by da', an element of the surface area of the sphere S' whose center is at the origin and whose radius is r'. It is then possible to write

$$\oint_S \frac{\mathbf{r} \cdot \mathbf{n}}{r^3} \, da = \oint_{S'} \frac{\mathbf{r}' \cdot \mathbf{n}}{r'^3} \, da' = 4\pi,$$

which shows that

$$\oint_S \mathbf{E} \cdot \mathbf{n} \, da = \frac{q}{4\pi\epsilon_0} 4\pi = \frac{q}{\epsilon_0} \tag{2-23}$$

in the special case described above. If q lies outside of S, it is clear from Fig. 2–4 that S can be divided into two areas S_1 and S_2 each of which subtends the same solid angle at the charge q. For S_2, however, the direction of the normal is toward q, while for S_1 it is away from q. Therefore the contributions of S_1 and S_2 to the surface integral are equal and opposite, and the total integral vanishes. Thus if the surface surrounds a point charge q, the surface integral of the normal component

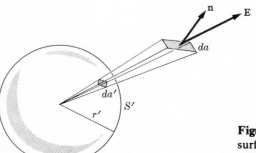

Figure 2–3 Construction of the spherical surface S' as an aid to evaluation of the solid angle subtended by da.

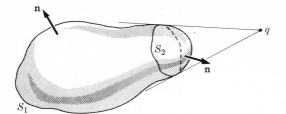

Figure 2–4 The closed surface S may be divided into two surfaces, S_1 and S_2, each of which subtend the same solid angle at q.

of the electric field is q/ϵ_0, while if q lies outside the surface the surface integral is zero. The preceding statement applies to any closed surface, even to so-called re-entrant ones. A study of Fig. 2–5 is sufficient to verify that this is indeed the case.

If several point charges $q_1, q_2, \ldots, q_N$ are enclosed by the surface S, then the total electric field is given by the first term of Eq. (2–8). Each charge subtends a full solid angle (4π); hence Eq. (2–23) becomes

$$\oint_S \mathbf{E} \cdot \mathbf{n} \, da = \frac{1}{\epsilon_0} \sum_{i=1}^{N} q_i. \tag{2–24}$$

This result can be readily generalized to the case of a continuous distribution of charge characterized by a charge density. If each element of charge $\rho \, dv$ is considered as a point charge, it contributes $\rho \, dv/\epsilon_0$ to the surface integral of the normal component of the electric field provided it is inside the surface over which we integrate. The total surface integral is then the sum of all contributions of this form due to the charge inside the surface. Thus if S is a closed surface which bounds the volume V,

$$\oint_S \mathbf{E} \cdot \mathbf{n} \, da = \frac{1}{\epsilon_0} \int_V \rho \, dv. \tag{2–25}$$

Equations (2–24) and (2–25) are known as Gauss's law. The term on the left, the integral of the normal component of the electric field over the surface S, is sometimes called the *flux* of the electric field through S.

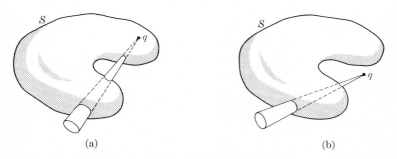

(a)   (b)

Figure 2–5 An element of solid angle cutting the surface S more than once.

Gauss's law may be expressed in yet another form by using the divergence theorem. The divergence theorem (1–37) states that

$$\oint_S \mathbf{F} \cdot \mathbf{n} \, da = \int_V \nabla \cdot \mathbf{F} \, dv.$$

If this theorem is applied to the surface integral of the normal component of **E**, it yields

$$\oint_S \mathbf{E} \cdot \mathbf{n} \, da = \int_V \nabla \cdot \mathbf{E} \, dv, \tag{2–26}$$

which, when substituted into Eq. (2–25), gives

$$\int_V \nabla \cdot \mathbf{E} \, dv = \frac{1}{\epsilon_0} \int_V \rho \, dv. \tag{2–27}$$

Equation (2–27) must be valid for all volumes, that is, for any choice of the volume V. The only way in which this can be true is if the integrands appearing on the left and on the right in the equation are equal. Thus the validity of Eq. (2–27) for any choice of V implies that

$$\nabla \cdot \mathbf{E} = \frac{1}{\epsilon_0} \rho. \tag{2–28}$$

This result may be thought of as a differential form of Gauss's law.

2–7 APPLICATION OF GAUSS'S LAW

Equation (2–28) or, more properly, a modified form of this equation which will be derived in Chapter 4, is one of the basic differential equations of electricity and magnetism. In this role it is important, of course; but Gauss's law also has practical utility. This practicality of the law lies largely in providing a very easy way to calculate electric fields in sufficiently symmetric situations. In other words, in certain highly symmetric situations of considerable physical interest, the electric field may be calculated by using Gauss's law instead of by the integrals given above or by the procedures of Chapter 3. When this can be done, it accomplishes a major saving in effort.

In order that Gauss's law be useful in calculating the electric field, it must be possible to choose a closed surface such that the electric field has a normal component that is either zero or a single fixed value at every point on the surface. As an example, consider a very long line charge of charge density λ per unit length, as shown in Fig. 2–6. The symmetry of the situation clearly indicates that the electric field is radial and independent of both position along the wire and angular position around the wire. These observations lead us to choose the surface shown in Fig. 2–6. For this surface it is easy to evaluate the integral of the normal component of the electric field. The circular ends contribute nothing, since the electric field is parallel to them. The cylindrical surface contributes $2\pi r l E_r$, since **E**

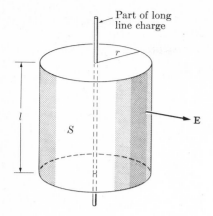

Part of long
line charge

r

l

S

E

Figure 2–6 A cylindrical surface to be used
with Gauss's law to find the electric field
produced by a long line charge.

is radial and independent of the position of the cylindrical surface. Gauss's law
then takes the form

$$2\pi r l E_r = \frac{\lambda l}{\epsilon_0}. \tag{2-29}$$

Equation (2–29) can be solved for E_r to give

$$E_r = \frac{\lambda}{2\pi\epsilon_0 r}. \tag{2-30}$$

The saving of effort accomplished by the use of Gauss's law will be more fully
appreciated by solving Problem 2–4, which involves direct application of Eq.
(2–8).

Another important result of Gauss's law is that the charge (net charge) of a
charged conductor resides on its outer surface. We saw in Section 2–5 that the
electric field inside a conductor vanishes. We may construct a gaussian surface
anywhere inside the conductor; by Gauss's law, the net charge enclosed by each of
these surfaces is zero. Finally, we construct the Gaussian surface S of Fig. 2–7;
again the net charge enclosed is zero. The only place left for the charge which is
not in contradiction with Gauss's law is for it to reside on the surface of the
conductor. Since there is no charge in the interior, part of the material could be
removed without changing anything. Thus the charge of a conducting *shell* must
reside entirely on the *outer* surface.

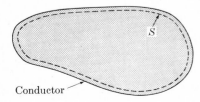

S

Conductor

Figure 2–7 A Gaussian surface S
constructed inside a charged conductor.

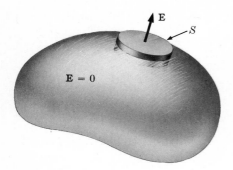

Figure 2–8 Application of Gauss's law to the closed, pillbox-shaped surface S that intersects the surface of a charged conductor.

The electric field just outside a charged conductor must be normal to the surface of the conductor. This follows because the surface is an equipotential, and $\mathbf{E} = -\nabla\varphi$. Let us assume that the charge on a conductor is given by the surface density function σ. If Gauss's law is applied to the small pillbox-shaped surface S of Fig. 2–8, then

$$E\,\Delta S = \left(\frac{\sigma}{\epsilon_0}\right)\Delta S,$$

where ΔS is the area of one of the pillbox bases. Hence, for the electric field just outside a conductor,

$$E = \frac{\sigma}{\epsilon_0}. \tag{2-31}$$

2–8 THE ELECTRIC DIPOLE

Two equal and opposite charges separated by a small distance form an electric dipole. The electric field and potential distribution produced by such a charge configuration can be investigated with the aid of the formulas of Sections 2–3 and 2–4. Suppose that a charge $-q$ is located at the point $\mathbf{r}'$ and a charge q is located at $\mathbf{r}' + \mathbf{l}$, as shown in Fig. 2–9; then the electric field at an arbitrary point $\mathbf{r}$ may be

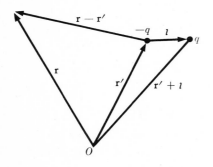

Figure 2–9 Geometry involved in calculating the electric field $\mathbf{E(r)}$ due to two point charges.

found by direct application of Eq. (2–8). The electric field at **r** is found to be

$$\mathbf{E(r)} = \frac{q}{4\pi\epsilon_0} \left\{ \frac{\mathbf{r} - \mathbf{r}' - \boldsymbol{l}}{|\mathbf{r} - \mathbf{r}' - \boldsymbol{l}|^3} - \frac{\mathbf{r} - \mathbf{r}'}{|\mathbf{r} - \mathbf{r}'|^3} \right\}. \tag{2-32}$$

This is the correct electric field for any value of q and any value of the separation $\boldsymbol{l}$; however, it is not easy to interpret. What we want is the dipole field, and in the dipole the separation $\boldsymbol{l}$ is small compared with $\mathbf{r} - \mathbf{r}'$; hence we may expand Eq. (2–32), keeping only the first nonvanishing term. Since this procedure is of general utility it will be considered in detail. The primary difficulty in making this expansion is caused by the denominator of the first term of Eq. (2–32). The reciprocal of this denominator can be rewritten as

$$|\mathbf{r} - \mathbf{r}' - \boldsymbol{l}|^{-3} = [(\mathbf{r} - \mathbf{r}')^2 - 2(\mathbf{r} - \mathbf{r}') \cdot \boldsymbol{l} + l^2]^{-3/2}$$

$$= |\mathbf{r} - \mathbf{r}'|^{-3} \left[1 - \frac{2(\mathbf{r} - \mathbf{r}') \cdot \boldsymbol{l}}{|\mathbf{r} - \mathbf{r}'|^2} + \frac{l^2}{|\mathbf{r} - \mathbf{r}'|^2} \right]^{-3/2}.$$

In the last form it is easy to expand by the binomial theorem, keeping only terms linear in $\boldsymbol{l}$. The result of this expansion is

$$|\mathbf{r} - \mathbf{r}' - \boldsymbol{l}|^{-3} = |\mathbf{r} - \mathbf{r}'|^{-3} \left\{ 1 + \frac{3(\mathbf{r} - \mathbf{r}') \cdot \boldsymbol{l}}{|\mathbf{r} - \mathbf{r}'|^2} + \cdots \right\}, \tag{2-33}$$

where terms involving l^2 have been dropped. Using Eq. (2–33) in Eq. (2–32) and again keeping only terms linear in $\boldsymbol{l}$ gives

$$\mathbf{E(r)} = \frac{q}{4\pi\epsilon_0} \left\{ \frac{3(\mathbf{r} - \mathbf{r}') \cdot \boldsymbol{l}}{|\mathbf{r} - \mathbf{r}'|^5} (\mathbf{r} - \mathbf{r}') - \frac{\boldsymbol{l}}{|\mathbf{r} - \mathbf{r}'|^3} + \cdots \right\}. \tag{2-34}$$

Equation (2–34) gives that part of the electric field, due to a finite electric dipole, which is proportional to the separation of the charges. There are other contributions proportional to the square, the cube, and higher powers of the separation. If, however, the separation is small, these higher powers contribute very little. In the limit as $\boldsymbol{l}$ goes to zero, all of the terms vanish unless the charge becomes infinite. In the limit as $\boldsymbol{l}$ goes to zero while q becomes infinite, in such a way that $q\boldsymbol{l}$ remains constant, all terms except the term linear in $\boldsymbol{l}$ vanish. In this limit a *point dipole* is formed. A point dipole has no net charge, no extent in space, and is completely characterized by its dipole moment, which is the limit of $q\boldsymbol{l}$ as $\boldsymbol{l}$ goes to zero. We use the symbol **p** to represent the electric dipole moment, and write

$$\mathbf{p} = q\boldsymbol{l}. \tag{2-35}$$

In terms of the dipole moment, Eq. (2–34) may be written .

$$\mathbf{E(r)} = \frac{1}{4\pi\epsilon_0} \left\{ \frac{3(\mathbf{r} - \mathbf{r}') \cdot \mathbf{p}}{|\mathbf{r} - \mathbf{r}'|^5} (\mathbf{r} - \mathbf{r}') - \frac{\mathbf{p}}{|\mathbf{r} - \mathbf{r}'|^3} \right\}. \tag{2-36}$$

The potential distribution produced by a point dipole is also important. This could be found by looking for a function with gradient equal to the right side of

Eq. (2–36). It is, however, easier to apply Eq. (2–15) to the charge distribution consisting of two point charges separated by a small distance. Using the notation of Eq. (2–32), the potential distribution is given by

$$\varphi(\mathbf{r}) = \frac{q}{4\pi\epsilon_0} \left[\frac{1}{|\mathbf{r} - \mathbf{r}' - \mathbf{l}|} - \frac{1}{|\mathbf{r} - \mathbf{r}'|} \right]. \tag{2–37}$$

By expanding the first term in exactly the same way as was done for the first term of Eq. (2–32) and retaining only the linear term in $\mathbf{l}$, Eq. (2–37) can be put in the form

$$\varphi(\mathbf{r}) = \frac{q}{4\pi\epsilon_0} \frac{(\mathbf{r} - \mathbf{r}') \cdot \mathbf{l}}{|\mathbf{r} - \mathbf{r}'|^3}. \tag{2–38}$$

This equation is valid to the same approximation as Eq. (2–34); namely, terms proportional to l^2 and to higher powers of l are neglected. For a point dipole, $\mathbf{p}$, Eq. (2–38) is exact; however, it is better written as

$$\varphi(\mathbf{r}) = \frac{1}{4\pi\epsilon_0} \frac{\mathbf{p} \cdot (\mathbf{r} - \mathbf{r}')}{|\mathbf{r} - \mathbf{r}'|^3}. \tag{2–39}$$

Equation (2–39) gives the potential $\varphi(\mathbf{r})$ produced by an electric dipole; from this potential the electric field (Eq. 2–36) may be determined. It is also interesting to inquire about the potential energy of an electric dipole which is placed in an *external electric field*. In the case of two charges, $-q$ at $\mathbf{r}$ and q at $\mathbf{r} + \mathbf{l}$, in an electric field described by the potential function $\varphi_{ext}(\mathbf{r})$, the potential energy is just

$$U = -q\varphi_{ext}(\mathbf{r}) + q\varphi_{ext}(\mathbf{r} + \mathbf{l}). \tag{2–40}$$

If $\mathbf{l}$ is small compared with $\mathbf{r}$, $\varphi_{ext}(\mathbf{r} + \mathbf{l})$ may be expanded in a power series in $\mathbf{l}$ and only the first two terms kept. The expansion gives

$$\varphi_{ext}(\mathbf{r} + \mathbf{l}) = \varphi_{ext}(\mathbf{r}) + \mathbf{l} \cdot \nabla\varphi_{ext}, \tag{2–41}$$

where the value of the gradient at point $\mathbf{r}$ is to be used. If this expansion is used in Eq. (2–40), the result is

$$U = q\mathbf{l} \cdot \nabla\varphi_{ext}. \tag{2–42}$$

Going to the limit of a point dipole gives simply

$$U(\mathbf{r}) = \mathbf{p} \cdot \nabla\varphi_{ext}, \tag{2–43}$$

which is, of course, exact. Since the electric field is the negative gradient of the electrostatic potential, an alternative form of Eq. (2–43) is

$$U(\mathbf{r}) = -\mathbf{p} \cdot \mathbf{E}_{ext}(\mathbf{r}). \tag{2–44}$$

This, then, is the potential energy of a dipole $\mathbf{p}$ in an external electric field $\mathbf{E}_{ext}$, where $\mathbf{E}_{ext}(\mathbf{r})$ is evaluated at the location of the dipole.

It is important to note that two potentials have been discussed in this section. In Eqs. (2–37), (2–38), and (2–39), the electrostatic potential *produced by* an electric dipole is considered. In Eqs. (2–40) through (2–43), the dipole is considered to be in an *existing electric field* described by a potential function $\varphi_{ext}(\mathbf{r})$. This electric field is due to charges other than those comprising the dipole; in fact, the dipole field must be excluded to avoid an infinite result. This statement could lead us to rather complicated questions concerning self-forces and self-energies which we cannot discuss here; however, it may be noted that the potential energy resulting from the interaction of an electric dipole with its own field arises from forces exerted on the dipole by itself. Such forces, known in dynamics as internal forces, do not affect the motion of the dipole as a whole. For our purposes further consideration of this question will be unnecessary.

2–9 MULTIPOLE EXPANSION OF ELECTRIC FIELDS

It is apparent from the definition of dipole moments given above that certain aspects of the potential distribution produced by a specified distribution of charge might well be expressed in terms of its electric dipole moment. In order to do so it is necessary, of course, to define the electric dipole moment of an arbitrary charge distribution. Rather than make an unmotivated definition, we shall consider a certain expansion of the electrostatic potential due to an arbitrary charge distribution. To reduce the number of position coordinates, a charge distribution in the neighborhood of the origin of coordinates will be considered. The further restriction will be made that the charge distribution can be entirely enclosed by a sphere of radius a which is small compared with the distance to the point of observation. An arbitrary point within the charge distribution will be designated by $\mathbf{r}'$, the charge density at that point by $\rho(\mathbf{r}')$, and the observation point by $\mathbf{r}$ (see Fig. 2–10). The potential at $\mathbf{r}$ is given by

$$\varphi(\mathbf{r}) = \frac{1}{4\pi\epsilon_0} \int_V \frac{\rho(\mathbf{r}')}{|\mathbf{r} - \mathbf{r}'|} \, dv', \tag{2–45}$$

where dv' is used to designate an element of volume in the charge distribution and V denotes the entire volume occupied by the charge distribution. In view of the restriction made above to points of observation which are remote from the origin, the quantity $|\mathbf{r} - \mathbf{r}'|^{-1}$ can be expanded in a series of ascending powers of $\mathbf{r}'/r$.

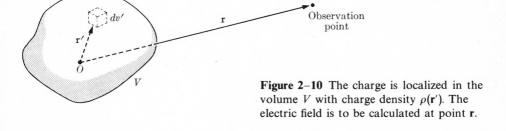

Figure 2–10 The charge is localized in the volume V with charge density $\rho(\mathbf{r}')$. The electric field is to be calculated at point $\mathbf{r}$.

The result of such an expansion is

$$|\mathbf{r} - \mathbf{r}'|^{-1} = (r^2 - 2\mathbf{r} \cdot \mathbf{r}' + r'^2)^{-1/2}$$

$$= \frac{1}{r} \left\{ 1 - \frac{1}{2} \left[-\frac{2\mathbf{r} \cdot \mathbf{r}'}{r^2} + \frac{r'^2}{r^2} \right] + \frac{1}{2} \frac{1}{2} \frac{3}{2} [\quad]^2 + \cdots \right\}, \qquad (2\text{--}46)$$

where only the first three terms are explicitly indicated. It should be noted that while $(r'/r)^2$ is negligible compared with $2\mathbf{r} \cdot \mathbf{r}/r^2$, it may not be dropped in the first set of brackets because it is of the same order as the dominant term in the second set of brackets. Using Eq. (2–46) in Eq. (2–45) and omitting terms involving the cube and higher powers of $\mathbf{r}'$ yields

$$\varphi(\mathbf{r}) = \frac{1}{4\pi\epsilon_0} \int_V \left\{ \frac{1}{r} + \frac{\mathbf{r} \cdot \mathbf{r}'}{r^3} + \frac{1}{2} \left[\frac{3(\mathbf{r} \cdot \mathbf{r}')^2}{r^5} - \frac{r'^2}{r^3} \right] + \cdots \right\} \rho(\mathbf{r}') \, dv'. \qquad (2\text{--}47)$$

Since $\mathbf{r}$ does not involve the variable of integration $\mathbf{r}'$, all of the $\mathbf{r}$ dependence may be taken from under the integral sign to obtain

$$\varphi(\mathbf{r}) = \frac{1}{4\pi\epsilon_0} \left\{ \frac{1}{r} \int_V \rho(\mathbf{r}') \, dv' + \frac{\mathbf{r}}{r^3} \cdot \int_V \mathbf{r}' \rho(\mathbf{r}') \, dv' \right.$$

$$\left. + \sum_{i=1}^{3} \sum_{j=1}^{3} \frac{1}{2} \frac{x_i x_j}{r^5} \int_V (3x_i' x_j' - \delta_{ij} r'^2) \rho(\mathbf{r}') \, dv' \right\}, \qquad (2\text{--}48)$$

where x_i, x_j are Cartesian components of $\mathbf{r}$, x_i', x_j' are the Cartesian components of $\mathbf{r}'$, and δ_{ij} is defined as follows:

$$\delta_{ij} = \begin{cases} 0, & i \neq j \\ 1, & i = j \end{cases}.$$

It is easy to interpret Eq. (2–48). The first integral in the equation is clearly the total charge, and the first term is the potential which would result if this total charge were concentrated at the origin. The second integral is very similar to the dipole moment defined in Section 2–8, and so it is called the *dipole moment of the charge distribution*. As a definition, this represents a generalization of the definition given for two equal and opposite point charges; it is easy to show, however, that both definitions give the same result for two equal and opposite point charges. The second term in Eq. (2–48) is the potential that would result if a point dipole equal to the dipole moment of the charge distribution were located at the origin of coordinates. It is interesting to note that the dipole moment of a charge distribution is independent of the origin of coordinates if the total charge is zero. To verify this, consider a new coordinate system with *origin* at $\mathbf{R}$ in the old system. Denoting a point with respect to the old system by $\mathbf{r}'$ and the same point with respect to the new system by $\mathbf{r}''$, we have

$$\mathbf{r}' = \mathbf{r}'' + \mathbf{R}. \qquad (2\text{--}49)$$

The dipole moment with respect to the old system is

$$\mathbf{p} = \int_V \mathbf{r}'\rho(\mathbf{r}')\, dv' = \int_V (\mathbf{r}'' + \mathbf{R})\rho(\mathbf{r}')\, dv' = \int_V \mathbf{r}''\rho\, dv' + \mathbf{R}Q, \qquad (2\text{-}50)$$

which proves the statement above.

The third term of Eq. (2–48) can be written

$$\sum_{i=1}^{3} \sum_{j=1}^{3} \frac{1}{2} \frac{x_i x_j}{r^5} Q_{ij}, \qquad (2\text{-}51)$$

where

$$Q_{ij} = \int_V (3x_i' x_j' - \delta_{ij} r'^2)\rho(\mathbf{r}')\, dv'. \qquad (2\text{-}52)$$

There are nine components of Q_{ij} corresponding to i, j equal to 1, 2, 3. Of these nine components six are equal in pairs, leaving six distinct components. This set of quantities forms the quadrupole moment tensor* and represents an extension of the dipole moment concept. There are, of course, higher-order moments which are generated by keeping higher-order terms in the expansion of Eq. (2–48). These higher-order multipoles are important in nuclear physics, but will not be considered further in this book.

The electric multipoles are used, as Eq. (2–48) indicates, to approximate the electric field of a charge distribution. There are, however, many other uses, all in the framework of approximating a real extended charge distribution by point charges, point dipoles, etc. These approximations often make it possible to solve extremely difficult problems.

2–10 THE DIRAC DELTA FUNCTION

In the general expressions for electric field and potential in Eqs. (2–8) and (2–15), we have distinguished between point charges and continuous charge distributions. For economy of notation, if for no other reason, it would be useful to be able to express point charges as a special case of a general charge-density function $\rho(\mathbf{r})$. The Dirac *delta function* $\delta(\mathbf{r})$ can serve this purpose, and in addition it is a valuable mathematical tool in many calculations. We write

$$\rho(\mathbf{r}) = q\delta(\mathbf{r}), \qquad \text{(point charge)} \qquad (2\text{-}53)$$

where

$$\delta(\mathbf{r}) = 0 \qquad \text{for } \mathbf{r} \neq 0,$$

$$\int \delta(\mathbf{r}')\, dv' = 1. \qquad (2\text{-}54)$$

---

* Tensors are a generalization of vectors and an elementary discussion is given in Appendix I.

Clearly this gives a mathematical expression to the physical idea of a point charge at $\mathbf{r} = 0$: the integrated charge density is q, but all of the charge is located exactly at the origin. The delta function is obviously a very highly singular mathematical function if it can be zero everywhere except at a single point and yet have a nonzero integral.* Nevertheless it is a legitimate mathematical object, which leads to no difficulties if one is cautious, not trying to differentiate it like a continuous function, for example. A variation of it can be made to represent a surface charge density $\sigma(\mathbf{r})$, that is, a charge distribution that vanishes everywhere except on a certain surface. With these extensions, the single integral over $\rho(\mathbf{r})$ is enough in Eqs. (2–8) and (2–15). For this application we note that

$$\int F(\mathbf{r}')\delta(\mathbf{r}') \, dv' = F(0), \tag{2–55}$$

where F is any scalar or vector function, since the integrand vanishes except at $\mathbf{r}' = 0$. Furthermore,

$$\int F(\mathbf{r}')\delta(\mathbf{r}' - \mathbf{r}_0) \, dv' = F(\mathbf{r}_0). \tag{2–56}$$

Thus, if $\rho(\mathbf{r}') = q_i \, \delta(\mathbf{r}' - \mathbf{r}_i)$,

$$\varphi(\mathbf{r}) = \frac{1}{4\pi\epsilon_0} \int \frac{q_i \delta(\mathbf{r}' - \mathbf{r}_i)}{|\mathbf{r} - \mathbf{r}'|} \, dv' = \frac{1}{4\pi\epsilon_0} \frac{q_i}{|\mathbf{r} - \mathbf{r}_i|}$$

for a point charge q_i at $\mathbf{r}_i$.

Some other properties of the delta function can be obtained as consequences of Gauss's law in differential form,

$$\mathbf{V} \cdot \mathbf{E} = \frac{1}{\epsilon_0} \rho. \tag{2–28}$$

For a point charge q at $\mathbf{r} = 0$, with (2–21)

$$\mathbf{V} \cdot \frac{q}{4\pi\epsilon_0} \frac{\mathbf{r}}{r^3} = \frac{1}{\epsilon_0} q\delta(\mathbf{r}),$$

or

$$\mathbf{V} \cdot \frac{\mathbf{r}}{r^3} = 4\pi\delta(\mathbf{r}). \tag{2–57}$$

Also, since

$$\mathbf{V} \frac{1}{r} = \frac{\mathbf{r}}{r} \frac{d}{dr}\left(\frac{1}{r}\right) = -\frac{\mathbf{r}}{r^3},$$

$$\nabla^2 \left(\frac{1}{r}\right) = -4\pi\delta(\mathbf{r}). \tag{2–58}$$

---

* The Riemann integral of such a function is zero if it exists at all, but the integration can be handled by the more general Lebesgue integral. See Appendix IV for other properties of the delta function.

Equations (2–57) and (2–58) are important enough that a more direct derivation is worthwhile. A straightforward differentiation shows that the divergence is zero everywhere but at the origin:

$$\nabla \cdot \frac{\mathbf{r}}{r^3} = \nabla\left(\frac{1}{r^3}\right) \cdot \mathbf{r} + \frac{1}{r^3}\nabla \cdot \mathbf{r}$$

$$= -\frac{3}{r^4}\frac{\mathbf{r}}{r} \cdot \mathbf{r} + \frac{3}{r^3} = 0, \qquad r \neq 0.$$

At $r = 0$ this gives $-\infty + \infty$, which is indeterminate. The divergence theorem, however, applied to a small sphere of radius R around the origin, gives

$$\int_V \nabla \cdot \frac{\mathbf{r}}{r^3}\, dv = \oint_S \frac{\mathbf{r} \cdot \mathbf{n}}{r^3}\, da = \frac{1}{R^2}\oint_S da = 4\pi.$$

Since the volume integral is 4π no matter how small the radius R, the integrand can be represented as $4\pi\delta(\mathbf{r})$, in agreement with Eq. (2–57). In other words, the delta function permits the divergence theorem to be applied to $\mathbf{r}/r^3$, even in a region that contains the singularity at the origin. The delta function is extremely useful whenever an integral over the divergence of $\mathbf{r}/r^3$, or over the Laplacian of $1/r$, is encountered.

2–11 SUMMARY

Electrostatics is based on Coulomb's law, which for a point charge q_1 at the origin and a point charge q at $\mathbf{r}$ gives the electrostatic force on q as

$$\mathbf{F}_e = \frac{1}{4\pi\epsilon_0}\frac{qq_1}{r^2}\frac{\mathbf{r}}{r},$$

where $\epsilon_0 = 8.854 \times 10^{-12}\ \mathrm{C^2/N \cdot m^2}$ in mks units and $\epsilon_0 = 1/4\pi$ in gaussian units. It is convenient to treat q as a test charge and abstract from it by defining the electric field $\mathbf{E}$ corresponding to the electric force $\mathbf{F}_e$,

$$\mathbf{F}_e = q\mathbf{E}.$$

The electrostatic field at $\mathbf{r}$ due to the source charge q_1 located at $\mathbf{r}_1 = 0$ is

$$\mathbf{E}(\mathbf{r}) = \frac{1}{4\pi\epsilon_0}\frac{q_1}{r^2}\frac{\mathbf{r}}{r}.$$

The curl and divergence of $\mathbf{E}$ are both of fundamental importance.

$$\nabla \times \frac{\mathbf{r}}{r^3} = 0.$$

$$\nabla \cdot \frac{\mathbf{r}}{r^3} = 4\pi\delta(\mathbf{r}),$$

where the Dirac delta function is defined by

$$\delta(\mathbf{r}) = 0, \qquad \mathbf{r} \neq 0,$$

$$\int \delta(\mathbf{r}) \, dv = 1.$$

The delta function has the property that for any function F

$$\int F(\mathbf{r})\delta(\mathbf{r} - \mathbf{r}_0) \, dv = F(\mathbf{r}_0).$$

Thus for a point charge

$$\nabla \times \mathbf{E} = 0.$$

$$\nabla \cdot \mathbf{E} = \frac{1}{\epsilon_0} q_1 \delta(\mathbf{r}).$$

1. Coulomb's law can be generalized to systems of many source charges, or to a continuous distribution of charge density $\rho(\mathbf{r})$ defined so that the element of charge in a volume element dv is

$$dq = \rho(\mathbf{r}) \, dv.$$

For a point charge q_i at $\mathbf{r}_i$,

$$\rho(\mathbf{r}) = q_i\delta(\mathbf{r} - \mathbf{r}_i).$$

Since the forces and fields are additive,

$$\mathbf{E}(\mathbf{r}) = \frac{1}{4\pi\epsilon_0} \int \frac{\mathbf{r} - \mathbf{r}'}{|\mathbf{r} - \mathbf{r}'|^3} \rho(\mathbf{r}') \, dv'.$$

Since ∇ is a linear operator,

$$\nabla \times \mathbf{E} = 0,$$

$$\nabla \cdot \mathbf{E} = \frac{1}{\epsilon_0} \rho(\mathbf{r}).$$

These are the fundamental differential equations that must be satisfied locally at every point by all electrostatic fields. (In fact, the divergence equation is satisfied even by time-dependent fields, and it is one of the four fundamental Maxwell equations.)

2. Gauss's law follows from the divergence equation by integrating both sides over an arbitrary volume V and applying the divergence theorem:

$$\int_S \mathbf{E} \cdot \mathbf{n} \, da = \frac{1}{\epsilon_0} Q,$$

where

$$Q = \int_V \rho(\mathbf{r}) \, dv$$

is the total charge inside V bounded by S. This has practical usefulness for calculating E in a few special situations where it can be argued on general grounds that E must be constant in magnitude and direction relative to some chosen surface S. It also shows that the charge on a conductor must lie on its outer surface.

3. The existence of an electrostatic potential function $\varphi(\mathbf{r})$ follows from the curl equation, such that

$$\mathbf{E} = -\nabla\varphi.$$

For a given E-field,

$$\varphi(\mathbf{r}) = -\int_{\mathbf{r}_0}^{\mathbf{r}} \mathbf{E} \cdot d\mathbf{l}.$$

For a specified charge distribution,

$$\varphi(\mathbf{r}) = \frac{1}{4\pi\epsilon_0} \int \frac{\rho(\mathbf{r}')}{|\mathbf{r} - \mathbf{r}'|} \, dv'.$$

This is easier to evaluate than the integral for $\mathbf{E}$. The scalar potential φ is related to the potential energy U of the conservative electrostatic force by

$$U = q\varphi.$$

4. At some distance away from the region where the source charges ρ are located, the multipole expansion of φ is useful:

$$\varphi(\mathbf{r}) = \frac{1}{4\pi\epsilon_0} \left[\frac{Q}{r} + \frac{\mathbf{p} \cdot \mathbf{r}}{r^3} + \cdots \right],$$

where

$$\mathbf{p} = \int_V \mathbf{r}'\rho(\mathbf{r}') \, dv'$$

is the dipole moment of the charge distribution. Usually the first nonvanishing term in the expansion is most important; we shall consider only the first two terms.

PROBLEMS

2–1 Two particles, each of mass m and having charge q, are suspended by strings of length l from a common point. Find the angle θ that each string makes with the vertical.

2–2 Two small identical conducting spheres have charges of 2.0×10^{-9} C and -0.5×10^{-9} C, respectively. When they are placed 4 cm apart, what is the force between them? If they are brought into contact and then separated by 4 cm, what is the force between them?

2–3 Point charges of 3×10^{-9} C are situated at each of three corners of a square whose side is 15 cm. Find the magnitude and direction of the electric field at the vacant corner point of the square.

2–4 Given an infinitely long line charge with uniform charge density λ per unit length. Using direct integration, find the electric field at a distance r from the line.

2–5 (a) A circular disk of radius R has a uniform surface charge density σ. Find the electric field at a point on the axis of the disk at a distance z from the plane of the disk. (b) A right circular cylinder of radius R and height L is oriented along the z-axis. It has a nonuniform volume density of charge given by $\rho(z) = \rho_0 + \beta z$ with reference to an origin at the center of the cylinder. Find the force on a point charge q placed at the center of the cylinder.

2–6 A thin, conducting, spherical shell of radius R is charged uniformly with total charge Q. By direct integration, find the potential at an arbitrary point (a) inside the shell, (b) outside the shell.

2–7 Two point charges, $-q$ and $+\frac{1}{2}q$, are situated at the origin and at the point $(a, 0, 0)$ respectively. At what point along the x-axis does the electric field vanish? In the x, y-plane, make a plot of the equipotential surface which goes through the point just referred to. Is this point a true minimum in the potential?

2–8 Show that the $\varphi = 0$ equipotential surface of the preceding problem is spherical in shape. What are the coordinates of the center of this sphere?

2–9 Given a right circular cylinder of radius R and length L containing a uniform charge density ρ. Calculate the electrostatic potential at a point on the cylinder axis but external to the distribution.

2–10 Given a region of space in which the electric field is everywhere directed parallel to the x-axis. Prove that the electric field is independent of the y- and z-coordinates in this region. If there is no charge in this region, prove that the field is also independent of x.

2–11 Given that the dielectric strength of air (i.e., the electric field above which the air becomes conducting) is 3×10^6 V/m, (a) what is the highest possible potential of an isolated spherical conductor of radius 10 cm? (b) What is the radius of a spherical conductor which could hold 1 coulomb of charge?

2–12 A conducting object has a hollow cavity in its interior. If a point charge q is introduced into the cavity, prove that the charge $-q$ is induced on the surface of the cavity. (Use Gauss's law.)

2–13 The electric field in the atmosphere at the earth's surface is approximately 200 V/m, directed downward. At 1400 m above the earth's surface, the electric field in the atmosphere is only 20 V/m, again directed downward. What is the average charge density in the atmosphere below 1400 m? Does this consist predominantly of *positive* or *negative* ions?

2–14 Two infinite parallel conducting plates are separated by the distance d. If the plates have uniform charge densities σ and $-\sigma$, respectively, on their inside surfaces, obtain an expression for the electric field between the plates. Prove that the electric field in the regions external to the plates is zero. [Two charged parallel conducting plates of finite area produce essentially the same electric field in the region between them as was found above provided the dimensions of the plates are large compared with the separation d; such an arrangement is called a *capacitor* (see Chapter 6).]

2–15 A spherical charge distribution has a volume charge density that is a function only of r, the distance from the center of the distribution. In other words, $\rho = \rho(r)$. If $\rho(r)$ is as given below, determine the electric field as a function of r. Integrate the result to obtain an expression for the electrostatic potential $\varphi(r)$, subject to the restriction that $\varphi(\infty) = 0$.

a) $\rho = A/r$ with A a constant for $0 \le r \le R$;
$\rho = 0$ for $r > R$.

b) $\rho = \rho_0$ (i.e., constant) for $0 \le r \le R$;
$\rho = 0$ for $r > R$.

2–16 An infinitely long circular rod of radius R contains a uniform charge density ρ. Use Gauss's law to find the electric field for $r > R$ and $r < R$.

2–17 Calculate the curl and divergence of $\mathbf{r}/r^a$. What charge density $\rho(r)$ would produce a field

$$\mathbf{E} = \frac{q}{4\pi\epsilon_0} \frac{\mathbf{r}}{r^a}?$$

What is the potential of this field?

2–18 Suppose the exponent in the Coulomb field were not exactly 3, but $a = 3 - \delta$, where $\delta \ll 1$. Calculate the integral of $\mathbf{V} \cdot \mathbf{E}$ over a spherical volume of radius R centered on the charge q.

2–19 The *screened Coulomb potential*

$$\varphi = \frac{q}{4\pi\epsilon_0} \frac{e^{-r/\lambda}}{r}$$

commonly occurs in a conducting medium. Calculate the corresponding electric field and charge density.

2–20 Using Eq. (2–39) for the potential produced by a dipole $\mathbf{p}$, make a plot of the traces of equipotential surfaces in a plane containing the dipole. For convenience, the dipole may be located at the origin. Use the results obtained to sketch in some of the lines of force. Compare the result with Fig. 2–1.

2–21 (a) Show that the force acting on a dipole $\mathbf{p}$ placed in an external electric field $\mathbf{E}_{ext}$ is $\mathbf{p} \cdot \mathbf{V}\mathbf{E}_{ext}$. (b) Show that the torque acting on the dipole in this field is

$$\tau = \mathbf{r} \times [\mathbf{p} \cdot \mathbf{V}\mathbf{E}_{ext}] + \mathbf{p} \times \mathbf{E}_{ext},$$

where $\mathbf{r}$ is the vector distance to the dipole from the point about which the torque is to be measured. The quantity $\mathbf{p} \times \mathbf{E}_{ext}$, which is independent of the point about which the torque is computed, is called the turning couple acting on the dipole.

2–22 Three charges are arranged in a linear array. The charge $-2q$ is placed at the origin, and two charges, each of $+q$, are placed at $(0, 0, l)$ and $(0, 0, -l)$ respectively. Find a relatively simple expression for the potential $\varphi(\mathbf{r})$ which is valid for distances $|\mathbf{r}| \gg l$. Make a plot of the equipotential surfaces in the x, z-plane.

2–23 What is the quadrupole moment tensor of the charge distribution discussed in Problem 2–22?

2–24 By using delta functions for the charge distribution of point charges, show that the dipole moment of a pair of point charges $\mathbf{p} = q\mathbf{l}$ follows from the general definition

$$\mathbf{p} = \int \mathbf{r}'\rho(\mathbf{r}') \, dv'.$$

2–25 Suppose a molecule is represented by a charge $-2q$ at the origin and charges $+q$ at l_1 and l_2, with $|l_1| = |l_2| = l$.

a) Find the dipole moment of the molecule.

b) For H_2O, $l = 0.958 \times 10^{-10}$ m and the angle between l_1 and l_2 is $\theta = 105°$. If $p = 6.14 \times 10^{-30}$ C · m, find the effective charge q.

2–26 Obtain the electric field of a point dipole by calculating the gradient of

$$\varphi = \frac{1}{4\pi\epsilon_0} \frac{\mathbf{p} \cdot \mathbf{r}}{r^3} .$$

CHAPTER 3 Solution of Electrostatic Problems

The solution to an electrostatic problem is straightforward for the case in which the charge distribution is everywhere specified, for then, as we have seen, the potential and electric field are given directly as integrals over this charge distribution:

$$\varphi(\mathbf{r}) = \frac{1}{4\pi\epsilon_0} \int \frac{dq'}{|\mathbf{r} - \mathbf{r}'|}, \tag{3-1}$$

$$\mathbf{E}(\mathbf{r}) = \frac{1}{4\pi\epsilon_0} \int \frac{(\mathbf{r} - \mathbf{r}') \, dq'}{|\mathbf{r} - \mathbf{r}'|^3}. \tag{3-2}$$

However, many of the problems encountered in practice are not of this type. If the charge distribution is not specified in advance, it may be necessary to determine the electric field *first*, before the charge distribution can be calculated. For example, an electrostatic problem may involve several conductors, with either the potential or total charge of each conductor given, but the distribution of surface charge will not be known in general, and cannot be obtained until a complete solution to the problem is effected.

Our aim in this chapter is to develop an alternative approach to electrostatic problems, and to accomplish this we first derive the fundamental differential equation that must be satisfied by the potential φ. For the present we shall disregard problems involving dielectric bodies; problems of this type will be solved in Chapter 4.

3–1 POISSON'S EQUATION

All of the basic relationships which we shall need here were developed in the preceding chapter. First, we have the differential form of Gauss's law,

$$\nabla \cdot \mathbf{E} = \frac{1}{\epsilon_0} \rho. \tag{3-3}$$

Furthermore, in a purely electrostatic field, $\mathbf{E}$ may be expressed as minus the gradient of the potential φ:

$$\mathbf{E} = -\nabla\varphi. \tag{3-4}$$

Combining Eqs. (3–3) and (3–4), we obtain

$$\nabla \cdot \nabla\varphi = -\frac{\rho}{\epsilon_0}. \tag{3-5a}$$

It is convenient to think of the divergence of the gradient as a single differential operator, $\nabla \cdot \nabla$ or ∇^2. The latter notation is preferred, and the operator is called the *Laplacian*:

$$\nabla^2\varphi = -\frac{\rho}{\epsilon_0}. \tag{3-5b}$$

It is evident that the Laplacian is a scalar differential operator, and (3–5b) is a differential equation. This is *Poisson's equation*. The operator ∇^2 involves differentiation with respect to more than one variable; hence Poisson's equation is a *partial differential equation* which may be solved once we know the functional dependence of $\rho(x, y, z)$ and the appropriate boundary conditions.

The operator ∇^2, just like the ∇, $\nabla\cdot$, and $\nabla\times$, makes no reference to any particular coordinate system. In order to solve a specific problem, we must write ∇^2 in terms of x, y, z or r, θ, ϕ, or etc. The choice of the particular set of coordinates is arbitrary, but substantial simplification of the problem is usually achieved by choosing a set compatible with the symmetry of the electrostatic problem. The form taken by $\nabla^2\varphi$ in various coordinate systems is easily found by first taking the gradient of φ, and then operating with $\nabla\cdot$, using specific expressions from Chapter 1:

Rectangular coordinates:

$$\nabla^2\varphi \equiv \frac{\partial^2\varphi}{\partial x^2} + \frac{\partial^2\varphi}{\partial y^2} + \frac{\partial^2\varphi}{\partial z^2}. \tag{3-6}$$

Spherical coordinates:

$$\nabla^2\varphi \equiv \frac{1}{r^2}\frac{\partial}{\partial r}\left(r^2\frac{\partial\varphi}{\partial r}\right) + \frac{1}{r^2\sin\theta}\frac{\partial}{\partial\theta}\left(\sin\theta\frac{\partial\varphi}{\partial\theta}\right) + \frac{1}{r^2\sin^2\theta}\frac{\partial^2\varphi}{\partial\phi^2}. \tag{3-7}$$

Cylindrical coordinates:

$$\nabla^2\varphi \equiv \frac{1}{r}\frac{\partial}{\partial r}\left(r\frac{\partial\varphi}{\partial r}\right) + \frac{1}{r^2}\frac{\partial^2\varphi}{\partial\theta^2} + \frac{\partial^2\varphi}{\partial z^2}. \tag{3-8}$$

For the form of the Laplacian in other, more complicated coordinate systems, the reader is referred to the references at the end of this chapter. It should be noted that r and θ have different meanings in Eqs. (3–7) and (3–8); in spherical coordinates r is the magnitude of the radius vector from the origin and θ is the polar angle. In cylindrical coordinates, r is the perpendicular distance from the cylinder axis and θ is the azimuthal angle about this axis.

3-2 LAPLACE'S EQUATION

In a certain class of electrostatic problems involving conductors, all of the charge is found either on the surface of the conductors or in the form of fixed point charges. In these cases ρ is zero at most points in space. And where the charge density vanishes, the Poisson equation reduces to the simpler form

$$\nabla^2 \varphi = 0, \qquad (3\text{-}9)$$

which is *Laplace's equation*.

Suppose we have a set of N conductors maintained at the potentials φ_I, $\varphi_{II}, \ldots, \varphi_N$. Our problem is to find the potential at all points in space outside of the conductors. This may be accomplished by finding a solution to Laplace's equation which reduces to $\varphi_I, \varphi_{II}, \ldots, \varphi_N$ on the surfaces of the appropriate conductors. Such a solution to Laplace's equation may be shown to be unique, i.e., there is no other solution to Laplace's equation that satisfies the same boundary conditions. A proof of this statement will be given below. The solution to Laplace's equation that we find in this way is not applicable to the interior of the conductors, because the conductors have surface charge, and this implies a discontinuity in the gradient of φ across the surface (see Section 2-7). But we have already seen that the interior of each conductor is a region of constant potential, so the solution to our problem is complete.

We shall describe in some detail two methods for solution of Laplace's equation: the first is a method for compounding a general solution to (3-9) from particular solutions in a coordinate system dictated by the symmetry of the problem; the second is the method of images. In addition, a completely general solution to the problem in two dimensions will be found. Before taking up these specific procedures, however, we stop to prove some important properties of the solution to Laplace's equation.

Theorem I *If $\varphi_1, \varphi_2, \ldots, \varphi_n$ are all solutions of Laplace's equation, then*

$$\varphi = C_1 \varphi_1 + C_2 \varphi_2 + \cdots + C_n \varphi_n, \qquad (3\text{-}10)$$

where the C's are arbitrary constants, is also a solution.

The proof of this follows immediately from the fact that

$$\nabla^2 \varphi = \nabla^2 C_1 \varphi_1 + \nabla^2 C_2 \varphi_2 + \cdots + \nabla^2 C_n \varphi_n$$
$$= C_1 \nabla^2 \varphi_1 + C_2 \nabla^2 \varphi_2 + \cdots + C_n \nabla^2 \varphi_n$$
$$= 0.$$

Through the use of Theorem I we may superimpose two or more solutions of Laplace's equation in such a way that the resulting solution satisfies a given set of boundary conditions. Examples will be given in the following sections.

Theorem II (*Uniqueness theorem*) *Two solutions of Laplace's equation that satisfy the same boundary conditions differ at most by an additive constant.*

To prove this theorem we consider the closed region V_0 exterior to the surfaces S_I, S_II, ..., S_N of the various conductors in the problem and bounded on the outside by a surface S, the latter being either a surface at infinity or a real physical surface which encloses V_0. Let us assume that φ_1 and φ_2 are two solutions of Laplace's equation in V_0 which, in addition, have the *same* boundary conditions on S, S_I, S_II, ..., S_N. These boundary conditions may be specified by assigning values of either φ or $\partial\varphi/\partial n$ on the bounding surfaces.

We define a new function $\Phi = \varphi_1 - \varphi_2$. Obviously, $\nabla^2\Phi = \nabla^2\varphi_1 - \nabla^2\varphi_2 = 0$ in V_0. Furthermore, either Φ or $\mathbf{n}\cdot\nabla\Phi$ vanishes on the boundaries. Let us apply the divergence theorem to the vector $\Phi\nabla\Phi$:

$$\int_{V_0} \nabla\cdot(\Phi\nabla\Phi)\, dv = \int_{S+S_\mathrm{I}+\cdots S_\mathrm{N}} \Phi\nabla\Phi\cdot\mathbf{n}\, da$$
$$= 0,$$

since the second integral vanishes. The divergence may be expanded according to Eq. (1–1–7) of Table 1–1 to give

$$\nabla\cdot(\Phi\nabla\Phi) = \Phi\nabla^2\Phi + (\nabla\Phi)^2.$$

But $\nabla^2\Phi$ vanishes at all points in V_0, so that the divergence theorem reduces in this case to

$$\int_{V_0} (\nabla\Phi)^2\, dv = 0.$$

Now $(\nabla\Phi)^2$ must be either positive or zero at each point in V_0, and since its integral is zero, it is evident that $(\nabla\Phi)^2 = 0$ is the only possibility.

The theorem is essentially proved. A function whose gradient is zero at all points cannot change; hence at all points in V_0, Φ has the same value that it has on the bounding surfaces. If the boundary conditions have been given by specifying φ_1 and φ_2 on the surfaces S, S_I, ..., S_N, then since $\Phi = 0$ on these surfaces, it vanishes throughout V_0. If the boundary conditions are given in terms of $\partial\varphi_1/\partial n$ and $\partial\varphi_2/\partial n$, then $\nabla\Phi$ equals zero at all points in V_0 and $\nabla\Phi\cdot\mathbf{n} = 0$ on the boundaries. The only solution compatible with the last statement is Φ equal to a constant.

3–3 LAPLACE'S EQUATION IN ONE INDEPENDENT VARIABLE

If φ is a function of one variable only, Laplace's equation reduces to an ordinary differential equation. Consider the case where φ is $\varphi(x)$, a function of the single rectangular coordinate x. Then

$$\frac{d^2\varphi}{dx^2} = 0 \qquad \text{and} \qquad \varphi(x) = ax + b \tag{3–11}$$

is the general solution, where a and b are constants chosen to fit the boundary conditions. This is the result already found in the preceding chapter for the potential between two charged conducting plates oriented normal to the x-axis.

The situation is no more complicated in other coordinate systems where φ is a function of a single variable. In spherical coordinates where φ equals $\varphi(r)$, Laplace's equation and its general solution become

$$\frac{1}{r^2} \frac{d}{dr} \left(r^2 \frac{d\varphi}{dr} \right) = 0, \qquad \varphi(r) = -\frac{a}{r} + b. \tag{3–12}$$

The general solution to Laplace's equation in cylindrical coordinates for a function that is independent of θ and z, that is, for $\varphi(r)$, is left as an exercise for the reader.

3–4 SOLUTIONS TO LAPLACE'S EQUATION IN SPHERICAL COORDINATES. ZONAL HARMONICS

We next turn our attention to solutions of Laplace's equation where φ is a function of more than one variable. Many of the problems of interest to us deal with conductors in the shape of spheres or cylinders, and thus solutions of Laplace's equation in either spherical or cylindrical coordinates are called for. We first take up the spherical problem, but we shall find it expedient to *limit the discussion* to cases in which φ is independent of the azimuthal angle ϕ. This limitation restricts the class of problems that we shall be able to solve; nevertheless, many interesting physical problems fall into this restricted category, and more complicated problems are really beyond the scope of this book.

For the spherical case, φ is $\varphi(r, \theta)$, where r is the radius vector from a fixed origin O and θ is the polar angle (see Fig. 3–1). Using Eq. (3–7), Laplace's equation becomes in this case

$$\frac{1}{r^2} \frac{\partial}{\partial r} \left(r^2 \frac{\partial \varphi}{\partial r} \right) + \frac{1}{r^2 \sin\theta} \frac{\partial}{\partial \theta} \left(\sin\theta \frac{\partial \varphi}{\partial \theta} \right) = 0. \tag{3–13}$$

This partial differential equation will be solved by a technique known as *separation of variables*. A solution of the form $\varphi(r, \theta) = Z(r)P(\theta)$ is substituted into Eq. (3–13), yielding

$$\frac{1}{r^2} P(\theta) \frac{d}{dr} \left(r^2 \frac{dZ}{dr} \right) + \frac{Z(r)}{r^2 \sin\theta} \frac{d}{d\theta} \left(\sin\theta \frac{dP}{d\theta} \right) = 0. \tag{3–14}$$

Note that the partial derivatives have been replaced by total derivatives, since Z and P are each functions of one variable only. Dividing through by $\varphi(r, \theta)$ and multiplying through by r^2, we transform Eq. (3–14) into

$$\frac{1}{Z} \frac{d}{dr} \left(r^2 \frac{dZ}{dr} \right) = -\frac{1}{P \sin\theta} \frac{d}{d\theta} \left(\sin\theta \frac{dP}{d\theta} \right). \tag{3–15}$$

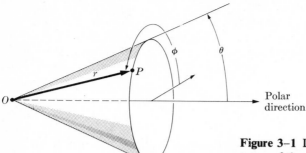

Figure 3–1 Location of the point P in terms of the spherical coordinates r, θ, ϕ.

The left side of this equation is a function of r only and the right side is a function of θ; the only way in which a function of r can equal a function of θ for all values of r and θ is for both functions to be constant. Hence let each side of Eq. (3–15) equal k, where k is the "separation constant."

Not all values of k necessarily yield solutions which are acceptable on physical grounds. Consider the θ equation first:

$$\frac{1}{\sin \theta} \frac{d}{d\theta} \left(\sin \theta \frac{dP}{d\theta} \right) + kP = 0. \tag{3–16}$$

This is Legendre's equation, and the only physically acceptable solutions which are defined over the full range of θ, from 0 to π, correspond to $k = n(n + 1)$, where n is a positive integer. The solution for a particular n will be denoted by $P_n(\theta)$. Solutions of Eq. (3–16) for other values of k are ill-behaved in the vicinity of $\theta = 0$ or $\theta = \pi$ radians, becoming infinite or even undefined at these values of θ.* These solutions cannot be made to fit physical boundary conditions and hence must be discarded.†

The acceptable solutions, the $P_n(\theta)$, are polynomials in $\cos \theta$, and are usually referred to as Legendre polynomials. The first four Legendre functions are given in Table 3–1. It is evident from Eq. (3–16) that the P_n may be multiplied by any arbitrary constant.

We now return to the radial equation

$$\frac{d}{dr} \left(r^2 \frac{dZ}{dr} \right) = n(n + 1)Z, \tag{3–17}$$

---

* The discussion here has been all too brief. The interested reader is referred to more mathematical texts for a detailed treatment of Legendre's equation. See, e.g., the books listed at the end of this chapter. Legendre's equation is usually written in a different form by substituting $x = \cos \theta$, and its solutions are then denoted by $P_n(x)$ or $P_n(\cos \theta)$.

† This statement requires some qualification. In some electrostatic problems the regions around $\theta = 0$ and $\theta = \pi$ may be naturally excluded, for example, by conducting conical surfaces; under these conditions solutions of Eq. (3–16) with other values of k could be used. Problems of this type will not be considered here.

Table 3–1 Legendre Polynomials for $n = 0, 1, 2,$ and 3

n	$P_n(\theta)$
0	1
1	$\cos \theta$
2	$\frac{1}{2}(3 \cos^2 \theta - 1)$
3	$\frac{1}{2}(5 \cos^3 \theta - 3 \cos \theta)$

where we have used the explicit form of k which gave acceptable θ solutions. Inspection of Eq. (3–17) shows that two independent solutions are

$$Z_n = r^n \quad \text{and} \quad Z_n = r^{-(n+1)}.$$

Solutions of Laplace's equation are obtained as the product $\varphi_n(r, \theta) = Z_n(r) \times P_n(\theta)$, where particular care must be exercised to have Z and P correspond to the same value of n. This is mandatory, since both sides of Eq. (3–15) are equal to the same constant, namely, $n(n + 1)$.

As a result of the above discussion we have solved Laplace's equation in spherical coordinates and have obtained the solutions that are known as *zonal harmonics*:

$$\varphi_n = r^n P_n(\theta) \quad \text{or} \quad \varphi_n = r^{-(n+1)} P_n(\theta), \tag{3–18}$$

where $P_n(\theta)$ is one of the polynomials listed in Table 3–1, and n is a positive integer or zero. The zonal harmonics form a complete set of functions, i.e., a general solution of Laplace's equation may be constructed as a superposition of these solutions according to Theorem I provided the physical problem shows the appropriate azimuthal symmetry. Several of the zonal harmonics are already well known to us: one of the $n = 0$ solutions, namely $\varphi = $ constant, is a trivial solution of Laplace's equation, valid in any coordinate system; the zonal harmonic r^{-1} is the potential of a point charge; and $r^{-2} \cos \theta$ is the potential of a dipole.

3–5 CONDUCTING SPHERE IN A UNIFORM ELECTRIC FIELD

We shall illustrate the usefulness of zonal harmonics for electrostatic problems having spherical symmetry by solving the problem of an uncharged conducting sphere placed in an *initially* uniform electric field $\mathbf{E}_0$. The lines of a uniform electric field are parallel, but the presence of the conductor alters the field in such a way that the field lines strike the surface of the conductor, which is an equipotential surface, normally. If we take the direction of the initially uniform electric field as the polar direction (z-direction), and if we make the origin of our coordinate system coincide with the center of the sphere, then from the symmetry of the problem it is clear that the potential will be independent of azimuthal angle ϕ, and may be expressed as a sum of zonal harmonics.

The spherical conductor, of radius a, is an equipotential surface; let us denote its potential by φ_0. Our problem is to find a solution to Laplace's equation in the region outside the sphere which reduces to φ_0 on the sphere itself, and which has the correct limiting form at large distances away. The solution may be formally written as

$$\varphi(r, \theta) = A_1 + C_1 r^{-1} + A_2 r \cos \theta + C_2 r^{-2} \cos \theta$$
$$+ \tfrac{1}{2} A_3 r^2 (3 \cos^2 \theta - 1) + \tfrac{1}{2} C_3 r^{-3} (3 \cos^2 \theta - 1) + \cdots, \qquad (3\text{-}19)$$

where the A's and C's are arbitrary constants. At large r, the electric field will be only slightly distorted from its initial form, and the potential will be that appropriate to a uniform electric field.

$$[\mathbf{E}(r, \theta)]_{r \to \infty} = \mathbf{E}_0 = E_0 \mathbf{k},$$
$$[\varphi(r, \theta)]_{r \to \infty} = - E_0 z + \text{constant},$$
$$= - E_0 r \cos \theta + \text{constant}. \qquad (3\text{-}20)$$

Hence, in order to make Eqs. (3–19) and (3–20) agree at large r, $A_2 = - E_0$; furthermore, all the A's from A_3 up must be set equal to zero.

The term $C_1 r^{-1}$ produces a radial field which, as we might expect, is compatible only with a spherical conductor bearing net total charge. Since our problem deals with an uncharged conductor, the constant C_1 must be set equal to zero. At the surface of the sphere $\varphi = \varphi_0$, and the potential must become independent of angle θ. The two terms involving $\cos \theta$ may be made to cancel each other, but the terms with higher inverse powers of r cannot be canceled one against the other because they contain *different* Legendre functions. The only possibility is to set all the C_i's with $i \geq 3$ equal to zero. Equation (3–19) now becomes

$$\varphi(r, \theta) = A_1 - E_0 r \cos \theta + C_2 r^{-2} \cos \theta, \qquad \text{for} \quad r \geq a,$$
$$\varphi(a, \theta) = \varphi_0. \qquad (3\text{-}21)$$

Since the two expressions must be equal at $r = a$, $A_1 = \varphi_0$ and $C_2 = E_0 a^3$.

From the final expression for the potential, we may calculate not only the electric field at all points in space (see Fig. 3–2) but also the surface density of charge on the conducting sphere:

$$E_r = - \frac{\partial \varphi}{\partial r} = E_0 \left(1 + 2 \frac{a^3}{r^3} \right) \cos \theta, \ \Bigg\}$$
$$\hspace{4cm} \text{for} \quad r \geq a, \qquad (3\text{-}22)$$
$$E_\theta = - \frac{1}{r} \frac{\partial \varphi}{\partial \theta} = - E_0 \left(1 - \frac{a^3}{r^3} \right) \sin \theta, \ \Bigg\}$$

$$\sigma(\theta) = \epsilon_0 E_r |_{r=a} = 3 \epsilon_0 E_0 \cos \theta. \qquad (3\text{-}23)$$

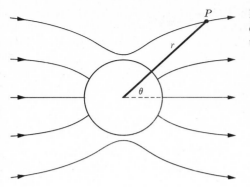

Figure 3–2 Lines of electric flux for the case of a conducting sphere placed in a uniform electric field.

The total charge on the sphere,

$$Q = a^2 \int_0^\pi \sigma(\theta)2\pi \sin \theta \, d\theta,$$

is obviously zero, which agrees with our initial assumption.

3–6 CYLINDRICAL HARMONICS

Laplace's equation in cylindrical coordinates may also be solved by the method of separation of variables. Here again it will be expedient to work out solutions for only a restricted class of problems, namely, those in which the potential is independent of the coordinate z. These solutions are appropriate for certain problems involving a long straight cylindrical conductor or wire, but *not* for those dealing with a short cylindrical segment.

If the potential is independent of z, Laplace's equation in cylindrical coordinates becomes

$$\frac{1}{r} \frac{\partial}{\partial r}\left(r \frac{\partial \varphi}{\partial r}\right) + \frac{1}{r^2} \frac{\partial^2 \varphi}{\partial \theta^2} = 0. \tag{3–24}$$

Substitution of $\varphi = Y(r)S(\theta)$ reduces the equation to

$$\frac{r}{Y} \frac{d}{dr}\left(r \frac{dY}{dr}\right) = -\frac{1}{S} \frac{d^2 S}{d\theta^2} = k, \tag{3–25}$$

where k again plays the role of a separation constant. The θ-equation is particularly simple; it has the solutions $\cos k^{1/2}\theta$ and $\sin k^{1/2}\theta$. But if these solutions are to make sense physically, each must be a single-valued function of θ; thus

$$\cos k^{1/2}(\theta + 2\pi) = \cos k^{1/2}\theta,$$

$$\sin k^{1/2}(\theta + 2\pi) = \sin k^{1/2}\theta.$$

Or, to put it differently, after θ has gone through its full range from 0 to 2π the function must join *smoothly* to its value at $\theta = 0$. This can be the case only if $k = n^2$, n being an integer. We may further require n to be positive (or zero) without losing any of these solutions.

Returning now to the r-equation, we are able to verify easily that $Y(r)$ is r^n or r^{-n}; unless $n = 0$ when $Y(r) = \ln r$ or $Y(r) = $ constant. Hence the required solutions to Laplace's equation, the so-called *cylindrical harmonics*, are

$$1, \qquad\qquad \ln r,$$

$$r^n \cos n\theta, \qquad r^{-n} \cos n\theta,$$

$$r^n \sin n\theta, \qquad r^{-n} \sin n\theta.$$

These functions form a complete set for the variables r, θ in cylindrical coordinates, and the potential $\varphi(r, \theta)$ may be developed as a superposition of cylindrical harmonics in accordance with Theorem I.

*3–7 LAPLACE'S EQUATION IN RECTANGULAR COORDINATES

In rectangular coordinates, the variables may be separated by making the substitution

$$\varphi(x, y, z) = f_1(x) f_2(y) f_3(z),$$

whereby Laplace's equation reduces to

$$\frac{1}{f_1(x)} \frac{d^2 f_1}{dx^2} + \frac{1}{f_2(y)} \frac{d^2 f_2}{dy^2} = -\frac{1}{f_3(z)} \frac{d^2 f_3}{dz^2}. \tag{3–26a}$$

The left side of this equation is a function of x and y, and the right side is a function of z only; hence both sides must be equal to the same constant, k. This is the first separation constant. The two equations obtained from Eq. (3–26a) are

$$\frac{d^2 f_3}{dz^2} + kf_3 = 0, \tag{3–26b}$$

$$\frac{1}{f_2} \frac{d^2 f_2}{dy^2} = k - \frac{1}{f_1} \frac{d^2 f_1}{dx^2}.$$

The latter equation has been written such that the variables x and y are separated; each side of this equation is now set equal to $-m$ (the second separation constant). Thus,

$$\frac{d^2 f_2}{dy^2} + mf_2 = 0, \tag{3–26c}$$

$$\frac{d^2 f_1}{dx^2} - (k + m)f_1 = 0. \tag{3–26d}$$

---

* Starred sections may be omitted without loss of continuity.

Equations (3–26b), (3–26c), and (3–26d) are easily solved. One of the typical solutions for $\varphi(x, y, z)$ is

$$\varphi(x, y, z) = Ae^{-(k+m)^{1/2}x} \cos m^{1/2}y \cos k^{1/2}z. \tag{3–27}$$

The other seven independent solutions for a pair of separation constants (k, m) are obtained by making one or more of the following substitutions: $+(k + m)^{1/2}x$ for $-(k + m)^{1/2}x$, $\sin m^{1/2}y$ for $\cos m^{1/2}y$, and $\sin k^{1/2}z$ for $\cos k^{1/2}z$.

Thus far there are no restrictions on k or m, but boundary conditions on the problem usually restrict k (or m) to a discrete set of positive or negative values. It is worth while making the point that it is the boundary conditions that really pick out the *pertinent solutions* to a partial differential equation; the function

$$\varphi(x, y, z) = \sum_p \sum_q A_{pq} e^{-(p^2+q^2)^{1/2}x} \cos py \cos qz$$

for fixed x and y is just the Fourier series expansion for an *arbitrary* even function of z.

The individual solutions, Eq. (3–27), do not represent particularly simple potentials, and we shall not try to correlate them with physical situations. The case where both separation constants are zero is more interesting; hence we turn our attention to this case. From Eq. (3–26d), it is evident that $f_1(x) = a_1 x$, or $f_1(x) = $ constant, is a solution; from Eq. (3–26c), we obtain $f_2(y)$, etc. Thus,

$$\varphi(x, y, z) = A_1 xyz + A_2 xy + A_3 yz + A_4 xz$$
$$+ A_5 x + A_6 y + A_7 z + A_8, \tag{3–28a}$$

where the A's are arbitrary constants. This solution may be applied to the case where three conducting planes intersect at right angles. If these planes are the coordinate planes xy, yz, and zx, and are all at the same potential, then

$$\varphi(x, y, z) = A_1 xyz + A_8. \tag{3–28b}$$

It is left as an exercise for the reader to determine the surface charge density on the coordinate planes that is compatible with Eq. (3–28b).

*3–8 LAPLACE'S EQUATION IN TWO DIMENSIONS. GENERAL SOLUTION

If the potential is a function of only two rectangular coordinates, Laplace's equation is written

$$\frac{\partial^2 \varphi}{\partial x^2} + \frac{\partial^2 \varphi}{\partial y^2} = 0. \tag{3–29a}$$

It is possible to obtain the general solution to this equation by means of a transformation to a new set of independent variables; nevertheless, it should be emphasized that such a transformation leads to a simplification of the original equation only in the two-dimensional case. Let

$$\xi = x + iy, \qquad \eta = x - iy,$$

where $i = \sqrt{-1}$ is the unit imaginary number. In terms of these relationships,

$$\frac{\partial^2}{\partial x^2} = \frac{\partial^2}{\partial \xi^2} + 2 \frac{\partial^2}{\partial \xi \, \partial \eta} + \frac{\partial^2}{\partial \eta^2},$$

$$\frac{\partial^2}{\partial y^2} = -\frac{\partial^2}{\partial \xi^2} + 2 \frac{\partial^2}{\partial \xi \, \partial \eta} - \frac{\partial^2}{\partial \eta^2},$$

and

$$\nabla^2 \varphi = 4 \frac{\partial^2 \varphi}{\partial \xi \, \partial \eta} = 0. \qquad (3\text{–}29\text{b})$$

It is evident that the general solution to (3–29b) is

$$\varphi = F_1(\xi) + F_2(\eta) = F_1(x + iy) + F_2(x - iy), \qquad (3\text{–}30)$$

where F_1 and F_2 are *arbitrary* functions. The functions F_1 and F_2 are complex quantities in general, but two real functions may be constructed in the following way. First let $F_2(x - iy) = F_1(x - iy)$, that is, let the two functions F_1 and F_2 have the same dependence on their arguments; then

$$\varphi_1 = F_1(x + iy) + F_1(x - iy) = 2 \text{ Re } [F_1(x + iy)],$$

where Re stands for *real part of*. Furthermore, the second real potential function is

$$\varphi_2 = -i[F_1(x + iy) - F_1(x - iy)] = 2 \text{ Im } [F_1(x + iy)],$$

where Im stands for *imaginary part of*. Thus the real and imaginary parts of any complex function $F(x + iy)$ are both solutions of Laplace's equation.

The solutions found in this way are not restricted to any particular coordinate system. For example, the cylindrical harmonics of Section 3–7 are obtained from the complex functions* $(x + iy)^n = r^n e^{in\theta}$, and $\ln (x + iy) = \ln r + i\theta$. On the other hand, when it comes to solving a particular two-dimensional problem, there is no standard procedure for finding the appropriate complex function. This method generates so many solutions that it is not possible to enumerate them all and cast out those that do not agree with boundary conditions on the problem. In simple cases, the required functions may be found by trial and error; in other cases, the method of *conformal mapping* (which is beyond the scope of this book) may be useful.

3–9 ELECTROSTATIC IMAGES

For a given set of boundary conditions the solution to Laplace's equation is unique, so that if one obtains a solution $\varphi(x, y, z)$ by any means whatever, and if this φ satisfies all boundary conditions, then a complete solution to the problem

---

* The cylindrical and rectangular coordinates are related in the usual way: $x = r \cos \theta$, $y = r \sin \theta$.

has been effected. The method of images is a procedure for accomplishing this result without specifically solving a differential equation. It is not universally applicable to all types of electrostatic problems, but enough interesting problems fall into this category to make it worth while discussing the method here.

Suppose the potential may be written in the following way:

$$\varphi(\mathbf{r}) = \varphi_1(\mathbf{r}) + \frac{1}{4\pi\epsilon_0} \int_S \frac{\sigma(\mathbf{r}') \, da'}{|\mathbf{r} - \mathbf{r}'|}, \tag{3-31}$$

where φ_1 is either a specified function or easily calculable, and the integral represents the contribution to the potential from surface charge on all conductors appearing in the problem. The function σ is not known. It may happen, and this is the essence of the image-charge method, that the last term in Eq. (3–31) can be replaced by a potential φ_2 that is due to a *specified* charge distribution. This is possible so long as the surfaces of all conductors coincide with equipotential surfaces of the combined $\varphi_1 + \varphi_2$. The specified charges producing φ_2 are called *image charges*. They do not really exist, of course. Their apparent location is *inside* the various conductors, and the potential $\varphi = \varphi_1 + \varphi_2$ is a valid solution to the problem only in the exterior region.

As an example of this method, we shall solve the problem of a point charge q placed near a conducting plane of infinite extent. To formulate the problem mathematically, let the conducting plane coincide with the yz-plane, and let the point charge lie on the x-axis at $x = d$ (see Fig. 3–3a). The potential fits the prescription Eq. (3–31), with

$$\varphi_1(x, y, z) = \frac{q}{4\pi\epsilon_0 r_1} = \frac{q}{4\pi\epsilon_0 \sqrt{(x - d)^2 + y^2 + z^2}}. \tag{3-32}$$

Consider now a different problem, that of two point charges (q and $-q$) a distance $2d$ apart, as shown in Fig. 3–3(b). The potential of these two charges,

$$\varphi(x, y, z) = \frac{q}{4\pi\epsilon_0 r_1} - \frac{q}{4\pi\epsilon_0 r_2}, \tag{3-33}$$

not only satisfies Laplace's equation at all points exterior to the charges, but also reduces to a constant (namely, zero) on the plane that perpendicularly bisects the segment joining the two charges. Thus Eq. (3–33) satisfies the boundary conditions of the original problem. Because solutions to Laplace's equation are unique, Eq. (3–33) is the correct potential in the entire half-space exterior to the conducting plane. The charge $-q$, which gives rise to the potential

$$\varphi_2(x, y, z) = -\frac{q}{4\pi\epsilon_0 r_2} = -\frac{q}{4\pi\epsilon_0 \sqrt{(x + d)^2 + y^2 + z^2}}, \tag{3-34}$$

is called the *image* of the point charge q. Naturally, the image does not really exist, and Eq. (3–32) does *not* give correctly the potential inside or to the left of the conducting plane in Fig. 3–3(a).

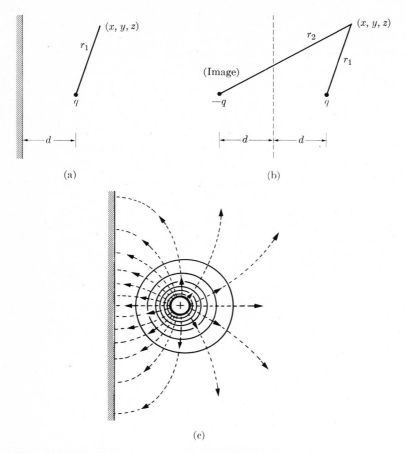

Figure 3–3 Problem of a point charge and conducting plane solved by means of the image-charge method: (a) original problem, (b) location of image charge, (c) lines of force (dashed) and equipotential surfaces (solid).

The electric field **E** in the exterior region may be obtained as the negative gradient of Eq. (3–33). Since the surface of the conducting plane represents an interface joining two solutions of Laplace's equation, namely, $\varphi = 0$ and Eq. (3–33), the discontinuity in the electric field is accommodated by a surface charge density σ on the plane:

$$\sigma(y, z) = \epsilon_0 E_x|_{x=0} = -\frac{qd}{2\pi(d^2 + y^2 + z^2)^{3/2}}. \tag{3–35}$$

The lines of force and equipotential surfaces appropriate to the original problem are shown in Fig. 3–3(c). These are the same lines of force and equipotential surfaces appropriate to the two point charge problem in Fig. 3–3(b) except that in the latter case the flux lines would continue into the left half-plane. It is evident from the figure that *all* of the electric flux lines which would normally converge on the image charge are intercepted by the plane in Fig. 3–3(c). Hence the total charge on the plane is equal to that of the image charge, $-q$. This same result may be obtained mathematically by integrating (3–35) over the entire surface (see Problem 3–14).

It is evident that the point charge q exerts an attractive force on the plane, because the induced surface charge is of the opposite sign. By Newton's law of action and reaction, this force is equal in magnitude to the force exerted on q by the plane. Since the point charge experiences no force due to its own field,

$$\mathbf{F} = -q\nabla\varphi_2, \tag{3–36}$$

which is just the force exerted on it by the image charge.

Another problem which may be solved simply in terms of images is that of determining the electric field of a point charge q in the vicinity of a right-angle intersection of two conducting planes (see Fig. 3–4a). The positions of the necessary image charges are shown in Fig. 3–4(b). It is readily seen that the two planes shown dotted in the figure are surfaces of zero potential due to the combined potentials of q and the three image charges.

(a)   (b)

Figure 3–4 Point charge in a right-angle corner.

3–10 POINT CHARGE AND CONDUCTING SPHERE

The principal difficulty in solving a problem by image technique is that of finding a group of image charges that, together with the originally specified charges, produce equipotential surfaces at the conductors. The problem is straightforward only in cases where the geometry is simple. Such is the case, however, for a point charge q in the vicinity of a conducting sphere; it requires a single image charge to make the sphere a surface of *zero potential*. An additional image charge is needed to change the potential of the sphere to some other constant value.

We shall first determine the magnitude and location of the image q' which together with the point charge q produces zero potential at all points on the sphere. The geometry of the situation is shown in Fig. 3–5. The point charge q is a distance d from the center of the sphere, and the radius of the sphere is a. It is

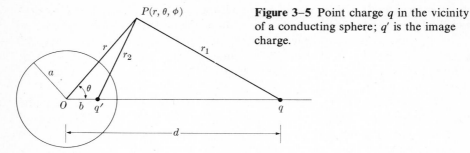

Figure 3–5 Point charge q in the vicinity of a conducting sphere; q' is the image charge.

apparent from the symmetry of the problem that the image charge q' will lie on the line passing through q and the center of the sphere.

The desired results are most easily obtained by means of spherical coordinates, with the origin of coordinates at the center of the sphere. Let the polar axis be taken as the line joining q to the origin. The distance b and the magnitude of q' are to be determined in terms of the specified quantities: q, d, a. The potential at an arbitrary point P due to q and q' is given by

$$\varphi(r, \theta, \phi) = \frac{q}{4\pi\epsilon_0 r_1} + \frac{q'}{4\pi\epsilon_0 r_2}$$

$$= \frac{1}{4\pi\epsilon_0}\left[\frac{q}{\sqrt{r^2 + d^2 - 2rd\cos\theta}} + \frac{q'}{\sqrt{r^2 + b^2 - 2rb\cos\theta}}\right]. \qquad (3\text{--}37)$$

On the surface of the sphere, $r = a$, and $\varphi(a, \theta, \phi) = 0$ for all θ and ϕ. But from expression (3–37), $\varphi(a, \theta, \phi)$ can equal zero for all θ *only* if the two square roots are proportional to each other. This is the case if $b = a^2/d$, for then

$$\sqrt{a^2 + b^2 - 2ab\cos\theta} = \frac{a}{d}\sqrt{d^2 + a^2 - 2ad\cos\theta}.$$

Hence,

$$b = \frac{a^2}{d}, \qquad (3\text{--}38)$$

and furthermore,

$$q' = -\frac{a}{d}q. \qquad (3\text{--}39)$$

These equations serve to specify the location and magnitude of the first image charge.

A second image charge q'' may be placed at the center of the sphere without destroying the equipotential nature of the spherical surface. The magnitude of q'' is arbitrary; it may be adjusted to fit the boundary conditions on the problem. Thus a complete solution to the point charge-conducting sphere problem has been effected; the potential at all points exterior to the sphere is

$$\varphi(r, \theta, \phi) = \frac{1}{4\pi\epsilon_0} \left[\frac{q}{r_1} + \frac{q'}{r_2} + \frac{q''}{r} \right]. \tag{3-40}$$

The potential of the spherical conductor itself is

$$\varphi(a, \theta, \phi) = \frac{q''}{4\pi\epsilon_0 a}; \tag{3-41}$$

and the surface density of charge on the sphere is

$$\sigma(\theta, \phi) = -\epsilon_0 \left. \frac{\partial\varphi}{\partial r} \right|_{r=a}. \tag{3-42}$$

All the lines of force which would normally converge on the image charges are intercepted by the sphere; hence the total charge on the sphere, Q, is equal to the sum of the image charges:

$$Q = q' + q''. \tag{3-43}$$

This result may be verified by direct integration of Eq. (3–42).

Special cases of interest are the *grounded* sphere: $\varphi(a) = 0$, $q'' = 0$; and the *uncharged spherical conductor*: $q'' = -q'$.

3–11 LINE CHARGES AND LINE IMAGES

Thus far, our image technique has been limited to problems involving point charges, and hence point images. In this section we shall take up several problems that may be solved by means of line image charges. Consider two infinitely long, parallel, line charges, with charges λ and $-\lambda$ per unit length, respectively, as shown in Fig. 3–6. The potential at any point is given by

$$\varphi = -\frac{\lambda}{2\pi\epsilon_0} \left[\ln r_1 - \ln r_2 \right] = -\frac{\lambda}{2\pi\epsilon_0} \ln \frac{r_1}{r_2}, \tag{3-44}$$

where r_1 and r_2 are the perpendicular distances from the point to the two line charges. The equipotentials are obtained by setting Eq. (3–44) equal to a constant, a procedure which is equivalent to requiring that

$$\frac{r_1}{r_2} = M, \tag{3-45}$$

where M is constant. Hence the equipotentials may be specified by Eq. (3–45).

The equipotential corresponding to $M = 1$ is the plane located halfway between the two line charges, shown as equipotential surface I in Fig. 3–6. The potential of the plane is zero. Hence the problem of a long line charge oriented parallel to a conducting plane has been effectively solved. The potential in the half-space is given correctly by Eq. (3–44). Let us assume that the line charge

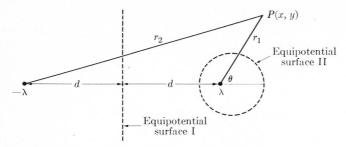

Figure 3–6 Two infinitely long, parallel line charges (of charge λ and $-\lambda$ per unit length) are shown cutting the plane of the paper.

shown on the right side of the figure is the specified charge, which is at a distance d from the conducting plane. Then the line charge on the left side of the figure plays the role of an image. Again, the total charge on the plane is equal to that of the image charge.

Let us next consider equipotential surfaces corresponding to other values of M. The general form of the surface may be found by expressing r_1 and r_2 in rectangular coordinates. For convenience, we choose the origin of the coordinate system on the positive line charge, and make this charge coincide with the z-axis; we let the second line charge be located at $x = -2d$, $y = 0$. Now

$$r_1^2 = x^2 + y^2$$

and

$$r_2^2 = (x + 2d)^2 + y^2,$$

so that Eq. (3–45) becomes, after a little algebraic manipulation,

$$x^2 + y^2 - \frac{4M^2xd}{1 - M^2} = \frac{4M^2d^2}{1 - M^2}. \tag{3–46}$$

This is the equation of a circular cylinder extending parallel to the z-axis. If M is less than one, the cylinder surrounds the positive line charge, as does equipotential surface II of the figure. The axis of the cylinder goes through the point

$$x = \frac{2M^2d}{1 - M^2}, \qquad y = 0; \tag{3–47}$$

and the radius of the cylinder is

$$R_c = \frac{2Md}{1 - M^2}. \tag{3–48}$$

We are now in a position to solve a number of interesting problems involving cylindrical conductors, but only one of this type will be discussed. Consider the problem of a long cylindrical conductor in the vicinity of a conducting plane, and oriented parallel to it. The cylinder bears the charge λ per unit length. Figure 3–6 may serve to illustrate the problem, the two conductors coinciding with the dotted surfaces. Both of the line charges are images in this case, and the potential in the region surrounding the cylinder and to the right of the plane is given by Eq. (3–44). It is evident that the charge induced on the plane is equal to $-\lambda$ per unit distance in the z-direction.

3–12 SYSTEM OF CONDUCTORS. COEFFICIENTS OF POTENTIAL

In the preceding sections several important methods for obtaining solutions to Laplace's equation have been discussed. Although general in scope, these methods are limited by practical considerations to problems in which the conductors have rather simple shapes. When their shapes are complicated, complete mathematical solution is out of the question; nevertheless, certain conclusions can be drawn about the system just because the potential satisfies Laplace's equation. In fact, we shall prove here that a linear relationship exists between the potential of one of the conductors and the charges on the various conductors in the system. The coefficients in this relationship, the so-called *coefficients of potential*, are functions only of the geometry and, although not always calculable analytically, may be determined numerically or directly from experiment.

Suppose there are N conductors in fixed geometry. Let all of the conductors be uncharged except conductor j, which bears the charge Q_{j0}. The appropriate solution to Laplace's equation in the space exterior to the conductors will be given the symbol $\varphi^{(j)}(x, y, z)$, and the potential of each of the conductors will be indicated by $\varphi_1^{(j)}, \varphi_2^{(j)}, \ldots, \varphi_j^{(j)}, \ldots, \varphi_N^{(j)}$. Now let us change the charge of the jth conductor to λQ_{j0}. The function $\lambda \varphi^{(j)}(x, y, z)$ satisfies Laplace's equation, since λ is a constant; that the new boundary conditions are satisfied by this function may be seen from the following argument. The potential at all points in space is multiplied by λ; thus all derivatives (and in particular the gradient) of the potential are multiplied by λ. Because $\sigma = \epsilon_0 E_n$, it follows that all charge densities are multiplied by λ. Thus the charge of the jth conductor is λQ_{j0} and all other conductors remain uncharged.

A solution of Laplace's equation which fits a particular set of boundary conditions is unique; therefore we have found *the correct solution*, $\lambda \varphi^{(j)}(x, y, z)$, to our modified problem. The interesting conclusion we draw from this discussion is that the potential of each conductor is proportional to the charge Q_j of conductor j, that is,

$$\varphi_i^{(j)} = p_{ij} Q_j, \qquad (i = 1, 2, \ldots, N), \tag{3–49}$$

where p_{ij} is a constant which depends only on the geometry.

The same argument may be applied to the case where conductor k is charged: $Q_k = v Q_{k0}$, all other conductors being uncharged. Here the appropriate solution to Laplace's equation is $v \varphi^{(k)}(x, y, z)$, where $\varphi^{(k)}$ is the solution for $v = 1$. It is apparent, then, that

$$\lambda \varphi^{(j)}(x, y, z) + v \varphi^{(k)}(x, y, z) \tag{3–50}$$

is a solution appropriate to the case where both conductors are charged. Again we appeal to the uniqueness of a solution for a given set of boundary conditions. Thus Eq. (3–50) is *the solution* for this case, and the potential of each conductor may be written as

$$\varphi_i = p_{ij}Q_j + p_{ik}Q_k, \qquad (i = 1, 2, \ldots, N). \tag{3–51}$$

This result may be generalized immediately to the case where all N conductors are charged:

$$\varphi_i = \sum_{j=1}^{N} p_{ij}Q_j. \tag{3–52}$$

This is the linear relationship between potential and charge which we have been seeking; the coefficients p_{ij} are called the *coefficients of potential*. In Chapter 6 it will be shown that the array of these coefficients is symmetrical, i.e., that $p_{ij} = p_{ji}$.

3–13 SOLUTIONS OF POISSON'S EQUATION

In the preceding sections, we have dealt exclusively with Laplace's equation and its solution. Laplace's equation is applicable to those electrostatic problems in which all the charge resides on surfaces of conductors or is concentrated in the form of point or line charges. (We shall see in the next chapter that if the region between the conductors is filled with one or more simple dielectric media, then Laplace's equation still holds in these media.)

Let us consider, now, an electrostatic problem in which part of the charge (the prescribed charge) is given by $\rho(x, y, z)$, a known function, and the rest of the charge (the induced charge) resides on the surfaces of conductors. Such a problem requires the solution of Poisson's equation. The general solution to this problem may be written as an integral of the type Eq. (3–1) over the *prescribed charge* plus a general solution to Laplace's equation. The solution to Laplace's equation must be chosen, however, so that the entire potential satisfies all boundary conditions.

When all of the charge is prescribed, i.e., when $dq = \rho(x, y, z)\, dv$ is known at all points in space, then Eq. (3–1) represents the entire solution to Poisson's equation, and this integral may be performed (either analytically or numerically). There is one case, however, where the solution to Poisson's equation may be obtained more directly than by means of the formal solution Eq. (3–1); this occurs when both ρ and φ are functions of only one independent variable. As an example of this case, let ρ be a function of the spherical coordinate, r, only, and let the entire charge be distributed in a spherically symmetric way. Then Eq. (3–5b) becomes

$$\frac{1}{r^2} \frac{d}{dr}\left(r^2 \frac{d\varphi}{dr}\right) = -\frac{1}{\epsilon_0}\, \rho(r). \tag{3–53}$$

We shall assume that the total charge is bounded, i.e., that either the charge does not extend to infinity or the charge density drops off sufficiently rapidly at large radii. Equation (3–53) may then be integrated directly, assuming the function $\rho(r)$ given, and the two constants of integration may be determined (1) from Gauss's Law for the electric field at some radius, and (2) from the fact that $\varphi \to 0$ as $r \to \infty$.

3–14 SUMMARY

The fundamental first-order vector differential equations for the electrostatic field, $\nabla \times \mathbf{E} = 0$ and $\nabla \cdot \mathbf{E} = \rho/\epsilon_0$, can be combined into a single scalar second-order differential equation, Poisson's equation

$$\nabla^2 \varphi = -\frac{\rho}{\epsilon_0},$$

where $\mathbf{E} = -\nabla\varphi$. If $\rho(\mathbf{r})$ is a given function in a region V, Poisson's equation has the particular solution

$$\varphi(\mathbf{r}) = \frac{1}{4\pi\epsilon_0} \int_V \frac{\rho(\mathbf{r}')}{|\mathbf{r} - \mathbf{r}'|} \, dv',$$

as can be seen by operating on the right side of this expression with ∇^2 inside the integral. To this particular solution can be added any solution of the corresponding homogeneous equation, Laplace's equation

$$\nabla^2 \varphi = 0.$$

The appropriate solution of Laplace's equation is one which will cause given boundary conditions to be satisfied on the boundary of V. If φ or $\partial\varphi/\partial n$ is specified on a closed boundary, the solution is unique.

1. If the region V encompasses all space, the appropriate boundary condition is $\varphi = 0$ at $r = \infty$. The solution of Laplace's equation that satisfies this condition is 0 so that the particular integral is the complete solution.

2. If $\rho = 0$ everywhere inside V, the complete solution is the solution of Laplace's equation that is continuous inside V and satisfies the imposed conditions at the boundary of V. The latter are $\varphi = constant$ if the boundary is composed of conductors.

3. Analytical solutions for Laplace's equation can be composed of linear combinations from a set of basis functions, which is chosen by considering any symmetry that appears in the particular problem. The set of zonal harmonics is useful for spherical boundaries and azimuthal symmetry, and the cylindrical harmonics for cylindrical boundaries and axial symmetry. The uniqueness theorem signals when enough functions from the infinite set have been used.

4. The method of images is a technique where one devises a configuration of fictitious image charges outside of V such that the potential field of these image charges plus the real charges inside V satisfies the boundary conditions for φ on the given boundaries of V. The uniqueness theorem assures that the field inside V is the correct one. The technique can be applied in only a few situations where the symmetry is adequate. The simplest example is a point charge in front of a conducting plane.

5. When all the charges reside on conducting surfaces, Laplace's equation requires that the potentials of the conductors are linear functions of their charges:

$$\varphi_i = \sum_j p_{ij} Q_j.$$

REFERENCES

The following texts are recommended for (1) a more complete discussion of Legendre's equation, (2) the general form of Laplace's equation in orthogonal, curvilinear coordinates, and (3) a more complete discussion of the solution to Laplace's equation:

Mathematical Physics by E. Butkov (Reading, Mass: Addison-Wesley, 1968).

Electromagnetic Theory by J. A. Stratton (New York: McGraw-Hill, 1941).

Classical Electricity and Magnetism, Second Edition, by W. Panofsky and M. Phillips (Reading, Mass.: Addison-Wesley, 1962).

PROBLEMS

3–1 Two spherical conducting shells of radii r_a and r_b are arranged concentrically and are charged to the potentials φ_a and φ_b, respectively. If $r_b > r_a$, find the potential at points between the shells, and at points $r > r_b$.

3–2 Two long cylindrical shells of radii r_a and r_b are arranged coaxially and are charged to the potentials φ_a and φ_b, respectively. Find the potential at points between the cylindrical shells.

3–3 If φ_1 is a solution to Laplace's equation, prove that the partial derivative of φ_1 with respect to one or more of the rectangular coordinates (e.g., $\partial\varphi_1/\partial x$, $\partial^2\varphi_1/\partial x^2$, $\partial^2\varphi_1/\partial x\partial y$, etc.) is also a solution.

3–4 Suppose φ satisfies Laplace's equation throughout a region V_0. Prove that the value of φ at any point O is the average of its values over the surface of any sphere centered at O that lies entirely in V_0:

$$\varphi(O) = \frac{1}{4\pi R^2} \oint \varphi \, da,$$

where R is the radius of the sphere. (*Hint:* Let $\psi = 1/r$ in Eq. (1–57).) Show that consequently φ has no maxima or minima inside V_0.

3–5 Expand the function

$$F(u) = (1 - 2xu + u^2)^{-1/2}$$

in Taylor series up to the term in u^3. Note that the coefficients are the first four Legendre polynomials $P_n(x)$. In fact, $F(u)$ is a *generating function* for all the Legendre polynomials:

$$F(u) = \sum_{n=0}^{\infty} P_n(x)u^n.$$

3–6 Show that half the zonal harmonics are generated by differentiating r^{-1} successively with respect to the rectangular coordinate z $(z = r \cos \theta)$.

3–7 Obtain $\nabla^2 \varphi$ in cylindrical coordinates, Eq. (3–8), from the rectangular form, Eq. (3–6), by direct substitution: $x = r \cos \theta$, $y = r \sin \theta$.

3–8 Find the potential of an axial quadrupole: point charges q, $-2q$, q placed on the z-axis at distances l, 0, $-l$ from the origin. Find the potential only at distances $r \gg l$, and show that this potential is proportional to one of the zonal harmonics.

3–9 Suppose a point dipole is located at the center of a grounded spherical conducting shell. Find the potential inside the shell. (*Hint:* Use zonal harmonics that are regular at the origin to satisfy the boundary conditions on the shell.)

3–10 For an uncharged conducting sphere placed in an initially uniform electric field, show that the potential due to the sphere is that of a point dipole and find the induced dipole moment.

3–11 A conducting sphere of radius a bearing total charge Q is placed in an initially uniform electric field E_0. Find the potential at all points exterior to the sphere.

3–12 A long cylindrical conductor of radius a bearing no net charge is placed in an initially uniform electric field $\mathbf{E}_0$. The direction of $\mathbf{E}_0$ is perpendicular to the cylinder axis. Find the potential at points exterior to the cylinder, and find also the charge density on the cylindrical surface.

*3–13 Show that Im $A[(x + iy)]^{1/2} = Ar^{1/2} \sin \frac{1}{2}\theta$ satisfies Laplace's equation, but that the electric field derived from this function has a discontinuity at $\theta = 0$. (Note that r and θ are cylindrical coordinates here.) The function may be used to describe the potential at the edge of a charged conducting plane. The conducting plane coincides with the xz-plane, but only for positive values of x. Find the charge density on the plane. Make a sketch showing several equipotential surfaces and several lines of force.

3–14 A point charge q is situated a distance d from a grounded conducting plane of infinite extent. Obtain the total charge induced on the plane by direct integration of the surface charge density.

3–15 Two point charges, q_1 and q_2, are located near a grounded conducting plane of infinite extent. Find the image charges which are needed to make the plane a surface of constant potential. From the result just obtained, can you predict the image charge distribution required for the case of a body of arbitrary shape with charge density ρ situated near a conducting plane of infinite extent?

3–16 Two grounded conducting planes intersect at $60°$ and a point charge q lies between them. Determine the positions of the image charges that will give the electric field between the planes.

3–17 A point charge q is located between two parallel, grounded, conducting planes which are separated by a distance d. Find the locations of the infinite number of image charges. Express the force on the charge q by an infinite series.

3–18 Find the force between a point charge q and an *uncharged* conducting sphere of radius a. The point charge is located a distance r from the center of the sphere, where $r > a$. Find an approximate expression valid for $r \gg a$.

3–19 Show that the problem of an uncharged conducting sphere in an initially uniform electric field $\mathbf{E}_0$ may be solved by means of images. (*Hint:* A uniform electric field in the vicinity of the origin may be approximated by the field of two point charges Q and $-Q$

---

* Starred problems are more difficult.

placed on the z-axis at $z = -L$ and $z = +L$, respectively. The field becomes more nearly uniform as $L \to \infty$. It is evident that $Q/2\pi\epsilon_0 L^2 = E_0$.)

3–20 A point charge q is located inside and at distance r from the center of a spherical conducting shell. The inner radius of the shell is a. Show that this problem can be solved by the image technique, and find the charge density σ induced on the inside surface of the shell. (The potential of the spherical shell cannot be completely specified in terms of q and its image, because exterior fixed charges can also contribute. Nevertheless, these exterior charges will add only a constant term to the potential.) Find the total charge induced on the inside surface of the shell (a) by physical arguments, and (b) by integration of σ over the surface.

3–21 A long conducting cylinder bearing a charge λ per unit length is oriented parallel to a grounded conducting plane of infinite extent. The axis of the cylinder is at distance x_0 from the plane, and the radius of the cylinder is a. Find the location of the line image, and find also the constant M (which determines the potential of the cylinder) in terms of a and x_0.

3–22 A spherical distribution of charge is characterized by a constant charge density ρ for $r \le R$. For radii greater than R, the charge density is zero. Find the potential $\varphi(r)$ by integrating Poisson's equation. Check this result by evaluating the integral (3–1). (*Hint:* To perform (3–1), divide the charge region into spherical concentric shells of thickness dr.)

3–23 A dipole $\mathbf{p}$ is oriented normal to and at distance d from an infinite conducting plane. The plane is grounded (i.e., at zero potential). Calculate the force exerted on the plane by the dipole.

3–24 A thunderstorm contains a charge $+Q$ at altitude h_1 and, directly below this, a charge $-Q$ at altitude h_2. Find an expression for the vertical electric field E_v at the earth's surface at distance d from the storm. For $h_1 = 5000$ m, $h_2 = 3000$ m, and $Q = 15$ C, make a graph showing how E_v varies, from $d = 0$ to $d = 20$ km.

3–25 Suppose $\varphi(x, y, z)$ satisfies Laplace's equation. Show that the value of φ at (x, y, z) is approximately equal to the average of its values at the six surrounding points $(x \pm d, y, z)$, $(x, y \pm d, z)$, $(x, y, z \pm d)$. (*Hint:* Calculate the Taylor series expansion of $\varphi(x + d, y, z)$ up to the term in d^3, and similarly for the other five points; add the results.) Computer programs for the numerical calculation of φ can be based on this result.

CHAPTER 4   The Electrostatic Field in Dielectric Media

Thus far, we have ignored problems involving dielectric media, and have dealt with cases in which the electric field is produced exclusively either by charges in a specified distribution or by free charge on the surface of conductors. We now wish to remedy this situation and take up the more general case.

An ideal dielectric material is one which has no free charges. Nevertheless, all material media are composed of molecules, these in turn being composed of charged entities (atomic nuclei and electrons), and the molecules of the dielectric are certainly affected by the presence of an electric field. The electric field causes a force to be exerted on each charged particle, positive particles being pushed in the direction of the field, negative particles oppositely, so that the positive and negative parts of each molecule are displaced from their equilibrium positions in opposite directions. These displacements, however, are limited (in most cases to very small fractions of a molecular diameter) by strong restoring forces which are set up by the changing charge configuration in the molecule. The term "bound charge," in contrast to the "free charge" of a conductor, is sometimes used to emphasize that such molecular charges are not free to move very far or to be extracted from the dielectric material. The overall effect from the macroscopic point of view is most easily visualized as a displacement of the entire positive charge in the dielectric relative to the negative charge. The dielectric is said to be *polarized*.

A polarized dielectric, even though it is electrically neutral on the average, produces an electric field, both at exterior points and inside the dielectric as well. As a result, we are confronted with what appears to be an awkward situation: the polarization of the dielectric depends on the total electric field in the medium, but a part of the electric field is produced by the dielectric itself. Furthermore, the distant electric field of the dielectric may modify the free charge distribution on conducting bodies, and this in turn will change the electric field *in the dielectric*. It is the main purpose of this chapter to develop general methods for handling this curious situation.

4–1 POLARIZATION

Consider a small volume element Δv of a dielectric medium which, as a whole, is electrically neutral. If the medium is polarized, then a separation of positive and negative charge has been effected, and the volume element is characterized by an electric dipole moment

$$\Delta \mathbf{p} = \int_{\Delta v} \mathbf{r} \, dq. \tag{4-1}$$

According to Section 2–9, this quantity determines the electric field produced by Δv at distant points (i.e., at distances from Δv large compared with the dimensions of the volume element).

Since $\Delta \mathbf{p}$ depends on the size of the volume element, it is more convenient to work with $\mathbf{P}$, the electric dipole moment per unit volume:

$$\mathbf{P} = \frac{\Delta \mathbf{p}}{\Delta v}. \tag{4-2}$$

Strictly speaking, $\mathbf{P}$ must be defined as the limit of this quantity as Δv becomes very small from the macroscopic viewpoint. In this way $\mathbf{P}$ becomes a point function, $\mathbf{P}(x, y, z)$. $\mathbf{P}$ is usually called the *electric polarization*, or simply the *polarization*, of the medium. Its dimensions are charge per unit area; in mks units, C/m^2.

It is apparent that $\mathbf{P}(x, y, z)$ is a vector quantity that, in each volume element, has the direction of $\Delta \mathbf{p}$. This, in turn, has the direction of displacement of positive charge relative to negative charge (see Fig. 4–1).

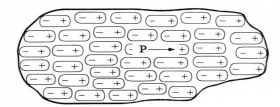

Figure 4–1 A piece of polarized dielectric material. Each volume element is represented as a dipole $\Delta \mathbf{p}$.

Although Δv is assumed very small from the macroscopic point of view, it still contains many molecules. It is sometimes desirable to speak about the electric dipole moment of a single molecule, that is,

$$\mathbf{p}_m = \int_{\text{molecule}} \mathbf{r} \, dq, \tag{4-3}$$

since a molecule is one of the small, electrically neutral entities that make up the dielectric material. It is evident from Eq. (4–1) that the dipole moment associated with Δv is given by $\Delta \mathbf{p} = \sum \mathbf{p}_m$, where the summation extends over all molecules inside the element Δv. Hence

$$\mathbf{P} = \frac{1}{\Delta v} \sum_m \mathbf{p}_m. \tag{4-4}$$

This approach will be developed further in Chapter 5.

Although Fig. 4–1 represents each volume element of the polarized dielectric as a small dipole, it may be more instructive to visualize the dielectric in terms of its molecules, and to imagine that each dipole of Fig. 4–1 represents a single molecule.

4–2 FIELD OUTSIDE OF A DIELECTRIC MEDIUM

Consider now a finite piece of dielectric material that is polarized, i.e., it is characterized at each point $\mathbf{r}'$ by a polarization, $\mathbf{P}(\mathbf{r}')$. The polarization gives rise to an electric field, and our problem is to calculate this field at point $\mathbf{r}$, which is outside of the dielectric body (see Fig. 4–2). As in Chapter 2, we shall find it more convenient to calculate first the potential $\varphi(\mathbf{r})$, and obtain the electric field as minus the gradient of φ.

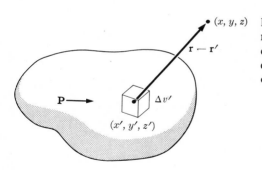

Figure 4–2 The electric field at (x, y, z) may be calculated by summing up the contributions due to the various volume elements $\Delta v'$ in V_0. The surface of V_0 is denoted by S_0.

Each volume element $\Delta v'$ of the dielectric medium is characterized by a dipole moment $\Delta \mathbf{p} = \mathbf{P} \, \Delta v'$, and since the distance between (x, y, z) and $\Delta v'$ is large compared with the dimensions of $\Delta v'$, this quantity (the dipole moment) completely determines $\Delta v'$'s contribution to the potential:

$$\Delta \varphi(\mathbf{r}) = \frac{\Delta \mathbf{p} \cdot (\mathbf{r} - \mathbf{r}')}{4\pi \epsilon_0 |\mathbf{r} - \mathbf{r}'|^3} = \frac{\mathbf{P}(\mathbf{r}') \cdot (\mathbf{r} - \mathbf{r}') \, \Delta v'}{4\pi \epsilon_0 |\mathbf{r} - \mathbf{r}'|^3}. \tag{4-5}$$

Here $\mathbf{r} - \mathbf{r}'$ is the vector, directed out from $\Delta v'$, whose magnitude is given by

$$|\mathbf{r} - \mathbf{r}'| = \sqrt{(x - x')^2 + (y - y')^2 + (z - z')^2}. \tag{4-6}$$

The entire potential at point $\mathbf{r}$ is obtained by summing the contributions from all parts of the dielectric:

$$\varphi(\mathbf{r}) = \frac{1}{4\pi\epsilon_0} \int_{V_0} \frac{\mathbf{P}(\mathbf{r}') \cdot (\mathbf{r} - \mathbf{r}')\, dv'}{|\mathbf{r} - \mathbf{r}'|^3}. \tag{4-7}$$

This result is correct, and φ may be evaluated directly from Eq. (4–7) if the functional form of $\mathbf{P}$ is known. It will be to our advantage, however, to express Eq. (4–7) in a rather different way by means of a simple mathematical transformation.

If $|\mathbf{r} - \mathbf{r}'|$ is given by Eq. (4–6), then

$$\nabla'\left(\frac{1}{|\mathbf{r} - \mathbf{r}'|}\right) = +\frac{\mathbf{r} - \mathbf{r}'}{|\mathbf{r} - \mathbf{r}'|^3}, \tag{4-8}$$

as may be seen by direct application of the gradient operator in Cartesian coordinates. The ∇' operator involves derivatives with respect to the primed coordinates. In certain circumstances it may be desirable to perform a gradient operation with respect to the unprimed coordinates; this will be indicated in the usual way by ∇. Evidently, ∇' operating on a function of $|\mathbf{r} - \mathbf{r}'|$ is equal to $-\nabla$ operating on the same function. We shall require the ∇ operator later in order to get the electric field at point $\mathbf{r}$. However, in performing the integral Eq. (4–7) over the dielectric volume V_0, the point $\mathbf{r}$ is held fixed; hence the integrand of Eq. (4–7) may be transformed by means of Eq. (4–8):

$$\frac{\mathbf{P} \cdot (\mathbf{r} - \mathbf{r}')}{|\mathbf{r} - \mathbf{r}'|^3} = \mathbf{P} \cdot \nabla'\left(\frac{1}{|\mathbf{r} - \mathbf{r}'|}\right). \tag{4-9}$$

Equation (4–9) may be further transformed by means of the vector identity (1–1–7) of Table 1–1:

$$\nabla' \cdot (f\mathbf{F}) = f\nabla' \cdot \mathbf{F} + \mathbf{F} \cdot \nabla' f, \tag{4-10}$$

where f is any scalar point function and $\mathbf{F}$ is an arbitrary vector point function. Here again the prime indicates differentiation with respect to the primed coordinates. Letting $f = (1/|\mathbf{r} - \mathbf{r}'|)$ and $\mathbf{F} = \mathbf{P}$, the integrand, Eq. (4–9), becomes

$$\frac{\mathbf{P} \cdot (\mathbf{r} - \mathbf{r}')}{|\mathbf{r} - \mathbf{r}'|^3} = \nabla' \cdot \left(\frac{\mathbf{P}}{|\mathbf{r} - \mathbf{r}'|}\right) - \frac{1}{|\mathbf{r} - \mathbf{r}'|} \nabla' \cdot \mathbf{P}. \tag{4-11}$$

Finally, the potential, Eq. (4–7), may be written as

$$\varphi(\mathbf{r}) = \frac{1}{4\pi\epsilon_0} \oint_{S_0} \frac{\mathbf{P} \cdot \mathbf{n}\, da'}{|\mathbf{r} - \mathbf{r}'|} + \frac{1}{4\pi\epsilon_0} \int_{V_0} \frac{(-\nabla' \cdot \mathbf{P})\, dv'}{|\mathbf{r} - \mathbf{r}'|}, \tag{4-12}$$

where the volume integral of $\nabla' \cdot (\mathbf{P}/|\mathbf{r} - \mathbf{r}'|)$ has been replaced by a surface integral through application of the divergence theorem, and $\mathbf{n}$, of course, is the outward normal to the surface element da' (outward means out of the dielectric).

The quantities $\mathbf{P} \cdot \mathbf{n}$ and $-\nabla \cdot \mathbf{P}$ that appear in the integrals of Eq. (4–12) are two scalar functions obtained from the polarization $\mathbf{P}$. It seems expedient to give these quantities special symbols, and since they have the dimensions charge per unit area and charge per unit volume, respectively, we write

$$\sigma_P \equiv \mathbf{P} \cdot \mathbf{n} = P_n, \tag{4-13}$$

and

$$\rho_P \equiv -\mathbf{\nabla} \cdot \mathbf{P}, \tag{4-14}$$

and call σ_P and ρ_P *polarization charge densities*. The surface density of polarization charge is given by the component of polarization normal to the surface, and the volume density of polarization charge is a measure of the nonuniformity of the polarization inside the material.

The potential due to the dielectric material,

$$\varphi(\mathbf{r}) = \frac{1}{4\pi\epsilon_0} \left[\oint_{S_0} \frac{\sigma_P \, da'}{|\mathbf{r} - \mathbf{r}'|} + \int_{V_0} \frac{\rho_P \, dv'}{|\mathbf{r} - \mathbf{r}'|} \right],$$

$$= \frac{1}{4\pi\epsilon_0} \int \frac{dq'_P}{|\mathbf{r} - \mathbf{r}'|}, \tag{4-15}$$

is now written in such a way that it is evident that it arises from a charge distribution. In other words, the dielectric material has been replaced by an appropriate distribution of polarization charge.

Although Eq. (4–15) has been obtained by means of a mathematical transformation, it should be possible to understand σ_P and ρ_P on purely physical grounds. That a surface charge density σ_P exists is evident from Fig. 4–1, where it is seen that this charge is made up from the ends of similarly oriented dipoles. In this way a charge density is developed on every surface that is not parallel to the polarization vector. Turning now to ρ_P we expect that $\rho_P \Delta v'$ represents the *excess* or *net charge* in the volume element $\Delta v'$. That this is truly the case may be seen in the following way: let us define two charge densities ρ^+ and ρ^- as representing the total positive charge and the total negative charge per unit volume, respectively. That is, ρ^+ represents all the atomic nuclei in unit volume of the dielectric and, similarly, ρ^- counts all the electrons. In the unpolarized state, each volume element of the dielectric is electrically neutral; hence

$$\rho_0^+ (x', y', z') + \rho_0^- (x', y', z') = 0, \tag{4-16}$$

where the subscript zero denotes densities in the unpolarized configuration. Let us assume that as a consequence of polarization the positive charge is displaced by $\mathbf{\delta}^+ (x, y, z)$ and the negative charge by $\mathbf{\delta}^- (x, y, z)$. The positive charge crossing an element of area da' is $\rho_0^+ \mathbf{\delta}^+ \cdot \mathbf{n} \, da'$, and thus the *gain of positive charge* by the volume element $\Delta v'$ during the polarization process is

$$-\oint_{\Delta S} \rho_0^+ \mathbf{\delta}^+ \cdot \mathbf{n} \, da', \tag{4-17}$$

where ΔS is the surface bounding $\Delta v'$. Similarly, the displacement of negative charge increases the charge (decreases the negative charge) in $\Delta v'$ by

$$\oint_{\Delta S} (-\rho_0^-)\, \boldsymbol{\delta}^- \cdot \mathbf{n}\, da. \tag{4-18}$$

The total gain in charge by the volume element $\Delta v'$ is the sum of Eqs. (4–17) and (4–18), and as a consequence of Eq. (4–16) may be written as

$$-\oint_{\Delta S} \rho_0^+ (\boldsymbol{\delta}^+ - \boldsymbol{\delta}^-) \cdot \mathbf{n}\, da' = -\nabla \cdot [\rho_0^+ (\boldsymbol{\delta}^+ - \boldsymbol{\delta}^-)]\, \Delta v'. \tag{4-19}$$

But $\boldsymbol{\delta}^+ - \boldsymbol{\delta}^-$ is just the relative displacement of positive and negative charge densities, and $\rho_0^+ (\boldsymbol{\delta}^+ - \boldsymbol{\delta}^-)$ is equivalent, therefore, to what we have called the polarization $\mathbf{P}$. Thus $\rho_P\, \Delta v'$ is the net charge in a volume element of the polarized dielectric.

At first sight it may seem rather strange that having started with electrically neutral volume elements of dielectric material, we end up with volume elements that bear a net charge. According to our original point of view, the dielectric is composed of elemental dipoles $\Delta \mathbf{p}$, and it was essential that each $\Delta \mathbf{p}$ be electrically neutral in order that Eq. (4–5) give the potential correctly. Now we find that so long as $\nabla \cdot \mathbf{P}$ does not vanish, the individual volume elements appear to be charged. The origin of this seeming paradox is found in the mathematical transformation Eq. (4–11); the contribution from each volume element is transformed to a different volume term and a surface term. The total charge in the volume and surface of the element is still zero; but when we stack various volume elements together to form a macroscopic piece of dielectric material, we find that the contributions to the potential from the various "internal surfaces" cancel out. We are left with effectively charged volume elements and a surface contribution from the actual boundary of the dielectric body.

The total polarization charge of a dielectric body,

$$Q_P = \int_{V_0} (-\nabla' \cdot \mathbf{P})\, dv' + \oint_{S_0} \mathbf{P} \cdot \mathbf{n}\, da', \tag{4-20}$$

must equal zero, since it was our premise that the dielectric, as a whole, is electrically neutral. This result is immediately obvious from the form of Eq. (4–20), which clearly vanishes as a consequence of the divergence theorem.

We now have two distinct expressions for the electrostatic potential $\varphi(\mathbf{r})$ due to a polarized dielectric specimen, namely, Eqs. (4–7) and (4–15). Both are correct, but we shall find the latter expression more convenient in most cases. The electric field $\mathbf{E}$ may be obtained as minus the gradient of Eq. (4–15). Since φ is a function of the coordinates (x, y, z), the appropriate gradient is $-\nabla$. The unprimed coordinates appear only in the function $1/|\mathbf{r} - \mathbf{r}'|$. Hence, noting that $\nabla(1/|\mathbf{r} - \mathbf{r}'|) = -\nabla'(1/|\mathbf{r} - \mathbf{r}'|)$ and using Eq. (4–8), we obtain

$$\mathbf{E}(\mathbf{r}) = \frac{1}{4\pi\epsilon_0} \left[\int_{S_0} \frac{\sigma_P(\mathbf{r} - \mathbf{r}')\, da'}{|\mathbf{r} - \mathbf{r}'|^3} + \int_{V_0} \frac{\rho_P(\mathbf{r} - \mathbf{r}')\, dv'}{|\mathbf{r} - \mathbf{r}'|^3} \right]. \tag{4-21}$$

4-3 THE ELECTRIC FIELD INSIDE A DIELECTRIC

Before we can write an expression for the electric field inside a polarized medium, it is necessary to define this electric field precisely. What we are interested in, of course, is the *macroscopic electric field*, i.e., the average electric field in a small region of the dielectric which, nevertheless, contains a large number of molecules. An alternative and perhaps preferable approach is to define the electric field directly in terms of a macroscopic experiment: *the (macroscopic) electric field is the force per unit charge on a test charge embedded in the dielectric, in the limit where the test charge is so small that it does not itself affect the charge distribution.* This test charge must be dimensionally small from the macroscopic point of view (what we shall call a "point" charge), but it will be large compared with the size of a molecule.

Although the above statement is the fundamental definition of the macroscopic electric field $\mathbf{E}$, it is difficult to use this definition directly to obtain an expression for the field, since we would have to calculate the force on a charged body of extended size, and then go to the limit as the size of the object decreased. Hence we find it expedient to use another property of the electric field to help us obtain the analytic expression we are seeking, and in this way we shall get $\mathbf{E}$ in terms of the polarization charges of the medium. Later, in Section 4–10, it will be shown that the quantity we have called $\mathbf{E}$ is indeed in agreement with the fundamental "force definition."

The electrostatic field in a dielectric must have the same basic properties that we found applied to $\mathbf{E}$ in vacuum; in particular, $\mathbf{E}$ is a conservative field, and hence derivable from a scalar potential. Thus

$$\nabla \times \mathbf{E} = 0$$

or, equivalently,

$$\oint \mathbf{E} \cdot d\mathbf{l} = 0.$$

Let us apply the last equation to the path $ABCD$ shown in Fig. 4–3, where the segment AB lies in a needle-shaped cavity cut out of the dielectric, and the segment CD lies in the dielectric proper. Since the segments AD and BC may be made arbitrarily small, the line integral reduces to

$$\mathbf{E}_v \cdot \mathbf{l} - \mathbf{E}_d \cdot \mathbf{l} = 0$$

or, equivalently,

$$E_{vt} = E_{dt}, \tag{4–22}$$

where the subscripts v and d refer to vacuum and dielectric respectively, and the subscript t stands for tangential component.

Equation (4–22) is valid regardless of the orientation of the needle-shaped cavity. If the "needle" is oriented along the direction of $\mathbf{E}$, then $E_{dt} = E_d$; further-

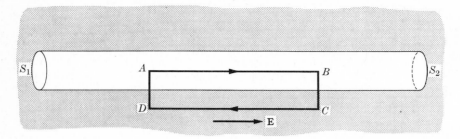

Figure 4–3 The path $ABCD$ lies partly in the needle-shaped cavity, and partly in the dielectric. In an isotropic dielectric (see Section 4–5) the polarization $\mathbf{P}$ has the direction of $\mathbf{E}$, so that, for the orientation of the needle shown, $\sigma_P = 0$ on the cylindrical walls. In an anisotropic dielectric, σ_P is not necessarily zero, but its value does not affect the longitudinal component of electric field in the cavity.

more, by symmetry, the field in the cavity is along the direction of the needle, that is, $E_{vt} = E_v$. We are thus led to an important conclusion:*

> *The electric field in a dielectric is equal to the electric field inside a needle-shaped cavity in the dielectric provided the cavity axis is oriented parallel to the direction of the electric field.*

Evidently, the problem of calculating the electric field inside a dielectric reduces to calculating the electric field inside a needle-shaped cavity in the dielectric. But the electric field in the cavity is an external field, and hence may be determined by means of the results of Section 4–2. Just as in Section 4–2, we assume here that the polarization of the dielectric is a given function $\mathbf{P}(x', y', z')$, and we calculate the potential and electric field arising from this polarization. Taking the field point $\mathbf{r}$ at the center of the cavity and using Eq. (4–15), we obtain for the potential

$$\varphi(\mathbf{r}) = \frac{1}{4\pi\epsilon_0} \int_{V_0 - V_1} \frac{\rho_P(x', y', z')\, dv'}{|\mathbf{r} - \mathbf{r}'|} + \frac{1}{4\pi\epsilon_0} \int_{S_0 + S'} \frac{\sigma_P(x', y', z')\, da'}{|\mathbf{r} - \mathbf{r}'|}, \quad (4\text{–}23)$$

where $V_0 - V_1$ is the volume of the dielectric excluding the "needle," S_0 is the exterior surface of the dielectric, and $S' = S_1 + S_2 + S_c$ are the needle surfaces. But from Fig. 4–3 it is seen that $\sigma_P = 0$ on the cylindrical surface S_c of the needle; furthermore, the needle may be made arbitrarily thin so that the surfaces S_1 and S_2 have negligible area. Thus only the exterior surfaces of the dielectric contribute, and the surface integral of Eq. (4–23) becomes identical in form to the surface

---

* This statement is strictly true only for isotropic dielectrics (see Section 4–5). For anisotropic dielectrics the symmetry argument fails, and our conclusion must be generalized: the electric field in a dielectric is equal to the longitudinal component of the electric field inside a needle-shaped cavity in the dielectric provided the cavity axis is oriented parallel to the direction of the electric field in the dielectric.

integral of Eq. (4–15). The volume integral of Eq. (4–23) excludes the cavity; however, the contribution of the cavity to this integral is negligible, as may readily be seen. The charge density ρ_P is bounded; the quantity $dv'/|\mathbf{r} - \mathbf{r}'|$ does not diverge at the field point (i.e., when $\mathbf{r}' = \mathbf{r}$) because the volume of a point is a higher-order zero than the $\lim |\mathbf{r} - \mathbf{r}'|$; and finally volume V_1 of the needle may be made arbitrarily small by making the cavity thin. Thus we need not exclude the volume V_1, and Eq. (4–23) becomes similar in form to Eq. (4–15). In other words, Eq. (4–15) gives the potential $\varphi(\mathbf{r})$ regardless of whether the point $\mathbf{r}$ is located inside or outside the dielectric.

The electric field $\mathbf{E}(\mathbf{r})$ may be calculated as minus the gradient of Eq. (4–23). But this differs only by a negligible amount from Eq. (4–21). *Thus Eq. (4–21) gives the medium's contribution to the electric field at* $\mathbf{r}$, *independently of whether* $\mathbf{r}$ *is inside or outside the medium.*

The calculations indicated in Eqs. (4–15) and (4–21) are straightforward for cases in which $\mathbf{P}(x, y, z)$ is a known function of position. (Some examples of this type are to be found among the problems at the end of this chapter.) In most cases, however, the polarization arises in response to an electric field which has been imposed on the dielectric medium [that is, $\mathbf{P}(x', y', z')$ is a function of the *total macroscopic electric field* $\mathbf{E}(x', y', z')$], and under these conditions the situation is much more complicated. First, it is necessary to know the functional form of $\mathbf{P}(\mathbf{E})$; but this is known experimentally in most cases and hence is not a source of difficulty. The real complication arises because $\mathbf{P}$ depends on the *total electric field*, including the contribution from the dielectric itself, and it is this contribution that we are in the process of evaluating. Thus we cannot determine $\mathbf{P}$ because we don't know $\mathbf{E}$, and conversely.

It is evident that a different approach to the problem is needed, and this will be provided in the following sections.

4–4 GAUSS'S LAW IN A DIELECTRIC.
THE ELECTRIC DISPLACEMENT

In Chapter 2 we derived an important relationship between electric flux and charge, namely, Gauss's law. This law states that the electric flux across an arbitrary closed surface is proportional to the total charge enclosed by the surface. In applying Gauss's law to a region containing charges embedded in a dielectric, we must be careful to include all of the charge inside the gaussian surface, polarization charge as well as the charge that we embedded.

In Fig. 4–4 the dashed surface S is an imaginary closed surface located inside a dielectric medium. We embed a certain amount of charge, Q, in the volume bounded by S, and we shall assume that this charge exists on the surfaces of three conductors in amounts q_1, q_2, and q_3. By Gauss's law,

$$\oint_S \mathbf{E} \cdot \mathbf{n} \, da = \frac{1}{\epsilon_0} (Q + Q_P), \qquad (4\text{–}24)$$

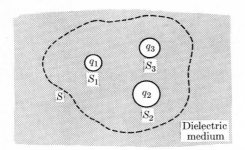

Figure 4–4 Construction of a Gaussian surface S in a dielectric medium.

where Q is the net embedded charge, i.e.,

$$Q = q_1 + q_2 + q_3,$$

and Q_P is the net polarization charge:

$$Q_P = \int_{S_1 + S_2 + S_3} \mathbf{P} \cdot \mathbf{n} \, da + \int_V (-\nabla \cdot \mathbf{P}) \, dv. \qquad (4\text{–}25a)$$

Here V is volume of the dielectric enclosed by S. There is no boundary of the dielectric material at S, so that the surface integral in Eq. (4–25a) *does not* contain a contribution from S.

If we transform the volume integral in Eq. (4–25a) to a surface integral by means of the divergence theorem, we must be careful to include contributions from all surfaces bounding V, namely, S, S_1, S_2, and S_3. It is evident that the last three contributions will cancel the first term of Eq. (4–25a), so that

$$Q_P = -\oint_S \mathbf{P} \cdot \mathbf{n} \, da. \qquad (4\text{–}25b)$$

Combining this result with (4–24), we obtain

$$\oint_S (\epsilon_0 \mathbf{E} + \mathbf{P}) \cdot \mathbf{n} \, da = Q. \qquad (4\text{–}26)$$

Equation (4–26) states that the flux of the vector $\epsilon_0 \mathbf{E} + \mathbf{P}$ through a closed surface is equal to the net charge that we embedded in the volume enclosed by the surface. This vector quantity is important enough to warrant a name and a separate symbol. We define, therefore, a new macroscopic field vector $\mathbf{D}$, the *electric displacement*:

$$\mathbf{D} = \epsilon_0 \mathbf{E} + \mathbf{P}, \qquad (4\text{–}27)$$

which evidently has the same units as $\mathbf{P}$, charge per unit area.*

---

* In Gaussian units, $\mathbf{D}$ is defined as $\mathbf{D} = \mathbf{E} + 4\pi\mathbf{P}$; $\mathbf{D}$, $\mathbf{E}$, and $\mathbf{P}$ all have the same units of charge per unit area, and in vacuum $\mathbf{D} = \mathbf{E}$.

In terms of **D**, Eq. (4–26) becomes

$$\oint_S \mathbf{D} \cdot \mathbf{n} \, da = Q, \tag{4–28}$$

and this result is usually referred to as Gauss's law for the electric displacement, or simply Gauss's law. Equation (4–28) is applicable to a region of space bounded by any closed surface S; if we apply it to a small region in which the charge enclosed is distributed as a charge density ρ, then Gauss's law becomes

$$\oint_S \mathbf{D} \cdot \mathbf{n} \, da = \rho \, \Delta V.$$

Dividing this equation by ΔV and proceeding to the limit, we obtain

$$\nabla \cdot \mathbf{D} = \rho, \tag{4–29}$$

a result which is sometimes called the differential form of Gauss's law.

The advantage of expressing the integral and differential forms of Gauss's law, Eqs. (4–28) and (4–29), in terms of the vector **D** is that only the charge Q or the charge density ρ that we embedded in the dielectric medium appears explicitly. This is what henceforth we shall usually call simply the charge (or the charge density). When it is necessary to distinguish it from the *polarization* charge Q_P of the medium or the *total* charge $Q + Q_P$, the charge Q will be called the *external charge*. By "external" we do not mean that it necessarily lies outside the physical boundary of the piece of material; we mean that it is in addition to the charges that make up the atomic constitution of the neutral material.* Since in many problems the external charges are given, it is an advantage that the total electrostatic field at each point in the dielectric medium is expressed as the sum of two parts,

$$\mathbf{E}(x, y, z) = \frac{1}{\epsilon_0} \mathbf{D}(x, y, z) - \frac{1}{\epsilon_0} \mathbf{P}(x, y, z), \tag{4–30}$$

where the first term, $(1/\epsilon_0)\mathbf{D}$, is related to the external charge density through its divergence, and the second term, $(-1/\epsilon_0)\mathbf{P}$, is proportional to the polarization of the medium. In a vacuum the electric field is given entirely by the first term in Eq. (4–30).

---

* The external charge is very often called the "free" charge, and polarization charge is sometimes used synonymously with bound charge. In electrostatics this confusion does not cause too much trouble because the external charge on a conductor is free (i.e., free to move around) and the polarization charge in a dielectric is bound. The external charge in a dielectric is not free, however—if it were it would quickly move to the surface and leak off. Also, most conducting media contain, in addition to the free charges that determine the electrostatic behavior of a conducting object, some bound charges, which in other circumstances contribute to polarization (Chapter 7). With time-dependent fields (Chapter 19), it is very important not to confound the distinction between external and polarization charges with that between free and bound charges. Thus we shall exclusively use the term *external* charge in this context.

4–5 ELECTRIC SUSCEPTIBILITY AND DIELECTRIC CONSTANT

In the introduction to this chapter it was stated that the polarization of a dielectric medium occurs in response to the electric field in the medium. The degree of polarization depends not only on the electric field, but also on the properties of the molecules which make up the dielectric material. From the macroscopic point of view, the behavior of the material is completely specified by an experimentally determined relationship, called a *constitutive equation*, $\mathbf{P} = \mathbf{P}(\mathbf{E})$, where $\mathbf{E}$ is the macroscopic electric field. This is a point relationship, and if $\mathbf{E}$ varies from point to point in the material, then $\mathbf{P}$ will vary accordingly.

For most materials, $\mathbf{P}$ vanishes when $\mathbf{E}$ vanishes. Since this is the usual behavior, we shall limit our discussion here to materials of this type. (Dielectrics with a permanent polarization will be discussed briefly in Section 5–4.) Furthermore, if the material is isotropic, then the polarization should have the same direction as the electric field that is causing it. These results are summarized by the constitutive equation

$$\mathbf{P} = \chi(E)\mathbf{E}, \tag{4-31}$$

where the scalar quantity $\chi(E)$ is called the *electric susceptibility* of the material. A great many materials are electrically isotropic; this category includes fluids, polycrystalline and amorphous solids, and some crystals. A treatment of the electrical properties of anisotropic materials is beyond the scope of this text.

Combining Eq. (4–31) with Eq. (4–27), we obtain an expression for $\mathbf{D}$ in isotropic media:

$$\mathbf{D} = \epsilon(E)\mathbf{E}, \tag{4-32}$$

$$\epsilon(E) = \epsilon_0 + \chi(E), \tag{4-33}$$

where $\epsilon(E)$ is the *permittivity* of the material. It is evident that ϵ, ϵ_0, and χ all have the same units.

Although we have been careful to write χ and ϵ in the form $\chi(E)$ and $\epsilon(E)$, nevertheless it is found experimentally that χ and ϵ are frequently independent of the electric field, except perhaps for very intense fields. In other words, χ and ϵ are constants characteristic of the material. Materials of this type will be called *linear dielectrics*, and they obey the relations

$$\mathbf{P} = \chi\mathbf{E}, \tag{4-31a}$$

$$\mathbf{D} = \epsilon\mathbf{E}. \tag{4-32a}$$

The electrical behavior of a material is now completely specified by either the permittivity ϵ or the susceptibility χ. It is more convenient, however, to work with a dimensionless quantity K defined by

$$\epsilon = K\epsilon_0. \tag{4-34}$$

K is called the dielectric coefficient, or simply the *dielectric constant*. From Eq. (4–33) it is evident that

$$K = \frac{\epsilon}{\epsilon_0} = 1 + \frac{\chi}{\epsilon_0}. \tag{4–35}$$

The dielectric constants for some commonly encountered materials are given in Table 4–1. Except for a few examples in which the polarization **P** of the material is specified, the problems in this book deal with linear dielectrics.

If the electric field in a dielectric is made very intense, it will begin to pull electrons completely out of the molecules, and the material will become conducting. The maximum electric field which a dielectric can withstand without breakdown is called its *dielectric strength*. The dielectric strengths E_{max}, of a few substances are also given in Table 4–1.

Table 4–1 Properties of Dielectric Materials* (Dielectric constant K and dielectric strength E_{max})

Material	K	E_{max}, V/m
Aluminium oxide	4.5	6×10^6
Glass†	5–10	9×10^6
Nylon	3.5	19×10^6
Polyethylene	2.3	18×10^6
Quartz (SiO_2)	4.3	
Sodium chloride	6.1	
Sulfur	4.0	
Wood†	2.5–8.0	
Alcohol, ethyl (0°C)	28.4	
Benzene (0°C)	2.3	
Water (distilled, 0°C)	87.8	
Water (distilled, 20°C)	80.1	
Air (1 atm)	1.00059	3×10^6
Air (100 atm)	1.0548	
CO_2 (1 atm)	1.000985	

* Data from the *Handbook of Chemistry and Physics*, 58th edition, CRC Press, Inc., Cleveland, Ohio.
† For materials such as glass and wood, the chemical composition varies; hence the range of dielectric constants. It is not to be inferred that the material is nonlinear.

4–6 POINT CHARGE IN A DIELECTRIC FLUID

One of the simplest problems involving a dielectric that we might consider is that of a point charge q in a homogeneous isotropic medium of infinite extent. The dielectric medium will be assumed to be linear and characterized by a dielectric constant K. Although this problem is quite simple, it will nevertheless prove instructive.

If the point charge q were situated in a vacuum, the electric field would be a pure radial field. But since **E**, **D**, and **P** are all parallel to one another at each point, the radial nature of the field is not changed by the presence of the medium. Furthermore, from the symmetry of the problem, **E**, **D**, and **P** can depend only on the distance from the point charge, not on any angular coordinate. Let us apply Gauss's law, Eq. (4–28), to a spherical surface of radius r which is located concentrically about q. For convenience, q will be located at the origin. Then

$$4\pi r^2 D = q$$

and

$$D = \frac{q}{4\pi r^2},$$

or

$$\mathbf{D} = \frac{q}{4\pi r^3}\,\mathbf{r}. \tag{4–36}$$

The electric field and polarization may now be evaluated quite easily:

$$\mathbf{E} = \frac{q}{4\pi K\epsilon_0 r^3}\,\mathbf{r}, \tag{4–37}$$

$$\mathbf{P} = \frac{(K-1)q}{4\pi K r^3}\,\mathbf{r}. \tag{4–38}$$

Thus the electric field is smaller by the factor K than would be the case if the medium were absent.

At this point, it will be instructive to look at the problem in more detail, and try to see why the dielectric has weakened the electric field. The electric field has its origin in all of the charge, polarization and external. The external charge is just the point charge q. The polarization charge, however, is made up from two contributions, a volume density $\rho_P = -\mathbf{\nabla} \cdot \mathbf{P}$, and a surface density $\sigma_P = \mathbf{P} \cdot \mathbf{n}$ on the surface of the dielectric in contact with the point charge. Using Eq. (4–38), we find that $\mathbf{\nabla} \cdot \mathbf{P}$ vanishes for $r \neq 0$, so there is no volume density of polarization charge in this case.

Our point charge q is a point in the macroscopic sense. Suppose it is large on a molecular scale, and we can assign to it a radius b which eventually will be made to approach zero. The total surface polarization charge is then given by

$$Q_P = \lim_{b \to 0} 4\pi b^2 (\mathbf{P} \cdot \mathbf{n})_{r=b} = -\frac{(K-1)q}{K}. \tag{4–39}$$

The total charge,

$$Q_P + q = \frac{1}{K}\,q, \tag{4–40}$$

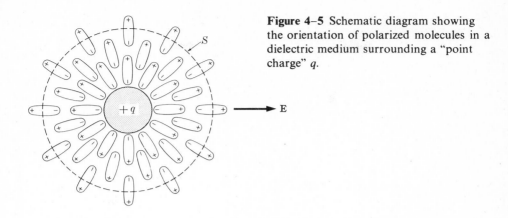

Figure 4–5 Schematic diagram showing the orientation of polarized molecules in a dielectric medium surrounding a "point charge" q.

appears as a point charge from the macroscopic point of view, and it is now clear why the electric field is a factor K smaller than it would be if the medium were absent. A schematic diagram of the *point charge q* in a dielectric medium is shown in Fig. 4–5.

4–7 BOUNDARY CONDITIONS ON THE FIELD VECTORS

Before we can solve more complicated problems, we must know how the field vectors **E** and **D** change in passing an interface between two media. The two media may be two dielectrics with different properties, or a dielectric and a conductor. Vacuum may be treated as a dielectric with permittivity ϵ_0.

Consider two media, 1 and 2, in contact as shown in Fig. 4–6. We shall assume that there is a surface density of external charge, σ, which may vary from point to point on the interface. Let us construct the small pillbox-shaped surface S that intersects the interface and encloses an area ΔS of the interface, the height of the

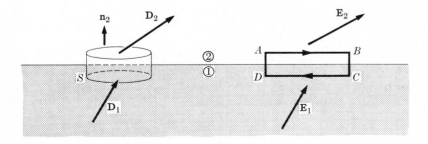

Figure 4–6 Boundary conditions on the field vectors at the interface between two media may be obtained by applying Gauss's law to S, and integrating $\mathbf{E} \cdot d\mathbf{l}$ around the path $ABCDA$.

pillbox being negligibly small in comparison with the diameter of the bases. The charge enclosed by S is

$$\sigma \, \Delta S + \tfrac{1}{2}(\rho_1 + \rho_2) \times \text{volume},$$

but the volume of the pillbox is negligibly small, so that the last term may be neglected. Applying Gauss's law to S, we find

$$\mathbf{D}_2 \cdot \mathbf{n}_2 \, \Delta S + \mathbf{D}_1 \cdot \mathbf{n}_1 \, \Delta S = \sigma \, \Delta S,$$

or

$$(\mathbf{D}_2 - \mathbf{D}_1) \cdot \mathbf{n}_2 = \sigma. \tag{4–41a}$$

Since $\mathbf{n}_2$ may serve as the normal to the interface,

$$D_{2n} - D_{1n} = \sigma. \tag{4–41b}$$

Thus the discontinuity in the normal component of $\mathbf{D}$ is given by the surface density of external charge on the interface. Or, to put it another way, if there is no charge on the interface between two media, the normal component of $\mathbf{D}$ is continuous.

Because the electrostatic field $\mathbf{E}$ may be obtained as minus the gradient of a potential, the line integral of $\mathbf{E} \cdot d\mathbf{l}$ around any closed path vanishes. Let us apply this result to the rectangular path $ABCD$ of Fig. 4–6. On this path, the lengths AB and CD will be taken equal to Δl and the segments AD and BC will be assumed to be negligibly small. Therefore

$$\mathbf{E}_2 \cdot \Delta \mathbf{l} + \mathbf{E}_1 \cdot (-\Delta \mathbf{l}) = 0,$$

or

$$(\mathbf{E}_2 - \mathbf{E}_1) \cdot \Delta \mathbf{l} = 0. \tag{4–42a}$$

Hence the desired result:

$$E_{2t} = E_{1t}, \tag{4–42b}$$

that is, the tangential component of the electric field is continuous across an interface.

The above results have been obtained for two arbitrary dielectric media, but it is interesting to see what the equations would predict if one of the media were a conductor. Since there is no molecular restoring force on the free charges of a conductor, it would seem that we would have $\chi = \infty$ for a conductor in Eq. (4–31), and $\epsilon = \infty$ according to Eq. (4–33). If medium 1 is taken as the conductor, then we conclude that $\mathbf{E}_1 = 0$ whatever the (finite) values of $\mathbf{P}_1$ and $\mathbf{D}_1$, as we have already inferred by different reasoning in Chapter 2. Since $\mathbf{E}_1$ vanishes, Eq. (4–42b) becomes

$$E_{2t} = 0. \tag{4–43}$$

The displacement $\mathbf{D}_1$ is not determined by these considerations, however; if for our present purposes we arbitrarily take it to be zero, Eq. (4–41b) becomes

$$D_{2n} = \sigma, \qquad (4\text{–}44)$$

where σ then represents the *total* surface charge on the conductor but does not include the polarization surface charge on the dielectric. An alternative approach is to solve the dielectric problem and then let K_1 go to infinity. (See Problems 4–12 and 4–14.) Then σ represents polarization charge of the conductor (plus any external charge on the conductor). Physically the result is the same. Note that the field in the dielectric is always perpendicular to the surface of the conductor according to Eq. (4–43).

It is evident on purely physical grounds that the potential φ must be continuous across an interface. This follows because the difference in potential, $\Delta\varphi$, between two closely spaced points is $-\mathbf{E} \cdot \Delta\mathbf{l}$, where $\Delta\mathbf{l}$ is the separation of the two points, and from what has been said above there is no reason to expect $\mathbf{E}$ to become infinite at an interface. Actually, the continuity of the potential is a boundary condition, but not independent of those already derived. It is equivalent in most cases to Eq. (4–42b).

From the discussion above and in preceding sections, it may be inferred that the electric displacement $\mathbf{D}$ is closely related to external charge. We should now like to prove an important property of $\mathbf{D}$, namely, that the flux of $\mathbf{D}$ is continuous in regions containing no external charge. To do so, we again resort to Gauss's law. Let us focus our attention on a region of space and construct *lines of displacement*, which are imaginary lines drawn in such a way that the direction of a line at any point is the direction of $\mathbf{D}$ at that point. Next we imagine a tube of displacement, a volume bounded on the sides by lines of $\mathbf{D}$ but not cut by them (see Fig. 4–7). The tube is terminated at its ends by the surfaces S_1 and S_2. Applying Gauss's law, we obtain,

$$\int_{S_2} \mathbf{D} \cdot \mathbf{n}\, da - \int_{S_1} \mathbf{D} \cdot \mathbf{n}'\, da = Q. \qquad (4\text{–}45)$$

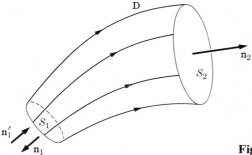

Figure 4–7 A tube of displacement flux.

If there is no external charge in the region, then $Q = 0$, and the same amount of flux enters the tube through S_1 as leaves through S_2. When external charge is present, it determines the discontinuity in displacement flux; thus lines of displacement terminate on external charges. The lines of force, on the other hand, terminate on either external or polarization charges.

4–8 BOUNDARY-VALUE PROBLEMS INVOLVING DIELECTRICS

The fundamental equation that has been developed in this chapter is

$$\mathbf{V} \cdot \mathbf{D} = \rho, \tag{4–46}$$

where ρ is the external charge density. If the dielectrics with which we are concerned are linear, isotropic, and homogeneous, then $\mathbf{D} = \epsilon\mathbf{E}$, where ϵ is a constant characteristic of the material, and we may write

$$\mathbf{V} \cdot \mathbf{E} = \frac{1}{\epsilon} \rho. \tag{4–47}$$

But the electrostatic field $\mathbf{E}$ is derivable from a scalar potential φ, i.e.,

$$\mathbf{E} = -\mathbf{V}\varphi;$$

so that

$$\nabla^2 \varphi = -\frac{1}{\epsilon} \rho. \tag{4–48}$$

Thus the potential in the dielectric satisfies Poisson's equation; the only difference between Eq. (4–48) and the corresponding equation for the potential in vacuum is that ϵ replaces ϵ_0 (and ρ means the external instead of the total charge density).

In most cases of interest the dielectric contains no charge distributed throughout its volume, that is, $\rho = 0$ inside the dielectric material. The charge exists on the surfaces of conductors or is concentrated in the form of point charges, which may, to be sure, be embedded in the dielectric. In these circumstances, the potential satisfies Laplace's equation throughout the body of the dielectric:

$$\nabla^2 \varphi = 0. \tag{4–49}$$

In some problems there may be a surface density of charge, σ, on the surface of a dielectric body or on the interface between two dielectric materials, but this does not alter the situation, and Eq. (4–49) still applies so long as $\rho = 0$.

An electrostatic problem involving linear, isotropic, and homogeneous dielectrics reduces, therefore, to finding solutions of Laplace's equation in each medium, and joining the solutions in the various media by means of the boundary conditions of the preceding section. There are many problems that may be solved by this method; one example will be discussed here and additional examples will be found in the problems at the end of the chapter.

4–9 DIELECTRIC SPHERE IN A UNIFORM ELECTRIC FIELD

We should like to determine how the lines of force are modified when a dielectric sphere of radius a is placed in a region of space containing an *initially* uniform electric field, E_0. Let us assume the dielectric to be linear, isotropic, and homogeneous, and to be characterized by the dielectric constant K. Furthermore, it bears no charge. The origin of our coordinate system may be taken at the center of the sphere, and the direction of E_0 as the polar direction (z-direction); the potential may then be expressed as a sum of zonal harmonics. Just as in Section 3–5, all boundary conditions can be satisfied by means of the two lowest-order harmonics, and we write

$$\varphi_1(r, \theta) = A_1 r \cos \theta + C_1 r^{-2} \cos \theta \qquad (4\text{--}50)$$

for the vacuum region (1) outside the sphere, and

$$\varphi_2(r, \theta) = A_2 r \cos \theta + C_2 r^{-2} \cos \theta \qquad (4\text{--}51)$$

for the dielectric region (2). The constants A_1, A_2, C_1, and C_2 are unknown and must be determined from the boundary conditions. The harmonic r^{-1} is not required, since its presence implies a net charge on the sphere. A constant term may be added to Eqs. (4–50) and (4–51), but since it turns out that the same constant is required in both equations, we may, without loss of generality, take it to be zero.

At distances far from the sphere, the electric field will retain its uniform character, and $\varphi_1 \rightarrow -E_0 r \cos \theta$. Hence $A_1 = -E_0$. Furthermore, unless $C_2 = 0$, the potential and associated electric field would become infinite at the center of the sphere, and this would imply the existence of a point dipole at the center, i.e., a dipole whose moment is not proportional to ΔV. Certainly, this is not the case; as has already been discussed in Section 4–3, the potential and macroscopic field do not become infinite in a dielectric devoid of point charges. Hence $C_2 = 0$, and the remaining constants, A_2 and C_1, may be obtained from the boundary conditions of Section 4–7.

Continuity of the potential across the interface between the dielectric and vacuum requires that $\varphi_1 = \varphi_2$ at $r = a$, or

$$-E_0 a + C_1 a^{-2} = A_2 a. \qquad (4\text{--}52)$$

Since the normal component of D at the interface is $D_r = -\epsilon(\partial\varphi/\partial r)$, the continuity of D_r (there is no charge on the surface of the dielectric) requires that $D_{1r} = D_{2r}$ at $r = a$, or

$$E_0 + 2C_1 a^{-3} = -KA_2. \qquad (4\text{--}53)$$

Continuity of E_t at $r = a$ is equivalent to Eq. (4–52).

Combining Eqs. (4–52) and (4–53), we obtain

$$A_2 = -\frac{3E_0}{K+2} \qquad (4\text{--}54)$$

and

$$C_1 = \frac{(K-1)a^3 E_0}{K+2}. \tag{4-55}$$

Thus the problem has been solved. The potential is given by Eqs. (4–50) or (4–51), and the constants A_1, C_1, A_2, and C_2 are all known. The components of **E** and **D** may be obtained at any point (r, θ, ϕ) by differentiation. It is evident from Eq. (4–54) and because $C_2 = 0$ that the electric field inside the sphere has the direction of $\mathbf{E}_0$ and is given by

$$\mathbf{E}_2 = \frac{3}{K+2}\,\mathbf{E}_0. \tag{4-56}$$

The lines of displacement and the lines of force are shown in Fig. 4–8.

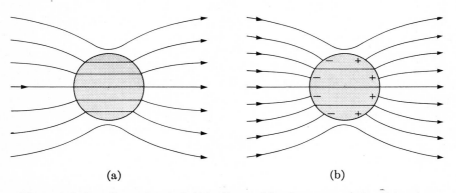

(a)   (b)

Figure 4–8 A uniform electric field is distorted by the presence of a dielectric sphere: (a) lines of electric displacement, (b) lines of the electric field.

*4–10 FORCE ON A POINT CHARGE EMBEDDED IN A DIELECTRIC

We are now in a position to determine the force on a small, spherical, charged conductor embedded in a linear, isotropic dielectric. In the limit in which the conductor is negligibly small from the macroscopic viewpoint, this calculation gives the force on a point charge.

The electric field and surface charge density at a representative point of the conductor surface will be obtained by the boundary-value procedure of the preceding section, and the force **F** may then be obtained from the integral over the surface:

$$\mathbf{F} = \oint_S \mathbf{E}'\sigma\,da. \tag{4-57}$$

Here $\mathbf{E}'$ stands for the electric field at the surface element da minus that part of the field produced by the element itself. In other words,

$$\mathbf{E}' = \mathbf{E} - \mathbf{E}_s, \tag{4-58}$$

where $\mathbf{E}_s$ is the electric field produced by the surface element of charge, $\sigma\, da$. It is important that $\mathbf{E}_s$ not be included in the field $\mathbf{E}'$, because the quantity $\mathbf{E}_s \sigma\, da$ represents the interaction of the charge element $\sigma\, da$ with its *own* field; this self-interaction clearly produces no net force on the element, but gives rise to a surface stress

$$\mathscr{F}_s = \sigma \mathbf{E}_s, \tag{4-59}$$

which is due to the mutual repulsion of the electrons (or of the excess positive ions) in the surface layer. This stress is balanced by strong cohesive forces in the material of which the element is composed. It should be pointed out that when calculating forces on charged objects in Chapters 2 and 3, we implicitly subtracted the self field $\mathbf{E}_s$; thus, when calculating the force on a point charge, the field produced *by* the point charge was not included. A further discussion of the forces on charged objects will be taken up in Section 6–8.

It may appear that the self field of the charged surface element $\sigma\, da$ is negligible because the element is of infinitesimal size. This, however, is not the case. The element is small from the macroscopic point of view, to be sure, but one never quite goes to the limit. At a point directly on its surface, the element appears to be an infinite plane, i.e., the element subtends an angle of 2π; hence,

$$\mathbf{E}_s = \frac{\sigma}{2\epsilon}\, \mathbf{n}, \tag{4-60}$$

where $\mathbf{n}$ is a normal to the element and ϵ is the permittivity of the dielectric in contact with it. Thus the stress $\mathscr{F}_s$ is proportional to σ^2, and is always a tension regardless of the sign of σ.

It is our purpose here to calculate the force on a *conductor*. Using the boundary conditions of Section 4–7, the total electric field at the conductor is given by

$$\mathbf{E} = \frac{\sigma}{\epsilon}\, \mathbf{n}. \tag{4-61}$$

Combining Eqs. (4–58), (4–60), and (4–61), we obtain

$$\mathbf{E}' = \tfrac{1}{2}\mathbf{E},$$

and the force on the conductor becomes

$$\mathbf{F} = \tfrac{1}{2} \oint_S \mathbf{E}\sigma\, da. \tag{4-57a}$$

Let us now fix our attention on a small spherical conductor embedded in a dielectric of infinite extent. The total charge on the conductor is Q; its radius is a.

Since we shall eventually go to the limit in which a becomes very small, and since variations in the electric field (if they exist) are on a macroscopic scale, it is sufficient to consider the case in which the electric field is initially uniform in the neighborhood of the conductor. Let us denote this uniform field by the symbol $\mathbf{E}_0$. The picture is similar to that of the boundary-value problem we solved in Section 3–5, except that here the conducting sphere is embedded (or immersed) in a dielectric of permittivity ϵ, and in addition bears a net charge Q.

By analogy with Section 3–5 we easily determine:

the potential,

$$\varphi(r, \theta) = \varphi_0 - E_0 r \cos \theta + \frac{E_0 a^3}{r^2} \cos \theta + \frac{Q}{4\pi\epsilon r} ; \qquad (4\text{–}62)$$

the electric field,

$$E_r = E_0(1 + 2a^3/r^3) \cos \theta + Q/4\pi\epsilon r^2, \qquad (4\text{–}63)$$
$$E_\theta = -E_0(1 - a^3/r^3) \sin \theta;$$

and the surface charge density on the surface of the sphere,

$$\sigma(\theta) = \epsilon E_r|_{r=a} = 3\epsilon E_0 \cos \theta + Q/4\pi a^2. \qquad (4\text{–}64)$$

The force may now be determined from Eq. (4–57a). By symmetry, the only nonzero component of force is that in the direction $\theta = 0$, i.e., in the z-direction:

$$F_z = \tfrac{1}{2} \int_0^\pi (E_r)_{r=a} \cos \theta \sigma(\theta) 2\pi a^2 \sin \theta \, d\theta$$
$$= E_0 Q, \qquad (4\text{–}65a)$$

or

$$\mathbf{F} = Q\mathbf{E}_0. \qquad (4\text{–}65b)$$

This result is unchanged as we go to the limit of small a. Thus the electric field in the dielectric, $\mathbf{E}_0$, is in agreement with the fundamental definition, namely, the force on a small test charge Q divided by the magnitude of Q.

4–11 SUMMARY

The electrostatic behavior of a dielectric medium is completely characterized by its dipole moment per unit volume or polarization,

$$\mathbf{P} = \frac{d\mathbf{p}}{dv}.$$

This produces the polarization charge density

$$\rho_P = -\nabla \cdot \mathbf{P}, \qquad (\sigma_P = \mathbf{n} \cdot \mathbf{P}),$$

giving rise to the potential

$$\varphi(\mathbf{r}) = \frac{1}{4\pi\epsilon_0} \left[\int \frac{\rho_P(\mathbf{r}') \, dv'}{|\mathbf{r} - \mathbf{r}'|} + \oint \frac{\sigma_P(\mathbf{r}') \, da'}{|\mathbf{r} - \mathbf{r}'|} \right].$$

The total E-field due to external charges plus the polarization charge satisfies

$$\nabla \cdot \mathbf{E} = \frac{1}{\epsilon_0} (\rho + \rho_P).$$

It is convenient to define the vector field

$$\mathbf{D} = \epsilon_0 \mathbf{E} + \mathbf{P}$$

such that

$$\nabla \cdot \mathbf{D} = \rho$$

with only the external charges as sources. The curl equation

$$\nabla \times \mathbf{E} = 0$$

is unchanged because it does not contain the charge density. In order to solve the equations for the fields, the constitutive equation

$$\mathbf{P} = \mathbf{P}(\mathbf{E})$$

must also be known for the particular material. Then the last four equations, subject to the boundary conditions

$$D_{2n} - D_{1n} = \sigma,$$

$$E_{2t} - E_{1t} = 0,$$

are sufficient to determine $\mathbf{E}$ and $\mathbf{D}$ inside and outside dielectrics.

1. The integral form of Gauss's law becomes

$$\oint_S \mathbf{D} \cdot \mathbf{n} \, da = Q,$$

where Q includes only the *external* charge located inside the surface S. The curl equation still allows the definition of the potential by

$$\mathbf{E} = -\nabla\varphi.$$

2. Most dielectric materials are linear, with constant susceptibility,

$$\mathbf{P} = \chi\mathbf{E}.$$

This constitutive equation combined with the definition of $\mathbf{D}$ gives

$$\mathbf{D} = \epsilon\mathbf{E},$$

where

$$\epsilon = \epsilon_0 + \chi.$$

The dielectric constant

$$K = \frac{\epsilon}{\epsilon_0}$$

is between 1 and 100 for most common dielectrics; for all dielectrics $K \geq 1$ ($\chi \geq 0$). For vacuum, $K = 1$ ($\chi = 0$). The electrostatic behavior of a conductor can be obtained by letting K be infinite.

   3. In a linear medium

$$\mathbf{V} \cdot \mathbf{E} = \frac{1}{\epsilon} \rho$$

and

$$\nabla^2 \varphi = -\frac{\rho}{\epsilon}.$$

The mathematical techniques for solving Poisson's and Laplace's equations are similar to those of Chapter 3, with the appropriate boundary conditions at a dielectric interface,

$$K_2 \frac{\partial \varphi_2}{\partial n} = K_1 \frac{\partial \varphi_1}{\partial n}$$

and

$$\varphi_2 = \varphi_1$$

(equivalent to $E_{2t} = E_{1t}$).

PROBLEMS

4–1 A thin dielectric rod of cross section A extends along the x-axis from $x = 0$ to $x = L$. The polarization of the rod is along its length, and is given by $P_x = ax^2 + b$. Find the volume density of polarization charge and the surface polarization charge on each end. Show *explicitly* that the total polarization charge vanishes in this case.

4–2 A dielectric cube of side L has a radial polarization given by $\mathbf{P} = A\mathbf{r}$, where A is a constant, and $\mathbf{r} = \mathbf{i}x + \mathbf{j}y + \mathbf{k}z$. The origin of coordinates is at the center of the cube. Find all polarization charge densities, and show *explicitly* that the total polarization charge vanishes.

4–3 A dielectric rod in the shape of a right circular cylinder of length L and radius R is polarized in the direction of its length. If the polarization is uniform and of magnitude P, calculate the electric field resulting from this polarization at a point on the axis of the rod.

4–4 Prove the following relationship between the polarization, $\mathbf{P}$, and the polarization charge densities ρ_P and σ_P, for a dielectric specimen of volume V and surface S.

$$\int_V \mathbf{P} \, dv = \int_V \rho_P \mathbf{r} \, dv + \int_S \sigma_P \mathbf{r} \, da.$$

Here, $\mathbf{r} = \mathbf{i}x + \mathbf{j}y + \mathbf{k}z$ is the position vector from any fixed origin. [*Hint:* Expand $\mathbf{V} \cdot (x\mathbf{P})$ according to Eq. (4–10).]

4–5 Two semi-infinite blocks of dielectric are placed almost in contact so that there exists a narrow gap of constant separation between them. The polarization $\mathbf{P}$ is constant throughout all of the dielectric material, and it makes the angle γ with the normal to the planes bounding the gap. Determine the electric field in the gap.

4–6 A long cylindrical conductor of radius a, bearing the charge λ per unit length, is immersed in a dielectric medium of constant permittivity ϵ. Find the electric field at distance $r > a$ from the axis of the cylinder.

4–7 Two dielectric media with dielectric constants K_1 and K_2 are separated by a plane interface. There is no external charge on the interface. Find a relationship between the angles θ_1 and θ_2, where these are the angles which an arbitrary line of displacement makes with the normal to the interface: θ_1 in medium 1, θ_2 in medium 2.

4–8 A coaxial cable of circular cross section has a compound dielectric. The inner conductor has an outside radius a; this is surrounded by a dielectric sheath of dielectric constant K_1 and of outer radius b. Next comes another dielectric sheath of dielectric constant K_2 and outer radius c. If a potential difference φ_0 is imposed between the conductors, calculate the polarization at each point in the two dielectric media.

***4–9** Two dielectric media with constant permittivities ϵ_1 and ϵ_2 are separated by a plane interface. There is no external charge on the interface. A point charge q is embedded in the medium characterized by ϵ_1, at a distance d from the interface. For convenience, we may take the yz-plane through the origin to be the interface, and we locate q on the x-axis at $x = -d$. If

$$r = \sqrt{(x + d)^2 + y^2 + z^2}, \quad \text{and} \quad r' = \sqrt{(x - d)^2 + y^2 + z^2},$$

then it is easily demonstrated that $(1/4\pi\epsilon_1)[(q/r) + (q'/r')]$ satisfies Laplace's equation at all points in medium 1 except at the position of q. Furthermore, $q''/4\pi\epsilon_2 r$ satisfies Laplace's equation in medium 2. Show that all boundary conditions can be satisfied by these potentials, and in so doing determine q' and q''. (Refer to Fig. 4–9.)

Figure 4–9

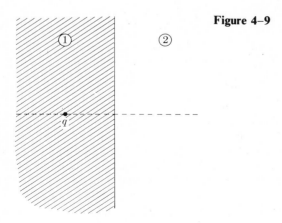

4–10 A long dielectric cylinder of radius a and dielectric constant K is placed in a uniform electric field $\mathbf{E}_0$. The axis of the cylinder is oriented at right angles to the direction of $\mathbf{E}_0$. The cylinder contains no external charge. Determine the electric field at points inside and outside the cylinder.

4-11 A point dipole **p** is placed at the center of a solid dielectric sphere of radius a and dielectric constant K. Find the electric field at points inside and outside the sphere. (*Hint:* The field outside is a dipole field; inside another term is needed in addition to a dipole field.)

4-12 Show that if K goes to infinity in the solution for a dielectric sphere in an initially uniform field, derived in Section 4-9, the result is the same as that for a conducting sphere derived in Section 3-5.

4-13 A plane slab of material with dielectric constant K_1 is bounded on both sides by material of dielectric constant K_2. The electric field in medium 2, $\mathbf{E}_2$, is given to be uniform and perpendicular to the boundaries. Find the field $\mathbf{E}_1$, the polarization $\mathbf{P}_1$, and the polarization charge in medium 1.

4-14 Show that if K_1 goes to infinity in Problem 4-13 the result agrees physically with Eq. (4-44).

4-15 Two parallel conducting plates are separated by the distance d and maintained at the potential difference $\Delta\varphi$. A dielectric slab, of dielectric constant K and of uniform thickness $t < d$, is inserted between the plates. Determine the field vectors **E** and **D** in the dielectric and also in the vacuum between dielectric and one plate. Neglect edge effects due to the finite size of the plates.

4-16 Two parallel conducting plates are separated by the distance d and maintained at the potential difference $\Delta\varphi$. A dielectric slab, of dielectric constant K and of uniform thickness d, is inserted snugly between the plates; however, the slab does not completely fill the volume between the plates. Find the electric field (a) in the dielectric, and (b) in the vacuum region between the plates. Find the charge density σ on that part of the plate (c) in contact with the dielectric, and (d) in contact with vacuum. (e) Find σ_P on the surface of the dielectric slab.

4-17 A conducting sphere of radius R floats half submerged in a liquid dielectric medium of permittivity ϵ_1. The region above the liquid is a gas of permittivity ϵ_2. The total charge on the sphere is Q. Find a radial inverse-square electric field satisfying all boundary conditions, and determine the free, bound, and total charge densities at all points on the surface of the sphere. Formulate an argument to show that this electric field is the actual one.

4-18 A uniform electric field $\mathbf{E}_0$ is set up in a medium of dielectric constant K. Prove that the field inside a spherical cavity in the medium is

$$\mathbf{E} = \frac{3K\mathbf{E}_0}{2K + 1}.$$

*4-19 A dielectric sphere of radius R has a permanent polarization **P**, which is uniform in direction and magnitude. The polarized sphere gives rise to an electric field. Determine this field both inside and outside the sphere. Inside the sphere the electric field, which is in the opposite direction to the polarization, is called a depolarizing field. (*Hint:* Since $\nabla \cdot \mathbf{P}$ vanishes at all points, the electrostatic potential satisfies Laplace's equation both inside and outside the sphere. Do *not* assume that the dielectric is characterized by a dielectric constant.)

4-20 In the text, it was shown that the polarization $\mathbf{P} = \rho_0^+(\delta^+ - \delta^-)$. Use this relation for the uniformly polarized sphere of Problem 4-19 to determine the external dipole field directly.

CHAPTER 5 Microscopic Theory of Dielectrics

In the preceding chapter we were concerned with the macroscopic aspects of dielectric polarization, and it was shown how in many cases the polarization could be taken into account through the introduction of a dielectric constant. In this way the electric field could be computed directly from a consideration of the external charge distribution. Although reference was made to the molecules of the dielectric several times in Chapter 4, a microscopic treatment of the material was not carried through in detail, and the over-all picture that was presented was certainly from the macroscopic point of view. We should now like to examine the molecular nature of the dielectric, and see how the electric field responsible for polarizing the molecule is related to the macroscopic electric field. Furthermore, on the basis of a simple molecular model it is possible to understand the linear behavior that is characteristic of a large class of dielectric materials.

5–1 MOLECULAR FIELD IN A DIELECTRIC

The electric field that is responsible for polarizing a molecule of the dielectric is called the molecular field, $\mathbf{E}_m$. This is the electric field at a molecular position in the dielectric; it is produced by all external sources and by all polarized molecules in the dielectric *with the exception* of the one molecule at the point under consideration. It is evident that $\mathbf{E}_m$ need not be the same as the macroscopic electric field because, as was discussed in Section 4–3, the latter quantity is related to the force on a test charge that is large in comparison with molecular dimensions.

The molecular field may be calculated in the following way. Let us cut out a small piece of the dielectric, leaving a spherical cavity surrounding the point at which the molecular field is to be computed. The dielectric that is left will be treated as a continuum, i.e., from the macroscopic point of view. Now we put the dielectric back into the cavity, molecule by molecule, except for the molecule at the center of the cavity where we wish to compute the molecular field. The molecules that have just been replaced are to be treated, not as a continuum, but as individual dipoles. The procedure just outlined can be justified only if the result of the calculation is independent of the size of the cavity; we shall see that under certain conditions this is indeed the case.

Let us suppose that the thin dielectric sample has been polarized by placing it in the uniform electric field between two parallel plates which are oppositely charged, as shown in Fig. 5–1(a). It will be assumed that the polarization is uniform on a macroscopic scale (i.e., $\mathbf{V} \cdot \mathbf{P} = 0$), and that $\mathbf{P}$ is parallel to the field producing it. The part of the dielectric outside the cavity may be replaced by a system of polarization charges as shown in Fig. 5–1(b), whence the electric field at the center of the cavity may be written as

$$\mathbf{E}_m = \mathbf{E}_x + \mathbf{E}_d + \mathbf{E}_s + \mathbf{E}'. \tag{5–1}$$

Here $\mathbf{E}_x$ is the primary electric field due to the charged parallel plates, $\mathbf{E}_d$ is the depolarizing field due to polarization charge on the outside surfaces of the dielectric, $\mathbf{E}_s$ is due to polarization charge on the cavity surface S, and $\mathbf{E}'$ is due to all of the dipoles inside of S. Although we are not concerned with the explicit form of $\mathbf{E}_x$, it is evident that if the dimensions of the plates are large compared with their separation, $E_x = (1/\epsilon_0)\sigma$, where σ is the surface charge density. The depolarizing field is also produced by two parallel planes of charge, this time with the density σ_P. Since $\sigma_P = P_n = \pm P$,

$$\mathbf{E}_d = -\frac{1}{\epsilon_0} \mathbf{P}. \tag{5–2}$$

Let us write the macroscopic electric field *in the dielectric* without a subscript, that is, $\mathbf{E}$. Since the normal component of the electric displacement $\mathbf{D}$ is continuous across the vacuum-dielectric interface, and since $\mathbf{D} = \epsilon_0\mathbf{E}_x$ in the vacuum just outside the dielectric slab,

$$\epsilon_0\mathbf{E}_x = \epsilon_0\mathbf{E} + \mathbf{P}. \tag{5–3}$$

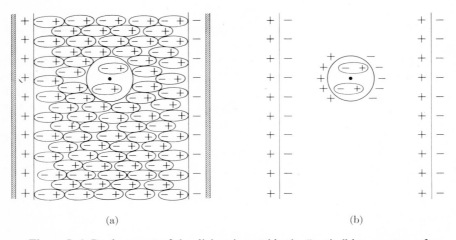

(a)   (b)

Figure 5–1 Replacement of the dielectric outside the "cavity" by a system of polarization charges.

Combining Eqs. (5–1), (5–2), and (5–3) yields

$$\mathbf{E}_m = \mathbf{E} + \mathbf{E}_s + \mathbf{E}', \tag{5-4}$$

which is an equation relating the molecular field to the macroscopic electric field in the dielectric material. This result is quite general, and not restricted to the geometry of Fig. 5–1; nevertheless, the above derivation is instructive and will be useful to the subject discussed in Section 5–4.

The field $\mathbf{E}_s$ arises from polarization charge density, $\sigma_P = P_n$, on the spherical surface S. Using spherical coordinates, and taking the polar direction along the direction of $\mathbf{P}$, as in Fig. 5–2, we obtain

$$d\mathbf{E}_s = \frac{(-P \cos \theta)}{4\pi\epsilon_0 r^3} \mathbf{r} \, da, \tag{5-5}$$

where $\mathbf{r}$ is the vector from the surface to the center of the sphere. From symmetry, it is evident that only the component of $d\mathbf{E}_s$ along the direction of $\mathbf{P}$ will contribute to the integral of Eq. (5–5) over the complete surface. Since $da = r^2 \sin \theta \, d\theta \, d\phi$,

$$\mathbf{E}_s = \frac{1}{4\pi\epsilon_0} \mathbf{P} \int_0^{2\pi} d\phi \int_0^{\pi} \cos^2 \theta \sin \theta \, d\theta$$

$$= \frac{1}{3\epsilon_0} \mathbf{P}. \tag{5-6}$$

Finally, we come to the last term in Eq. (5–4), that due to the electric dipoles inside S. There is a number of important cases for which this term vanishes. If there are a great many dipoles in the cavity, if they are oriented parallel but randomly distributed in position, and if there are no correlations between the positions of the dipoles, then $\mathbf{E}' = 0$. This is the situation that might prevail in a gas or a liquid. Similarly, if the dipoles in the cavity are located at the regular atomic positions of a cubic crystal,* then again $\mathbf{E}' = 0$. In this connection, the reader is referred to Problem 5–2.

In the general case, $\mathbf{E}'$ is not zero, and if the material contains several species of molecule, $\mathbf{E}'$ may differ at the various molecular positions. It is this term that

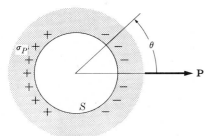

Figure 5–2 Calculation of the "cavity" surface contribution to $\mathbf{E}_m$.

---

* Crystals with the highest symmetry belong to the cubic system.

gives rise to the anisotropic electrical behavior of calcite, for example. It is not our purpose, however, to develop a theory of anisotropic materials; hence we restrict further discussion to the rather large class of materials in which $\mathbf{E}' = 0$. Thus, Eq. (5-4) reduces to

$$\mathbf{E}_m = \mathbf{E} + \frac{1}{3\epsilon_0}\,\mathbf{P}. \qquad (5-7)$$

It is interesting to note that this result would be obtained directly by the above method if the spherical cavity were created by removing just one molecule. But under these conditions the cavity would be so small that the replacement of the rest of the dielectric by a system of polarization charges could not be justified.

The dipole moment of a molecule per unit polarizing field is called its *polarizability*, α. In other words,

$$\mathbf{p}_m = \alpha \mathbf{E}_m. \qquad (5-8)$$

If there are N molecules per unit volume, then the polarization $\mathbf{P} = N\mathbf{p}_m$, and combining this result with Eqs. (5-7) and (5-8), we obtain

$$\mathbf{P} = N\alpha\left(\mathbf{E} + \frac{1}{3\epsilon_0}\,\mathbf{P}\right). \qquad (5-9)$$

This equation may be rewritten in terms of the dielectric constant, K, since $\mathbf{P} = (K-1)\epsilon_0\mathbf{E}$. In this way, Eq. (5-9) becomes

$$\alpha = \frac{3\epsilon_0}{N}\,\frac{(K-1)}{(K+2)}, \qquad (5-10)$$

which is known as the Clausius-Mossotti equation. It is evident that Eq. (5-10) defines a molecular property, namely, the molecular polarizability, in terms of quantities that can be determined on a macroscopic basis.

5-2 INDUCED DIPOLES. A SIMPLE MODEL

The molecules of a dielectric may be classified as *polar* or *nonpolar*. A polar molecule is one that has a permanent dipole moment, even in the absence of a polarizing field $\mathbf{E}_m$. In the next section the response of a polar dielectric to an external electric field will be studied, but here we deal with the somewhat simpler problem involving nonpolar molecules, in which the "centers of gravity" of the positive and negative charge distributions normally coincide. Symmetrical molecules such as H_2, N_2, and O_2, or monatomic molecules such as He, Ne, and Ar, fall into this category.

The application of an electric field causes a relative displacement of the positive and negative charges in nonpolar molecules, and the molecular dipoles so created are called *induced dipoles*. The simplest type of molecule that can be envisaged is that composed of a single neutral atom. It is possible to construct a simple classical model for the atom and from this model derive an expression for the induced dipole moment, and hence for its polarizability. Although specifically designed to treat monatomic molecules, the model may be used for symmetrical diatomic molecules by applying it separately to each of the atoms in the molecule to obtain the atomic polarizabilities. The molecular polarizability is then the sum of these, or twice the atomic polarizability.

An atom consists of an extremely small positively charged nucleus surrounded by orbital electrons that are in a state of continual motion. Since the electrons traverse their orbits in an exceedingly short time, of the order of 10^{-15} second, it is evident that in the equivalent "static" atom each electronic charge is smeared over its orbit. Quantum mechanics tells us that although this picture is essentially correct, it is somewhat naive; the electrons are not really localized on orbits, but have a finite probability of being situated in any part of the atom. Thus the response of the atom to an electrostatic field or to slowly varying electric fields may be treated by considering the electron to be distributed over its orbit in the atom, and each orbit to be smeared over a substantial part of the atomic volume. In short, a simple classical model of the atom consistent with this picture is a point positive charge (the nucleus) surrounded by a spherically symmetric cloud of negative charge in which the density is essentially uniform out to the atomic radius R_0, and zero at larger radii.

We are now in a position to compute the polarizability of this "atom." The nucleus will be assigned the charge Ze, where e is the absolute value of the electronic charge and Z is the atomic number. Since the atom is electrically neutral, the total charge in the electron cloud is $-Ze$. If the atom is placed in a polarizing field $\mathbf{E}_m$, the nucleus will be displaced *relative* to the center of the charge cloud by a distance that we shall call x. This displacement will be in the direction of $\mathbf{E}_m$. We shall assume that the charge cloud moves rigidly during this displacement, i.e., there is no distortion of the cloud by the polarizing field. The displacement x may be determined from the equilibrium of forces on the nucleus; the force ZeE_m acts in the direction of the field, whereas an electrostatic force between the nucleus and charge cloud tends to restore the initial configuration. By Gauss's law, the negative charge attracting the nucleus is that part of the cloud within the sphere of radius x, and if the electronic density in the cloud is uniform, then this charge is Zex^3/R_0^3. Hence

$$\frac{(Ze)(Zex^3/R_0^3)}{4\pi\epsilon_0 x^2} = ZeE_m, \tag{5–11}$$

or

$$Zex = 4\pi\epsilon_0 R_0^3 E_m. \tag{5–12}$$

Since the atomic dipole created in this process is $\mathbf{p}_m = Ze\mathbf{x}$, the last equation may be compared with Eq. (5–8), whence

$$\alpha = 4\pi\epsilon_0 R_0^3. \tag{5–13}$$

The atomic model just described may be tested by comparing results obtained from it with results derived from other sources. For example, Eq. (5–13) may be combined with the Clausius-Mossotti equation (5–10) to eliminate α; the resulting equation predicts the atomic radius R_0 in terms of experimentally determined quantities. R_0 obtained in this way agrees reasonably well with results from other experiments in those cases for which the model is particularly suited; R_0 is of the order of magnitude of 1 angstrom unit, i.e., 10^{-10} m. (See Problem 5–1.)

The polarizability derived in Eq. (5–13) is a constant, independent of the polarizing field. Hence Eq. (5–13) leads to a constant value of K, and the dielectric so described is linear.

5–3 POLAR MOLECULES. THE LANGEVIN-DEBYE FORMULA

As mentioned in the preceding section, a polar molecule has a permanent dipole moment. A polar molecule consists of at least two different species of atoms; during molecule formation some of the electrons may be completely or partially transferred from one atomic species to the other, the resulting electronic arrangement being such that positive and negative charge centers do not coincide in the molecule. In the absence of an electric field a macroscopic piece of the polar dielectric is not polarized, since the individual dipoles are randomly oriented, as shown in Fig. 5–3. The polarization has been defined as

$$\mathbf{P} = \frac{1}{\Delta v} \sum \mathbf{p}_m, \qquad (5\text{–}14)$$

where the summation extends over all molecules in the volume element Δv. When the $\mathbf{p}_m$ are oriented at random, the summation vanishes.

Figure 5–3 A random distribution of permanent dipoles.

If the polar dielectric is subjected to an electric field, the individual dipoles experience torques which tend to align them with the field. If the field is strong enough, the dipoles may be completely aligned, and the polarization achieves the saturation value

$$\mathbf{P}_s = N\mathbf{p}_m, \qquad (5\text{–}15)$$

where N is the number of molecules per unit volume. This orientation effect is in addition to the induced dipole effects, which are usually present also. For the moment, we shall ignore the induced dipole contribution, but its effect will be added in later.

At field strengths normally encountered, the polarization of a polar dielectric is usually far from its saturation value, and if the temperature of the specimen is raised the polarization becomes even smaller. The lack of complete dipole alignment is due to the thermal energy of the molecules, which tends to produce random dipole orientations. The average effective dipole moment per molecule may be calculated by means of a principle from statistical mechanics which states that at temperature T the probability of finding a particular molecular energy E is proportional to

$$e^{-E/kT}, \tag{5-16}$$

where k is Boltzmann's constant and T is the absolute temperature. A complete discussion of the basis for this principle will not be given here; the reader familiar with the Maxwell velocity distribution in a perfect gas has already encountered the principle. According to the Maxwell distribution law, the probability of a molecular velocity v is proportional to $e^{-mv^2/2kT}$. But in Maxwell's perfect gas the molecules have only kinetic energy, $\frac{1}{2}mv^2$; in the general case E in Eq. (5–16) must include both kinetic energy E_k and potential energy U, and the factor becomes

$$e^{-E_k/kT}e^{-U/kT}. \tag{5-17}$$

The potential energy of a permanent dipole p_0 in an electric field $\mathbf{E}_m$ is

$$U = -\mathbf{p}_0 \cdot \mathbf{E}_m = -p_0 E_m \cos \theta, \tag{5-18}$$

where θ is the angle between $\mathbf{p}_0$ and the electric field. Since the molecular kinetic energies do not depend on the electric field, we can ignore the velocity distribution completely in the following calculation. The effective dipole moment of a molecular dipole is its component along the field direction, i.e., $p_0 \cos \theta$. Using the above principle, the average value of this quantity is found to be

$$\langle p_0 \cos \theta \rangle = \frac{\int p_0 \cos \theta e^{+p_0 E_m \cos \theta/kT} \, d\Omega}{\int e^{+p_0 E_m \cos \theta/kT} \, d\Omega}, \tag{5-19}$$

where $d\Omega$ is an element of solid angle that may be replaced by $2\pi \sin \theta \, d\theta$ and the limits on θ are 0 and π. Since p_0, E_m, and kT are constants in the integration, the integrals may be readily performed. It is convenient to define

$$y = \frac{p_0 E_m}{kT}. \tag{5-20}$$

Equation (5–19) then yields:

$$\langle p_0 \cos \theta \rangle = p_0 \left[\coth y - \frac{1}{y} \right], \tag{5-21}$$

which is known as the Langevin formula. A plot of this function is given in Fig. 5–4.

It can be seen from the figure that Eq. (5–21) does indeed give a saturation effect at large field strengths. At small values of y, however, the curve is linear, and it is this linear region which is important at ordinary temperatures. The molecular

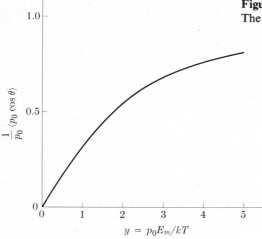

Figure 5–4 Plot of the Langevin function. The asymptotic value as $y \rightarrow \infty$ is one.

dipole moment p_0 of most polar materials is such that $y \ll 1$ for a full range of field strengths, even for those approaching the dielectric strength of the material, so long as the temperature is above about 250 K. Thus a dielectric material containing polar molecules is, in general, *linear*.

Since it is the linear region of Eq. (5–21) that is important, it is appropriate to expand coth y in a power series and keep only leading terms (see Problem 5–4). The first term cancels the last term in Eq. (5–21), with the result that

$$\langle p_0 \cos \theta \rangle \approx \tfrac{1}{3} p_0 y = \frac{p_0^2 E_m}{3kT}. \tag{5–22a}$$

The term $\langle p_0 \cos \theta \rangle$ is the average effective dipole moment; therefore the polarization $P = N \langle p_0 \cos \theta \rangle$ and has the direction of $\mathbf{E}_m$. Hence Eq. (5–22a) may be written in the form

$$\frac{1}{N} \mathbf{P} = \frac{p_0^2}{3kT} \mathbf{E}_m. \tag{5–22b}$$

From a comparison of this equation with Eq. (5–8), it is evident that the polarizability α (i.e., the molecular dipole moment per unit polarizing field) is

$$\alpha = \frac{p_0^2}{3kT}. \tag{5–23}$$

This result has been derived by neglecting induced dipole moments, and represents what we might call an *orientational* polarizability. Induced dipole effects, such as have been considered in the previous section, give rise to what might be termed a *deformation* polarizability, α_0. In the general case, then, the total molecular polarizability is

$$\alpha = \alpha_0 + \frac{p_0^2}{3kT}, \tag{5-24}$$

an expression that is known as the Langevin-Debye equation and has been of great importance in interpreting molecular structures.

*5–4 PERMANENT POLARIZATION. FERROELECTRICITY

It was seen in Section 5–1 that it is the molecular field $\mathbf{E}_m$ that is responsible for polarizing the individual molecules. The relationship between $\mathbf{E}_m$ and the macroscopic electric field $\mathbf{E}$ was given in Eq. (5–7). In most cases the polarization is proportional to $\mathbf{E}$, so that $\mathbf{E}_m$ vanishes when $\mathbf{E}$ goes to zero. But under certain conditions Eq. (5–7) is also compatible with a permanent (or spontaneous) polarization. When $\mathbf{E}$ is set equal to zero,

$$\mathbf{E}_m = \frac{1}{3\epsilon_0} \mathbf{P}_0, \tag{5-25}$$

or, in words, if a polarization $\mathbf{P}_0$ exists, it will create an electric field at the molecule, which tends to polarize the molecule. To be sure, a polarizing field exists; but if this field gives rise to a polarization different from $\mathbf{P}_0$, then the solution is not self-consistent. Therefore, if N is the number of molecules per unit volume,

$$\mathbf{P}_0 = N\alpha\mathbf{E}_m = \frac{N\alpha}{3\epsilon_0} \mathbf{P}_0, \tag{5-26}$$

which is satisfied when either

$$\mathbf{P}_0 = 0$$

or

$$\frac{N\alpha}{3\epsilon_0} = 1. \tag{5-27}$$

Thus the condition for a permanent polarization is Eq. (5–27).†

For most materials $N\alpha/3\epsilon_0$ is less than one, and ordinary dielectric behavior results. In a few crystalline solids, however, condition Eq. (5–27) is met. Such materials are called *ferroelectric* because their electrical properties are analogous

---

† Strictly speaking, Eq. (5–27) has been derived for materials that are composed of only one species of molecule, and for which the term $\mathbf{E}'$ of Section 5–1 vanishes. In a quantitative theory applicable to the general case, Eq. (5–27) is replaced by a set of simultaneous equations. Such complications are not necessary for a fundamental understanding of the origin of ferroelectricity, and consequently will not be discussed here.

to the magnetic properties of ferromagnetic materials. The best known example of a ferroelectric material is barium titanate, $BaTiO_3$, which exhibits a spontaneous dipole moment at temperatures below 120°C. This temperature is called the *Curie point* of the material.

The polarized state of a ferroelectric material is a relatively stable one, and one that can persist for long periods of time. This statement may surprise us to some extent because a polarized specimen is subjected to its own depolarizing field and, depending on the geometry of the specimen, this depolarizing field may be rather large. The depolarizing field is largest for a specimen in the shape of a flat slab, polarized in a direction normal to its faces. As was seen in Section 5–1, if the dimensions of the slab face are large compared with the slab thickness, then

$$\mathbf{E}_d = -\frac{1}{\epsilon_0}\,\mathbf{P}. \tag{5–28}$$

Actually, the high stability of a polarized ferroelectric is due to the fact that there is *no* depolarizing field on the specimen, even for the case of slab geometry. The specimen is polarized by placing it between parallel conducting plates that subsequently have a large potential difference applied to them. In this process the *free* charge from the plates will, to a large extent, be neutralized by surface polarization charge, as is also the case during polarization of a conventional dielectric. If the parallel plates are now brought to the same potential by short-circuiting them, the polarized state of the ferroelectric is still energetically favorable, so the *free charge stays in place*, still neutralizing the polarization charge. The situation is something like that shown in Fig. 5–5; the *free charge* is held in place by the surface polarization charge. The macroscopic field inside the ferroelectric is zero; furthermore, the external electric field is zero, and it is difficult to distinguish the polarized specimen from a conventional unpolarized dielectric material.

If a large potential difference of the opposite sign is now applied to the plates surrounding the polarized ferroelectric, the specimen will change its polarization, and free charge of the opposite sign will flow to the plates from the external circuit, sufficient not only to neutralize the free charge already there, but also to neutralize the new polarization charge. Thus a ferroelectric slab between two parallel plates

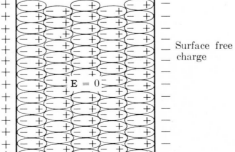

Surface free charge

Figure 5–5 A polarized piece of ferroelectric material.

may serve as the basic element of a memory device; it is capable of storing $\pm$ or $\mp$, and its polarization persists in the absence of an external electric field. The number $\pm$ or $\mp$ may be read by applying a potential difference across the specimen. If the applied field is in the direction of the original polarization, no charge will pass through the external circuit; if the potential difference is opposite to the original polarization, a charge will flow through the external circuit as the polarization of the ferroelectric changes its direction.

A polarized ferroelectric is stable against a reversed electric field provided this electric field is not too large. Figure 5–6 shows the complete curve of polarization versus electric field; it is evident that for low fields there are two values of P for each value of E. A curve such as that in Fig. 5–6 is called a *hysteresis loop.* Hysteresis means "to lag behind," and it is apparent that the polarization vector lags the electric field vector. Points b and a are the stable configurations at $E = 0$; they represent the polarizations $\pm$ and $\mp$, respectively. Point c is the electric field which must be exceeded in order to reverse the polarization.

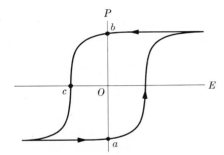

Figure 5–6 Hysteresis curve for a ferroelectric specimen.

5–5 SUMMARY

The macroscopic polarization P of an isotropic dielectric material depends on the molecular dipole moment (or its component) p_m, which arises in response to the local electric field at the molecule—the molecular field E_m:

$$P = Np_m(E_m).$$

Usually p_m is proportional to E_m to a good approximation,

$$p_m = \alpha E_m.$$

The molecular field depends on the applied field E and also on the polarization itself (i.e., the dipole fields of all the other molecules). In the simplest cases

$$E_m = E + \frac{1}{3\epsilon_0} P.$$

In the usual case E_m can be eliminated from these equations to give

$$P = \chi E$$

with the constant susceptibility

$$\chi = \frac{N\alpha}{1 - \frac{N\alpha}{3\epsilon_0}}.$$

But if the molecules are very polarizable ($N\alpha > 3\epsilon_0$), another solution is possible with $E = 0$, $P \neq 0$ (i.e., the material can be spontaneously polarized in zero applied field, as in a ferroelectric).

1. All molecules exhibit an induced dipole moment in an electric field due to the deformation of the electronic charge distribution. A simple linear model leads to a constant atomic polarizability proportional to the atomic volume,

$$\alpha = 4\pi\epsilon_0 R_0^3.$$

2. Polar molecules, which have a permanent dipole moment p_0, display an additional orientational polarizability that is described by the Langevin function derived by statistical mechanics. At high temperature T, this contribution is also linear, with

$$\alpha = \frac{p_0^2}{3kT}.$$

3. A few materials, for example barium titanate, exhibit ferroelectricity.

PROBLEMS

5-1 Use the Clausius-Mossotti equation to determine the polarizability of *atoms* in the air molecules: N_2, O_2. (Note that only the weighted average of the polarizabilities for nitrogen and oxygen may be obtained from Eq. (5-10).) Combine this result with the theory of Section 5-2 to determine the average radius of the atom in an air molecule.

5-2 Figure 5-7 shows a simple cubic lattice of molecules all of which have the same (in direction and magnitude) dipole moment $\mathbf{p}_m$. Let us fix our attention on one particular molecule, call it j. It is evident that j has six nearest neighbors at distance a, twelve next-nearest neighbors at distance $\sqrt{2}\,a$, etc. Find the electric field at j due to the six $\mathbf{p}_m$ on nearest-neighbor molecules for an arbitrary orientation of $\mathbf{p}_m$. (Let the lines joining j to its nearest neighbors define the x-, y-, and z-axes. For simplicity, take $\mathbf{p}_m$ in the xz-plane, making an angle θ with the x-axis.)

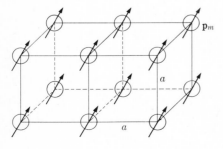

Figure 5-7 Part of a simple cubic array of molecules, each with dipole moment $\mathbf{p}_m$.

5-3 Using the result of Problem 5–1 for the atomic polarizability of nitrogen, compute the relative displacement of the nitrogen nucleus and electron cloud at a field strength $E_m = 3 \times 10^6$ V/m. Compare this displacement with the radius of the atom found in Problem 5–1.

5-4 By using the well-known series expansions for e^y, expand coth y, and obtain Eq. (5–22a) from Eq. (5–21). Go one step further and obtain another term in the series Eq. (5–22a).

5-5 Water is a polar molecule for which the Clausius-Mossotti equation is, strictly speaking, not applicable. Assume its validity, however, and determine p_0 for the water molecule.

CHAPTER 6   Electrostatic Energy

Many problems in mechanics are greatly simplified by means of energy considerations. Hence, when the mechanical behavior of an electrical system is to be studied, it may prove advantageous to use energy methods. In general, the energy of a system of charges, just like that of any other mechanical system, may be divided into its potential and kinetic contributions. Under static conditions, however, the entire energy of the charge system exists as potential energy, and we are particularly concerned with that potential energy that arises from electrical interaction of the charges, the so-called *electrostatic energy*.

In Section 2–4, it was shown that the electrostatic energy U of a point charge is closely related to the electrostatic potential φ at the position of the point charge. In fact, if q is the magnitude of a particular point charge, then the work done by the force on the charge when it moves from position A to position B is

$$\text{Work} = \int_A^B \mathbf{F} \cdot d\mathbf{l} = q \int_A^B \mathbf{E} \cdot d\mathbf{l}$$

$$= -q \int_A^B \nabla\varphi \cdot d\mathbf{l} = -q(\varphi_B - \varphi_A). \tag{6-1}$$

Here the force $\mathbf{F}$ has been assumed to be only the electric force $q\mathbf{E}$ at each point along the path. Under these conditions the charged particle would accelerate. If it does not accelerate, the electric force must be balanced at each point by an equal and opposite force applied by some other agency, so that the *total* work is zero and the kinetic energy does not change. The work done by this *other* force is

$$W = q(\varphi_B - \varphi_A), \tag{6-2}$$

which is equal to the increase in electrostatic energy of the charge over the path interval $A \to B$.

Similar considerations may be applied to more complicated systems of charges; in fact, the electrostatic energy of an arbitrary charge distribution may be calculated as the work required to assemble this distribution of charge against the Coulomb interaction of the charges without imparting to it other forms of energy.

6–1 POTENTIAL ENERGY OF A GROUP OF POINT CHARGES

By the electrostatic energy of a group of m point charges, we mean the potential energy of the system relative to the state in which all point charges are infinitely separated from one another. This energy may be obtained rather easily by calculating the work to assemble the charges, according to (6–2), bringing in one at a time. The first charge q_1 may be placed in position without any work, $W_1 = 0$; to place the second, q_2, requires

$$W_2 = \frac{q_2 q_1}{4\pi\epsilon_0 r_{21}},$$

where $r_{21} = |\mathbf{r}_2 - \mathbf{r}_1|$. For the third charge, q_3,

$$W_3 = q_3 \left[\frac{q_1}{4\pi\epsilon_0 r_{31}} + \frac{q_2}{4\pi\epsilon_0 r_{32}} \right].$$

The work required to bring in the fourth charge, fifth charge, etc., may be written down in a similar fashion. The total electrostatic energy of the assembled m-charge system is the sum of the W's, namely,

$$U = \sum_{j=1}^{m} W_j = \sum_{j=1}^{m} \left(\sum_{k=1}^{j-1} \frac{q_j q_k}{4\pi\epsilon_0 r_{jk}} \right). \tag{6–3}$$

Let us abbreviate this result for U as

$$U = \sum_{j=1}^{m} \sum_{k=1}^{j-1} W_{jk}.$$

Now if we arrange the W_{jk} in the form of a matrix, noting that $W_{jk} = W_{kj}$ and putting $W_{jj} = 0$, it will be apparent that U can also be written as

$$U = \frac{1}{2} \sum_{j=1}^{m} \sum_{k=1}^{m} W_{jk}, \qquad (W_{jj} = 0).$$

That is, a factor of $\frac{1}{2}$ appears in this form of the summation, which is more symmetric, in order that the interaction between each pair of charges not be counted twice. Thus an alternative and more convenient form of Eq. (6–3) is

$$U = \frac{1}{2} \sum_{j=1}^{m} \sum_{k=1}^{m}{}' \frac{q_j q_k}{4\pi\epsilon_0 r_{jk}}, \tag{6–4}$$

where the prime on the second summation means that the term $k = j$ is specifically excluded.

Equation (6–4) may be written in a somewhat different way by noting that the final value of the potential φ at the jth point charge due to the other charges of the system is

$$\varphi_j = \sum_{k=1}^{m}{}' \frac{q_k}{4\pi\epsilon_0 r_{jk}}. \tag{6–5}$$

Thus the electrostatic energy of the system is

$$U = \frac{1}{2} \sum_{j=1}^{m} q_j \varphi_j. \qquad (6\text{-}6)$$

If the point charges had been assembled in a linear dielectric medium of infinite extent, instead of in vacuum, then the permittivity ϵ would replace ϵ_0 in Eqs. (6–4) and (6–5), but Eq. (6–6) would remain unchanged. In the following section it will be shown that this last equation has rather general validity. It applies to a group of point charges that are located in more than one dielectric medium; it even applies to conductors of finite size. The only limitation on the validity of Eq. (6–6) is that all dielectrics in the electrical system be linear.

6–2 ELECTROSTATIC ENERGY OF A CHARGE DISTRIBUTION

In this section we shall calculate the electrostatic energy of an arbitrary charge distribution with volume density ρ and surface density σ. Some of the charge may reside on the surfaces of conductors; in fact, it will be explicitly assumed that there are conductors in the system. In addition, it will be assumed that the dielectrics in the system are *linear*; this restriction is necessary in order that the work expended in bringing the system to its final charged state shall be independent of the way in which this final state is reached.

Suppose we assemble the charge distribution by bringing in charge increments δq from a reference potential $\varphi_A = 0$. If the charge distribution is partly assembled and the potential at a particular point in the system is $\varphi'(x, y, z)$, then, from Eq. (6–2), the work required to place δq at this point is

$$\delta W = \varphi'(x, y, z) \, \delta q. \qquad (6\text{-}7)$$

The charge increment δq may be added to a volume element located at (x, y, z), such that $\delta q = \delta\rho \, \Delta v$, or δq may be added to a surface element at the point in question, whereby $\delta q = \delta\sigma \, \Delta a$. The total electrostatic energy of the assembled charge distribution is obtained by summing contributions of the form Eq. (6–7).

Since the work required to assemble the charges is independent of the order in which things are done, we may choose a particular assembly procedure for which the summation of the δW's is conveniently calculated. This procedure is one in which all parts of the system are brought to their final charge values in concert, i.e., at any stage of the charging process all charge densities will be at the same fraction of their final values. Let us call this fraction α. If the final values of the charge densities are given by the functions, $\rho(x, y, z)$ and $\sigma(x, y, z)$, then the charge densities at an arbitrary stage are $\alpha\rho(x, y, z)$ and $\alpha\sigma(x, y, z)$. Furthermore, the increments in these densities are $\delta\rho = \rho(x, y, z) \, \delta\alpha$ and $\delta\sigma = \sigma(x, y, z) \, \delta\alpha$. The total electrostatic energy, which is obtained by summing Eq. (6–7), is

$$U = \int_0^1 \delta\alpha \int_V \rho(x, y, z)\varphi'(\alpha; x, y, z) \, dv + \int_0^1 \delta\alpha \int_S \sigma(x, y, z)\varphi'(\alpha; x, y, z) \, da.$$

But since all charges are at the same fraction, α, of their final values, the potential $\varphi'(\alpha; x, y, z) = \alpha\varphi(x, y, z)$, where φ is the final value of the potential at (x, y, z). Making this substitution, we find that the integration over α is readily done, and yields

$$U = \tfrac{1}{2} \int_V \rho(\mathbf{r})\varphi(\mathbf{r})\, dv + \tfrac{1}{2} \int_S \sigma(\mathbf{r})\varphi(\mathbf{r})\, da, \qquad (6\text{–}8)$$

the desired result for the energy of a charge distribution. It is important to note that the volume of integration V must be large enough to include *all* of the charge density in the problem, and that the potential φ is just that due to the charge density ρ (and σ) itself. Normally the charge density vanishes outside some bounded region, in which case V may be taken to include all space. If all space is filled with a single dielectric medium except for certain conductors, the potential is given by

$$\varphi(\mathbf{r}) = \frac{1}{4\pi\epsilon} \int_V \frac{\rho(\mathbf{r}')\, dv'}{|\mathbf{r} - \mathbf{r}'|} + \frac{1}{4\pi\epsilon} \int_S \frac{\sigma(\mathbf{r}')\, da'}{|\mathbf{r} - \mathbf{r}'|}. \qquad (6\text{–}9)$$

If several dielectrics are present, the proper boundary conditions must be satisfied, for example, by adding suitable solutions of Laplace's equation to Eq. (6–9). Equations (6–8) and (6–9) are generalizations of Eqs. (6–6) and (6–5) for point charges. The latter can be recovered as a special case by letting

$$\rho(\mathbf{r}) = \sum_{j=1}^{m} q_j\, \delta(\mathbf{r} - \mathbf{r}_j),$$

$$\rho(\mathbf{r}') = \sum_{k=1}^{m}{}' q_k\, \delta(\mathbf{r}' - \mathbf{r}_k),$$

where one must remember specifically to throw out the term $k = j$. When ρ is a continuous distribution, the vanishing of the denominator in Eq. (6–9) does not cause the integral to diverge, and so it is unnecessary to exclude the point $\mathbf{r}' = \mathbf{r}$.

It was stipulated that conductors are present in the system. Although Eq. (6–8) covers this case very well, it is convenient to separate out the contribution from the conductors explicitly. The last integral involves, in part, integrations over the surface of these conductors; since a conductor is an equipotential region, each of these integrations may be done:

$$\tfrac{1}{2} \int_{\text{conductor } j} \sigma\varphi\, da = \tfrac{1}{2} Q_j\varphi_j, \qquad (6\text{–}10)$$

where Q_j is the charge on the jth conductor. Hence Eq. (6–8) becomes

$$U = \tfrac{1}{2} \int_V \rho\varphi\, dv + \tfrac{1}{2} \int_{S'} \sigma\varphi\, da + \tfrac{1}{2} \sum_j Q_j\varphi_j, \qquad (6\text{–}11)$$

where the last summation is over all conductors, and the surface integral is restricted to nonconducting surfaces. As we have seen in Chapter 3, in many problems of practical interest all of the charge resides on the surfaces of conductors. In these circumstances Eq. (6–11) reduces to

$$U = \tfrac{1}{2} \sum_j Q_j \varphi_j. \tag{6–12}$$

We shall have occasion to develop this equation in a later section of this chapter.

For the present, we should like to compare Eq. (6–12) with Eq. (6–6), which was derived for an assembly of point charges. It appears at first sight that the two equations are identical; however, there is an important difference. Equation (6–12) was derived by starting with uncharged macroscopic conductors which were gradually charged by bringing in charge increments; thus the energy described by Eq. (6–12) includes both interaction energy between different conductors and the self-energies of the charge on each individual conductor. If there is only a single conductor, its self-energy $U = \tfrac{1}{2} Q_1 \varphi_1$ is due to the interaction energy of the charges assembled on that conductor. In deriving Eq. (6–6), however, each point charge was brought in as a unit; hence the energy to assemble the point charge from *smaller* charge increments, the so-called self-energy of the point charge, is not included. An attempt to calculate it would give an infinite result if the charge is a mathematical point; but it is not included in the formulation of Coulomb's law of the force between point charges, and it should not be considered now. In order to see that Eq. (6–12) gives the same result in the limit where the conductors become very small, approaching "point" charges, the potential of the jth conductor may be written as the sum of two terms,

$$\varphi_j = \varphi_{j1} + \varphi_{j2}, \tag{6–13}$$

where φ_{j1} is the contribution to the potential due to the charge on conductor j itself, and φ_{j2} is the contribution from charge on other conductors. Thus Eq. (6–12) becomes

$$U = \tfrac{1}{2} \sum_j Q_j \varphi_{j1} + \tfrac{1}{2} \sum_j Q_j \varphi_{j2}. \tag{6–14}$$

The first term of this equation represents the various self-energies of the conductors. Each self-energy, $\tfrac{1}{2} Q_j \varphi_{j1}$, depends on the environment of the conductor (since the charge distribution on each conductor adjusts itself to its environment); furthermore, the only physically meaningful potential associated with conductor j is the total potential φ_j. Thus the decomposition, Eq. (6–14), does *not* make a great deal of sense in general. However, if the conductors are so small that they may be treated as point charges from the macroscopic point of view, then redistribution of charge on the "point" cannot be important, and each self-energy may be taken to be independent of its environment. In addition, since by potential at the point charge j we mean φ_{j2}, the energy required to place a group of *previously charged, very small conductors* in position is the second summation in Eq. (6–14), and this is equivalent to Eq. (6–6).

6–3 ENERGY DENSITY OF AN ELECTROSTATIC FIELD

In the preceding section an expression was developed for the electrostatic energy of an arbitrary distribution of charge. This expression, Eq. (6–8), involves an explicit integration over the charge distribution. It is possible, however, to express the electrostatic energy of the system in a different way, and this alternate form is frequently rather useful. By means of a mathematical transformation (integration by parts), therefore, we convert Eq. (6–8) to an integral involving the field vectors **E** and **D** of the system.

We again consider an arbitrary distribution of charge characterized by the densities ρ and σ. For convenience, it will be assumed that the charge system is bounded, i.e., it is possible to construct a closed surface of finite dimensions which encloses all of the charge. In addition, all surface densities of charge σ will be assumed to reside on conductor surfaces. The last statement is really no restriction at all, since a surface charge density on a dielectric-dielectric interface may be spread out slightly and then treated as a volume density, ρ. The densities ρ and σ are related to the electric displacement;

$$\rho = \mathbf{V} \cdot \mathbf{D}$$

throughout the dielectric regions, and

$$\sigma = \mathbf{D} \cdot \mathbf{n}$$

on the conductor surfaces.* Hence Eq. (6–8) becomes

$$U = \tfrac{1}{2} \int_V \varphi\, \mathbf{V} \cdot \mathbf{D}\, dv + \tfrac{1}{2} \int_S \varphi \mathbf{D} \cdot \mathbf{n}\, da. \qquad (6\text{–}15)$$

The volume integral here refers to the region where $\mathbf{V} \cdot \mathbf{D}$ is different from zero, and this is the region external to the conductors. The surface integral is over the conductors.

The integrand in the first integral of Eq. (6–15) may be transformed by means of a vector identity which we have had occasion to use several times before, Eq. (1–1–7) of Table 1–1:

$$\varphi\, \mathbf{V} \cdot \mathbf{D} = \mathbf{V} \cdot \varphi \mathbf{D} - \mathbf{D} \cdot \mathbf{V}\varphi.$$

Of the two volume integrals resulting from this transformation, the first may be converted to a surface integral through the use of the divergence theorem. Finally, using the fact that $\mathbf{E} = -\mathbf{V}\varphi$, we may write Eq. (6–15) as

$$U = \tfrac{1}{2} \int_{S+S'} \varphi \mathbf{D} \cdot \mathbf{n}'\, da + \tfrac{1}{2} \int_V \mathbf{D} \cdot \mathbf{E}\, dv + \tfrac{1}{2} \int_S \varphi \mathbf{D} \cdot \mathbf{n}\, da. \qquad (6\text{–}16)$$

---

* Here we are taking the first of the alternative viewpoints of a conductor described in Section 4–7.

This equation may be simplified substantially. The surface $S + S'$ over which the first integral of Eq. (6–16) is to be evaluated is the entire surface bounding the volume V. It consists, in part, of S (the surfaces of all conductors in the system), and also of S' (a surface which bounds our system from the outside, and which we may choose to locate at infinity). In both cases the normal $\mathbf{n}'$ is directed *out of* the volume V. In the last integral the normal $\mathbf{n}$ is directed out of the conductor, hence *into* V. Thus the two surface integrals over S cancel each other. It remains to show that the integral over S' vanishes.

If our charge distribution, which is arbitrary but bounded, bears a net charge, then at large distances from the charge system the potential falls off inversely as the distance, i.e., as r^{-1}. $\mathbf{D}$ falls off as r^{-2}. The area of a closed surface that passes through a point at distance r is proportional to r^2. Hence the value of the integral over S', which bounds our system at distance r, is proportional to r^{-1}, and when S' is moved out to infinity, its contribution vanishes.

If the charge distribution bears zero net charge, then the potential at large distances acts like some multipole and falls off *more* rapidly than r^{-1}. Again the contribution from S' may be seen to vanish. Thus, for the electrostatic energy, we have

$$U = \tfrac{1}{2} \int_V \mathbf{D} \cdot \mathbf{E} \, dv, \qquad (6\text{–}17)$$

where the integration is over the volume of the system external to the conductors, i.e., over the various dielectrics in the system. The integration may, of course, be extended to include all space, since the electric field $\mathbf{E}$ equals zero inside a conductor. If this formulation is applied to fields that are produced in part by point charges, it is essential to subtract their infinite "self-energy" explicitly. (See Problem 6–7.)

Where is the electrostatic energy of the electrical system located? This is a question whose precise meaning is difficult to pin down; nevertheless, it is convenient to imagine the energy to be stored in the electric field. Equation (6–17) shows that such a procedure is at least not unreasonable, and in addition it prescribes that the energy be distributed with a density $\tfrac{1}{2}\mathbf{D} \cdot \mathbf{E}$ per unit volume. Hence we are led to the concept of *energy density* in an electrostatic field:

$$u = \tfrac{1}{2}\mathbf{D} \cdot \mathbf{E}. \qquad (6\text{–}18a)$$

Since Eq. (6–17) was derived on the basis of linear dielectrics, each dielectric is characterized by a constant permittivity ϵ. Furthermore, the discussion in preceding chapters has been limited to isotropic dielectrics. Thus Eq. (6–18a) is equivalent to

$$u = \tfrac{1}{2}\epsilon E^2 = \frac{1}{2}\frac{D^2}{\epsilon}. \qquad (6\text{–}18b)$$

6–4 ENERGY OF A SYSTEM OF CHARGED CONDUCTORS. COEFFICIENTS OF POTENTIAL

In Section 3–12 it was shown that a linear relationship exists between the potentials and charges on a set of conductors. In fact, in a system composed of N conductors, the potential of one of them is given by

$$\varphi_i = \sum_{j=1}^{N} p_{ij} Q_j. \tag{3–51}$$

The derivation of Eq. (3–51) was carried out for N conductors in vacuum; however, it is clear that this derivation also holds when dielectrics are present in the system, so long as these dielectrics are both linear and devoid of external charge. The coefficient p_{ij} is the potential of the ith conductor due to a unit charge on conductor j. These coefficients are usually referred to as *coefficients of potential*.

In Section 6–2 an expression was developed for the electrostatic energy of a set of N charged conductors, namely, Eq. (6–12). Combining this result with Eq. (3–51), we obtain

$$U = \tfrac{1}{2} \sum_{i=1}^{N} \sum_{j=1}^{N} p_{ij} Q_i Q_j. \tag{6–19}$$

Thus the energy is a quadratic function of the charges on the various conductors.

A number of general statements can be made about the coefficients p_{ij}, the most important being that (1) $p_{ij} = p_{ji}$, (2) all of the p_{ij} are positive, and (3) $p_{ii} - p_{ij} \geq 0$ for all j. The first of these statements follows from Eq. (6–19), which expresses U as $U(Q_1 \cdots Q_N)$; thus

$$dU = \left(\frac{\partial U}{\partial Q_1}\right) dQ_1 + \cdots + \left(\frac{\partial U}{\partial Q_N}\right) dQ_N.$$

If dQ_1 only is changed, then

$$dU = \left(\frac{\partial U}{\partial Q_1}\right) dQ_1 = \tfrac{1}{2} \sum_{j=1}^{N} (p_{1j} + p_{j1}) Q_j \, dQ_1. \tag{6–20}$$

This increment in the electrostatic energy may also be calculated directly from Eq. (6–2). Bringing in dQ_1 from a zero potential reservoir, we obtain

$$dU = dW = \varphi_1 \, dQ_1 = \sum_{j=1}^{N} p_{1j} Q_j \, dQ_1. \tag{6–21}$$

Equations (6–20) and (6–21) must be equivalent for all possible values of the Q_j, which implies that

$$\tfrac{1}{2}(p_{1j} + p_{j1}) = p_{1j},$$

or

$$p_{j1} = p_{1j}. \tag{6–22}$$

The second statement above, that the potential produced by a net positive charge is positive, is almost intuitively obvious but difficult to prove in a rigorous way. That the third statement is true may be seen from the following argument: let conductor i bear a positive charge Q_i, all other conductors being uncharged. Since conductor j $(j \neq i)$ is uncharged, the net number of lines of displacement leaving this conductor is zero. We distinguish two cases: (a) there are no lines of displacement leaving or impinging upon conductor j, whence we infer that the conductor is in an equipotential region, i.e., it is shielded by another conductor. For example, it could be located inside conductor i, and its potential might be φ_i. In this circumstance, $p_{ij} = p_{ii}$. If conductor j is inside conductor k, then $p_{ik} = p_{ij}$; we immediately transfer our attention to conductor k. (b) Lines of displacement flux leaving conductor j are balanced in number by lines impinging on it. The origin of the displacement flux is the charge on i; hence it must be possible to trace a flux line, which impinges on j, back (perhaps via other conductors) to i. Thus i is at a higher potential than j:

$$\varphi_i > \varphi_j, \quad (Q_i \text{ is positive})$$

or

$$p_{ii} > p_{ij}. \tag{6-23}$$

We must, however, add the equality sign to cover case (a).

The usefulness of the coefficients p_{ij} may be illustrated by means of a simple example. Problem: to find the potential of an *uncharged* spherical conductor in the presence of a point charge q at distance r from the center of the sphere, where $r > R$, and R is the radius of the spherical conductor. The point charge and sphere are taken to be a system of two conductors, and use is made of the equality $p_{12} = p_{21}$. If the sphere is charged (Q) and the "point" uncharged, then the potential of the "point" is $Q/4\pi\epsilon_0 r$; thus

$$p_{12} = p_{21} = \frac{1}{4\pi\epsilon_0 r}.$$

Evidently, when the "point" has charge q and the sphere is uncharged, the potential of the latter is $q/4\pi\epsilon_0 r$.

6-5 COEFFICIENTS OF CAPACITANCE AND INDUCTION

Equation (3–51), which was derived in Chapter 3 and discussed again in Section 6–4, is a set of N linear equations giving the potentials of the conductors in terms of their charges. This set of equations may be solved for the Q_i's, yielding

$$Q_i = \sum_{j=1}^{N} c_{ij}\varphi_j, \tag{6-24}$$

where c_{ii} is called a *coefficient of capacitance* and c_{ij} $(i \neq j)$ is a *coefficient of induction*. The actual solution of Eq. (3–51), expressing each c in terms of the p_{ij}, can be done by matrix inversion, using determinants, for example.

Properties of the c's follow from those of the p's, which we have already discussed. Thus: (1) $c_{ij} = c_{ji}$, (2) $c_{ii} > 0$, (3) the coefficients of induction are negative or zero. (See Problem 6–10.)

Equation (6–24) may be combined with Eq. (6–12) to give an alternative expression for the electrostatic energy of an N-conductor system:

$$U = \tfrac{1}{2} \sum_{i=1}^{N} \sum_{j=1}^{N} c_{ij} \varphi_i \varphi_j. \tag{6–25}$$

6–6 CAPACITORS

Two conductors that can store equal and opposite charges ($\pm Q$), with a potential difference between them which is independent of whether other conductors in the system are charged, form what is called a *capacitor*. This independence of other charges implies that one of the pair of conductors is shielded by the other; in other words, the potential contributed to *each* of the pair by other charges must be the same. Such a situation is depicted in Fig. 6–1 where conductors 1 and 2 form a device of this type. In general, if two conductors, 1 and 2, form a capacitor, we may write

$$\varphi_1 = p_{11} Q + p_{12}(-Q) + \varphi_x,$$
$$\varphi_2 = p_{12} Q + p_{22}(-Q) + \varphi_x, \tag{6–26}$$

where $\pm Q$ are the charges stored and φ_x is the common potential contributed by other charges.

If Eqs. (6–26) are subtracted, we find

$$\Delta\varphi = \varphi_1 - \varphi_2 = (p_{11} + p_{22} - 2p_{12})Q. \tag{6–27}$$

Thus the difference in potential between the conductors of a capacitor is proportional to the charge stored, Q. (Obviously, the total charge stored is zero but, by convention, the absolute value of the charge on one of the two conductors is called the *charge* on the capacitor.) Equation (6–27) may be written

$$Q = C\,\Delta\varphi, \tag{6–28}$$

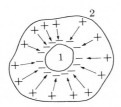

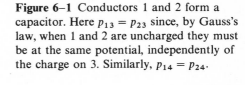

Figure 6–1 Conductors 1 and 2 form a capacitor. Here $p_{13} = p_{23}$ since, by Gauss's law, when 1 and 2 are uncharged they must be at the same potential, independently of the charge on 3. Similarly, $p_{14} = p_{24}$.

where $C = (p_{11} + p_{22} - 2p_{12})^{-1}$ is called the *capacitance* of the capacitor. Evidently C is the charge stored per unit of potential difference; in the mks system C is measured in C/V, or farads (1 F $\equiv$ 1 C/V).

Using the results of previous sections in this chapter, the energy of a charged capacitor may be written as

$$U = \tfrac{1}{2}C(\Delta\varphi)^2 = \tfrac{1}{2}Q\,\Delta\varphi = \frac{1}{2}\frac{Q^2}{C}. \tag{6-29}$$

If the two conductors making up the capacitor have simple geometrical shapes, the capacitance may be obtained analytically. Thus, for example, it is easy to calculate the capacitance of two parallel plates, two coaxial cylinders, two concentric spheres, or that of a cylinder and a plane. The capacitance of a parallel-plate capacitor (Fig. 6–2) will be derived here; other simple cases are taken up in the exercises at the end of the chapter.

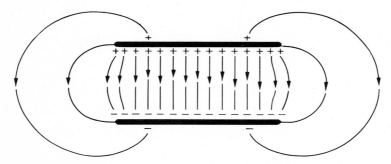

Figure 6–2 The electric field between oppositely charged parallel plates of finite area.

Except for the fringing field at the edge of the parallel plates, the electric field between them is uniform. An ideal parallel-plate capacitor is one in which the plate separation d is very small compared with dimensions of the plate; thus the fringing field may be neglected in the ideal case. If the region between the plates is filled with dielectric of permittivity ϵ, then the electric field between the plates is

$$E = \frac{1}{\epsilon}\sigma = \frac{Q}{\epsilon A}, \tag{6-29}$$

where A is the area of one plate. The potential difference $\Delta\varphi = Ed$. Therefore

$$C = \frac{Q}{\Delta\varphi} = \frac{\epsilon A}{d} \tag{6-30}$$

is the capacitance of this capacitor.

When a capacitor is depicted as part of an electric circuit, it is usually indicated by the symbol ⊣⊢ . Two or more capacitors may be joined together by

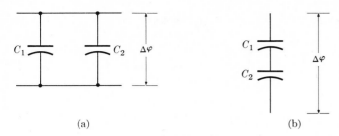

Figure 6–3 (a) Parallel and (b) series connection of two capacitors.

connecting one of the conductors of the first capacitor to a conductor of the second, etc. Possible ways of joining two capacitors are by parallel connection (Fig. 6–3a), or by series connection (Fig. 6–3b). After the capacitors are joined, it is usually desirable to talk about the equivalent capacitance of the combination. In the parallel case, the same voltage $\Delta\varphi$ that appears across each capacitor also appears across the combination; hence the equivalent capacitance is given by

$$C = \frac{Q_{\text{total}}}{\Delta\varphi} = C_1 + C_2. \tag{6–31a}$$

If two uncharged capacitors are connected in series and subsequently charged, *conservation of charge* requires that each capacitor acquire the same charge. Thus the equivalent capacitance C of the combination is related to C_1 and C_2 by the expression

$$\frac{1}{C} = \frac{1}{C_1} + \frac{1}{C_2}. \tag{6–31b}$$

6–7 FORCES AND TORQUES

Thus far in this chapter we have developed a number of alternative procedures for calculating the electrostatic energy of a charge system. We shall now show how the force on one of the objects in the charge system may be calculated from a knowledge of this electrostatic energy. Let us suppose we are dealing with an isolated system composed of a number of parts (conductors, point charges, dielectrics), and we allow one of these parts to make a small displacement $d\mathbf{r}$ under the influence of the electrical forces $\mathbf{F}$ acting upon it. The work performed by the electrical force on the system in these circumstances is

$$dW = \mathbf{F} \cdot d\mathbf{r}$$
$$= F_x \, dx + F_y \, dy + F_z \, dz. \tag{6–32}$$

Because the system is isolated, this work is done at the expense of the electrostatic energy U; in other words, according to Eq. (6–1),

$$dW = -dU. \tag{6–33}$$

Combining Eqs. (6–32) and (6–33) yields

$$-dU = F_x\, dx + F_y\, dy + F_z\, dz$$

and

$$F_x = -\frac{\partial U}{\partial x}, \tag{6–34}$$

with similar expressions for F_y and F_z. That is, in this case $\mathbf{F}$ is a conservative force, and $\mathbf{F} = -\nabla U$.

If the object under consideration is constrained to move in such a way that it rotates about an axis, then Eq. (6–32) may be replaced by

$$dW = \boldsymbol{\tau} \cdot d\boldsymbol{\theta}, \tag{6–35}$$

where $\boldsymbol{\tau}$ is the electrical torque and $d\boldsymbol{\theta}$ is the angular displacement. Writing $\boldsymbol{\tau}$ and $d\boldsymbol{\theta}$ in terms of their components (τ_1, τ_2, τ_3), $(d\theta_1, d\theta_2, d\theta_3)$, and combining Eqs. (6–33) and (6–35), we obtain

$$\tau_1 = -\frac{\partial U}{\partial \theta_1}, \tag{6–36}$$

etc.

Thus our goal has been achieved:

$$F_x = -\left(\frac{\partial U}{\partial x}\right)_Q, \tag{6–34a}$$

$$\tau_1 = -\left(\frac{\partial U}{\partial \theta_1}\right)_Q, \tag{6–36a}$$

where the subscript Q has been added to denote that the system is isolated, and hence its total charge remains constant during the displacement $d\mathbf{r}$ or $d\boldsymbol{\theta}$. To exploit this method, it is necessary to express U in analytic form, and the specific dependence of U on the coordinate x, or θ_1, must be given. An example showing the usefulness of the method will be presented shortly.

Equations (6–34a) and (6–36a) do not, however, cover all cases of interest because, as was mentioned in their derivation, they are limited to isolated systems in which the charge in the system remains constant. In another important class of problems all of the charge exists on the surfaces of conductors, and these are maintained at fixed potentials by means of external sources of energy (e.g., by means of batteries). Here again we may allow one of the parts of the system to move under the influence of electrical forces acting upon it, and the work performed (this time by the system and the batteries) will still be related to the force by Eq. (6–32). But the work becomes, in this instance,

$$dW = dW_b - dU, \tag{6–37}$$

where dW_b is the work supplied by the batteries. Before we can proceed to an expression linking U and the force on some part of the system for this case it will be necessary to eliminate dW_b from Eq. (6–37).

The electrostatic energy U of a system of charged conductors has been given earlier, in Eq. (6–12). If, now, some part of the system is displaced while at the same time the potentials of all conductors remain fixed,

$$dU = \tfrac{1}{2} \sum_j \varphi_j \, dQ_j. \tag{6–38}$$

Furthermore, the work supplied by the batteries, dW_b, is the work required to move each of the charge increments dQ_j from zero potential to the potential of the appropriate conductor; by Eq. (6–2) this is

$$dW_b = \sum_j \varphi_j \, dQ_j. \tag{6–39}$$

Thus

$$dW_b = 2 \, dU. \tag{6–40}$$

Using this equation to eliminate dW_b from Eq. (6–37) and combining the result with Eq. (6–32), we obtain

$$dU = F_x \, dx + F_y \, dy + F_z \, dz$$

or

$$F_x = \left(\frac{\partial U}{\partial x}\right)_\varphi. \tag{6–41}$$

Here the subscript φ is used to denote the fact that all potentials are maintained constant during the virtual displacement $d\mathbf{r}$. In a similar fashion, we may derive

$$\tau_1 = \left(\frac{\partial U}{\partial \theta_1}\right)_\varphi. \tag{6–42}$$

As an example of the energy method, let us consider the following problem. A parallel-plate capacitor of plate separation d has the region between its plates filled by a block of solid dielectric of permittivity ϵ. The dimensions of each plate are length l, width w. The plates are maintained at the constant potential difference $\Delta\varphi$. If the dielectric block is withdrawn along the l dimension until only the length x remains between the plates (see Fig. 6–4), calculate the force tending to pull the block back into place.

Solution. The energy of the system may be calculated by any of several methods. Thus, for example, since $E = \Delta\varphi/d$ is the same everywhere between the plates, let us use

$$U = \tfrac{1}{2} \int_V \epsilon E^2 \, dv,$$

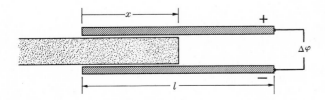

Figure 6–4 Dielectric slab partially withdrawn from between two charged plates.

where the region of integration need include only those parts of space where $E \neq 0$. Neglecting fringing effects at the edge of the capacitor, we find

$$U = \tfrac{1}{2}\epsilon \left(\frac{\Delta\varphi}{d}\right)^2 dwx + \tfrac{1}{2}\epsilon_0 \left(\frac{\Delta\varphi}{d}\right)^2 dw(l - x).$$

The force may be calculated from Eq. (6–41):

$$F_x = \tfrac{1}{2}(\epsilon - \epsilon_0)w \frac{(\Delta\varphi)^2}{d} = \tfrac{1}{2}(K - 1)\epsilon_0 E^2 A.$$

in the direction of increasing x.

The case in which the plates are isolated (constant charge Q) is treated in Problems 6–19 and 6–24.

*6–8 FORCE ON A CHARGE DISTRIBUTION

This chapter would not be complete without a brief discussion of the calculation of electrical force from first principles, i.e., by direct integration, although this procedure has been discussed at some length in an earlier chapter (see Section 4–10). The important thing to remember is that when calculating the force on a charge element dq, the electric field produced by this element, $\mathbf{E}_s$, must be subtracted from the total electric field:

$$d\mathbf{F} = (\mathbf{E} - \mathbf{E}_s)\, dq. \tag{6–43}$$

Thus, for example, when we calculate the force on a point charge, the infinite electric field produced *by* the point charge itself must be excluded from the effective electric field acting at the point. The effect of an extended charge distribution interacting with its own electric field is such as to produce internal stresses in the charge, but these stresses can never combine in such a way that they would tend to produce a rigid displacement of the charge.

The force on an object bearing the surface charge $\sigma(x, y, z)$ is obtained by combining Eqs. (4–57) and (4–58):

$$\mathbf{F} = \oint_S (\mathbf{E} - \mathbf{E}_s)\sigma\, da, \tag{6–44}$$

where the integral is taken over the entire surface of the object. The field $\mathbf{E}_s$ is given by Eq. (4–60):

$$\mathbf{E}_s = \frac{\sigma}{2\epsilon}\, \mathbf{n}. \tag{6–45}$$

If the object is a conductor, there is a simple relationship between the total electric field at the surface, $\mathbf{E}$, and $\mathbf{E}_s$. Thus the force on a conductor, as we have already found in Section 4–10, is

$$\mathbf{F} = \tfrac{1}{2}\oint_S \sigma\mathbf{E}\, da, \tag{6–46a}$$

or

$$\mathbf{F} = \oint_S \frac{\sigma^2}{2\epsilon} \, \mathbf{n} \, da. \tag{6–46b}$$

Finally, let us determine the force on a volume charge distribution. The force on a charge element $\rho \, dv$ is

$$d\mathbf{F} = (\mathbf{E} - \mathbf{E}_s)\rho \, dv. \tag{6–47}$$

But the field $\mathbf{E}_s$ produced by the volume element dv is proportional to the volume divided by the square of some relevant dimension of the element, and this ratio approaches zero in the limit where $dv \to 0$. Thus $\mathbf{E}_s$ is a negligible fraction of $\mathbf{E}$, and we may write

$$\mathbf{F} = \int_{V_0} \rho \mathbf{E} \, dv \tag{6–48}$$

for the force on the charge contained in the volume V_0.

*6–9 THERMODYNAMIC INTERPRETATION OF ELECTROSTATIC ENERGY

The electrostatic energy of a system of charged conductors and dielectrics has been obtained in a variety of forms, in particular we have

$$U = \tfrac{1}{2} \int_V \mathbf{D} \cdot \mathbf{E} \, dv, \tag{6–17}$$

where the integration extends over all dielectrics (including vacuum). The question naturally arises whether U can be interpreted thermodynamically, i.e., does it form part of the internal energy of the system? To answer this question, we must go back to the derivation of U, where we showed that U was the work done on the system in bringing it to its charged condition. Thus U is really a work term, and the problem at hand is to determine under what conditions a work increment may be identified with a thermodynamic property of the system.

From the first law of thermodynamics (which expresses conservation of energy), for a reversible process

$$dE = T \, dS + dW, \tag{6–49}$$

where dE represents the change in internal energy of the system, dS represents the change in entropy, dW is the work done on the system, and T is the absolute temperature. The quantity $T \, dS$, of course, is the heat added to the system during the process.

It is evident that the work increment dW may be identified with the change in internal energy dE only for an adiabatic process, that is, a process in which $dS = 0$. But the temperature of the system will change in general during an adiabatic process, and the dielectric coefficients, which are functions of the temperature, will change also. Recall that Eq. (6–17) was derived from Eq. (6–8), and the latter equation was obtained on the assumption that the various dielectric coefficients remained constant during the charging process. Thus $dW = dU$ does not hold for an adiabatic process. Hence we must restrict our interest to isothermal processes, and here it is not possible to identify dW with dE.

The thermodynamic quantity called the Helmholtz *free energy* of the system is defined by $F = E - TS$. Differentiating and combining the result with Eq. (6–49) yields

$$dF = dE - T\,dS - S\,dT$$
$$= -S\,dT + dW. \tag{6–50}$$

This is just the equation we need. For an isothermal process, dF is equal to dW and dW is equal to dU; thus we may say that the electrostatic energy is the *free energy* of the electrostatic system: $dU = dF$. This energy represents the maximum work that can be extracted at a later time from the electrostatic field.

For a system kept at constant temperature, the free energy plays the same role as does the potential energy for a mechanical (i.e., a temperature-independent) system.

6–10 SUMMARY

The electrostatic potential energy of a system of point charges is calculated as the work that would have to be done by an outside agency against the Coulomb forces between the charges in order to assemble the given configuration. It is expressible as

$$U = \tfrac{1}{2} \sum q_j \varphi_j,$$

where φ_j, the potential at the position of q_j due to all the *other* charges, is

$$\varphi_j = \sum{}' \frac{q_k}{4\pi\epsilon_0 r_{jk}},$$

with the term $k = j$ excluded. For a general charge distribution the electrostatic energy becomes, provided that all dielectrics present are linear,

$$U = \tfrac{1}{2} \int \rho\varphi\,dv,$$

where the potential φ is that produced by the external charge density ρ in the presence of the dielectric media. (ρ may include charge concentrated in a surface distribution or point charges.) Integration by parts transforms the energy in linear dielectrics into an integral,

$$U = \int u\,dv,$$

over the energy density of the electric field,

$$u = \tfrac{1}{2}\mathbf{E} \cdot \mathbf{D} = \tfrac{1}{2}\epsilon E^2 = \frac{1}{2}\frac{D^2}{\epsilon}.$$

When this formulation is applied to point charges, their infinite "self-energy" must be subtracted.

1. When all the charge is a surface distribution on conductors, whose surfaces are equipotentials, the electrostatic energy specializes to

$$U = \tfrac{1}{2}\sum Q_j \varphi_j.$$

Then the coefficients in the linear functions

$$\varphi_i = \sum p_{ij} Q_j,$$

and in the inverse functions

$$Q_i = \sum c_{ij} \varphi_j,$$

are found to satisfy the conditions

$$p_{ij} = p_{ji}, \qquad c_{ij} = c_{ji}.$$

(In addition, $p_{ii} \geq p_{ij} > 0$ and $c_{ii} > 0 \geq c_{ij}$.)

2. In the special case where two conductors form a capacitor,

$$U = \tfrac{1}{2}Q \, \Delta\varphi,$$

with

$$Q = C \, \Delta\varphi.$$

For a parallel-plate capacitor,

$$C = \frac{\epsilon A}{d}.$$

3. The electric force on a part of an isolated system, with constant charge on each conductor, is the negative gradient of the electrostatic energy,

$$F_x = -\left(\frac{\partial U}{\partial x}\right)_Q.$$

If the system is not isolated, but instead the potential of each conductor is maintained constant by an outside agency (battery), the force is given by

$$F_x = +\left(\frac{\partial U}{\partial x}\right)_\varphi.$$

PROBLEMS

6–1 A fast electron (kinetic energy $= 3.0 \times 10^{-17}$ J) enters a region of space containing a uniform electric field $E = 1000$ V/m. The field is parallel to the electron's motion, and in a direction such as to decelerate it. How far does the electron travel before it is brought to rest? (Charge of electron $= 1.60 \times 10^{-19}$ C.)

6–2 Given a spherical dielectric shell (inner radius a, outer radius b, dielectric constant K) and a point charge q, infinitely separated. Now let the point charge be placed at the center of the dielectric shell. Determine the change in energy of the system.

6–3 A dipole ql is positioned perpendicular to a conducting plane, so that the charge $-q$ is at a distance d and the charge $+q$ is at $d + l$. Calculate the electrostatic energy of the system of charges. (*Hint:* Consider the energy of the system consisting of the true charges plus the image charges in vacuum, where the image charges are chosen to give the correct E-field in front of the plane.)

6–4 Given a spherical shell of charge, radius R, uniform surface charge density σ_0. Determine the self energy of the distribution in two ways: (a) by direct integration of Eq. (6–8); (b) by integration over the field, $\frac{1}{2} \int \mathbf{E} \cdot \mathbf{D} \, dv$.

6–5 Given a spherical charge distribution of radius R, uniform charge density ρ_0. Determine the self-energy of the distribution in two ways: (a) by direct integration of Eq. (6–8); (b) by an integration over the field, $\frac{1}{2} \int \mathbf{E} \cdot \mathbf{D} \, dv$.

6–6 Let us assume that an electron is a uniformly charged, spherical particle of radius R. Assume further that the rest energy, mc^2 (where m is the mass of the electron, and c is the velocity of light), is electrostatic in origin and given by the result of Problem 6–5. By putting in appropriate numerical values for the charge and mass of the electron, determine its "classical radius" R.

6–7 Two point charges q_1 and q_2 are separated by a distance d. If their respective fields at a point $\mathbf{r}$ are $\mathbf{E}_1$ and $\mathbf{E}_2$,

$$E^2 = E_1^2 + E_2^2 + 2\mathbf{E}_1 \cdot \mathbf{E}_2.$$

(a) Show that the integrals of E_1^2 and E_2^2 over space are divergent. This is the infinite "self-energy" which must be subtracted from the energy U. (b) Set up an integral for the contribution of $2\mathbf{E}_1 \cdot \mathbf{E}_2$ to U, and show that it is not divergent.

6–8 (a) What is the capacity of a capacitor that can store 1000 J at 1000 V? (b) Assuming the capacitor has parallel plates separated by 10^{-5} m and filled with a material of dielectric constant 2, what is the necessary area of the plates?

6–9 An electrophorus consists of a flat circular plate of wax and a similar plate of metal with an insulating handle. The wax plate is given a bound charge Q by rubbing it with fur. The metal plate is laid on the wax plate and temporarily grounded, so that it acquires a charge $-Q$. The metal plate is finally removed from the wax plate, retaining its charge $-Q$. Suppose the radius of the plates is 10 cm, $Q = 0.3$ μC, and the initial separation of the two plates is 10^{-5} m. Find the potential difference between the plates and the stored energy when the separation is (a) $d = 10^{-5}$ m, (b) $d = 0.02$ m.

6–10 A system of conductors consists of only two conductors. Find the coefficients of capacity and induction explicitly in terms of the coefficients of potential p_{ij}.

6–11 Two spherical conductors are located in vacuum. Conductor 1, of radius R, is grounded (i.e., at zero potential). Conductor 2 is so small that it may be treated as a point charge. It bears the charge q and is located at distance d from the grounded sphere. What is the charge induced on the grounded sphere? (Use the concept of coefficient of potential.)

6–12 Given a system of two conducting objects in a linear dielectric medium. Conductor 1 is uncharged, and conductor 2 is grounded. Prove that conductor 1 is also at ground potential.

6–13 A parallel-plate capacitor is made with a composite dielectric. A sheet of dielectric of permittivity ϵ_1, thickness d_1, is placed on top of a second dielectric sheet (permittivity ϵ_2, thickness d_2). The combination is placed between parallel conducting plates which are separated by the distance $d_1 + d_2$. What is the capacitance per unit plate area of the capacitor?

6–14 A long, conducting cylinder of radius a is oriented parallel to and at distance h from an infinite conducting plane. Show that the capacitance of the system, per unit length of the cylinder, is given by

$$C = 2\pi\epsilon_0/\cosh^{-1}(h/a).$$

(See Section 3–11.)

6–15 Two identical air capacitors are connected in series, and the combination is maintained at the constant potential difference of 50 V. If a dielectric sheet, of dielectric constant 10 and thickness equal to one-tenth of the air gap, is inserted into one of the capacitors, calculate the voltage across this capacitor.

6–16 The capacitance of a gold-leaf electroscope is not quite constant because the leaf moves closer to the case as $\Delta\varphi$ increases. The expected form for the capacitance is

$$C = a + b(\Delta\varphi)^2.$$

How would you determine the constants a and b for a particular instrument? What is the energy of the electroscope when it is charged? Is the energy entirely electrical?

6–17 Two concentric, spherical, conducting shells of radii r_1 and r_2 are maintained at potentials φ_1 and φ_2, respectively. The region between the shells is filled with a dielectric medium. Show by direct calculation that the energy stored in the dielectric is equal to $C(\varphi_1 - \varphi_2)^2/2$, and thereby determine C, the capacitance of the system.

6–18 Two coaxial, cylindrical conductors of approximately the same radius are separated in the radial dimension by the distance d. The cylinders are inserted normally into a liquid dielectric of susceptibility χ and mass density ζ. The cylinders are maintained at the potential difference $\Delta\varphi$. To what height h does the dielectric rise between the conductors? (Neglect surface tension.)

6–19 A parallel-plate capacitor has the region between its plates filled with a dielectric slab of dielectric constant K. The plate dimensions are width w and length l, and the plate separation is d. The capacitor is charged while it is connected to a potential difference $(\Delta\varphi)_0$, after which it is disconnected. The dielectric slab is now partially withdrawn in the l dimension until only the length x remains between the plates. (a) What is the potential difference across the capacitor? (b) What is the force tending to pull the dielectric slab back to its original position?

6–20 The capacitance of a variable air capacitor changes linearly from 50 to 364 pF during a rotation from 0° to 180°. When set at 75°, a potential difference of 400 V is maintained across the capacitor. What is the direction and magnitude of electrostatic torque experienced by the capacitor?

*6–21 An uncharged, conducting, spherical shell of mass m floats with one fourth of its volume submerged in a liquid dielectric of dielectric constant K. To what potential must the sphere be charged to float half submerged? (*Hint:* Assume the electric field of the half-submerged, charged shell to be a purely radial field, and show later that the sum of $\sigma + \sigma_P$ over the spherical surface is such as to justify this assumption.)

6–22 A dielectric slab of thickness d and dielectric constant K fills the region between the plates of a parallel-plate capacitor. The plate area is A. Calculate the electrostatic force on one of the capacitor plates (a) on the assumption that the dielectric is in direct contact with the plate, (b) on the assumption that there is a narrow air space between dielectric and plate. The plates are maintained at the potential difference $\Delta\varphi$ in both cases.

6–23 Solve the example of Fig. 6–4 with the electrostatic energy in the form $U = \frac{1}{2}C(\Delta\varphi)^2$, where $C = C(x)$ is the capacitance of the capacitor with the dielectric block inserted a distance x.

6–24 Suppose the electrostatic energy of a system is $U = \frac{1}{2}Q\,\Delta\varphi$, with $Q = C\,\Delta\varphi$. Let $C = C(x)$ depend on some geometrical parameter x which specifies the position of a part of the system. Show that the force on that part, for a particular value of Q and $\Delta\varphi$, is the same according to Eq. (6–34a) for constant Q as according to Eq. (6–41) for constant $\Delta\varphi$.

6–25 Solve the example of Fig. 6–4 if the thickness of the dielectric slab t is much less than the separation of the plates d. (*Hint:* In this case, according to the result of Problem 4–15, it is D rather than E that is nearly the same everywhere between the plates.)

CHAPTER 7 Electric Current

Up to this point we have been dealing with charges at rest; now we wish to consider charges in uniform motion. This statement implies that we shall be dealing with conductors of electricity because, by definition, a conductor is a material in which the charge carriers are free to move under steady electric fields. (See Section 2–5.) The preceding definition includes not only the conventional conductors, such as metals and alloys, but also semiconductors, electrolytes, ionized gases, imperfect dielectrics, and even vacuums in the vicinity of a thermionic emitting cathode. In many conductors the charge carriers are electrons; in other cases the charge may be carried by positive or negative ions.

Moving charge constitutes a *current*, and the process whereby charge is transported is called *conduction*. To be precise: the current, I, is defined as the rate at which charge is transported through a given surface in a conducting system (e.g., through a given cross section of a wire). Thus

$$I = \frac{dQ}{dt},\tag{7–1}$$

where $Q = Q(t)$ is the net charge transported in time t. The unit of current in the mks system is the ampere (A), named for the French physicist, André Marie Ampère. Evidently,

$$1 \text{ ampere} = 1 \frac{\text{coulomb}}{\text{second}}.$$

7–1 NATURE OF THE CURRENT

In a metal, current is carried entirely by electrons, while the heavy positive ions are fixed at regular positions in the crystal structure (Fig. 7–1). Only the valence (outermost) atomic electrons are free to participate in the conduction process; the other electrons are tightly bound to their ions. Under steady-state conditions, electrons may be fed into the metal at one point and removed at another, produc-

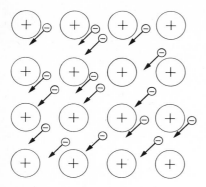

Figure 7–1 Schematic diagram of the motion of conduction electrons in a metal.

ing a current, but the metal as a whole is electrostatically neutral. Strong electrostatic forces keep excess electrons from accumulating at any point in the metal. Similarly, a deficiency of electrons is remedied by electrostatic forces of the opposite sign. We shall see later that excess charge is dissipated extremely rapidly in a conductor. Thus we note that it is possible to study the subject of electric current without taking into account detailed electrostatic effects associated with the charge carriers.

In an electrolyte, the current is carried by both positive and negative ions, although, because some ions move faster than others, conduction by one type of ion usually predominates. It is important to note that positive and negative ions traveling in *opposite* directions (Fig. 7–2) contribute to the current in the *same* direction. The basis for this fact is evident from Eq. (7–1), since the net charge transported through a given surface depends on both the sign of the charge carrier and the direction in which it is moving. Thus, in Fig. 7–2, both the positive and negative carrier groups produce currents to the right; by convention, the direction in which the positive carrier moves (or, equivalently, the direction opposite to that in which the negative carrier moves) is taken as the direction, or *sense*, of the current. In general, an electric current arises in response to an electric field. If an electric field is imposed on a conductor, it will cause positive charge carriers to move in the general direction of the field and negative carriers in a direction

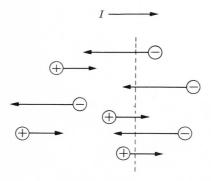

Figure 7–2 Current produced by the motion of both positive and negative charge carriers.

opposite to the field; hence all currents produced in the process have the same direction as the field.

In a gas discharge, the current is carried by both electrons and positive ions, but because the electrons are so much more mobile than the heavy ions, practically all of the current is carried by electrons. Gas conduction is somewhat complicated, because the electronic and ionic populations vary greatly with the experimental conditions (they are determined primarily by the gas pressure and the potential drop across the gas). Under certain conditions *cascading* occurs, in which process the few ions that are initially present accelerate and make inelastic collisions with neutral atoms, thereby producing additional ions and electrons. The additional ions can also give rise to ionizing collisions, with the result that the carrier density builds up enormously.

In Figs. 7–1 and 7–2 we have pictured the charge carriers as falling into groups, each of which has a common motion, called the *drift motion* of the group. The picture has been greatly oversimplified, however. Each group of charge carriers actually represents an assembly of particles in thermal equilibrium with its environment, and so each particle has thermal motion as well as drift motion. But the thermal motion, although it may be large, is also random, and hence gives rise to no organized transport of charge. The drift motion, on the other hand, is not random. In considering the conduction process, then, it is permissible to forget about the random motion, which in the end adds up to nothing, and to use the simple picture presented in Figs. 7–1 and 7–2. For certain other transport processes, however, such as conduction in a thermal gradient (which gives rise to the thermoelectric effect), it is necessary to take the thermal motion into account in a detailed way in order to understand the phenomena fully.

The currents we have described thus far in this section are known as *conduction currents*. These currents represent the drift motion of charge carriers in a *neutral* medium; the medium as a whole may be, and usually is, at rest. Liquids and gases may also undergo hydrodynamic motion, and if the medium has a charge density, this hydrodynamic motion will produce currents. Such currents, arising from mass transport of a *charged* medium are called *convection currents*. Convection currents are important to the subject of atmospheric electricity; in fact, the upward convection currents in thunderstorms are sufficient to maintain the normal potential gradient of the atmosphere above the earth. The motion of charged particles in vacuum (such as electrons in a vacuum diode) also constitutes a convection current. A characteristic feature of the convection current is that it is not electrostatically neutral, and its electrostatic charge must usually be taken into account.

In the rest of this chapter we shall deal exclusively with conduction currents.

7–2 CURRENT DENSITY. EQUATION OF CONTINUITY

We shall now consider a conducting medium which has only one type of charge carrier, of charge q. The number of these carriers per unit volume will be denoted by N. In accordance with the preceding section, we ignore their random thermal

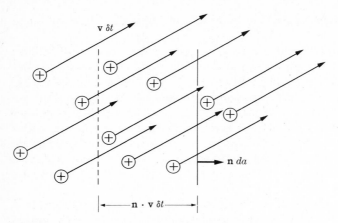

Figure 7–3 The drift motion of charge carriers across the plane *da* in time δt.

motion, and assign the same drift velocity **v** to each carrier. We are now in a position to calculate the current through an element of area *da* such as is shown in Fig. 7–3. During the time δt each carrier moves a distance **v** δt; from the figure it is evident that the charge δQ that crosses *da* during time δt is *q* times the sum of all charge carriers in the volume **v** · **n** δt *da*, where **n** is a unit vector normal to the area *da*. From Eq. (7–1), the current

$$dI = \frac{\delta Q}{\delta t} = \frac{qN\mathbf{v} \cdot \mathbf{n}\, \delta t\, da}{\delta t}$$

$$= Nq\mathbf{v} \cdot \mathbf{n}\, da. \tag{7–2}$$

If there is more than one kind of charge carrier present, there will be a contribution of the form (7–2) from each type of carrier. In general,

$$dI = \left[\sum_i N_i q_i \mathbf{v}_i \right] \cdot \mathbf{n}\, da \tag{7–3}$$

is the current through the area *da*. The summation is over the different carrier types. The quantity in brackets is a vector that has dimensions of current per unit area; this quantity is called the current density, and is given the symbol **J**:

$$\mathbf{J} = \sum_i N_i q_i \mathbf{v}_i. \tag{7–4}$$

The current density may be defined at each point in a conducting medium and is, therefore, a vector point function. It is a useful quantity, one that enters directly into the fundamental differential equations of electromagnetic theory. The mks unit of **J** is ampere per meter^2 (A/m^2).

Equation (7–3) may be written as

$$dI = \mathbf{J} \cdot \mathbf{n} \, da,$$

and the current through the surface S, an arbitrarily shaped surface area of macroscopic size, is given by the integral

$$I = \int_S \mathbf{J} \cdot \mathbf{n} \, da. \tag{7–5}$$

The current density $\mathbf{J}$ and the charge density ρ are not independent quantities, but are related at each point through a differential equation, the so-called *equation of continuity*. This relationship has its origin in the fact that charge can neither be created nor destroyed; the equation is most easily derived by applying Eq. (7–5) to an arbitrary *closed* surface S. The electric current entering V, the volume enclosed by S, is given by

$$I = -\oint_S \mathbf{J} \cdot \mathbf{n} \, da = -\int_V \mathbf{\nabla} \cdot \mathbf{J} \, dv, \tag{7–6}$$

the last integral being obtained through the use of the divergence theorem. The minus sign in Eq. (7–6) comes about because $\mathbf{n}$ is the outward normal and we wish to call I positive when the net flow of charge is from the outside of V to within. But from Eq. (7–1), I is equal to the rate at which charge is transported into V:

$$I = \frac{dQ}{dt} = \frac{d}{dt} \int_V \rho \, dv. \tag{7–7a}$$

Since we are dealing with a fixed volume V, the time derivative operates only on the function ρ. However, ρ is a function of position as well as of time, so that the time derivative becomes the partial derivative with respect to time when it is moved inside the integral. Hence

$$I = \int_V \frac{\partial \rho}{\partial t} \, dv. \tag{7–7b}$$

Equations (7–6) and (7–7b) may now be equated:

$$\int_V \left(\frac{\partial \rho}{\partial t} + \mathbf{\nabla} \cdot \mathbf{J} \right) dv = 0. \tag{7–8}$$

But V is completely arbitrary, and the only way that Eq. (7–8) can hold for an *arbitrary volume segment* of the medium is for the integrand to vanish at each point. Hence, the equation of continuity:

$$\frac{\partial \rho}{\partial t} + \mathbf{\nabla} \cdot \mathbf{J} = 0. \tag{7–9}$$

7–3 OHM'S LAW. CONDUCTIVITY

It is found experimentally that in a metal at constant temperature the current density $\mathbf{J}$ is linearly proportional to the electric field (Ohm's law). Thus

$$\mathbf{J} = g\mathbf{E}. \tag{7–10}$$

The constant of proportionality g is called the *conductivity*. Equation (7–10) has approximate validity for a large number of the common conducting materials; in the general case, however, Eq. (7–10) must be replaced by

$$\mathbf{J} = g(\mathbf{E})\mathbf{E},$$

where $g(\mathbf{E})$ is a function of the electric field. Materials for which Eq. (7–10) holds are called *linear media* or *ohmic media*. Here again, as with dielectrics, we shall be most concerned with the linear case.

The reciprocal of the conductivity is called the *resistivity η*; thus*

$$\eta = \frac{1}{g}. \tag{7–11}$$

The unit of η in the mks system is volt-meter per ampere, or simply ohm-meter, where the ohm (Ω) is defined by

$$1 \text{ ohm} = \frac{1 \text{ volt}}{1 \text{ ampere}}.$$

The unit of conductivity g is $\Omega^{-1}\,\mathrm{m}^{-1}$, sometimes written mho/meter.

Resistivities of a number of common materials are given in Table 7–1. It is apparent from this table that all materials conduct electricity to some extent, but that the materials we have called insulators (dielectrics) are much poorer conductors than the metals by a tremendous factor (as much as 10^{23}). The distinction between a conductor and an insulator will be discussed in a more quantitative way in Section 7–7.

Consider a conducting specimen obeying Ohm's law, in the shape of a straight wire of uniform cross section whose ends are maintained at a constant potential difference $\Delta\varphi$. The wire is assumed to be homogeneous and characterized by the constant conductivity g. Under these conditions an electric field will exist in the wire, the field being related to $\Delta\varphi$ by the relation

$$\Delta\varphi = \int \mathbf{E} \cdot d\mathbf{l}. \tag{7–12a}$$

It is evident that there can be no steady-state component of electric field at right angles to the axis of the wire, since by Eq. (7–10) this would produce a continual charging of the wire's surface. Thus the electric field is purely longitudinal.

---

* The common symbols for resistivity and conductivity are ρ and σ, respectively, but to avoid the possibility of confusion with volume charge density ρ and surface charge density σ, we shall use the symbols η and g.

Table 7–1 Resistivity η and Temperature Coefficient of Resistance α of Some Common Materials at 20°C*

Material	η, Ω m	$\alpha = \dfrac{1}{\eta}\dfrac{d\eta}{dT}$, $(°C)^{-1}$
Aluminum	2.65×10^{-8}	0.0043
Copper	1.67×10^{-8}	0.0068
Gold	2.35×10^{-8}	0.004
Iron	9.71×10^{-8}	0.0065
Nickel	6.84×10^{-8}	0.0069
Silver	1.59×10^{-8}	0.0041
Mercury	95.8×10^{-8}	0.0009
Tungsten	5.51×10^{-8}	0.0045
Constantin (Cu 60, Ni 40)	49.0×10^{-8}	0.0000
Nichrome	100.0×10^{-8}	0.0004
Germanium (pure)	0.46	-0.048
Germanium ($5 \times 10^{-6}\%$ As)	0.011	
Graphite	1.4×10^{-5}	
NaCl Solution (saturated)	0.044	-0.005
Aluminum oxide	1×10^{14}	
Glass	10^{10}–10^{14}	
Iodine	1.3×10^{7}	
Quartz (SiO_2)	1×10^{13}	
Sulfur	2×10^{15}	
Wood	10^{8}–10^{11}	

* Data from the *Handbook of Chemistry and Physics*, 58th edition, CRC Press, Inc., Cleveland, Ohio.

Furthermore, because of the geometry, the electric field must be the same at all points along the wire. Therefore Eq. (7–12a) reduces to

$$\Delta\varphi = El, \qquad (7\text{–}12b)$$

where l is the length of the wire. But an electric field implies a current of density $\mathbf{J} = g\mathbf{E}$. The current through any cross section of the wire is

$$I = \int_A \mathbf{J} \cdot \mathbf{n}\, da = JA, \qquad (7\text{–}13)$$

where A is the cross-sectional area of the wire. Combining Eq. (7–13) with Eqs. (7–10) and (7–12b), we obtain

$$I = \frac{gA}{l}\,\Delta\varphi, \qquad (7\text{–}14)$$

which provides a linear relationship between I and $\Delta\varphi$.

The quantity l/gA is called the *resistance* of the wire; resistance will be denoted by the symbol R. Using R, we may rewrite Eq. (7–14):

$$\Delta\varphi = RI, \qquad (7\text{–}15)$$

which is the familiar form of Ohm's law (R is evidently measured in units of ohms). In the next section it will be inferred that Eq. (7–10) implies Eq. (7–15), independently of the shape of the conductor. Equation (7–15) may be considered to be a definition of the resistance of an object or device that is passing a constant current. In the general case, R will depend upon the value of this current. However, as was mentioned earlier, we are primarily interested in linear materials, and here R is independent of the current.

The work done by the field when a charge dQ moves through the potential difference $\Delta\varphi$ is $dW = dQ \, \Delta\varphi$. The corresponding power is

$$P = I \, \Delta\varphi = I^2 R = (\Delta\varphi)^2/R,$$

where the last two forms are obtained by combining with Ohm's law. This power is dissipated as Joule heating of the material.

7–4 STEADY CURRENTS IN CONTINUOUS MEDIA

There is a very close analogy between an electrostatic system of conductors and dielectrics, on the one hand, and a system that conducts a steady current, on the other. This analogy is the subject of the present section.

Let us consider a homogeneous, ohmic, conducting medium under conditions of steady-state conduction. Since we are dealing specifically with the steady state, the local charge density $\rho(x, y, z)$ is at its equilibrium value, and $\partial\rho/\partial t = 0$ for each point in the medium. Hence, the equation of continuity (Eq. 7–9) reduces to

$$\mathbf{V} \cdot \mathbf{J} = 0, \qquad \text{(steady currents).} \tag{7–16}$$

Using Ohm's law in combination with (7–16), we obtain

$$\mathbf{V} \cdot g\mathbf{E} = 0,$$

which for a homogeneous medium reduces to

$$\mathbf{V} \cdot \mathbf{E} = 0.$$

But since $\mathbf{V} \times \mathbf{E} = 0$ for a static field, $\mathbf{E}$ is derivable from a scalar potential:

$$\mathbf{E} = -\mathbf{V}\varphi.$$

Combination of the last two equations yields

$$\mathbf{V}^2\varphi = 0, \tag{7–17}$$

which is Laplace's equation.

We see, therefore, that a steady-state conduction problem may be solved in the same way as electrostatic problems. Laplace's equation is solved by one of the techniques discussed in Chapter 3, the appropriate solution being determined, as always, by the boundary conditions. Boundary conditions that are sufficient to solve the problem are those that specify either φ or $\mathbf{J}$ at each point on the surface of the conducting medium. Specifying $\mathbf{J}$ at the surface is equivalent to specifying $\mathbf{E}$, since the two vectors are connected by Ohm's law. Once the appropriate solution to Laplace's equation has been found, **E** (and hence **J**) may be determined at each point inside the medium from the gradient operation.

Under steady-state conduction the current which crosses an interfacial area between two conducting media may be computed in two ways: in terms of the current density in medium 1, or in terms of the current density in medium 2. Since the two procedures must yield the same result, the normal component of **J** must be continuous across the interface:

$$J_{1n} = J_{2n}, \tag{7-10a}$$

or

$$g_1 E_{1n} = g_2 E_{2n}. \tag{7-18b}$$

This equation is the analog to the equation for the continuity of D_n across dielectric interfaces in electrostatic problems.

Since the field is static in each medium,

$$\oint \mathbf{E} \cdot d\mathbf{l} = 0$$

for a closed path that links both media, and

$$E_{1t} = E_{2t} \tag{7-19}$$

by the derivation of Section 4–7. This equation is evidently the same for both types of problems (electrostatic and steady conduction).

An example of the ideas presented above is found in the "electrolytic tank" shown in Fig. 7–4. Here a number of metallic conductors, which are connected to external sources of potential, are placed in a liquid conducting medium (ideally of infinite extent) with moderate conductivity (such as a salt solution). Since the conductivity of the salt solution is *much* smaller than that of a metal (see Table 7–1), the electric field in the metal (for the same current density) is *much* smaller than that in the solution. The ratio of fields is so small that **E** in the metal may be neglected, and each metallic conductor may be assumed to be an equipotential volume. A small conducting probe may be used, as shown in Fig. 7–4, to explore the potential in the solution, and in this way a plot of the equipotential surfaces can be made. A possible use of this experimental approach is that it provides a numerical solution to Laplace's equation that, in the case of complicated geometry, might be more difficult to determine theoretically. The solution found is not limited to the conduction problem but applies equally well to the equivalent electrostatic problem in which the same metallic conductors are surrounded by a dielectric medium (Fig. 7–5).

As a second example of the relation between conduction and electrostatics, we consider two metallic conductors in an infinite, homogeneous, ohmic medium of moderate conductivity g. If the metallic conductors are maintained at the potentials φ_1 and φ_2, the current I between them is

$$I = \frac{\varphi_1 - \varphi_2}{R},$$

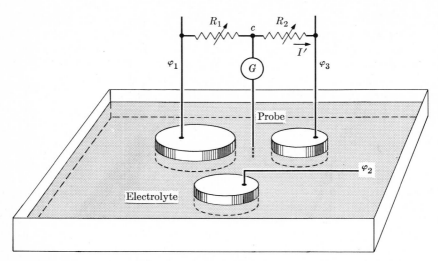

Figure 7–4 Two-dimensional electrolytic tank. The three metallic conductors are maintained at potentials φ_1, φ_2, and φ_3, where for convenience it is assumed that $\varphi_1 > \varphi_2 > \varphi_3$. The symbol ⌇⌇ stands for a resistor whose resistance may be varied, and G is a galvanometer. The wires are taken to be of negligible resistance. If the resistors R_1 and R_2 are adjusted so that there is no current through G, then $\varphi_{\text{probe}} = \varphi_c$, and the same current I' passes through both R_1 and R_2. In these circumstances $\varphi_{\text{probe}} = \varphi_1 - I'R_1 = \varphi_3 + I'R_2$, or $\varphi_{\text{probe}} = \varphi_1 - (\varphi_1 - \varphi_3)R_1/(R_1 + R_2)$.

where R is the resistance of the medium.* This current may be written in terms of the current density **J** in the medium:

$$I = \oint_S \mathbf{J} \cdot \mathbf{n} \, da,$$

where S is any closed surface which completely surrounds one of the conductors (except for an insulated metal wire to lead current onto the conductor so as to maintain its potential constant). But

$$\mathbf{J} = g\mathbf{E}.$$

Combining the last three equations, we obtain

$$\frac{\varphi_1 - \varphi_2}{R} = g \oint_S \mathbf{E} \cdot \mathbf{n} \, da. \qquad (7\text{–}20)$$

If the identical electric field were produced by electrostatic charges on the two metallic conductors in a *dielectric* medium, then by Gauss's law

$$\oint_S \mathbf{E} \cdot \mathbf{n} \, da = \frac{1}{\epsilon} Q, \qquad (7\text{–}21)$$

---

* Since I is analogous to Q in the electrostatic problem, I is proportional to $\Delta\varphi$, and $1/R$ is defined as the proportionality constant. See Eq. (7–22).

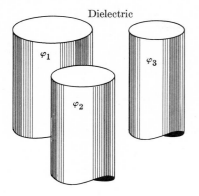

Dielectric

Figure 7–5 The equivalent electrostatic problem to the conduction problem of the preceding figure. Since Fig. 7–4 depicted two-dimensional conduction, the electrostatic problem is also two-dimensional, and each conductor is an infinitely long cylinder.

where Q is the charge on the metallic conductor surrounded by the surface S and ϵ is the permittivity of the medium. In these circumstances, the two conductors form a capacitor:

$$Q = C(\varphi_1 - \varphi_2). \qquad (7\text{–}22)$$

Insertion of Eqs. (7–21) and (7–22) into Eq. (7–20) yields

$$RC = \frac{\epsilon}{g}, \qquad (7\text{–}23)$$

which is a relation between the resistance of the medium and the capacitance of the equivalent electrostatic problem.

This relation is in fact more than an analogy between conducting and dielectric media; it also holds for any single medium with conductivity g and dielectric constant ϵ. Since there is no ideal dielectric, every real dielectric has some nonvanishing g, no matter how small. From the other extreme, even a good conductor has its own value of ϵ, whatever it may be.* Thus a capacitor has a leakage resistance, and a resistor has a small associated capacitance; in each case R and C are related by Eq. (7–23) (approximately, since the medium is not infinite).

7–5 APPROACH TO ELECTROSTATIC EQUILIBRIUM

In Chapter 2 it was shown that the excess charge on a conductor resides on its surface. This, of course, is the equilibrium situation. The approach to equilibrium was not studied, but it was stated that for good (metallic) conductors the attainment of equilibrium is extremely rapid. The poorer the conductor, the slower is the approach to electrostatic equilibrium; in fact, if the conductivity of the mate-

---

* It is not obvious how the dielectric constant of a reasonably good conductor could be measured, but it will become clear in Chapter 13. The time scale within which a "static" measurement would have to be made will be found in the following section. The value $\epsilon = \infty$ suggested in Chapter 4 is applicable only in the absence of current; with a steady current only the bound charges contribute to ϵ, not the free charges.

rial is extremely low, it may take years or even longer for electrostatic equilibrium to obtain.

Consider a homogeneous, isotropic medium characterized by conductivity g and permittivity ϵ, which has a volume density of prescribed charge $\rho_0(x, y, z)$. If this conducting system is suddenly isolated from applied electric fields, it will tend toward the equilibrium situation where there is no excess charge in the interior of the system. According to the equation of continuity,

$$\frac{\partial \rho}{\partial t} + \mathbf{V} \cdot \mathbf{J} = 0, \tag{7-9}$$

which, with the aid of Ohm's law, becomes

$$\frac{\partial \rho}{\partial t} + g\mathbf{V} \cdot \mathbf{E} = 0. \tag{7-24}$$

But $\mathbf{V} \cdot \mathbf{E}$ is related to the sources of the field; in fact, $\mathbf{V} \cdot \mathbf{E} = \rho/\epsilon$, so that

$$\frac{\partial \rho}{\partial t} + \frac{g}{\epsilon} \rho = 0. \tag{7-25}$$

The solution to this partial differential equation is, for constant g and ϵ,

$$\rho(x, y, z, t) = \rho_0(x, y, z)e^{-gt/\epsilon}, \tag{7-26}$$

and it is seen that the equilibrium state is approached exponentially.

From Eq. (7–26) it is evident that the quantity ϵ/g has the dimensions of time; it is called the *time constant* or relaxation time t_c of the medium:

$$t_c = \frac{\epsilon}{g} = \epsilon\eta. \tag{7-27}$$

The time constant is a measure of how fast the conducting medium approaches electrostatic equilibrium; precisely, it is the time required for the charge in a specified region to decrease to $1/e$ of its original value.

A material will reach its equilibrium charge distribution in a specific application when its time constant is much shorter than the characteristic time required to make the pertinent measurement. For some applications a time constant of less than 0.1 second is sufficient to ensure conductorlike behavior; since most non-metallic permittivities* fall into the range ϵ_0 to $10\epsilon_0$, this requires a material with resistivity less than 10^9 or 10^{10} $\Omega \cdot$ m. For high-frequency applications a shorter time constant, and a correspondingly smaller resistivity, is required for true conductorlike behavior; in fact,

$$t_c \ll \frac{1}{f},$$

where f is the highest frequency involved in the experiment. Just the opposite condition holds for dielectriclike behavior.

---

* We cannot apply this relation to a metal since we do not know the proper value of ϵ to use; in fact, $t_c \approx \tau \approx 10^{-14}$ s, where τ is the collision time to be discussed in Section 7–7. As will be seen, for times shorter than τ the assumption that $\mathbf{J} = g\mathbf{E}$ is not valid.

7–6 RESISTANCE NETWORKS AND KIRCHHOFF'S LAWS

Thus far we have discussed conduction primarily from the point of view of charge transport in a conducting medium, and have approached the problem in terms of the differential equations that must apply at each point. In these cases the important quantity to be determined is the current density, $\mathbf{J}$. But in many problems of practical interest the electric charge carriers are constrained to follow a high conduction path called a *circuit*, and then the quantities of interest are the *currents* in each part of the circuit. In this section we shall limit the discussion to circuits carrying steady currents, i.e., to *direct current circuits*. A circuit may consist of several different branches; in fact, as a possible definition, a circuit is a network of conducting paths, each of which may contain applied voltages. The central problem of circuit analysis is: *given the resistance and applied voltage in each circuit branch, find the current in each of these branches.*

In Chapter 2 it was shown that the integral of the tangential component of an electrostatic field around any closed path vanishes; i.e.,

$$\oint \mathbf{E} \cdot d\mathbf{l} = 0 \tag{7–28}$$

for an electrostatic field. For an ohmic material, $\mathbf{J} = g\mathbf{E}$. In the general case this is modified to $\mathbf{J} = g(\mathbf{E})\,\mathbf{E}$, but $g(\mathbf{E})$ is always a positive quantity. Thus it follows that a purely electrostatic force cannot cause a current to circulate in the same sense around an entire circuit. Or, in other words, a steady current cannot be maintained solely by means of electrostatic forces. A charged particle q may, however, experience other forces (mechanical, "chemical," etc.) in addition to the macroscopic electrostatic force, so that in part of the circuit the charges move opposite to the direction of $\mathbf{E}$. In the previous sections we sidestepped the question concerning the cause of electric current by assuming that two points on a conducting object were maintained at the constant potential difference $\Delta\varphi$ by means of external energy sources. It is still sufficient for our purpose here to assume the existence of such *applied voltages*,* but we shall make a brief digression to discuss how they may in fact be produced.

In the laboratory a steady voltage is usually produced either by a battery or by an electronic power supply (which rectifies and smooths the line voltage), but it could be produced by a variety of other means, e.g., a Van de Graaff generator. The last is conceptually the simplest case to analyze. In the Van de Graaff generator charges are literally put onto a conveyor belt at one terminal and forcibly carried to another location of *higher* potential energy at the other terminal where they are removed from the belt. In steady-state operation, $\oint \mathbf{E} \cdot d\mathbf{l} = 0$ around any closed path; e.g., the integral is negative along the belt and equally positive along an outside path between the terminals. A steady outside current can be allowed to

---

* The applied voltage is usually called an *electromotive force* (or *emf*) in other books, although applied voltage is the term used in the laboratory for this potential difference. The historical term emf and the concept itself are rather confusing and unnecessary, so they will not be used here. We reserve the term emf for a somewhat different concept to be introduced later (Chapter 11).

flow through a resistance connected between the terminals if the belt is made to move fast enough; the power input is simply the mechanical power necessary to crank the belt that is transporting charges in opposition to an electric field. The operation of a battery is similar (except that the "forces" at work in a battery depend on the quantum mechanics of electrochemistry), and $\oint \mathbf{E} \cdot d\mathbf{l} = 0$ around any closed path, even one passing through the electrolyte in the battery. What is important for the purpose of circuit analysis, however, is simply that $\oint \mathbf{E} \cdot d\mathbf{l} = 0$ around a closed path containing the terminals of the voltage source—one leg of the path through the resistance network and the other directly across the terminals but external to the source. The purpose of electric circuit theory is to develop a procedure for analyzing the first leg, through the resistance network; we do not need to analyze the cause (mechanical, chemical, or whatever) of the voltage difference between the terminals of the power source, but simply refer to it as an applied voltage, $\mathscr{V}$. An ideal source would provide an applied voltage $\mathscr{V}_0$ that is independent of the current drawn from the source; but the terminal voltage of a real source depends to some degree on the current, $\mathscr{V} = \mathscr{V}(I)$. The simplest assumption, which is usually applicable, is that the dependence is linear:

$$\mathscr{V} = \mathscr{V}_0 - R_I I.$$

The coefficient R_I is called the *internal resistance* and $\mathscr{V}_0$ is called the *open-circuit voltage* (or the emf, in most other books).

Before proceeding to the general network problem, we first review the elementary series and parallel connections of resistors. The resistance defined in Section 7–3 is a property of the material object under consideration, and it depends upon both the nature of the material from which the object is composed and its geometry. (The resistivity, on the other hand, depends only upon the nature of the conducting material.) A conducting object of convenient shape that is characterized primarily by its resistance is called a *resistor*; it is usually denoted by the symbol ⌁ . Resistors may be connected to form a resistance network; the ways in which two resistors may be combined are illustrated in Fig. 7–6. Part (a) shows a *series* connection; here the same current I passes through both resistors. Applying Eq. (7–15) to each resistor, and noting that the potential difference* $V = V_1 + V_2$, we find that

$$V = R_1 I + R_2 I = (R_1 + R_2)I.$$

Thus the equivalent resistance of the combination is

$$R = R_1 + R_2 \qquad \text{(series connection)}. \tag{7–29}$$

In the *parallel* connection (Fig. 7–6b) the potential difference across each resistor is the same, and the total current through the combination is $I = I_1 + I_2$. Applying Eq. (7–15), we find

$$I = \frac{1}{R_1} V + \frac{1}{R_2} V = \left(\frac{1}{R_1} + \frac{1}{R_2} \right) V,$$

---

* In this section we shall use the symbol V instead of $\Delta \varphi$ for a potential difference, so as to accord with the most common electric circuit notation.

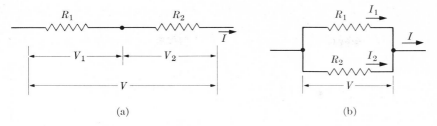

Figure 7–6 (a) Series and (b) parallel connection of two resistors.

and the equivalent resistance R of the combination is obtained from

$$\frac{1}{R} = \frac{1}{R_1} + \frac{1}{R_2} \quad \text{(parallel combination)}. \tag{7–30}$$

The equivalent resistance of a more complicated network like that in Fig. 7–7 may be determined by combining the resistors in pairs according to Eqs. (7–29) or (7–30), and then repeating the process until only one equivalent resistance remains. This procedure is not possible for every network; however, every two-terminal network can be reduced to one equivalent resistance by the procedure of the following paragraph.

Any network problem can be solved in a systematic way by means of two rules known as Kirchhoff's laws.* Before stating these laws, we define two terms. A *branch point* is a point of the circuit where three or more conductors are joined together, such as point a, b, c, or d in Fig. 7–8. A *loop* is any closed conducting path in the network. Kirchhoff's laws may now be stated:

I. *The algebraic sum of the currents flowing toward a branch point is zero*; i.e.,

$$\sum I_j = 0. \tag{I}$$

II. *The algebraic sum of the voltage differences around any loop of the network is zero*; i.e.,

$$\sum V_j = 0. \tag{II}$$

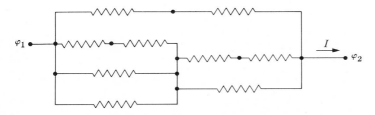

Figure 7–7 A resistor network.

---

* Named for Gustav Robert Kirchhoff (1824–1887).

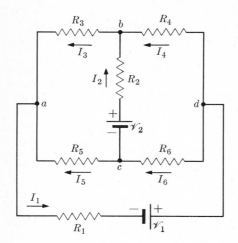

Figure 7–8 A typical circuit requiring the application of Kirchhoff's laws. The symbol —$\underset{-}{\vert}\vert\underset{+}{\vert}$— is used to designate an applied voltage. In a typical circuit problem, the $\mathscr{V}$'s and R's are specified, the currents are to be found. Two of the six equations for the currents in the above circuit are $-I_1 + I_3 + I_5 = 0$ and $\mathscr{V}_1 = I_6 R_6 + I_5 R_5 + I_1 R_1$.

The first law is just a formal statement of the fact that charge does not accumulate at a branch point in the circuit as a result of the steady current. It is a restatement of the equation of continuity in the form of Eqs. (7–6) and (7–7), and as such it is equivalent to

$$\mathbf{V} \cdot \mathbf{J} = 0. \qquad \text{(steady currents)} \qquad (7\text{–}16)$$

The second law is merely a restatement of

$$\oint \mathbf{E} \cdot d\mathbf{l} = 0. \qquad \text{(static fields)} \qquad (7\text{–}28)$$

To apply Kirchhoff's laws, we need to recall Ohm's law:
    The potential drop in a resistance R_j is

$$V_j = I_j R_j, \qquad \text{(resistor)} \qquad (7\text{–}15)$$

where the higher potential is taken to be at the end at which the assumed current enters the resistance. This is the integral form of

$$\mathbf{J} = g\mathbf{E}. \qquad \text{(linear medium)} \qquad (7\text{–}10)$$

Finally we label the applied voltages

$$V_j = -\mathscr{V}_j. \qquad \text{(applied voltage)}$$

Combining this and Eq. (7–15), we can rewrite Kirchhoff's law II as

$$\sum \mathscr{V}_j = \sum I_j R_j. \qquad (\text{IIa})$$

If the internal resistances of the sources must be taken into account, they can be transferred to the right side of (IIa).

Before applying Kirchhoff's laws to a specific problem, it is necessary to assume directions for the currents in each of the branches. These directions should be indicated in the circuit diagram. The formulation of Eqs. (I) and (IIa) is then carried out on the basis of the assumed directions. If the numerical solution of these equations yields a negative value for a particular current, the correct direction of this current is opposite to that assumed. In the problem illustrated in Fig. 7–8, there are six unknown currents; these are designated by the symbols $I_1, I_2, I_3, I_4, I_5,$ and I_6, each having been given an assumed direction.

Kirchhoff's Law I may be applied at each branch point of the circuit, but the equations so obtained are not all independent. The general rule is that if there are n branch points, only $n - 1$ of these will produce independent equations. In the problem shown in Fig. 7–8, there are six unknown currents; the solution requires three branch-point equations and three loop equations.

The summations in (I) and (IIa) are algebraic sums. In (I) the current is considered positive if its assumed direction points toward the branch point in question, or is taken with the negative sign if its assumed direction points away from the junction. In applying the loop equations, some direction (either clockwise or counterclockwise) must be taken as the traversal direction. An applied voltage is taken with the positive sign if the voltage (by itself) would produce a positive current in the traversal direction; an IR term is taken with the positive sign if the current through the resistor in question is in the direction of traversal of the loop.

7–7 MICROSCOPIC THEORY OF CONDUCTION

On the basis of a simple microscopic model of a conductor it is possible to understand the linear behavior that is stated as Ohm's law, as well as some other experimental features of conduction. We consider a free particle of the medium, with charge q and mass m. Under the influence of the local electric force, $q\mathbf{E}$, its drift velocity will increase according to $m\,d\mathbf{v}/dt = q\mathbf{E}$. If the charged particle were in a vacuum, it would continue to accelerate. In a material medium passing a steady current, however, the drift velocity is constant, and therefore the total force on the particle must be zero. Another force, due to the medium, must be acting in addition to the electric force. The simplest possible assumption is that this retarding force is proportional to the velocity, so that the equation of motion is

$$m\frac{d\mathbf{v}}{dt} = q\mathbf{E} - G\mathbf{v}. \tag{7–31}$$

It can be seen immediately that when $d\mathbf{v}/dt = 0$,

$$\mathbf{v} = \frac{q}{G}\mathbf{E} \tag{7–32}$$

is the steady-state solution for the drift velocity. It is interesting to examine the complete solution of Eq. (7–31), however; this is

$$\mathbf{v}(t) = \frac{q}{G}\mathbf{E}(1 - e^{-Gt/m}) \tag{7–33}$$

if one makes the initial condition $\mathbf{v}(0) = 0$. This shows that the local drift velocity approaches its steady value exponentially, like $e^{-t/\tau}$, where the *relaxation time* τ is

$$\tau = \frac{m}{G}. \tag{7-34}$$

Eliminating G between Eqs. (7–32) and (7–34), we find the steady-state drift velocity to be

$$\mathbf{v} = \frac{q\tau}{m}\,\mathbf{E}. \tag{7-35}$$

Combining this with Eq. (7–4) for a single kind of charge carrier, we get the current density

$$\mathbf{J} = N q \mathbf{v} = \frac{N q^2 \tau}{m}\,\mathbf{E}, \tag{7-36}$$

proportional to the field according to Ohm's law. Comparison with Eq. (7–10) gives the conductivity

$$g = \frac{N q^2 \tau}{m}, \tag{7-37}$$

or in case there are several kinds of charge carriers,

$$g = \sum \frac{N_i q_i^2 \tau_i}{m_i}.$$

For a reasonably good electronic conductor such as a semiconductor or a metal (but not an electrolyte), we can interpret τ physically as the *mean time between collisions* of a conduction electron. In such a material the electron accelerates for a short period, after which it makes a collision with one of the atoms of the material. As a result of this collision the electron is thrown off in a random direction, so that the average effect of a collision is to reduce the drift velocity of the electron to zero again. If the mean collision time is τ, and the mean net velocity is v, then the electrons lose the momentum mv after each time τ. In the steady state the rate of momentum loss mv/τ is equated to the rate of momentum gain qE, and the result is identical to Eq. (7–35). The mean time τ is related to the *mean free path* of the electron by

$$l = v_T \tau, \tag{7-38}$$

where v_T is the *thermal* velocity of the electrons. It is important to reemphasize that v_T is much greater* than the drift velocity v (though random in direction). For most metals v_T is of the order of 10^6 m/s (nearly independent of temperature), and for a semiconductor it is about an order of magnitude smaller at room temperature; on the other hand the mean drift velocity v is not greater than about

---

* It is only because of this fact that τ can be taken to be independent of the accelerating field E.

10^{-2} m/s in normal metals. In metals and semiconductors the mean free path is typically a few hundred angstroms (10^{-8} m) at room temperature, so that $\tau \approx 10^{-14}$ s in metals; in semiconductors τ may be an order of magnitude longer. In either case, since τ is also the time for the start up or decay of an Ohmic current, this is practically instantaneous after the field is applied or removed in resistors made of these materials.* We note that in a metal the relaxation time for current decay τ, the collision time, and the time constant for dissipation of excess charge density t_c all happen to be the same, although they are conceptually different.†

It is evident from Table 7–1 that the group of materials with the highest electrical conductivity is that of the *metals*. These materials have high conductivity both because they contain a large density of charge carriers, of the order of one for each atom of the metal, and because the drift velocity per unit electric field is high. In metals we deal with only one type of charge carrier, the electron. Hence the conduction equations are simple in this case:

$$\mathbf{J} = -N e \mathbf{v}, \tag{7–39}$$

$$g = Ne(v/E) = Ne^2\tau/m, \tag{7–40}$$

where e is the absolute value of the electronic charge. The drift velocity of the electron per unit electric field (v/E) is called the *mobility* of the electron. A large mobility implies a long collision time τ or, equivalently, a long mean free path. In order to get some feeling for the mean free path of electrons in a metal, we have to appeal to the dynamics of electron collisions. We know that the conductor is electrostatically neutral only on the average, that there are large variations in potential over distances of the order of one angstrom unit, and that a charged particle, such as an electron, ought to collide or be scattered by variations in potential. But we know also that the wave nature of the electron plays an important role in its motion on an atomic scale. A solution to the electron collision problem using wave-mechanical concepts is beyond the scope of this book; we merely state a result. *In a perfect crystal with a three-dimensional periodic potential, an electron wave makes no collision; its collision time τ is infinite.* Thus the finite conductivity of metals arises from imperfections in the perfectly periodic structure. These imperfections are of two types: (1) impurities and geometric imperfections (such as grain boundaries in polycrystalline material), and (2) thermally induced imperfections arising from the thermal motion of the atoms in the structure. Both types often contribute independently to the resistivity (Matthiessen's rule), so that

$$\eta = \eta_1 + \eta_2(T), \tag{7–41}$$

where T is the absolute temperature.

---

* On the macroscopic distance scale, the limitation is the longer time required for the propagation of the field at the velocity of light—e.g., 10^{-10} s over a 3 cm distance in vacuum.

† In a poor conductor, the collision time may have no meaning, or t_c may be *inversely* proportional to τ according to Eqs. (7–27) and (7–37).

In very pure metals the dominant contribution to the resistivity at ordinary temperatures is the scattering of electron waves by thermally displaced atoms. Thus $\eta \approx \eta_2(T)$. The scattering cross section of a displaced atom is proportional to the square of its vibration amplitude (x^2), in other words, to its maximum potential energy. Assuming elastic restoring forces operating on the displaced atoms,

$$(\text{Potential energy})_{max} = (\text{Kinetic energy})_{max} \propto kT,$$

so that

$$\eta \approx \eta_2 \propto (\tau_2)^{-1} \propto x^2 \propto T, \tag{7-42}$$

or, in words, the resistivity of a pure metal is proportional to the absolute temperature. The temperature coefficient of resistance, $(1/\eta)\, d\eta/dT$, for a very pure metal is, therefore,

$$\alpha = \frac{1}{\eta} \frac{d\eta}{dT} \approx \frac{1}{T}, \tag{7-43}$$

in approximate agreement with the metal entries of Table 7–1. Strictly speaking, the preceding argument is valid only for temperatures above the Debye temperature of the metal (the temperature above which all the atomic vibration modes are excited). At temperatures somewhat below the Debye temperature, η drops below the linear relationship predicted by Eq. (7–42). At very low temperatures the contribution from η_1 cannot be neglected.

The addition of small amounts of a soluble impurity always increases resistivity. An alloy, which may be regarded as an impure metal, always has a higher resistivity than that of the lower resistivity parent metal (Fig. 7–9). The temperature coefficient α of an alloy is obviously lower than that of a pure metal just because its resistivity is higher, but certain alloys that have extremely small temperature coefficients have been developed.

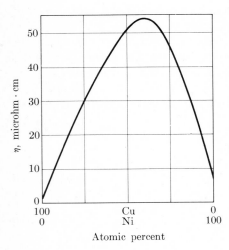

Figure 7–9 Resistivity of copper-nickel alloys as a function of composition at 20°C.

7–8 SUMMARY

The most important technological uses of electricity depend on currents of moving charges; these are also of fundamental importance to magnetism, as described in the next chapter. Defined locally at a point in space, the current density is

$$\mathbf{J} = \sum N_i q_i \mathbf{v}_i.$$

Since the charge density is

$$\rho = \sum N_i q_i,$$

the latter may vanish even though the former does not. This case (conduction, as opposed to convection) is the one considered here. The total current through a surface S is

$$I = \int_S \mathbf{J} \cdot \mathbf{n}\, da,$$

and

$$I = \frac{dQ}{dt}.$$

The conservation of charge is expressed locally by the equation of continuity

$$\mathbf{V} \cdot \mathbf{J} + \frac{\partial \rho}{\partial t} = 0.$$

(For the present we consider mainly steady currents, for which $\mathbf{V} \cdot \mathbf{J} = 0$, as well as $\partial \mathbf{J}/\partial t = 0$.) The conduction current in a medium is given by a constitutive equation, which in the simplest linear case defines the conductivity g:

$$\mathbf{J} = g\mathbf{E}.$$

1. The integral form of the linear constitutive equation is Ohm's law

$$V = IR;$$

the resistance of a straight conductor of uniform cross section is

$$R = l/gA.$$

2. In a continuous conducting medium with steady currents the potential obeys the Laplace equation,

$$\nabla^2 \varphi = 0.$$

The boundary conditions on $\mathbf{E}$ are the same as in a dielectric medium, and those on $\mathbf{J}$ are similar to the ones on $\mathbf{D}$. Consequently, for two conductors embedded in an infinite medium,

$$RC = \frac{\epsilon}{g}. \qquad \text{(mks units)}$$

3. If the volume charge density in a conducting medium is not zero initially, it disappears with a time constant

$$t_c = \frac{\epsilon}{g}. \qquad \text{(mks units)}$$

For metals this time is of the order of 10^{-14} s; for poorer conductors it may be up to many months.

4. For electric circuits, the static equations $\mathbf{V} \cdot \mathbf{J} = 0$ and $\mathbf{V} \times \mathbf{E} = 0$ become Kirchhoff's two laws,

$$\sum I_j = 0 \text{ at a junction,}$$

$$\sum V_j = 0 \text{ around a loop.}$$

In order to supply the power dissipated in resistors by steady currents, certain voltages must be applied by devices, e.g., batteries, whose operation cannot be described within the framework of electrostatics. Then, with Ohm's law, a complete solution of the circuit problem is straightforward.

5. The microscopic theory of Ohmic conduction depends on the existence of a linear retarding force that acts on free charges in the medium, in addition to the accelerating electric force. Expressed in terms of a relaxation time τ, this gives

$$g = \frac{Nq^2\tau}{m}.$$

The time τ is the time constant for the local establishment of an Ohmic current after the application of the field; in practical cases τ is short (10^{-14} s for metals). For good electronic conductors (metals, semiconductors), τ is interpreted as the mean time between collisions. In these cases it depends on the electronic mean free path according to

$$\tau = l/v_T,$$

where v_T is the random thermal velocity (not the net drift velocity).

PROBLEMS

7–1 The maximum rated current for a copper wire of cross-sectional area 2 mm^2 is 20 A. (a) What is the corresponding current density in A/m^2? (b) On the assumption that each copper atom contributes one conduction electron, calculate the electronic drift velocity corresponding to this current density. (Avogadro's number: $N_0 = 6.02 \times 10^{23}$ atoms per mole; atomic weight of copper: 63.5; density of copper: 8.92 g/cm^3.) (c) Use the observed conductivity to calculate the average collision time for an electron in copper.

7–2 The conductivity of seawater is approximately 4.3 $(\Omega\text{m})^{-1}$. Find the current density in a cell 1 cm long with 1 cm^2 cross-sectional area when 3 V is applied. Calculate the average drift velocity, assuming that the concentration of ions is 2 percent.

7–3 Two infinite, plane, parallel plates of metal are separated by the distance d. The space between the plates is filled with two conducting media, the interface between the media being a plane which is parallel to the metal plates. The first medium (conductivity g_1, permittivity ϵ_1) is of thickness a, and the second (conductivity g_2, permittivity ϵ_2) is of thickness $d - a$. The metal plates are maintained at potentials φ_1 and φ_2, respectively. In the steady state, what is the potential of the interface separating the two media, and what is the surface density of charge on this interface?

7–4 A system of charges and currents is completely contained *inside* the fixed volume V. The dipole moment of the charge-current distribution (see Section 2–9) is defined by

$$\mathbf{p} = \int_V \mathbf{r}\rho \, dv,$$

where $\mathbf{r}$ is the position vector from a fixed origin. Prove that

$$\int_V \mathbf{J} \, dv = \frac{d}{dt} \mathbf{p}.$$

(*Hint:* First prove the identity

$$\int_V \mathbf{J} \, dv = \oint_S \mathbf{r}\mathbf{J} \cdot \mathbf{n} \, da - \int_V \mathbf{r}\nabla \cdot \mathbf{J} \, dv,$$

and note that $\mathbf{J}$ vanishes on the surface S.)

7–5 A conducting medium is in a uniform field $\mathbf{E}_0$. A spherical cavity of radius a is formed in the medium. (a) Find the potential inside and outside the cavity. (b) Find the surface charge that appears on the cavity. (c) Make a sketch of the field lines.

7–6 A parallel plate capacitor is filled with a material of dielectric constant K and conductivity g. It is charged with an initial charge Q. (a) Show that the charge leaks off the plates as an exponential function of time. (b) Show that the total Joule heat production equals the electrostatic energy stored initially. (c) What is the time constant for the discharge if the material is silicon oxide? (See Tables 4–1 and 7–1.)

7–7 Two long cylindrical shells of metal (radii r_1 and r_2, with $r_2 > r_1$) are arranged coaxially. The plates are maintained at the potential difference $\Delta\varphi$. (a) The region between the shells is filled with a medium of conductivity g. Use Ohm's law, $\mathbf{J} = g\mathbf{E}$, to calculate the electric current between unit lengths of the shells. (b) If the region between the shells is filled with a nonconducting medium of permittivity ϵ, the capacitance of the system may be computed from the definition $C = Q/\Delta\varphi$. Show explicitly for this geometry that the product of resistance per unit length and capacitance per unit length $= \epsilon/g$.

7–8 The leakage resistance of a rubber cable insulation is measured in the following way: a length l of the insulated cable is immersed in a salt-water solution, a potential difference is applied between the cable conductor and the solution, and the resulting cable current is measured. In a particular case 3 m of cable is immersed in the solution; with 200 volts between cable conductor and solution the current measured is 2×10^{-9} A. The insulation thickness is equal to the radius of the central conductor. What is the electrical resistivity of the insulation?

7–9 A long copper wire of radius a is stretched parallel to and at distance h from an infinite copper plate. The region above the plate and surrounding the wire is filled with a medium of conductivity g. Show that the electrical resistance between the two copper electrodes, per unit length of the wire, is given by

$$R = \frac{1}{2\pi g} \cosh^{-1} \frac{h}{a}.$$

7–10 A homogeneous, isotropic sphere of conductivity g is subjected to a potential $\varphi_0 \cos \theta$ at all points on its surface. Here θ is the usual polar angle measured with respect to an axis through the center of the sphere. Determine the current density $\mathbf{J}$ at all points inside the sphere.

7–11 Two cylindrical copper electrodes of radius a are oriented normal to a silicon disk of thickness s, and are separated axially by the distance b. The electrodes are embedded in the disk to the depth s; in other words, they go completely through the disk. The lateral dimensions of the disk are large compared with b, and may be considered infinite. Taking the conductivity of silicon to be g, find the current between the electrodes when their potential difference is $\Delta\varphi$.

*7–12 A square copper plate of length $20a$, thickness s, and conductivity g is subjected to a potential difference: two opposite edges of the plate are maintained at the potentials φ_0 and $-\varphi_0$, respectively. (a) What is the electrical resistance of the plate? (b) A small hole of radius a is drilled through the plate at its center. Determine the *approximate* fractional change in resistance. (*Hint:* Find the potential distribution in the plate with the aid of the cosine θ cylindrical harmonics. Unfortunately, this distribution is not quite correct, because the two opposite edges of the square are not exact equipotentials. An approximate solution is obtained by taking the average potential of the two edges equal to $\pm\varphi_0$.)

7–13 Calculate the ratio of power dissipation to surface area for the conductors described in (a) Problem 7–1 and (b) Problem 7–2.

7–14 Given three resistors of 1 Ω, 2 Ω, and 3 Ω. Find the sixteen different resistances that can be made with these resistors.

7–15 A 0.4-watt lamp bulb is designed for operation with 2 volts across its terminals. A resistance R is placed in parallel with the bulb, and the combination is put in series with a 3-ohm resistor and a 3-volt battery (internal resistance, $\frac{1}{3}$ ohm). What should the value of R be if the lamp is to operate at design voltage?

*7–16 A resistance line, total resistance nR, is connected between the potential φ_0 and ground (ground is reference potential). The line is supported by $n - 1$ poles at equal resistance intervals such that the line resistance between poles is R. The leakage resistance to ground at each pole is βR. If φ_m is the line potential at the mth pole, show that

$$\varphi_{m+1} - (2 + \beta^{-1})\varphi_m + \varphi_{m-1} = 0.$$

7–17 Two batteries with open circuit voltage $\mathscr{V}_1$ and $\mathscr{V}_2$, and with internal resistances R_1 and R_2, respectively, are connected in parallel with each other and with the load resistance R. (a) Find the current through the load. (b) If the load resistance is varied and other quantities kept fixed, what should R be in order that it dissipate maximum power?

7–18 A group of n identical cells of open circuit voltage $\mathscr{V}_0$ and internal resistance R_I is used to supply current to a load resistor R. Show that if the n cells are connected in series with each other and with R, then $I = n\mathscr{V}_0/(R + nR_I)$, whereas if the cells are connected in parallel and the combination put in series with R, then $I = \mathscr{V}_0/(R + R_I/n)$.

7–19 Six identical resistors (R) are joined to form a hexagon. Six more resistors (all again of the same resistance R) are connected between the six vertices and the center of the hexagon. (a) What is the equivalent resistance between opposite vertices? (b) between adjacent vertices?

7–20 Six resistors form the sides of a tetrahedron. Five of the resistors are identical (R), the sixth is R_1. A potential difference is applied across one of the resistors adjoining R_1. Show that the Joule heat production in R_1 is maximum when $R_1 = (3/5)R$.

7-21 A Wheatstone-bridge circuit is obtained from the circuit of Fig. 7-8 by making $\mathcal{V}_2 = 0$, and substituting a galvanometer R_g for R_2. We shall also take $R_1 = 0$. The balance condition of the bridge (no current through the galvanometer) is obtained when $R_3 R_6 = R_4 R_5$. Thus an unknown resistance, for example R_6, may be determined in terms of known resistances: $R_6 = R_4 R_5/R_3$ at balance. (a) Find the current through the galvanometer when the bridge is off balance. (b) Assume that the bridge is to be balanced by varying R_4. The sensitivity of the bridge is defined by $S = CR_4(\partial I_2/\partial R_4)_0$, where C is the galvanometer deflection per unit current, and the subscript zero means that the derivative is to be evaluated at balance. Show that

$$S = \frac{C\mathcal{V}_1}{R_3 + R_4 + R_5 + R_6 + R_g(1 + R_5/R_6)(1 + R_4/R_3)}.$$

*7-22 The Wheatstone bridge of the preceding problem is nearly balanced. Let $R_5/R_3 = \alpha$, and $R_6/R_4 = \alpha(1 - \epsilon)$, where $\epsilon \ll 1$. If the resistance R_g is negligible, show that $I_2/I_1 = \alpha\epsilon/(\alpha + 1)^2$.

*7-23 A resistance of approximately 10 ohms is to be measured in the Wheatstone-bridge circuit of Problem 7-21. A large selection of standard resistances is available. The maximum power allowed in the bridge is 5 W. If $R_g = 100\ \Omega$, and the galvanometer will just detect a signal of 4×10^{-9} A, what is the highest precision one can obtain in measuring the unknown resistor? Assume that the standard resistors are exact, and do not limit the accuracy.

*7-24 A linear, conducting medium is connected at n points to electrodes with the fixed potentials: $\varphi_1, \varphi_2, \ldots, \varphi_n$. Show that the Joule heat production in the medium is given by $\sum_{i=1}^{n} \varphi_i I_i$, where I_i is the current *entering* the medium through electrode i.

CHAPTER 8 The Magnetic Field of Steady Currents

The second kind of field that enters into the study of electricity and magnetism is, of course, the magnetic field. Such fields or, more properly, the effects of such fields have been known since ancient times, when the effects of the naturally occurring permanent magnet magnetite (Fe_3O_4) were first observed. The discovery of the north- and south-seeking properties of this material had a profound influence on early navigation and exploration. Except for this application, however, magnetism was a little used and still less understood phenomenon until the early nineteenth century, when Oersted discovered that an electric current produced a magnetic field. This work, together with the later work of Gauss, Henry, Faraday, and others, brought the magnetic field into prominence as a partner to the electric field. The theoretical work of Maxwell and others (see Chapters 11 and 16) has shown that this partnership is real, and that the electric and magnetic fields are inextricably intertwined. The efforts of practical men have resulted in the development of the electrical machinery, communications equipment, and computers which involve magnetic phenomena and play such an important role in our everyday life. In this chapter the basic definitions of magnetism will be given, the production of magnetic fields by steady currents will be studied, and some important groundwork for future developments will be laid.

8-1 THE DEFINITION OF MAGNETIC INDUCTION

In Chapter 2 the Coulomb force on a charge q located at $\mathbf{r}$ due to a charge q_1 at the origin was given by

$$\mathbf{F}_e = \frac{1}{4\pi\epsilon_0} \frac{qq_1}{r^2} \frac{\mathbf{r}}{r}. \tag{8-1}$$

Implicit in this discussion was the assumption that the two charges were at rest. If the charges are uniformly moving, with velocities $\mathbf{v}$ and $\mathbf{v}_1$, respectively, there is an additional *magnetic force* $\mathbf{F}_m$ exerted on q by q_1,

$$\mathbf{F}_m = \frac{\mu_0}{4\pi} \frac{qq_1}{r^2} \mathbf{v} \times \left(\mathbf{v}_1 \times \frac{\mathbf{r}}{r} \right). \tag{8-2}$$

The number $\mu_0/4\pi$ plays the same role here as $1/4\pi\epsilon_0$ played in electrostatics, i.e., it is the constant that is required to make an experimental law compatible with a set of units. In mks units, by definition

$$\frac{\mu_0}{4\pi} = 10^{-7} \text{N} \cdot \text{s}^2/\text{C}^2$$

exactly, and this leads to the primary definition of the coulomb. (See Section 8–3.) Also as in the case of the electrostatic force, it is convenient to abstract from the properties of the "test charge" by defining a magnetic *field*; in this case not only the test charge q but also its velocity $\mathbf{v}$ must be factored out:

$$\mathbf{F}_m = q\mathbf{v} \times \mathbf{B}, \tag{8–3}$$

where the *magnetic induction* $\mathbf{B}$ is

$$\mathbf{B} = \frac{\mu_0}{4\pi} \frac{q_1}{r^2} \mathbf{v}_1 \times \frac{\mathbf{r}}{r}. \tag{8–4}$$

If a number of moving source charges is present, the magnetic forces and fields are additive. Some sort of limiting process should be included in the definition of $\mathbf{B}$ also, to ensure that the test charge does not affect the sources of $\mathbf{B}$. The unit for magnetic induction in the mks system according to Eq. (8–3) is the newton-second per coulomb-meter, called the *tesla* (T). If both an electric field and a magnetic field are present, the total force on a moving charge is $\mathbf{F}_e + \mathbf{F}_m$,

$$\mathbf{F} = q(\mathbf{E} + \mathbf{v} \times \mathbf{B}), \tag{8–5}$$

known as the *Lorentz force*.

The magnetic force between two charges is more complicated than the electric force because of the velocity dependence and the cross products. First, the similarities are that the magnitude of both forces depends on the product of the charges and the inverse square of their separation (in addition to a dimensional constant). The direction of the magnetic force, however, is not along the line joining the particles (i.e., it is not a central force), unless $\mathbf{v}$ happens to be perpendicular to $\mathbf{r}$; the force is always in the plane defined by $\mathbf{r}$ and $\mathbf{v}_1$. More importantly, the force is always perpendicular to $\mathbf{v}$; from Eq. (8–3), $\mathbf{v} \cdot \mathbf{F}_m = 0$ for any field $\mathbf{B}$, so that a magnetic force never does any work on a charged particle. A further comparison between $\mathbf{F}_m$ and $\mathbf{F}_e$ is facilitated if we multiply numerator and denominator of Eq. (8–2) by ϵ_0. Comparison of the result with Eq. (8–1) shows that $\epsilon_0 \mu_0$ must have the dimensions of an inverse velocity squared. Let us write

$$\epsilon_0 \mu_0 = \frac{1}{c^2}, \tag{8–6}$$

where c has the dimensions of a velocity, so that

$$\mathbf{F}_m = \frac{1}{4\pi\epsilon_0} \frac{qq_1}{r^2} \frac{\mathbf{v}}{c} \times \left(\frac{\mathbf{v}_1}{c} \times \frac{\mathbf{r}}{r} \right).$$

Using the defined value of μ_0 and the experimental value of ϵ_0, one finds that

$$c = 2.9979 \times 10^8 \text{ m/s},$$

which is numerically just the experimental velocity of light.* In Chapter 16 we shall see that this numerical coincidence is not accidental, but is a necessary consequence if light is an electromagnetic wave. Here we need not delve into the meaning of the relationship, but simply use the experimental fact. It means that for a given pair of particles

$$\frac{F_m}{F_e} \lesssim \frac{v}{c} \frac{v_1}{c}.$$

That is, if the particle velocities are small compared to the velocity of light, the magnetic interaction is very much smaller than the electric interaction. In fact, Eqs. (8–1), (8–2), and (8–4) are only first approximations to the correct relativistic expressions to be derived in Chapter 21, and they hold only for $v_1 \ll c$. We may note that the fields produced by the uniformly moving charge q_1 are related by

$$\mathbf{B} = \frac{\mathbf{v}_1}{c} \times \frac{\mathbf{E}}{c}.$$

(This relation happens to hold for arbitrarily large velocities, even though $\mathbf{E}$ and $\mathbf{B}$ both become modified when v_1 is comparable to c.) Finally, it is remarkable that the magnetic force does not depend only on the *relative* velocity of the two charges, but is different in a moving coordinate system†; and that it does not simply change sign when the particle labels are exchanged. These aspects need not concern us now, however, since they cancel out in the applications that are to be made in this and the following chapters.

Since $F_m \ll F_e$, it might seem at first glance that the magnetic force could always be neglected in comparison to the electric one, but there are systems of particles where this is not so. Notably in a conduction current, where positive and negative charges are present in equal densities, the macroscopic electric field is zero, but the magnetic field of the moving charges is not. This is the case in

---

* Equation (8–6) must hold in any consistent system of units. In gaussian units, where $\epsilon_0 = 1/4\pi$ by definition, $\mu_0/4\pi = 1/c^2$ is an experimental value. A more troublesome difference between the two systems of units is that in gaussian units the two c's are split up with the two v's, so that one defines

$$\mathbf{B} = \frac{q_1}{r^2} \frac{\mathbf{v}_1}{c} \times \frac{\mathbf{r}}{r} \qquad \text{and} \qquad \mathbf{F}_m = q \frac{\mathbf{v}}{c} \times \mathbf{B}.$$

This has the advantage that $\mathbf{B}$ is dimensionally the same as $\mathbf{E}$ (and that the relativistic form v/c appears explicitly).

† In particular it vanishes in a coordinate system moving with $\mathbf{v}$. This dependence on the coordinate system contradicts the assumption of classical mechanics that forces are the same in all inertial coordinate systems. It is our first hint that the theory of relativity is necessary to accommodate electromagnetism.

electromagnets, motors, transformers, and other situations where magnetic forces have very great practical importance. We therefore begin with the study of magnetic interactions between conduction currents. In the next section we discuss the force on a conduction current in an existing magnetic field, and in Section 8–3 we discuss the production of a magnetic field by a given conduction current.

8–2 FORCES ON CURRENT-CARRYING CONDUCTORS

From the Lorentz force (Eq. 8–5), an expression for the force on an element $d\mathbf{l}$ of a current-carrying conductor can be found. If $d\mathbf{l}$ is an element of conductor with its sense taken in the direction of the current I that it carries, then $d\mathbf{l}$ is parallel to the velocity $\mathbf{v}$ of the charge carriers in the conductor. If there are N charge carriers per unit volume in the conductor, the force on the element $d\mathbf{l}$ is

$$dF = NA\,|d\mathbf{l}|\,q\mathbf{v} \times \mathbf{B}, \tag{8–7}$$

where A is the cross-sectional area of the conductor and q is the charge per charge carrier. If several kinds of charge carriers are involved, then a summation must be included in Eq. (8–7); however, the final result, Eq. (8–8), is unchanged. Since $\mathbf{v}$ and $d\mathbf{l}$ are parallel, an alternative form of Eq. (8–7) is

$$d\mathbf{F} = Nq\,|\mathbf{v}|\,A\,d\mathbf{l} \times \mathbf{B}; \tag{8–7}$$

however, $Nq\,|\mathbf{v}|\,A$ is just the current for a single species of carrier. Therefore the expression

$$d\mathbf{F} = I\,d\mathbf{l} \times \mathbf{B} \tag{8–8}$$

is written for the force on an infinitesimal element of a charge-carrying conductor.

Equation (8–8) can be integrated to give the force on a complete (or closed) circuit. If the circuit in question is represented by the contour C, then

$$\mathbf{F} = \oint_C I\,d\mathbf{l} \times \mathbf{B}. \tag{8–9}$$

So long as $\mathbf{B}$ depends on position, the only simplification that can be made in Eq. (8–9) is to factor I from under the integral sign. If, however, $\mathbf{B}$ is uniform, i.e., independent of position, then it too can be removed from under the integral, to give

$$\mathbf{F} = I\left\{\oint_C d\mathbf{l}\right\} \times \mathbf{B}.$$

The remaining integral is easy to evaluate. Since it is the sum of infinitesimal vectors forming a complete circuit, it must be zero. Thus

$$\mathbf{F} = \oint_C I\,d\mathbf{l} \times \mathbf{B} = 0 \qquad (\text{B uniform}). \tag{8–10}$$

Another interesting quantity is the torque on a complete circuit. Since torque is moment of force, the infinitesimal torque $d\tau$ is given by

$$d\tau = \mathbf{r} \times d\mathbf{F} = I\mathbf{r} \times (d\mathbf{l} \times \mathbf{B}).\qquad(8\text{--}11)$$

The torque on a complete circuit is

$$\tau = I \oint_C \mathbf{r} \times (d\mathbf{l} \times \mathbf{B}).\qquad(8\text{--}12)$$

Once again, unless $\mathbf{B}$ is uniform no further simplification can be made; however, if it is uniform a straightforward expansion is accomplished by writing

$$d\mathbf{l} \times \mathbf{B} = \mathbf{i}(dyB_z - dzB_y) + \mathbf{j}(dzB_x - dxB_z) + \mathbf{k}(dxB_y - dyB_x).\qquad(8\text{--}13)$$

From these components the components of $\mathbf{r} \times (d\mathbf{l} \times \mathbf{B})$ are readily found to be

$$[\mathbf{r} \times (d\mathbf{l} \times \mathbf{B})]_x = y\,dxB_y - y\,dyB_x - z\,dzB_x + z\,dxB_z,$$
$$[\mathbf{r} \times (d\mathbf{l} \times \mathbf{B})]_y = z\,dyB_z - z\,dzB_y - x\,dxB_y + x\,dyB_x,\qquad(8\text{--}14)$$
$$[\mathbf{r} \times (d\mathbf{l} \times \mathbf{B})]_z = x\,dzB_x - x\,dxB_z - y\,dyB_z + y\,dzB_y.$$

Since $\mathbf{B}$ is assumed to be independent of $\mathbf{r}$ (uniform field) the components of $\mathbf{B}$ may be factored out of the integrals appearing in the expansion of Eq. (8–12). The spatial integrations which must be performed are of two general forms:

$$\oint \xi\,d\xi\qquad(8\text{--}15a)$$

and

$$\oint \xi\,d\eta,\qquad(8\text{--}15b)$$

where ξ represents any coordinate and η represents any coordinate different from ξ. The first of these is trivial because it represents the integral from some lower limit ξ_1 to some upper limit ξ_2 of $\xi\,d\xi$, plus the integral from ξ_2 to ξ_1 of $\xi\,d\xi$. Since interchanging the limits introduces a minus sign, the result is zero, which eliminates six terms from Eq. (8–14). Integrals of the form (8–15b) involve only two variables, ξ, η, hence it makes no difference whether the integral is taken around the actual curve C or around its projection on the ξ, η-plane, as shown in Fig. 8–1. By using the projection on the ξ, η-plane it is easy to see what Eq. (8–15b) represents. In Fig. 8–2 the ξ, η-plane is shown along with the infinitesimal area $\xi\,d\eta$. The integral can be written

$$\oint \xi\,d\eta = \int_a^b \xi_1(\eta)\,d\eta + \int_b^a \xi_2(\eta)\,d\eta.\qquad(8\text{--}16)$$

This, of course, gives just the area enclosed by the projected curve, and in the figure is positive. If ξ and η appear in cyclic order for a right-hand coordinate system, then the direction in which the contour is circled would give a normal in

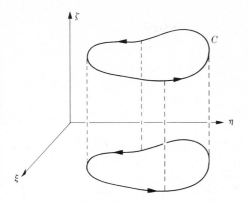

Figure 8–1 Projection of the curve C on the ξ, η-plane.

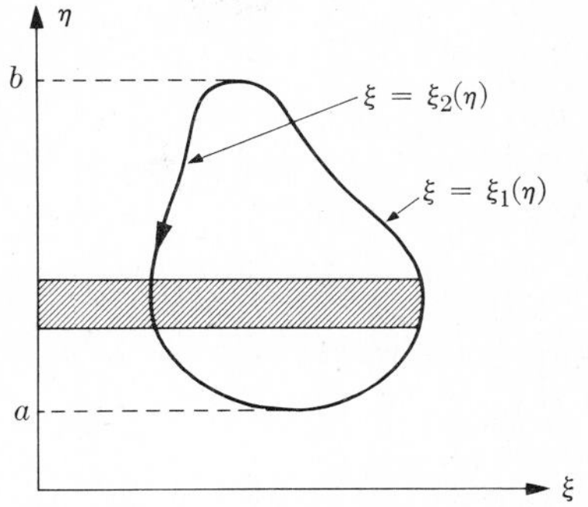

Figure 8–2 Evaluation of the integral $\int \xi \, d\eta$.

the positive ζ-direction. Thus we may write

$$\oint \xi \, d\eta = A_\zeta, \tag{8–17}$$

with ξ, η, ζ a cyclic permutation of x, y, z. Using this result to evaluate the integrals gives

$$\tau_x = I \oint_C [\mathbf{r} \times (d\mathbf{l} \times \mathbf{B})]_x = I(A_y B_z - A_z B_y), \tag{8–18}$$

with similar expressions for the y- and z-components. The three expressions are neatly summarized in the expression

$$\boldsymbol{\tau} = I\mathbf{A} \times \mathbf{B}, \tag{8–19}$$

where $\mathbf{A}$ is the vector whose components are the areas enclosed by projections of the curve C on the yz-, zx-, and xy-planes.* The quantity $I\mathbf{A}$ appears very frequently in magnetic theory and is referred to as the *magnetic moment* of the circuit. The symbol $\mathbf{m}$ will be used for magnetic moment:

$$\mathbf{m} = I\mathbf{A}, \tag{8–20}$$

with $\mathbf{A}$ defined as above.

It is easy to show, by the technique used above, that the integral of $\mathbf{r} \times d\mathbf{l}$ around a closed path gives twice the area enclosed by the curve. Thus

$$\tfrac{1}{2} \oint_C \mathbf{r} \times d\mathbf{l} = \mathbf{A}. \tag{8–21}$$

---

* Note that no restriction to plane curves has been imposed on C and that this definition of $\mathbf{A}$ makes any such restriction unnecessary.

This can be used to obtain

$$\mathbf{m} = \tfrac{1}{2}I \oint_C \mathbf{r} \times d\mathbf{l} \tag{8-22}$$

as an alternative expression for the magnetic moment. If, instead of being confined to wires, the current exists in a medium, then the identification

$$I\,d\mathbf{l} \to \mathbf{J}\,dv \tag{8-23}$$

is appropriate, as has been shown earlier. We then write

$$d\mathbf{m} = \tfrac{1}{2}\mathbf{r} \times \mathbf{J}\,dv, \tag{8-24}$$

which is useful in discussing the magnetic properties of matter.

8-3 THE LAW OF BIOT AND SAVART

In 1820, just a few weeks after Oersted announced his discovery that currents produce magnetic effects, Ampere presented the results of a series of experiments that may be generalized and expressed in modern mathematical language as

$$\mathbf{F}_2 = \frac{\mu_0}{4\pi} I_1 I_2 \oint_1 \oint_2 \frac{d\mathbf{l}_2 \times [d\mathbf{l}_1 \times (\mathbf{r}_2 - \mathbf{r}_1)]}{|\mathbf{r}_2 - \mathbf{r}_1|^3}. \tag{8-25}$$

This rather formidable expression can be understood with reference to Fig. 8–3. The force $\mathbf{F}_2$ is the force exerted on circuit 2 due to the influence of circuit 1, the $d\mathbf{l}$'s and $\mathbf{r}$'s are explained by the figure. By definition,

$$\frac{\mu_0}{4\pi} = 10^{-7}\ \text{N/A}^2$$

in mks units, and Eq. (8–25) serves as the primary definition of the ampere, in terms of which the coulomb is defined. Equation (8–25) appears, superficially, to

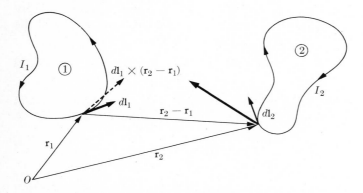

Figure 8–3 Magnetic interaction of two current circuits.

violate Newton's third law because of the lack of symmetry; however, by using some of the theorems of vector analysis it can be shown that it is actually symmetric, that is, $\mathbf{F}_2 = -\mathbf{F}_1$. (See Problem 8–4.)

From Eq. (8–9) it is apparent that Eq. (8–25) implies

$$\mathbf{B}(\mathbf{r}_2) = \frac{\mu_0}{4\pi} I_1 \oint_1 \frac{d\mathbf{l}_1 \times (\mathbf{r}_2 - \mathbf{r}_1)}{|\mathbf{r}_2 - \mathbf{r}_1|^3}. \tag{8–26}$$

This equation is a generalization of the Biot and Savart law,* which name will be used both for Eq. (8–26) and the differential form

$$d\mathbf{B}(\mathbf{r}_2) = \frac{\mu_0}{4\pi} \frac{I_1 \, d\mathbf{l}_1 \times (\mathbf{r}_2 - \mathbf{r}_1)}{|\mathbf{r}_2 - \mathbf{r}_1|^3}. \tag{8–27}$$

Equation (8–27) is an immediate consequence of Eq. (8–4) as applied to a conductor, if one uses the same argument that led to Eq. (8–7). As a last point, Eqs. (8–26) and (8–27) take the forms

$$\mathbf{B}(\mathbf{r}_2) = \frac{\mu_0}{4\pi} \int_V \frac{\mathbf{J}(\mathbf{r}_1) \times (\mathbf{r}_2 - \mathbf{r}_1)}{|\mathbf{r}_2 - \mathbf{r}_1|^3} \, dv_1 \tag{8–28}$$

and

$$d\mathbf{B}(\mathbf{r}_2) = \frac{\mu_0}{4\pi} \frac{\mathbf{J}(\mathbf{r}_1) \times (\mathbf{r}_2 - \mathbf{r}_1)}{|\mathbf{r}_2 - \mathbf{r}_1|^3} \, dv_1 \tag{8–29}$$

for a continuous distribution of current described by the current density $\mathbf{J}(\mathbf{r})$.

It is an experimental observation that all magnetic induction fields can be described in terms of a current distribution. That is, $\mathbf{B}$ always has the form of Eq. (8–28), with some $\mathbf{J}(\mathbf{r}_1)$. This implies that there are no isolated magnetic poles and that

$$\mathbf{\nabla} \cdot \mathbf{B} = 0. \tag{8–30}$$

Equation (8–30) is true for any $\mathbf{B}$ of the form (8–28) or (8–26), as can be verified mathematically: we take the divergence of Eq. (8–28). Using $\mathbf{\nabla} \cdot (\mathbf{F} \times \mathbf{G}) = -\mathbf{F} \cdot \mathbf{\nabla} \times \mathbf{G} + \mathbf{G} \cdot \mathbf{\nabla} \times \mathbf{F}$ gives

$$\mathbf{\nabla}_2 \cdot \mathbf{B}(\mathbf{r}_2) = -\frac{\mu_0}{4\pi} \int_V \mathbf{J}(\mathbf{r}_1) \cdot \mathbf{\nabla}_2 \times \frac{\mathbf{r}_2 - \mathbf{r}_1}{|\mathbf{r}_2 - \mathbf{r}_1|^3} \, dv_1.$$

However, $(\mathbf{r}_2 - \mathbf{r}_1)/|\mathbf{r}_2 - \mathbf{r}_1|^3$ is the gradient of $-1/|\mathbf{r}_2 - \mathbf{r}_1|$ with respect to $\mathbf{r}_2$. Since the curl of any gradient is zero, it follows that

$$\mathbf{\nabla}_2 \cdot \mathbf{B}(\mathbf{r}_2) = 0.$$

---

* In passing, we mention that there has been some controversy over the naming of various laws. We do not wish to enter into this controversy, but refer the interested reader to E. T. Whittaker's excellent history, *History of the Theories of Aether and Electricity*, Vol. I, Philosophical Library, New York, 1951.

8-4 ELEMENTARY APPLICATIONS OF THE BIOT AND SAVART LAW

The range of problems to which Eq. (8–28) (or Eq. 8–26) can be applied is limited primarily by the difficulty experienced in performing the integrations. Some of the tractable situations are considered in this section; in later sections other techniques for obtaining **B** will be considered.

As a first example, the magnetic field due to a long straight wire will be considered. The wire is imagined to lie along the x-axis from minus infinity to plus infinity and to carry a current I. The field will be computed at a typical point $\mathbf{r}_2$ on the y-axis. The geometry is best explained by Fig. 8–4. The magnetic induction is just

$$\mathbf{B}(\mathbf{r}_2) = \frac{\mu_0}{4\pi} I \int_{-\infty}^{\infty} \frac{dx\, \mathbf{i} \times (\mathbf{r}_2 - \mathbf{r}_1)}{|\mathbf{r}_2 - \mathbf{r}_1|^3}. \tag{8–31}$$

Since $\mathbf{r}_2 - \mathbf{r}_1$ lies in the xy-plane,

$$\mathbf{i} \times (\mathbf{r}_2 - \mathbf{r}_1) = |\mathbf{r}_2 - \mathbf{r}_1|\, \sin\theta\, \mathbf{k}. \tag{8–32}$$

Furthermore,

$$\frac{a}{x} = \tan(\pi - \theta) = -\tan\theta \tag{8–33}$$

and

$$|\mathbf{r}_2 - \mathbf{r}_1| = a \csc(\pi - \theta) = a \csc\theta. \tag{8–34}$$

Using these relationships to convert Eq. (8–31) to an integral on θ from 0 to π gives

$$\mathbf{B}(\mathbf{r}_2) = \frac{\mu_0}{4\pi} I\mathbf{k}\, \frac{1}{a} \int_0^{\pi} \sin\theta\, d\theta = \frac{\mu_0 I}{4\pi a} \mathbf{k}(-\cos\theta)\Big|_0^{\pi} = \frac{\mu_0 I}{2\pi a} \mathbf{k}. \tag{8–35}$$

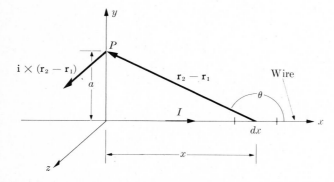

Figure 8–4 Magnetic field at point P due to a long straight wire.

To use this result more generally, it is only necessary to note that the problem exhibits an obvious symmetry about the x-axis. Thus we conclude that the lines of **B** are circles everywhere, with the conductor as center. This is in complete agreement with the elementary result that gives the direction of **B** by a right-hand rule.

As a second simple circuit, a circular turn will be considered. The magnetic field produced by such a circuit at an arbitrary point is very difficult to compute; however, if only points on the axis of symmetry are considered, the expression for **B** is relatively simple. In this example a complete vector treatment will be used to demonstrate the technique. Figure 8–5 illustrates the geometry and the coordinates to be used. The field is to be calculated at point $\mathbf{r}_2$ on the z-axis; the circular turn lies in the xy-plane. The magnetic induction is given by Eq. (8–26) in which, from Fig. 8–5, the following expressions are to be used:

$$d\mathbf{l} = a\, d\theta(-\mathbf{i} \sin\theta + \mathbf{j} \cos\theta),$$

$$\mathbf{r}_2 - \mathbf{r}_1 = -\mathbf{i}a \cos\theta - \mathbf{j}a \sin\theta + \mathbf{k}z, \tag{8-36}$$

$$|\mathbf{r}_2 - \mathbf{r}_1| = (a^2 + z^2)^{1/2}.$$

Substituting these into Eq. (8–26) yields

$$\mathbf{B}(z) = \frac{\mu_0 I}{4\pi} \int_0^{2\pi} \frac{(\mathbf{i}za \cos\theta + \mathbf{j}za \sin\theta + \mathbf{k}a^2)}{(z^2 + a^2)^{3/2}}\, d\theta. \tag{8-37}$$

The first two terms integrate to zero, leaving

$$\mathbf{B}(z) = \frac{\mu_0 I}{2} \frac{a^2}{(z^2 + a^2)^{3/2}}\, \mathbf{k}, \tag{8-38}$$

which is, of course, entirely along the z-axis.

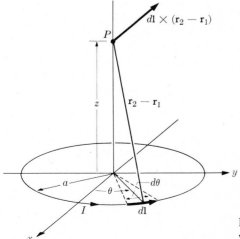

Figure 8–5 Axial field of a circular turn of wire.

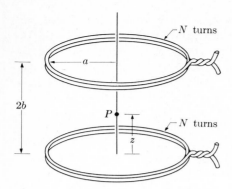

Figure 8–6 Axial field of a Helmholtz coil.

A frequently used current configuration is the Helmholtz coil, which consists of two circular coils of the same radius, with a common axis, separated by a distance chosen to make the second derivative of **B** vanish at a point on the axis halfway between the coils. Figure 8–6 shows such a configuration. The magnetic induction at point P is

$$B_z(z) = \frac{N\mu_0 I a^2}{2} \left\{ \frac{1}{(z^2 + a^2)^{3/2}} + \frac{1}{[(2b - z)^2 + a^2]^{3/2}} \right\}, \qquad (8\text{–}39)$$

which is obtained by applying Eq. (8–38) to each of the two coils. The factor N is included to handle the situation where each coil contains N turns. The first derivative of B_z with respect to z is

$$\frac{dB_z}{dz} = \frac{\mu_0 N I a^2}{2} \left\{ -\frac{3}{2} \frac{2z}{(z^2 + a^2)^{5/2}} - \frac{3}{2} \frac{2(z - 2b)}{[(2b - z)^2 + a^2]^{5/2}} \right\}. \qquad (8\text{–}40)$$

At $z = b$ this derivative vanishes. The second derivative with respect to z is

$$\frac{d^2 B_z}{dz^2} = -\frac{3\mu_0 N I a^2}{2} \left\{ \frac{1}{(z^2 + a^2)^{5/2}} - \frac{5}{2} \frac{2z^2}{(z^2 + a^2)^{7/2}} \right.$$

$$\left. + \frac{1}{[(2b - z)^2 + a^2]^{5/2}} - \frac{5}{2} \frac{2(z - 2b)^2}{[(2b - z)^2 + a^2]^{7/2}} \right\}.$$

At $z = b$ this reduces to

$$\frac{d^2 B_z}{dz^2}\bigg|_{z=b} = -\frac{3\mu_0 N I a^2}{2} \left\{ \frac{b^2 + a^2 - 5b^2 + b^2 + a^2 - 5b^2}{(b^2 + a^2)^{7/2}} \right\}, \qquad (8\text{–}41)$$

which vanishes if $a^2 - 4b^2 = 0$. Thus the appropriate choice for b is

$$2b = a, \qquad (8\text{–}42)$$

that is, the coil separation should equal the radius. With this separation, the magnetic induction at the midpoint is

$$B_z = \frac{\mu_0 N I}{a} \frac{8}{5^{3/2}}. \qquad (8\text{–}43)$$

Helmholtz coils play an important role in scientific research, where they are frequently used to produce a relatively uniform magnetic field over a small region of space. Let us consider the magnetic field at a point on the axis near the midpoint between the coils. The field $B_z(z)$ can be developed in a Taylor's series about the point $z = \frac{1}{2}a$:

$$B_z(z) = B_z(\tfrac{1}{2}a) + (z - \tfrac{1}{2}a) \left.\frac{\partial B_z}{\partial z}\right|_{z=\frac{1}{2}a} + \cdots$$

Since the first three derivatives vanish,

$$B_z(z) = B_z(\tfrac{1}{2}a) + \tfrac{1}{24}(z - \tfrac{1}{2}a)^4 \left.\frac{\partial^4 B_z}{\partial z^4}\right|_{z=\frac{1}{2}a} + \cdots$$

If the fourth derivative is evaluated explicitly, $B_z(z)$ can be written as

$$B_z(z) = B_z(a/2)\left\{1 - \frac{144}{125}\left(\frac{z - a/2}{a}\right)^4\right\}. \tag{8-44}$$

Thus for the region where $|z - a/2|$ is less than $a/10$, $B_z(z)$ deviates from $B_z(a/2)$ by less than one and a half parts in ten thousand.

The tesla is a rather large unit for measuring most laboratory fields; consequently the unit *gauss* for B from the Gaussian system* of units is commonly used: one gauss equals 10^{-4} tesla. For reference purposes we give

$$B_z = \frac{32\pi N}{5^{3/2}a}\frac{I}{10}, \qquad I \text{ in } \mathbf{ampere}, \qquad a \text{ in } \mathbf{cm}, \qquad B \text{ in } \mathbf{gauss}, \tag{8-43a}$$

for the induction at the midpoint of the Helmholtz coil. Of course N is still the number of turns in each of the two coils.

Another device to which Eq. (8–38) can be applied is the solenoid. A solenoid may be described as N turns uniformly wound on a cylindrical form of radius a and length L. Such a configuration is shown in Fig. 8–7. The magnetic induction at point z_0 is found by dividing the length L into elements dz, such as the one shown,

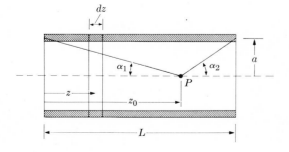

Figure 8–7 Axial magnetic field of a solenoid.

---

* This system of units is discussed in Appendix II.

and applying Eq. (8–38) to each element and summing the results. Noting that the element dz contains $N\,dz/L$ turns, we find that

$$B_z(z_0) = \frac{\mu_0 NI}{L}\frac{a^2}{2}\int_0^L \frac{dz}{[(z_0 - z)^2 + a^2]^{3/2}}. \tag{8-45}$$

The change of variable, $z - z_0 = a \tan\theta$, leads to

$$B_z(z_0) = \frac{\mu_0 NI}{2L}\int_{\theta_1}^{\theta_2}\cos\theta\,d\theta = \frac{\mu_0 NI}{L}\left[\frac{\sin\theta_2 - \sin\theta_1}{2}\right], \tag{8-46}$$

where $\theta_1 = -\tan^{-1}(z_0/a)$ and $\theta_2 = \tan^{-1}(L - z_0)/a$. The fact that sines appear rather than just ones, as in the elementary formula, represents end corrections. To help understand the approximation that is usually made, namely, $B_z = \mu_0 NI/L$, it is convenient to introduce the angles α_1 and α_2 (both positive) shown in Fig. 8–7. In terms of these angles, Eq. (8–46) becomes

$$B_z(z_0) = \frac{\mu_0 NI}{L}\left[\frac{\cos\alpha_1 + \cos\alpha_2}{2}\right]. \tag{8-47}$$

If the solenoid is long compared with its radius and z_0 is not too close to either zero or L, then α_1 and α_2 are both small angles and may be approximated by

$$\alpha_1 \cong \frac{a}{z_0}; \qquad \alpha_2 \cong \frac{a}{L - z_0}. \tag{8-48}$$

Maintaining quadratic terms in the expansions of $\cos\alpha_1$ and $\cos\alpha_2$, we obtain

$$B_z(z_0) \cong \frac{\mu_0 NI}{L}\left\{1 - \frac{a^2}{4z_0^2} - \frac{a^2}{4(L - z_0)^2}\right\}. \tag{8-49}$$

From this, we conclude that if $z_0 = L/2$ and $L/a = 10$, a 2 percent error results from neglecting the quadratic terms.

8–5 AMPERE'S CIRCUITAL LAW

For magnetic induction fields given by Eq. (8–26) or (8–28) that are due to steady currents, i.e., to currents that satisfy

$$\mathbf{V} \cdot \mathbf{J} = 0, \tag{8-50}$$

a very important equation for the curl of $\mathbf{B}$ can be derived. This is done by simply calculating the curl of Eq. (8–28). The curl involves differentiation with respect to $\mathbf{r}_2$, and hence operates only on the factor $(\mathbf{r}_2 - \mathbf{r}_1)/|\mathbf{r}_2 - \mathbf{r}_1|^3$:

$$\mathbf{V}_2 \times \mathbf{B}(\mathbf{r}_2) = \frac{\mu_0}{4\pi}\int_V \left[\mathbf{J}(\mathbf{r}_1)\left(\mathbf{V}_2 \cdot \frac{\mathbf{r}_2 - \mathbf{r}_1}{|\mathbf{r}_2 - \mathbf{r}_1|^3}\right) - \mathbf{J}(\mathbf{r}_1) \cdot \mathbf{V}_2 \frac{\mathbf{r}_2 - \mathbf{r}_1}{|\mathbf{r}_2 - \mathbf{r}_1|^3}\right] dv_1.$$

The derivative may now be changed to differentiation with respect to $\mathbf{r}_1$ (with a minus sign) in the second term because of the symmetry between $\mathbf{r}_2$ and $\mathbf{r}_1$.

$$\mathbf{V}_2 \times \mathbf{B}(\mathbf{r}_2) = \frac{\mu_0}{4\pi}\int_V \left[\mathbf{J}(\mathbf{r}_1)4\pi\,\delta(\mathbf{r}_2 - \mathbf{r}_1) - \mathbf{J}(\mathbf{r}_1) \cdot \mathbf{V}_1 \frac{\mathbf{r}_1 - \mathbf{r}_2}{|\mathbf{r}_1 - \mathbf{r}_2|^3}\right] dv_1.$$

The first term is expressed in terms of the Dirac delta function, as in Eq. (2–57); it integrates immediately to give $\mu_0 \mathbf{J}(\mathbf{r}_2)$. The second term can be shown to vanish by means of an integration by parts:

$$\mathbf{V}_1 \cdot \left(\mathbf{J} \frac{x_1 - x_2}{|\mathbf{r}_1 - \mathbf{r}_2|^3} \right) = \frac{x_1 - x_2}{|\mathbf{r}_1 - \mathbf{r}_2|^3} \mathbf{V}_1 \cdot \mathbf{J} + \mathbf{J} \cdot \mathbf{V}_1 \frac{x_1 - x_2}{|\mathbf{r}_1 - \mathbf{r}_2|^3}$$

for the x-component, and similarly for the other components. The term with $\mathbf{V} \cdot \mathbf{J}$ vanishes by the assumption (8–50), and the volume integral of the left side can be converted using the divergence theorem to a surface integral, which vanishes if the surface is chosen to lie outside a bounded region where $\mathbf{J}$ is nonvanishing. (The same result follows directly from the identity 1–2–4 of Table 1–2.) Thus the final result is

$$\mathbf{V} \times \mathbf{B}(\mathbf{r}_2) = \mu_0 \mathbf{J}(\mathbf{r}_2), \tag{8–51}$$

which will be called the *differential form* of Ampere's law. In Chapter 9 this will be modified so as to be more useful when magnetic materials are present; however, Eq. (8–51) is still valid so long as $\mathbf{J}$ is the *total* current and $\mathbf{V} \cdot \mathbf{J} = 0$.

Stokes's theorem can be used to transform Eq. (8–51) into an integral form, which is sometimes very useful. This application of Stokes's theorem is written

$$\int_S \mathbf{V} \times \mathbf{B} \cdot \mathbf{n} \, da = \oint_C \mathbf{B} \cdot d\mathbf{l}. \tag{1–45}$$

Using Eq. (8–51) for $\mathbf{V} \times \mathbf{B}$ gives

$$\oint_C \mathbf{B} \cdot d\mathbf{l} = \mu_0 \int_S \mathbf{J} \cdot \mathbf{n} \, da, \tag{8–52}$$

which simply says that the line integral of $\mathbf{B}$ around a closed path is equal to μ_0 times the total current through the closed path.

It is instructive to verify Eq. (8–52) for a simple case. The long straight wire provides a particularly good example. In this case $\mathbf{B}$ at a distance r from the conductor is given by $B(r) = \mu_0 I/2\pi r$, and it is tangential to a circle of radius r with center at the conductor. Figure 8–8 illustrates the geometry. The current is directed upward, and C is described in the counterclockwise direction. From the figure,

$$\mathbf{B} \cdot d\mathbf{l} = |\mathbf{B}| \, |d\mathbf{l}| \cos \chi = |\mathbf{B}| r \, d\theta. \tag{8–53}$$

With $|\mathbf{B}|$ as given above,

$$\oint_C \mathbf{B} \cdot d\mathbf{l} = \int_0^{2\pi} \frac{\mu_0 I}{2\pi r} r \, d\theta = \mu_0 I, \tag{8–54}$$

which represents a special case of Eq. (8–52).

Ampere's circuital law, as Eq. (8–52) is called, is in many ways parallel to Gauss's law in electrostatics. By this is meant that it can be used to obtain the magnetic field due to a certain current distribution of high symmetry without having to evaluate the complicated integrals that appear in the Biot law. As an

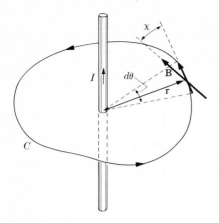

Figure 8–8 Verification of Ampère's circuital law for long, straight-wire geometry.

example, consider a coaxial cable consisting of a small center conductor of radius *a* and a coaxial cylindrical outer cable conductor of radius *b*, as shown in Fig. 8–9. Assume that the two conductors carry equal total currents of magnitude *I* in opposite directions, the center being directed out of the paper. From the symmetry of the problem it is clear that **B** must be everywhere tangent to a circle centered on the center conductor and drawn through the point at which **B** is being considered. Furthermore, **B** cannot depend on the azimuthal angle. The appropriate curves to use in the application of Eq. (8–52) are circles centered on the center conductor. For such a circle of radius *r*

$$\oint \mathbf{B} \cdot d\mathbf{l} = 2\pi r B, \tag{8–55}$$

which must equal μ_0 times the total current through the loop. Thus

$$2\pi r B = \mu_0 I, \qquad a < r < b$$

$$2\pi r B = 0, \qquad b < r. \tag{8–56}$$

This apparently trivial result can be obtained by integration of the Biot law only with considerable difficulty.

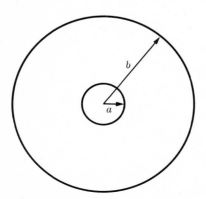

Figure 8–9 Cross section through a coaxial cable.

8–6 THE MAGNETIC VECTOR POTENTIAL

The calculation of electric fields was much simplified by the introduction of the electrostatic potential. The possibility of making this simplification resulted from the vanishing of the curl of the electric field. The curl of the magnetic induction does not vanish; however, its divergence does. Since the divergence of any curl is zero, it is reasonable to assume that the magnetic induction may be written

$$\mathbf{B} = \mathbf{V} \times \mathbf{A}. \tag{8–57}$$

The only other requirement placed on $\mathbf{A}$ is that

$$\mathbf{V} \times \mathbf{B} = \mathbf{V} \times \mathbf{V} \times \mathbf{A} = \mu_0 \mathbf{J}. \tag{8–58}$$

Using the identity

$$\mathbf{V} \times \mathbf{V} \times \mathbf{A} = \mathbf{V}\mathbf{V} \cdot \mathbf{A} - \mathbf{V}^2 \mathbf{A} \tag{8–59}$$

and specifying that $\mathbf{V} \cdot \mathbf{A} = 0$, yields

$$\mathbf{V}^2 \mathbf{A} = -\mu_0 \mathbf{J}. \tag{8–60}$$

Integrating each rectangular component and using the solution for Poisson's equation as a guide leads to

$$\mathbf{A}(\mathbf{r}_2) = \frac{\mu_0}{4\pi} \int_V \frac{\mathbf{J}(\mathbf{r}_1)}{|\mathbf{r}_2 - \mathbf{r}_1|} \, dv_1. \tag{8–61}$$

The integrals involved in this expression are much easier to evaluate than those involved in the Biot law; however, they are also more complicated than those used to obtain the electrostatic potential.

An alternative way of obtaining Eq. (8–61) is by the direct transformation of Eq. (8–28) to the form of Eq. (8–57). This is done by noting that

$$\frac{\mathbf{r}_2 - \mathbf{r}_1}{|\mathbf{r}_2 - \mathbf{r}_1|^3} = -\mathbf{V}_2 \frac{1}{|\mathbf{r}_2 - \mathbf{r}_1|}, \tag{8–62}$$

where $\mathbf{V}_2$ indicates that the differentiation is with respect to $\mathbf{r}_2$. The vector identity

$$\mathbf{V} \times (\varphi \mathbf{F}) = \varphi \mathbf{V} \times \mathbf{F} - \mathbf{F} \times \mathbf{V}\varphi, \tag{8–63}$$

which is valid for any vector $\mathbf{F}$ and any scalar φ, gives

$$\mathbf{V}_2 \times \left\{ \frac{1}{|\mathbf{r}_2 - \mathbf{r}_1|} \mathbf{J}(\mathbf{r}_1) \right\} = -\mathbf{J}(\mathbf{r}_1) \times \mathbf{V}_2 \frac{1}{|\mathbf{r}_2 - \mathbf{r}_1|}, \tag{8–64}$$

since $\mathbf{J}(\mathbf{r}_1)$ does not depend on $\mathbf{r}_2$. Combining these results in Eq. (8–28) leads to

$$\begin{aligned}
\mathbf{B}(\mathbf{r}_2) &= \frac{\mu_0}{4\pi} \int_V \mathbf{J}(\mathbf{r}_1) \times \frac{(\mathbf{r}_2 - \mathbf{r}_1)}{|\mathbf{r}_2 - \mathbf{r}_1|^3} \, dv_1 \\
&= -\frac{\mu_0}{4\pi} \int_V \mathbf{J}(\mathbf{r}_1) \times \mathbf{V}_2 \frac{1}{|\mathbf{r}_2 - \mathbf{r}_1|} \, dv_1 \\
&= \frac{\mu_0}{4\pi} \int_V \mathbf{V}_2 \times \frac{\mathbf{J}(\mathbf{r}_1)}{|\mathbf{r}_2 - \mathbf{r}_1|} \, dv_1.
\end{aligned} \tag{8–65}$$

The curl can be taken outside the integral, which puts Eq. (8–65) into exactly the form of Eq. (8–57). Thus

$$A(\mathbf{r}_2) = \frac{\mu_0}{4\pi} \int_{V_1} \frac{\mathbf{J}(\mathbf{r}_1)}{|\mathbf{r}_2 - \mathbf{r}_1|} \, dv_1 \tag{8–61}$$

results also from this approach.

To avoid leaving a false impression, namely, that the vector potential is as useful as the electrostatic potential in computing simple fields, it must be noted that there are essentially no cases where A can be computed in simple closed form (although it could always be done numerically for bounded current distributions). The long straight wire gives an infinite result for A when Eq. (8–61) is used.* The circular turn involves elliptic integrals, and so on. It should also be noted that evaluating the vector potential at a single point is not useful, because the magnetic induction is obtained by differentiation. The principal use of the vector potential is in approximations such as that discussed in the next section, and in problems involving electromagnetic radiation (see Chapters 16 and 20).

8–7   THE MAGNETIC FIELD OF A DISTANT CIRCUIT

The magnetic vector potential due to a small circuit at large distances can be evaluated relatively easily. The expression for the vector potential Eq. (8–61) may be applied to current circuits by making the substitution: $\mathbf{J} \, dv \rightarrow I \, d\mathbf{r}$. Thus

$$A(\mathbf{r}_2) = \frac{\mu_0 I}{4\pi} \oint \frac{d\mathbf{r}_1}{|\mathbf{r}_2 - \mathbf{r}_1|}. \tag{8–66}$$

For circuits whose dimensions are small compared with $\mathbf{r}_2$ the denominator can be approximated. To do so, we write, as in Eq. (2–46),

$$|\mathbf{r}_2 - \mathbf{r}_1|^{-1} = (r_2^2 + r_1^2 - 2\mathbf{r}_1 \cdot \mathbf{r}_2)^{-1/2} \tag{8–67}$$

and expand in powers of r_1/r_2 to get

$$|\mathbf{r}_2 - \mathbf{r}_1|^{-1} = \frac{1}{r_2} \left[1 + \frac{\mathbf{r}_1 \cdot \mathbf{r}_2}{r_2^2} + \cdots \right] \tag{8–68}$$

to first order in r_1/r_2. Using this in Eq. (8–66) gives

$$A(\mathbf{r}_2) = \frac{\mu_0 I}{4\pi} \left\{ \frac{1}{r_2} \oint d\mathbf{r}_1 + \frac{1}{r_2^3} \oint d\mathbf{r}_1(\mathbf{r}_1 \cdot \mathbf{r}_2) + \cdots \right\}. \tag{8–69}$$

The first integral vanishes; the second integrand is one term in the expansion

$$(\mathbf{r}_1 \times d\mathbf{r}_1) \times \mathbf{r}_2 = -\mathbf{r}_1(\mathbf{r}_2 \cdot d\mathbf{r}_1) + d\mathbf{r}_1(\mathbf{r}_1 \cdot \mathbf{r}_2). \tag{8–70}$$

---

* There is actually a finite vector potential for a long straight wire: $A = -(\mu_0 I/2\pi) \ln r \, \mathbf{k}$ in cylindrical coordinates for a wire on the z-axis carrying a current $I\mathbf{k}$. This can be verified by direct calculation of $\mathbf{V} \times \mathbf{A}$. (See Appendix III.)

To eliminate the first term on the right in Eq. (8–70), the differential of $\mathbf{r}_1(\mathbf{r}_2 \cdot \mathbf{r}_1)$ for a small change in $\mathbf{r}_1$ is written as

$$d[\mathbf{r}_1(\mathbf{r}_2 \cdot \mathbf{r}_1)] = \mathbf{r}_1(\mathbf{r}_2 \cdot d\mathbf{r}_1) + d\mathbf{r}_1(\mathbf{r}_2 \cdot \mathbf{r}_1), \tag{8–71}$$

which is of course exact. Adding Eqs. (8–70) and (8–71) and dividing by two yields

$$d\mathbf{r}_1(\mathbf{r}_1 \cdot \mathbf{r}_2) = \tfrac{1}{2}(\mathbf{r}_1 \times d\mathbf{r}_1) \times \mathbf{r}_2 + \tfrac{1}{2}\,d[\mathbf{r}_1(\mathbf{r}_2 \cdot \mathbf{r}_1)]. \tag{8–72}$$

Since the last term is an exact differential, it contributes nothing to the second integral in Eq. (8–69). Thus it follows that

$$\mathbf{A}(\mathbf{r}_2) = \frac{\mu_0}{4\pi} \left[\frac{I}{2} \oint \mathbf{r}_1 \times d\mathbf{r}_1 \right] \times \frac{\mathbf{r}_2}{r_2^3}. \tag{8–73}$$

Equation (8–22) defines the quantity in brackets as the magnetic moment, $\mathbf{m}$, of the circuit. Hence

$$\mathbf{A}(\mathbf{r}_2) = \frac{\mu_0}{4\pi} \frac{\mathbf{m} \times \mathbf{r}_2}{r_2^3}. \tag{8–74}$$

In this derivation it has been assumed that all $r_1 \ll r_2$; hence Eq. (8–74) is not valid for an arbitrary origin, but only for an origin close to the circuit.

The magnetic induction can be determined by taking the curl of Eq. (8–74). This is readily accomplished by using vector identities. First,

$$\mathbf{B}(\mathbf{r}_2) = \nabla \times \mathbf{A}(\mathbf{r}_2) = \frac{\mu_0}{4\pi} \nabla \times \left(\mathbf{m} \times \frac{\mathbf{r}_2}{r_2^3} \right)$$

$$= \frac{\mu_0}{4\pi} \left[-(\mathbf{m} \cdot \nabla) \frac{\mathbf{r}_2}{r_2^3} + \mathbf{m} \nabla \cdot \frac{\mathbf{r}_2}{r_2^3} \right]. \tag{8–75}$$

The first term in the brackets can be transformed by noting that

$$m_x \frac{\partial}{\partial x_2} \left(\frac{\mathbf{r}_2}{r_2^3} \right) = \frac{m_x \mathbf{i}}{r_2^3} - 3m_x x_2 \frac{\mathbf{r}_2}{r_2^5}; \tag{8–76}$$

hence

$$(\mathbf{m} \cdot \nabla) \frac{\mathbf{r}_2}{r_2^3} = \frac{\mathbf{m}}{r_2^3} - \frac{3(\mathbf{m} \cdot \mathbf{r}_2)\mathbf{r}_2}{r_2^5}. \tag{8–77}$$

The second term involves only the calculation of

$$\nabla \cdot \frac{\mathbf{r}_2}{r_2^3} = \frac{3}{r_2^3} - \mathbf{r}_2 \cdot \frac{3\mathbf{r}_2}{r_2^5} = 0. \qquad (r_2 \neq 0) \tag{8–78}$$

Finally,

$$\mathbf{B}(\mathbf{r}_2) = \frac{\mu_0}{4\pi} \left[-\frac{\mathbf{m}}{r_2^3} + \frac{3(\mathbf{m} \cdot \mathbf{r}_2)\mathbf{r}_2}{r_2^5} \right] \qquad \text{(magnetic dipole)}. \tag{8–79}$$

Equation (8–79) shows that the magnetic field of a distant circuit does not depend on its detailed geometry, but only on its magnetic moment **m**. Comparison with Eq. (2–36) shows that (8–79) is of the same form as the electric field due to an electric dipole, which explains the name *magnetic dipole* field. **m** is usually called the *magnetic dipole moment* of the circuit.

8–8 THE MAGNETIC SCALAR POTENTIAL

Equation (8–51) indicates that the curl of the magnetic induction is zero wherever the current density is zero. When this is the case, the magnetic induction in such regions can be written as the gradient of a scalar potential:

$$\mathbf{B} = -\mu_0 \nabla \varphi^*. \tag{8–80}$$

However, the divergence of **B** is also zero, which means that

$$\nabla \cdot \mathbf{B} = -\mu_0 \nabla^2 \varphi^* = 0. \tag{8–81}$$

Thus φ^*, which is called the magnetic scalar potential, satisfies Laplace's equation. Much of the work of electrostatics can be taken over directly and used to evaluate φ^* for various situations; however, care must be taken in applying the boundary conditions. (See Problem 8–25.)

The expression for the scalar potential of a magnetic dipole is particularly useful. If it is noted that Eq. (8–79) can be written

$$\mathbf{B}(\mathbf{r}_2) = -\mu_0 \nabla \left(\frac{\mathbf{m} \cdot \mathbf{r}_2}{4\pi r_2^3} \right), \tag{8–82}$$

then it is clear that

$$\varphi^*(\mathbf{r}_2) = \frac{\mathbf{m} \cdot \mathbf{r}_2}{4\pi r_2^3} \tag{8–83}$$

for a magnetic dipole **m**.

A large circuit C can be divided into many small circuits by means of a mesh, as shown in Fig. 8–10. If each small loop formed by the mesh carries the same

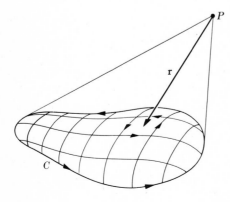

Figure 8–10 A macroscopic current circuit constructed from elemental magnetic dipoles.

current as originally was carried by the circuit C, then, because of the cancellation of currents in the common branch of adjacent loops, the net effect is the same as if the charge flowed only in the circuit C. For any one of the small loops, the magnetic moment may be written as

$$d\mathbf{m} = I\mathbf{n}\,da, \qquad (8\text{--}84)$$

since each of the loops is sufficiently small to be regarded as planar. Using this expression in Eq. (8-83) and integrating over the surface bounded by C gives

$$\varphi^*(P) = \frac{I}{4\pi}\int_S \frac{\mathbf{r}_2 \cdot \mathbf{n}\,da}{r_2^3}. \qquad (8\text{--}85)$$

In this equation $\mathbf{r}_2$ must be interpreted as the vector from da to the point P, that is, $-\mathbf{r}$, as shown in Fig. 8-10. Making the change $\mathbf{r}_2 = -\mathbf{r}$ results in

$$\varphi^*(P) = -\frac{I}{4\pi}\int_S \frac{\mathbf{r} \cdot \mathbf{n}\,da}{r^3}. \qquad (8\text{--}86)$$

The quantity $\mathbf{r} \cdot \mathbf{n}\,da$ is just r times the projection of da on a plane perpendicular to $\mathbf{r}$. Thus $\mathbf{r} \cdot \mathbf{n}\,da/r^3$ is the solid angle subtended by da at P. Equation (8-86) may then be written as

$$\varphi^*(P) = -\frac{I\Omega}{4\pi}, \qquad (8\text{--}87)$$

where Ω is the solid angle subtended by the curve C at the point P.

The magnetic scalar potential can be used for the calculation of the magnetic field due either to current-carrying circuits or to magnetic double layers (layers of dipoles). This procedure is occasionally useful in dealing with circuit problems; however, its principal use is in dealing with magnetic materials.

8–9 MAGNETIC FLUX

The quantity

$$\Phi = \int_S \mathbf{B} \cdot \mathbf{n}\,da \qquad (8\text{--}88)$$

is known as the *magnetic flux* and is measured in webers (Wb).† It is analogous to the electric flux discussed earlier, but it is of much greater importance. The flux through a closed surface is zero, as can be seen by computing

$$\oint_S \mathbf{B} \cdot \mathbf{n}\,da = \int_V \nabla \cdot \mathbf{B}\,dv = 0. \qquad (8\text{--}89)$$

From this it follows also that the flux through a circuit is independent of the particular surface used to compute the flux. Use will be made of these results in Chapter 11, when electromagnetic induction is discussed.

---

† Thus a tesla is equal to a weber/m^2, which was formerly used as the mks unit for $\mathbf{B}$.

8–10 SUMMARY

Magnetostatics is based on the addition of a magnetic force to the coulomb force when the charges are moving. In mks units, the Lorentz force on a test charge q with velocity $\mathbf{v}$ is

$$\mathbf{F} = q(\mathbf{E} + \mathbf{v} \times \mathbf{B}).$$

The magnetic field of a source charge q_1 moving uniformly with velocity $\mathbf{v}_1$ is

$$\mathbf{B} = \frac{\mathbf{v}_1}{c} \times \frac{\mathbf{E}}{c},$$

where $\mathbf{E}$ is the electric field produced by q_1 and

$$c = 1/\sqrt{\epsilon_0 \mu_0} \cong 3 \times 10^8 \text{ m/s}$$

is the velocity of light. (In gaussian units, B in these formulas is replaced by B/c.) The results are applied to conduction currents by writing

$$Nq \, dv \, \mathbf{v} = \mathbf{J} \, dv = I \, d\mathbf{l},$$

where $Nq = \rho$, $\rho\mathbf{v} = \mathbf{J}$ for the moving species of charged particle.

1. The force on an element of wire $d\mathbf{l}$ in a magnetic field $\mathbf{B}$ is

$$d\mathbf{F} = I \, d\mathbf{l} \times \mathbf{B}.$$

The torque on a circuit is for a constant B-field

$$\boldsymbol{\tau} = \mathbf{m} \times \mathbf{B},$$

where (mks units)

$$\mathbf{m} = \tfrac{1}{2} I \oint_C \mathbf{r} \times d\mathbf{l}$$

is the magnetic moment of the circuit. The integral $\frac{1}{2} \oint_C \mathbf{r} \times d\mathbf{l}$ is the vector whose components are the areas enclosed by the projections of the curve C on the coordinate planes.

2. The magnetic field produced by a current element $I \, d\mathbf{l}'$ is

$$d\mathbf{B}(\mathbf{r}) = \frac{\mu_0}{4\pi} \frac{I \, d\mathbf{l}' \times (\mathbf{r} - \mathbf{r}')}{|\mathbf{r} - \mathbf{r}'|^3},$$

where $\mu_0/4\pi = 10^{-7}$ N/A^2 in mks units; the field of a complete circuit is calculated by integrating this around the circuit. For a general current distribution $\mathbf{J}(\mathbf{r}')$ as source,

$$\mathbf{B}(\mathbf{r}) = \frac{\mu_0}{4\pi} \int_V \frac{\mathbf{J}(\mathbf{r}') \times (\mathbf{r} - \mathbf{r}')}{|\mathbf{r} - \mathbf{r}'|^3} \, dv'.$$

By differentiating this, we find there are no magnetic monopoles:

$$\mathbf{V} \cdot \mathbf{B} = 0.$$

$$\mathbf{V} \times \mathbf{B} = \mu_0 \mathbf{J}$$

for a steady current distribution with

$$\mathbf{V} \cdot \mathbf{J} = 0.$$

These are the fundamental differential equations that must be satisfied locally at every point by all magnetostatic fields. (The divergence equation is satisfied even by time dependent fields, and it is the second of the four fundamental Maxwell's equations.)

3. Ampere's law follows from the curl equation by integrating both sides over an arbitrary surface S and applying Stokes's theorem:

$$\oint_C \mathbf{B} \cdot d\mathbf{l} = \mu_0 I,$$

where

$$I = \int_S \mathbf{J} \cdot \mathbf{n} \, da$$

is the total current through S bounded by C. This has practical usefulness for calculating B in a few special situations of high symmetry, where it can be seen that $\mathbf{B}$ is constant in magnitude and direction relative to some suitable curve C.

4. The existence of the vector potential function $\mathbf{A}(\mathbf{r})$ follows from the divergence equation, such that

$$\mathbf{B} = \mathbf{V} \times \mathbf{A}.$$

For a specified current distribution,

$$\mathbf{A}(\mathbf{r}) = \frac{\mu_0}{4\pi} \int_V \frac{\mathbf{J}(\mathbf{r}')}{|\mathbf{r} - \mathbf{r}'|} \, dv'.$$

5. At a large distance from the region where the source currents $\mathbf{J}$ are located, the multipole expansion of $\mathbf{A}$ gives

$$\mathbf{A}(\mathbf{r}) = \frac{\mu_0}{4\pi} \frac{\mathbf{m} \times \mathbf{r}}{r^3} + \cdots.$$

(There is no monopole term.)

6. In regions where $\mathbf{J} = 0$, a scalar potential $\varphi^*(\mathbf{r})$ can be defined (since $\mathbf{V} \times \mathbf{B} = 0$), such that

$$\mathbf{B} = -\mu_0 \mathbf{V}\varphi^*.$$

Like the electrostatic potential, this satisfies Laplace's equation

$$\nabla^2 \varphi^* = 0,$$

but the different boundary conditions may bring different sets of solutions into play. The dipole solution is the same,

$$\varphi^*(\mathbf{r}) = \frac{\mathbf{m} \cdot \mathbf{r}}{4\pi r^3}.$$

PROBLEMS

8–1 A charged particle of mass m and charge q moves in a uniform magnetic induction field $\mathbf{B}_0$. Show that the most general motion of the particle traces out a helix, the cross section of which is a circle of radius $R = mv_\perp/qB$. (Here $v_\perp$ is the component of velocity of the particle which is perpendicular to $\mathbf{B}_0$.)

8–2 The Hamiltonian for a charged particle moving in a uniform magnetic induction field, $\mathbf{B}_0$, which is parallel to the z-axis, is given by

$$\mathscr{H} = \frac{1}{2m} p^2 - \frac{qB_0}{2m} (xp_y - yp_x) + \frac{q^2 B_0^2}{8m} (x^2 + y^2).$$

Show that the equations of motion that may be derived from $\mathscr{H}$ are consistent with the results of Problem 8–1.

8–3 A proton of velocity 10^7 m/s is projected at right angles to a uniform magnetic induction field of 0.1 T. (a) How much is the particle path deflected from a straight line after it has traversed a distance of 1 cm? (b) How long does it take the proton to traverse a $90°$ arc?

8–4 Show that the force law Eq. (8–25) can be transformed to

$$\mathbf{F}_2 = -\frac{\mu_0}{4\pi} I_1 I_2 \oint_1 \oint_2 d\mathbf{l}_2 \cdot d\mathbf{l}_1 \frac{\mathbf{r}_2 - \mathbf{r}_1}{|\mathbf{r}_2 - \mathbf{r}_1|^3},$$

which is clearly symmetric in that $\mathbf{F}_2 = -\mathbf{F}_1$.

8–5 Suppose a very long solenoid carries a (superconducting) current of 10 A and has 1000 turns per cm. Find the radial force per unit length f on one turn of the winding. Show that the tension in the wire T is $T = fa$, where a is the radius of the solenoid.

8–6 Show that the force between parallel wires carrying currents I_1 and I_2, both in the same direction, is one of attraction. If the two parallel wires are very long and separated by distance a, find the magnetic force on segment $d\mathbf{l}_2$ of wire 2.

8–7 Given a current circuit in the shape of a regular hexagon of side a. If the circuit carries the current I, find the magnetic induction at the center of the hexagon.

8–8 Given a thin strip of metal of width w and very long. The current in the strip is along its length; the total current is I. Find the magnetic induction in the plane of the strip at distance b from the nearer edge.

8–9 A large number N of closely spaced turns of fine wire is wound in a single layer upon the surface of a wooden sphere of radius a, with the planes of the turns perpendicular to the axis of the sphere and completely covering its surface. If the current in the winding is I, determine the magnetic field at the center of the sphere.

8–10 A solenoid 15 cm long is wound in two layers. Each layer contains 100 turns; the first layer is 2 cm in radius, the second 2.05 cm. If the winding carries a current of 3 A, find the magnetic induction at various points along the axis of the solenoid. Make a plot of the axial magnetic induction as a function of distance, from the center to one end of the solenoid.

8–11 A solenoid of square cross section (i.e., a solenoid in which the individual turns are in the shape of a square) has N turns per unit length and carries current I. The cross-sectional dimension is a. If the solenoid is very long, find the axial magnetic induction at its center.

8–12 The magnetic induction at a point on the axis (z-axis) of a circular turn of wire carrying current I is given in Eq. (8–38). Use the fact that $\mathbf{V} \cdot \mathbf{B} = 0$ to get an approximate expression for B_r (the radial component of the magnetic field) which is valid for points very near the axis.

8–13 The vertical component of the magnetic induction between the pole faces of a particle accelerator is given by $B_z = B_z(r, z)$, where $r = (x^2 + y^2)^{1/2}$ is the distance from the axis of the pole faces. (a) If $|B_z|$ is a decreasing function of r, show that the lines of magnetic intensity bow outward, as shown in Fig. 8–11, regardless of whether the upper pole is a north or south pole. (*Hint:* Use the fact that $\mathbf{V} \times \mathbf{B} = 0$, and that $B_r = 0$ on the median plane.) (b) If the lines of $\mathbf{B}$ bow as shown in the figure, show that accelerated particles that drift away from the median plane experience a force tending to restore them to the median plane, regardless of whether they are positively or negatively charged.

Figure 8–11

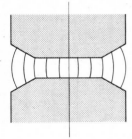

*8–14 It is evident from Eq. (8–30) that only a certain class of vector fields qualifies as a physically realizable magnetic induction field. (a) Verify that

$$\mathbf{B} = (\mathbf{r}/r) \times \nabla g(x, y, z),$$

with arbitrary $g(x, y, z)$ is an appropriate magnetic field; (b) find the current density $\mathbf{J}$ producing it, if g is a solution of Laplace's equation.

8–15 For a homogeneous, isotropic, nonmagnetic medium of conductivity g, in which there are steady currents, show that $\mathbf{B}$ satisfies the vector Laplace equation: $\nabla^2 \mathbf{B} = 0$.

8–16 By using Ampere's circuital law, find the magnetic induction at distance r from the center of a long wire carrying current I. Do so for both $r > R$ and $r < R$, where R is the radius of the wire. Show explicitly that the magnetic induction vanishes on the axis of the wire.

8–17 A cylindrical conductor of radius b contains a cylindrical hole of radius a; the axis of the hole is parallel to the axis of the conductor and at a distance s away from it,

$a < s < b - a$. The conductor carries a uniform current density **J**. Find the *B*-field in the hole on the diameter that coincides with a diameter of the conductor. (*Hint:* Treat an equivalent current distribution with density **J** throughout the hole as well as the conductor, plus $-$**J** in the hole.)

8–18 Assume that in a very long (infinite) solenoid the field is entirely in the *z*-direction, both inside and outside the solenoid. (a) Use Ampere's law to show that the field is uniform inside and outside. Thus, if **B** vanishes at an infinite distance from the axis, it is zero everywhere outside. (b) Use Ampere's law to find **B** inside. Show that the result agrees with Eq. (8–47) in the limit $a/L \to 0$.

8–19 A toroid is wound uniformly, as shown in Fig. 11–2. It has *N* turns of wire, which carry current *I*. The inner radius of the toroid is *a*, the outer *b*. Find the magnetic induction at various points inside the toroidal winding. Find the ratio b/a that will permit **B** in the ring to vary by no more than 25 percent.

8–20 Show that the magnetic vector potential for two long, straight, parallel wires carrying the same current, *I*, in opposite directions is given by

$$\mathbf{A} = \frac{\mu_0 I}{2\pi} \ln \left(\frac{r_2}{r_1} \right) \mathbf{n},$$

where r_2 and r_1 are the distances from the field point to the wires, and **n** is a unit vector parallel to the wires.

8–21 Given the following set of conductors: an infinitely long straight wire surrounded by a thin cylindrical shell of metal (at radius *b*) arranged coaxially with the wire. The two conductors carry equal but opposite currents, *I*. Find the magnetic vector potential for the system.

8–22 (a) Show that $\oint_C \mathbf{A} \cdot d\mathbf{l} = \Phi$, where Φ is the magnetic flux through a surface bounded by the circuit *C*. (b) Use this and the results of Problem 8–18 to find **A** at a distance *r* outside $(r > a)$ and inside $(r < a)$ a very long solenoid. (c) Check that $\nabla \times \mathbf{A} = \mathbf{B}$.

8–23 (a) Show by direct differentiation of Eq. (8–61) that $\nabla \cdot \mathbf{A} = 0$. (b) Show that $\mathbf{A} + \nabla \psi$, where ψ is an arbitrary function, is also a vector potential for the same *B*-field as **A**. (c) Show that, by a suitable choice of ψ, the vector potential of **B** can have any divergence desired.

8–24 Show that the following are all possible vector potentials of the uniform field, $\mathbf{B} = B\mathbf{k}$: $\mathbf{A}_1 = -By\mathbf{i}$, $\mathbf{A}_2 = Bx\mathbf{j}$, $\mathbf{A}_3 = -\frac{1}{2}\mathbf{r} \times \mathbf{B}$. For which of these is $\nabla \cdot \mathbf{A} = 0$? Show that $\mathbf{A}_1 - \mathbf{A}_2$ is the gradient of a function, $\nabla \psi$.

8–25 Show that the *B*-field outside of a long straight wire carrying a current *I* is derivable from the scalar potential

$$\varphi^* = -\frac{I}{2\pi} \theta$$

in cylindrical coordinates, and that φ^* satisfies Laplace's equation. Why is this φ^* not one of the cylindrical harmonics (as would be the case for the electrostatic potential of a line charge)?

8–26 The magnetic dip angle is defined as the angle between the direction of the magnetic induction and the tangent plane at the earth's surface. Derive an expression for the dip angle as a function of geomagnetic latitude, on the assumption that the induction is a dipole field.

*8–27 (a) Show that the magnetic scalar potential for a point on the axis (z-axis) of a circular loop, of radius a, is given by

$$\varphi^* = \tfrac{1}{2}I \left| 1 - \frac{z}{\sqrt{a^2 + z^2}} \right|.$$

(b) Expand this formula according to the binomial theorem to obtain a series expression valid for $z < a$.

(c) The magnetic scalar potential φ^* should satisfy Laplace's equation; furthermore, by symmetry, $\varphi^* = \varphi^*(r, \theta)$, where r is the distance from the center of the loop to the field point and θ is the angle between $\mathbf{r}$ and the z-axis. Show that by using the zonal harmonics, Eq. (3–18), a solution for φ^*, that reduces to the potential obtained in (b) on the symmetry axis, can be constructed.

(d) Use the φ^* obtained in (c) to find B_r and B_θ at points off the symmetry axis of the loop.

*8–28 A sphere of radius a carrying surface charge density σ (rigidly attached) is rotated about an axis through its center with constant angular velocity ω. Show that the magnetic field at an external point is a dipole field and find the equivalent dipole moment.

8–29 Two dipoles $\mathbf{m}_1$ and $\mathbf{m}_2$ are in the same plane; $\mathbf{m}_1$ is fixed but $\mathbf{m}_2$ is free to rotate about its center. Show that, for equilibrium, $\tan \theta_1 = -2 \tan \theta_2$, where θ_1, θ_2 are the angles between $\mathbf{r}$ and $\mathbf{m}_1, \mathbf{m}_2$ respectively ($\mathbf{r}$ is the vector displacement between $\mathbf{m}_2$ and $\mathbf{m}_1$).

CHAPTER 9 Magnetic Properties of Matter

In Chapter 8 we discussed techniques for finding the magnetic induction field due to a *specified* distribution of currents. Thus, for example, if we are dealing with a current-carrying circuit consisting of a closed loop of wire, the magnetic field in the vacuum region surrounding the wire may be calculated with the aid of Biot's Law. Now let the region surrounding the wire be filled with a material medium. Will the magnetic induction be altered by the presence of the matter? The answer is "yes."

All matter consists ultimately of atoms, and each atom consists of electrons in motion. These electron circuits, each of which is confined to a single atom, are what we shall call *atomic currents*. It thus appears that we have two kinds of current: (1) a conventional current, which consists of charge transport, i.e., the motion of free electrons or charged ions, and (2) atomic currents, which are pure circulatory currents and give rise to no charge transport. However, both kinds of current may produce magnetic fields.

9–1 MAGNETIZATION

Each atomic current is a tiny closed circuit of atomic dimensions, and it may therefore be appropriately described as a magnetic dipole. In fact, the dipole moment is the quantity of interest here, since the distant magnetic induction field due to a single atom is completely determined by specifying its magnetic dipole moment, $\mathbf{m}$.

Let the magnetic moment of the ith atom be $\mathbf{m}_i$. We now define a macroscopic vector quantity, the *magnetization* $\mathbf{M}$, by the same method used to define polarization in Chapter 4. We sum up vectorially all of the dipole moments in a small volume element Δv, and then divide the result by Δv; the resulting quantity,

$$\mathbf{M} = \lim_{\Delta v \to 0} \frac{1}{\Delta v} \sum_i \mathbf{m}_i, \qquad (9\text{–}1)$$

is called the magnetic dipole moment per unit volume, or simply the *magnetiza-tion*. The limit process in Eq. (9–1) is our usual macroscopic limit process; Δv is made very small from the macroscopic point of view, but not so small that it does not contain a statistically large number of atoms. The quantity $\mathbf{M}$ then becomes a vector point function. In the unmagnetized state, the summation $\sum \mathbf{m}_i$ will sum to zero as a result of random orientation of the $\mathbf{m}_i$, but in the presence of an external exciting field $\mathbf{M}$ usually depends on this field. The specific dependence of $\mathbf{M}$ on $\mathbf{B}$ will be taken up in Section 9–6.

For the moment, we shall assume that $\mathbf{M}(x, y, z)$ is a known function, and shall compute the magnetized material's contribution to the magnetic field from the equations developed in Section 8–7.

The vector function $\mathbf{M}$ provides us with a macroscopic description of the atomic currents inside matter. Specifically, $\mathbf{M}$ measures the number of atomic current circuits per unit volume times the average or effective magnetic moment of each circuit. From the purely macroscopic point of view all magnetic effects due to matter can be described adequately in terms of $\mathbf{M}$, or by its derivatives. One of these derivatives, $\mathbf{V} \times \mathbf{M}$, turns out to be the equivalent transport current density which would produce the same magnetic field as $\mathbf{M}$ itself; it is called the *magne-tization current density* $\mathbf{J}_M$. Before we derive this important relationship linking $\mathbf{J}_M$ and $\mathbf{M}$, let us look at a simplified model of magnetized matter as though it consisted of atomic loop currents circulating in the same direction, side by side (Fig. 9–1). If the magnetization is uniform, the currents in the various loops tend to cancel each other out, and there is no net effective current in the interior of the material. If the magnetization is nonuniform, the cancellation will not be complete. As an example of nonuniform magnetization, consider the abrupt change in magnetization shown in Fig. 9–2; if we focus our attention on the region between the dotted lines, it is evident that there is more charge moving down than there is moving up. This we call the *magnetization current*. Thus, even though there is no charge transport, there is an effective motion of charge downward, and this "current" can produce a magnetic field.

Figure 9–1 Simplified picture of magnetic material consisting of atomic loop currents circulating in the same direction.

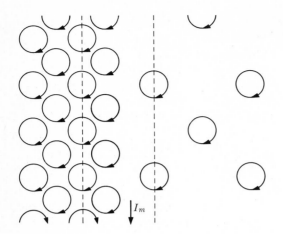

Figure 9–2 Example of abrupt change in magnetization.

It remains for us to derive the relationship between $\mathbf{J}_M$ and $\mathbf{M}$. Let us consider two small volume elements in a piece of magnetic material, each element of volume $\Delta x\, \Delta y\, \Delta z$, and located next to each other in the direction of the y-axis (Fig. 9–3). If the magnetization in the first volume element is $\mathbf{M}(x, y, z)$, then the magnetization in the second element is

$$\mathbf{M}(x,\ y,\ z) + \frac{\partial \mathbf{M}}{\partial y}\, \Delta y + \text{higher-order terms.}$$

The x-component of magnetic moment of the first element, $M_x\, \Delta x\, \Delta y\, \Delta z$, may be written in terms of a circulating current, I_c':

$$M_x\, \Delta x\, \Delta y\, \Delta z = I_c'\, \Delta y\, \Delta z. \tag{9–2}$$

Similarly, the x-component of magnetic moment of the second element, neglecting higher-order terms that vanish in the limit where each volume element becomes

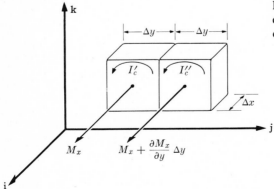

Figure 9–3 Replacement of volume elements of magnetized material by circulating currents I_c' and I_c''.

very small, is

$$\left(M_x + \frac{\partial M_x}{\partial y}\,\Delta y\right)\Delta x\,\Delta y\,\Delta z = I_c''\,\Delta y\,\Delta z. \tag{9-3}$$

The net upward current in the middle region of the two volume elements is

$$I_c' - I_c'' = -\frac{\partial M_x}{\partial y}\,\Delta x\,\Delta y. \tag{9-4}$$

We next consider two adjacent volume elements along the x-axis and focus our attention on the y-component of the magnetization in each cell. In the middle region of the two cells the net upward current due to the circulating currents that define the magnetic moments is

$$(I_c)_{\text{up}} = \frac{\partial M_y}{\partial x}\,\Delta x\,\Delta y. \tag{9-5}$$

These are the only circulating currents of a particular cell that give rise to a net current in the z-direction. This net current, which comes about from nonuniform magnetization, is called the magnetization current. This current is not a transport current but derives, as we have seen, from circulatory currents, i.e., from atomic currents in the material. The effective area for each of the currents in (9–4) and (9–5) is $\Delta x\,\Delta y$. Thus

$$(J_M)_z = \frac{\partial M_y}{\partial x} - \frac{\partial M_x}{\partial y} \tag{9-6a}$$

or

$$\mathbf{J}_M = \nabla \times \mathbf{M}. \tag{9-6b}$$

The magnetization current density is the curl of the magnetization.

9–2 THE MAGNETIC FIELD PRODUCED BY MAGNETIZED MATERIAL

According to Eq. (9–1), each volume element $\Delta v'$ of magnetized matter is characterized by a magnetic moment

$$\Delta \mathbf{m} = \mathbf{M}(x', y', z')\,\Delta v'. \tag{9-7}$$

Using the results of Section 8–7, we may write the contribution to the magnetic field at point (x, y, z) from each $\Delta \mathbf{m}$ (or, equivalently, from each $\Delta v'$). The magnetic field is then obtained as an integral over the entire volume of material, V_0. This procedure is indicated schematically in Fig. 9–4.

Instead of calculating $\mathbf{B}$ directly, we find it expedient to work with the vector potential $\mathbf{A}$, and to obtain $\mathbf{B}$ subsequently by means of the curl operation. Accord-

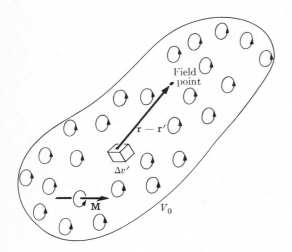

ing to Section 8–7, the vector potential at (x, y, z) is given by

$$A(x, y, z) = \frac{\mu_0}{4\pi} \int_{V_0} \frac{M(x', y', z') \times (r - r')}{|r - r'|^3} dv'$$

$$= \frac{\mu_0}{4\pi} \int_{V_0} M(x', y', z') \times \nabla' \frac{1}{|r - r'|} dv'. \tag{9–8}$$

By means of the vector identities (1–1–9) and (1–2–3) in Tables 1–1 and 1–2, this integral may be transformed to

$$A(x, y, z) = \frac{\mu_0}{4\pi} \int_{V_0} \frac{\nabla' \times M}{|r - r'|} dv' + \frac{\mu_0}{4\pi} \int_{S_0} \frac{M \times n}{|r - r'|} da', \tag{9–9}$$

where S_0 is the surface of V_0. Using Eq. (9–6b) and defining a surface magnetization current density j_M (i.e., a magnetization current per unit length flowing in the surface layer) by the relation

$$j_M = M \times n, \tag{9–10}$$

we may write Eq. (9–9) as

$$A(r) = \frac{\mu_0}{4\pi} \int_{V_0} \frac{J_M(r') dv'}{|r - r'|} + \frac{\mu_0}{4\pi} \int_{S_0} \frac{j_M da'}{|r - r'|}. \tag{9–11}$$

We might have ventured to predict the final expression, Eq. (9–11). Nevertheless, it is gratifying to see that it has come out of the mathematics in a natural way. Thus the vector potential produced by a distribution of atomic currents inside matter has the same form as that produced by a distribution of true transport currents. We should point out that Eq. (9–10) is the proper expression for the surface current density that is consistent with $J_M = \nabla' \times M$. The j_M must be introduced whenever **M** changes abruptly, as it might at the interface between two media, but if the region of discontinuity in **M** is imagined to be spread out over the distance $\Delta\xi$, then it can be shown that $\mathbf{j}_M$ is contained in the term $\mathbf{J}_M \, \Delta\xi$. (Or, if the region is very thin, $\mathbf{j}_M$ could be represented by a surface delta-function.)

Although Eq. (9–11) is both correct and of such form that it integrates nicely with the results of Chapter 8, it presents some practical difficulties when it comes to the task of computing **B** from a specified distribution of magnetization. First, there is the $\nabla \times \mathbf{M}$ operation to perform, and second, another curl operation is involved in obtaining **B** from the **A** field. It certainly is preferable to work with scalar quantities if possible, and the gradient of a scalar field (such as we encountered in electrostatics) is easier to compute than the curl of a vector field. For this reason we go back to Eq. (9–8) and try another approach. We are interested, after all, in **B**, not **A**, so let us formally take the curl:

$$\mathbf{B}(\mathbf{r}) = \nabla \times \mathbf{A}$$

$$= \frac{\mu_0}{4\pi} \int_{V_0} \nabla \times \left[\mathbf{M} \times \frac{(\mathbf{r} - \mathbf{r}')}{|\mathbf{r} - \mathbf{r}'|^3} \right] dv', \tag{9–12}$$

where the differential operators in the curl act on the unprimed coordinates.

As the reader may have anticipated, our next job is to transform the integrand of Eq. (9–12). To do so, we appeal to the vector identities of Table 1–1. According to (1–1–10),

$$\nabla \times (\mathbf{F} \times \mathbf{G}) = (\nabla \cdot \mathbf{G})\mathbf{F} - (\nabla \cdot \mathbf{F})\mathbf{G} + (\mathbf{G} \cdot \nabla)\mathbf{F} - (\mathbf{F} \cdot \nabla)\mathbf{G}.$$

Letting $\mathbf{F} = \mathbf{M}$ and $\mathbf{G} = (\mathbf{r} - \mathbf{r}')/|\mathbf{r} - \mathbf{r}'|^3$, and noting that the differentiations are with respect to the unprimed coordinates, we find that the identity reduces to

$$\nabla \times \left[\mathbf{M} \times \frac{(\mathbf{r} - \mathbf{r}')}{|\mathbf{r} - \mathbf{r}'|^3} \right] = \mathbf{M} \, \nabla \cdot \left[\frac{(\mathbf{r} - \mathbf{r}')}{|\mathbf{r} - \mathbf{r}'|^3} \right] - (\mathbf{M} \cdot \nabla) \frac{(\mathbf{r} - \mathbf{r}')}{|\mathbf{r} - \mathbf{r}'|^3}, \tag{9–13}$$

since $\nabla \cdot \mathbf{M}(x', y', z') = 0$, etc. Thus

$$\mathbf{B}(\mathbf{r}) = \mathbf{B}_\mathrm{I}(\mathbf{r}) + \mathbf{B}_\mathrm{II}(\mathbf{r}), \tag{9–14}$$

where

$$\mathbf{B}_\mathrm{I}(\mathbf{r}) = \frac{\mu_0}{4\pi} \int_{V_0} \mathbf{M} \, \nabla \cdot \left[\frac{(\mathbf{r} - \mathbf{r}')}{|\mathbf{r} - \mathbf{r}'|^3} \right] dv', \tag{9–14a}$$

$$\mathbf{B}_\mathrm{II}(\mathbf{r}) = -\frac{\mu_0}{4\pi} \int_{V_0} (\mathbf{M} \cdot \nabla) \frac{(\mathbf{r} - \mathbf{r}')}{|\mathbf{r} - \mathbf{r}'|^3} \, dv'. \tag{9–14b}$$

We consider the simpler $\mathbf{B}_\mathrm{I}$ integral first. Using Eq. (2–57), we obtain

$$\mathbf{B}_\mathrm{I}(\mathbf{r}) = \frac{\mu_0}{4\pi} \int_{V_0} \mathbf{M}(\mathbf{r}') 4\pi \, \delta(\mathbf{r} - \mathbf{r}') \, dv'$$

$$= \mu_0 \mathbf{M}(\mathbf{r}). \tag{9–15}$$

We next consider the $\mathbf{B}_{II}$ integral. The integrand may be transformed by means of a second identity (1–1–6), which becomes

$$\nabla \left[\mathbf{M} \cdot \frac{(\mathbf{r} - \mathbf{r}')}{|\mathbf{r} - \mathbf{r}'|^3} \right] = (\mathbf{M} \cdot \nabla) \frac{(\mathbf{r} - \mathbf{r}')}{|\mathbf{r} - \mathbf{r}'|^3} + \mathbf{M} \times \nabla \times \left[\frac{(\mathbf{r} - \mathbf{r}')}{|\mathbf{r} - \mathbf{r}'|^3} \right]. \quad (9\text{–}16)$$

The last term in (9–16) contains

$$\nabla \times \left[\frac{(\mathbf{r} - \mathbf{r}')}{|\mathbf{r} - \mathbf{r}'|^3} \right] = -\nabla \times \nabla \frac{1}{|\mathbf{r} - \mathbf{r}'|},$$

which vanishes identically. Hence

$$\mathbf{B}_{II}(\mathbf{r}) = -\mu_0 \nabla \frac{1}{4\pi} \int_{V_0} \mathbf{M}(\mathbf{r}') \cdot \frac{(\mathbf{r} - \mathbf{r}')}{|\mathbf{r} - \mathbf{r}'|^3} \, dv',$$

which may be written as in Eq. (8–80),

$$\mathbf{B}_{II}(\mathbf{r}) = -\mu_0 \nabla \varphi^*(\mathbf{r}). \quad (9\text{–}17)$$

The quantity $\varphi^*(\mathbf{r})$ is a scalar field, the magnetic scalar potential due to magnetic material:

$$\varphi^*(\mathbf{r}) = \frac{1}{4\pi} \int_{V_0} \mathbf{M}(\mathbf{r}') \cdot \frac{(\mathbf{r} - \mathbf{r}')}{|\mathbf{r} - \mathbf{r}'|^3} \, dv'. \quad (9\text{–}18)$$

Adding the two contributions, (9–15) and (9–17), we find for the magnetic induction field:

$$\mathbf{B}(\mathbf{r}) = -\mu_0 \nabla \varphi^*(\mathbf{r}) + \mu_0 \mathbf{M}(\mathbf{r}). \quad (9\text{–}19)$$

Thus the magnetic induction due to a magnetized distribution of matter may be expressed as the sum of two terms: the gradient of a scalar field plus a term proportional to the local magnetization. At an external point, i.e., in vacuum, $\mathbf{M}$ is zero, and the magnetic induction is then just the gradient of a scalar field which is the integral of the distant dipole fields given by Eq. (8–83).

9–3 MAGNETIC SCALAR POTENTIAL AND MAGNETIC POLE DENSITY

The expression for the magnetic scalar potential, Eq. (9–18), is similar in form to that for the electrostatic potential arising from polarized dielectric material. Here again mathematical transformation is suggested:

$$\frac{\mathbf{M} \cdot (\mathbf{r} - \mathbf{r}')}{|\mathbf{r} - \mathbf{r}'|^3} = \mathbf{M} \cdot \nabla' \frac{1}{|\mathbf{r} - \mathbf{r}'|}$$

$$= \nabla' \cdot \frac{\mathbf{M}}{|\mathbf{r} - \mathbf{r}'|} - \frac{1}{|\mathbf{r} - \mathbf{r}'|} \nabla' \cdot \mathbf{M}, \quad (9\text{–}20)$$

so that Eq. (9–18) becomes

$$\varphi^*(\mathbf{r}) = \frac{1}{4\pi} \int_{S_0} \frac{\mathbf{M} \cdot \mathbf{n} \, da'}{|\mathbf{r} - \mathbf{r}'|} - \frac{1}{4\pi} \int_{V_0} \frac{\nabla' \cdot \mathbf{M}}{|\mathbf{r} - \mathbf{r}'|} \, dv', \tag{9–21}$$

where S_0 is the surface of the region V_0.

By analogy with Section 4–2, it is expedient to define two scalar quantities:

$$\rho_M(\mathbf{r}') \equiv -\nabla' \cdot \mathbf{M}(\mathbf{r}'), \tag{9–22}$$

called the *magnetic pole density*, and

$$\sigma_M(\mathbf{r}') \equiv \mathbf{M}(\mathbf{r}') \cdot \mathbf{n}, \tag{9–23}$$

the *surface density of magnetic pole strength*. These quantities are quite useful even though rather artificial; they play the same role in the theory of magnetism that ρ_P and σ_P play in dielectric theory. The units of ρ_M and σ_M are A/m² and A/m, respectively.

Consider, for example, a uniformly magnetized bar magnet. Since the magnetization is uniform, $\rho_M = 0$. The only surface densities that do not vanish are on those surfaces that have a normal component of the magnetization; these are called the *poles* of the magnet. This is a somewhat idealized example, yet not too different from the common laboratory bar magnet familiar to the reader. (In practice, the poles of a magnet exert a demagnetizing influence that destroys the uniformity of $\mathbf{M}$ and thus spreads each pole over a somewhat larger region than just the surface.)

The total pole strength of every magnet is zero. This statement follows directly from the divergence theorem:

$$\int_{V_0} (-\nabla \cdot \mathbf{M}) \, dv + \int_{S_0} \mathbf{M} \cdot \mathbf{n} \, da = 0.$$

We now complete the derivation that was started earlier. Equation (9–18) becomes

$$\varphi^*(\mathbf{r}) = \frac{1}{4\pi} \int_{V_0} \frac{\rho_M \, dv'}{|\mathbf{r} - \mathbf{r}'|} + \frac{1}{4\pi} \int_{S_0} \frac{\sigma_M \, da'}{|\mathbf{r} - \mathbf{r}'|}, \tag{9–18a}$$

and $\mathbf{B}(x, y, z)$ is obtained as $-\mu_0$ times the gradient with respect to the unprimed coordinates, plus the term $\mu_0 \mathbf{M}$:

$$\mathbf{B}(\mathbf{r}) = \frac{\mu_0}{4\pi} \int_{V_0} \rho_M \frac{(\mathbf{r} - \mathbf{r}')}{|\mathbf{r} - \mathbf{r}'|^3} \, dv' + \frac{\mu_0}{4\pi} \int_{S_0} \sigma_M \frac{(\mathbf{r} - \mathbf{r}')}{|\mathbf{r} - \mathbf{r}'|^3} \, da' + \mu_0 \mathbf{M}(\mathbf{r}). \tag{9–19a}$$

This equation represents the contribution from the magnetized material in V_0 to the magnetic induction at (x, y, z).

9–4 SOURCES OF THE MAGNETIC FIELD. MAGNETIC INTENSITY

In the preceding sections we have seen how magnetized material produces a magnetic field. Furthermore, Chapter 8 dealt with the magnetic effects of conventional currents. In the general case, both types of magnetic sources are present: conventional or transport currents, which can be measured in the laboratory, and the atomic currents inside matter. It is important to realize that under certain conditions the same piece of matter may produce a magnetic field both because it is magnetized and because it is carrying a transport current of charge carriers. Thus, for example, one of our best magnetic materials, iron, may carry a transport current via its free electrons, but the fixed iron ions in the crystal contain atomic currents, which can be oriented to produce a strong magnetization.

In general, the expression for the magnetic field may be written as

$$\mathbf{B}(\mathbf{r}) = \frac{\mu_0}{4\pi} \int_V \frac{\mathbf{J} \times (\mathbf{r} - \mathbf{r}')}{|\mathbf{r} - \mathbf{r}'|^3} \, dv' - \mu_0 \nabla \varphi^*(\mathbf{r}) + \mu_0 \mathbf{M}(\mathbf{r}), \qquad (9\text{–}24)$$

where

$$\varphi^*(\mathbf{r}) = \frac{1}{4\pi} \int_V \frac{\rho_M \, dv'}{|\mathbf{r} - \mathbf{r}'|} + \frac{1}{4\pi} \int_S \frac{\sigma_M \, da'}{|\mathbf{r} - \mathbf{r}'|}. \qquad (9\text{–}25)$$

The volume V extends over all current-carrying regions and over all matter; the surface S includes all surfaces and interfaces between different media. The current density $\mathbf{J}$ includes only conventional currents of the charge transport variety, whereas the effect of atomic currents is found in the magnetization vector $\mathbf{M}$ (and potential φ^*).

Equation (9–24) may be solved for $\mathbf{B}$ if $\mathbf{M}$ and $\mathbf{J}$ are specified at all points. In most problems, however, $\mathbf{J}$ is specified but $\mathbf{M}(x', y', z')$ depends on $\mathbf{B}(x', y', z')$, so that even if the functional form of $\mathbf{M}(\mathbf{B})$ is known, (9–24) provides at best an integral equation for $\mathbf{B}$. To help get around this difficulty we introduce an auxiliary magnetic vector, the *magnetic intensity* $\mathbf{H}$, defined by

$$\mathbf{H} = \frac{1}{\mu_0} \mathbf{B} - \mathbf{M}. \qquad (9\text{–}26)$$

By combining (9–24) and (9–26) we obtain

$$\mathbf{H}(\mathbf{r}) = \frac{1}{4\pi} \int_V \frac{\mathbf{J} \times (\mathbf{r} - \mathbf{r}')}{|\mathbf{r} - \mathbf{r}'|^3} \, dv' - \nabla \varphi^*(\mathbf{r}). \qquad (9\text{–}27)$$

It appears that we have gained nothing by this maneuver because $\mathbf{H}$ still depends on $\mathbf{M}$ through ρ_M and σ_M; but in the next section we shall show how $\mathbf{H}$ is related to the conventional current density $\mathbf{J}$ through a differential equation. The situation is similar to the electrostatic case, where the auxiliary vector $\mathbf{D}$ is related to the prescribed charge density through its divergence.

The field vector $\mathbf{H}$ plays an important role in magnetic theory, particularly in problems involving permanent magnets. These will be discussed in later sections of the chapter. The units of $\mathbf{H}$ are the same as those of $\mathbf{M}$, namely, A/m.

9–5 THE FIELD EQUATIONS

In Chapter 8, the basic equations describing the magnetic effects of currents were expressed in differential form:

$$\mathbf{V} \cdot \mathbf{B} = 0, \qquad \mathbf{V} \times \mathbf{B} = \mu_0 \mathbf{J}.$$

We should now like to see how these equations are modified when the magnetic field $\mathbf{B}$ includes a contribution from magnetized material.

The reader will recall that the divergence equation ($\mathbf{V} \cdot \mathbf{B} = 0$) implied that $\mathbf{B}$ could be written as the curl of a vector function $\mathbf{A}$. But this result is not limited to magnetic fields produced by conventional currents. The field produced by magnetized matter is derivable from atomic currents; in fact, this approach was used in Section 9–2. Thus $\mathbf{B}$ may always be written as $\mathbf{V} \times \mathbf{A}$, and the divergence equation always holds:

$$\mathbf{V} \cdot \mathbf{B} = 0. \tag{9–28}$$

The "curl equation" is the differential form of Ampere's circuital law. Here we must be careful to include all types of currents that can produce a magnetic field. Hence, in the general case, this equation is properly written as

$$\mathbf{V} \times \mathbf{B} = \mu_0 (\mathbf{J} + \mathbf{J}_M), \tag{9–29}$$

where $\mathbf{J}$ is the transport current density, and $\mathbf{J}_M$ is the magnetization current density. Equation (9–6b) may be combined with (9–29) to yield

$$\mathbf{V} \times \left(\frac{1}{\mu_0} \mathbf{B} - \mathbf{M} \right) = \mathbf{J},$$

which, according to (9–26), is equivalent to

$$\mathbf{V} \times \mathbf{H} = \mathbf{J}. \tag{9–30}$$

Hence, the auxiliary magnetic vector $\mathbf{H}$ is related to the *transport current density* through its curl. This also follows by taking the curl of (9–27).

Equations (9–28) and (9–30) are the fundamental magnetic field equations. These equations, together with appropriate boundary conditions and an experimental relationship between $\mathbf{B}$ and $\mathbf{H}$, are sufficient to solve magnetic problems. In some instances it is preferable to use an integral formulation of the theory. With the aid of Stokes's theorem, (9–30) may be converted to

$$\int_S \mathbf{V} \times \mathbf{H} \cdot \mathbf{n} \, da = \oint_C \mathbf{H} \cdot d\mathbf{l}$$

$$= \int_S \mathbf{J} \cdot \mathbf{n} \, da,$$

or

$$\oint_C \mathbf{H} \cdot d\mathbf{l} = I. \tag{9-31}$$

In other words, the line integral of the tangential component of the magnetic intensity around a closed path C is equal to the entire transport current through the area bounded by the curve C.

Because of the divergence theorem, Eq. (9–28) is equivalent to

$$\oint_S \mathbf{B} \cdot \mathbf{n} \, da = 0. \tag{9-32}$$

The magnetic flux through any closed surface is zero.

9–6 MAGNETIC SUSCEPTIBILITY AND PERMEABILITY. HYSTERESIS

In order to solve problems in magnetic theory, it is essential to have a relationship between **B** and **H** or, equivalently, a relationship between **M** and one of the magnetic field vectors. These relationships depend on the nature of the magnetic material and are usually obtained from experiment.

In a large class of materials there exists an approximately linear relationship between **M** and **H**. If the material is isotropic as well as linear,*

$$\mathbf{M} = \chi_m \mathbf{H}, \tag{9-33}$$

where the dimensionless scalar quantity χ_m is called the *magnetic susceptibility*. If χ_m is positive, the material is called *paramagnetic*, and the magnetic induction is strengthened by the presence of the material. If χ_m is negative, the material is *diamagnetic*, and the magnetic induction is weakened by the presence of the material. Although χ_m is a function of the temperature, and sometimes varies quite drastically with temperature, it is generally safe to say that χ_m for paramagnetic and diamagnetic materials is quite small; i.e.,

$$|\chi_m| \ll 1 \quad \text{(for paramagnetic, diamagnetic materials).} \tag{9-34}$$

The susceptibilities of some common materials are given in Table 9–1.

In most handbooks and tabulations of physical data, χ_m is not listed directly, but instead is given as the *mass susceptibility*, $\chi_{m,\text{mass}}$, or the *molar susceptibility*, $\chi_{m,\text{molar}}$. These are defined by

$$\chi_m = \chi_{m,\text{mass}} \, d, \tag{9-35}$$

$$\chi_m = \chi_{m,\text{molar}} \, \frac{d}{A}, \tag{9-36}$$

---

* If the material is anisotropic but linear, Eq. (9–33) is replaced by the tensor relationships

$$M_x = \chi_{m,11} H_x + \chi_{m,12} H_y + \chi_{m,13} H_z,$$

etc. In these circumstances **M** does not necessarily have the same direction as **H**. We shall restrict ourselves in this book to isotropic media.

Table 9–1 Magnetic Susceptibility of Some Paramagnetic and Diamagnetic Materials at Room Temperature*

Material	χ_m	$\chi_{m,mass}$, m^3/kg
Aluminum	2.1×10^{-5}	0.77×10^{-8}
Bismuth	-16.4×10^{-5}	-1.68×10^{-8}
Copper	-0.98×10^{-5}	-0.11×10^{-8}
Diamond	-2.2×10^{-5}	-0.62×10^{-8}
Gadolinium chloride (GdCl$_3$)	603.0×10^{-5}	133.3×10^{-8}
Gold	-3.5×10^{-5}	-0.18×10^{-8}
Magnesium	1.2×10^{-5}	0.68×10^{-8}
Mercury	-2.8×10^{-5}	-0.21×10^{-8}
Silver	-2.4×10^{-5}	-0.23×10^{-8}
Sodium	0.84×10^{-5}	0.87×10^{-8}
Titanium	18.0×10^{-5}	4.01×10^{-8}
Tungsten	7.6×10^{-5}	0.40×10^{-8}
Carbon dioxide (1 atm)	-1.19×10^{-8}	-0.60×10^{-8}
Hydrogen (1 atm)	-0.22×10^{-8}	-2.48×10^{-8}
Nitrogen (1 atm)	-0.67×10^{-8}	-0.54×10^{-8}
Oxygen (1 atm)	193.5×10^{-8}	135.4×10^{-8}

* Data obtained from the *Handbook of Chemistry and Physics*, 58th edition, CRC Press, Inc., Cleveland, Ohio. Practically all sources of data list magnetic susceptibilities in gaussian (cgs) units; if the superscript (1) is used to indicate the constant in the gaussian system, then $\chi_m = 4\pi\chi_m^{(1)}$ and $\chi_{m,mass} = 4\pi \times 10^{-3}\chi_{m,mass}^{(1)}$.

where d is the mass density of the material and A is the molecular weight. Since **M** and **H** both have the dimensions of magnetic moment per unit volume, it is evident that $\chi_{m,mass}$ **H** and $\chi_{m,molar}$ **H** give magnetic moment per unit mass and magnetic moment per mole, respectively. For convenience, the mass susceptibility is also listed in Table 9–1.

A linear relationship between **M** and **H** implies also a linear relationship between **B** and **H**:

$$\mathbf{B} = \mu\mathbf{H}, \tag{9-37}$$

where the *permeability* μ is obtained from the combination of Eqs. (9–26) and (9–33);

$$\mu = \mu_0(1 + \chi_m). \tag{9-38}$$

The dimensionless quantity

$$K_m = \frac{\mu}{\mu_0} = 1 + \chi_m \tag{9-39}$$

is sometimes tabulated instead of χ_m. This quantity, K_m, is called the *relative permeability*. For the paramagnetic and diamagnetic materials of Table 9–1, it is evident that K_m is very close to unity.

Table 9–2. Properties of Ferromagnetic Materials at Room Temperature*

M_s = saturation magnetization, H_s = magnetic intensity
required for saturation, H_c = coercivity, B_r = remanence

Material	Composition, %	$\mu_0 M_s$, T	H_s, A/m	K_m, maximum
Iron (annealed)		2.15	1.6×10^5	5,500
Cobalt		1.79	7.0×10^5	
Nickel		0.61	5.5×10^5	
ALLOYS			H_c, A/m	
Iron-silicon	96 Fe, 3 Si	2.02	56	8,000
Permalloy	55 Fe, 45 Ni	1.60	5.6	50,000
Mumetal	5 Cu, 2 Cr, 77 Ni, 16 Fe	0.75	1.2	150,000
Permendur	50 Co, 50 Fe	2.40	159	6,000
Mn ferrite	$MnFe_2O_4$	0.49		2,500
Ni ferrite	$NiFe_2O_4$	0.32		2,500
		B_r, T		
Cobalt steel	52 Fe, 36 Co, 4 W, 6 Cr, 0.8 C	0.97	19×10^3	
Alnico V	51 Fe, 8 Al, 14 Ni 24 Co, 3 Cu	1.25	49×10^3	

* Data from *American Institute of Physics Handbook*, 3rd edition, McGraw-Hill, New York, 1972.

The *ferromagnetics* form another class of magnetic material. Such a material is characterized by a possible permanent magnetization and by the fact that its presence usually has a profound effect on the magnetic induction. Ferromagnetic materials are *not* linear, so that Eqs. (9–33) and (9–37) with constant χ and μ *do not* apply.* It has been expedient, however, to use Eq. (9–37) as the defining equation for μ, i.e., with $\mu = \mu(\mathbf{H})$, but the reader should be cautioned that this practice can lead to difficulty in certain situations. If the μ of a ferromagnetic material is defined by Eq. (9–37), then, depending on the value of $\mathbf{H}$, μ goes through an entire range of values from infinity to zero and may be either positive or negative. The best advice that can be given is to consider each problem involving ferromagnetism separately, try to determine which region of the **B-H** diagram is important for the particular problem, and make approximations appropriate to this region.

First, let us consider an unmagnetized sample of ferromagnetic material. If the magnetic intensity, initially zero, is increased *monotonically*, then the **B-H** relationship will trace out a curve something like that shown in Fig. 9–5. This is called the *magnetization curve* of the material. It is evident that μ's taken from the magnetiza-

---

* Certain kinds of iron, referred to as *soft iron*, can be treated as *approximately* linear nevertheless.

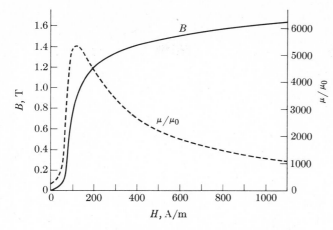

Figure 9–5 Magnetization curve and relative permeability of commercial iron (annealed).

tion curve, using the expression $\mu = B/H$, are always of the same sign (positive), but they show a rather large spectrum of values. The maximum permeability occurs at the "knee" of the curve; in some materials this maximum permeability is as large as $10^5 \mu_0$; in others it is much lower. The reason for the knee in the curve is that the magnetization **M** approaches a maximum value in the material, and

$$\mathbf{B} = \mu_0(\mathbf{H} + \mathbf{M})$$

continues to increase at very large **H** only because of the $\mu_0\mathbf{H}$ term. The maximum value of **M** is called the *saturation magnetization* of the material. (See Table 9–2.)

Next consider a ferromagnetic specimen magnetized by the above procedure. If the magnetic intensity **H** is decreased, the **B-H** relationship does not follow back down the curve of Fig. 9–5, but instead moves along the new curve of Fig. 9–6 to point *r*. The magnetization, once established, does not disappear with the removal of **H**; in fact, it takes a reversed magnetic intensity to reduce the magnetization to zero. If **H** continues to build up in the reversed direction, then **M** (and hence **B**) will establish itself in the reversed direction, and Fig. 9–6 begins to show a certain symmetry. Finally, when **H** once again increases, the operating point follows the lower curve of Fig. 9–6. Thus the **B-H** curve for increasing **H** is entirely different from that for decreasing **H**. This phenomenon is called *hysteresis*, from the Greek word meaning "to lag"; the magnetization literally lags the exciting field.

The curve of Fig. 9–6 is called the *hysteresis loop* of the material. The value of **B** at point *r* is known as the *retentivity* or *remanence*; the magnitude of **H** at point *c* is called the *coercive force* or *coercivity* of the material. From Fig. 9–6 it is evident that the value of μ, defined by Eq. (9–37), is negative in the second and fourth quadrants of the diagram. The shape of the hysteresis loop depends not only upon the nature of the ferromagnetic material (Fig. 9–7) but also on the

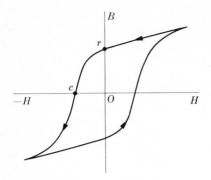

Figure 9-6 Typical hysteresis loop of a ferromagnetic material.

maximum value of **H** to which the material is subjected (Fig. 9–8). However, once $\mathbf{H_{max}}$ is sufficient to produce saturation in the material, the hysteresis loop does not change shape with increasing $\mathbf{H_{max}}$. (See Table 9–2.) For soft iron, the hysteresis is relatively small.

For certain applications it is desirable to know the effective permeability of a material to a small alternating H-field superposed on a large constant field. Thus if

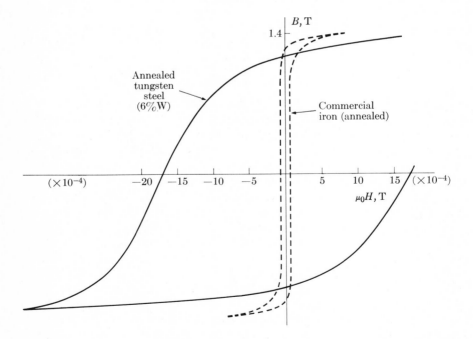

Figure 9-7 Comparison of the hysteresis curves of several materials. (Note that $\mu_0 H$ is plotted along the abscissa instead of just H. $\mu_0 = 4\pi \times 10^{-7}$ T·m/A.) Data from R. M. Bozorth, *Ferromagnetism*, Van Nostrand, N.Y., 1951.

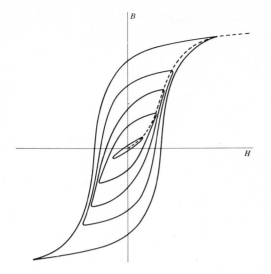

Figure 9–8 Major hysteresis loop and several minor hysteresis loops for a typical material.

ΔB is the change in magnetic field produced by a change ΔH in the magnetic intensity, the *incremental permeability* μ_{in} is defined by

$$\mu_{in} = \frac{\Delta B}{\Delta H},\qquad(9\text{--}40)$$

and is approximately equal to the slope of the hysteresis curve which goes through the point in question.

Ferromagnetic materials are used either (1) to increase the magnetic flux of a current circuit, or (2) as sources of the magnetic field (permanent magnets). For use as a permanent magnet, the material is first magnetized to saturation by placing it in a strong magnetic field (i.e., by placing it between the poles of an electromagnet or by placing it in a solenoid subjected to a momentary large current). However, when the permanent magnet is removed from the external field, it will in general be subject to a demagnetizing field; this will be discussed in detail in Sections 9–8 and 9–11. Thus the second quadrant of the hysteresis loop diagram is the important part of the **B-H** relationship for a permanent magnet material (Fig. 9–9).

9–7 BOUNDARY CONDITIONS ON THE FIELD VECTORS

Before we can solve magnetic problems, even simple ones, we must know how the field vectors **B** and **H** change in passing an interface between two media. The interface to be considered may be between two media with different magnetic properties, or between a material medium and vacuum.

Consider two media, 1 and 2, in contact, as shown in Fig. 9–10. Let us construct the small pillbox-shaped surface S, which intersects the interface, the

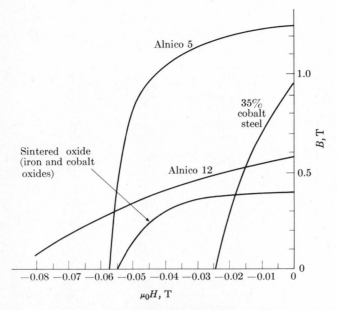

Figure 9–9 Hysteresis curves of permanent magnet materials. (Note that $\mu_0 H$ is plotted along the abscissa instead of just H.)

height of the pillbox being negligibly small in comparison with the diameter of the bases. Applying the flux integral, Eq. (9–32), to the surface S, we find

$$\mathbf{B}_2 \cdot \mathbf{n}_2 \, \Delta S + \mathbf{B}_1 \cdot \mathbf{n}_1 \, \Delta S = 0,$$

where $\mathbf{n}_2$ and $\mathbf{n}_1$ are the outward-directed normals to the upper and lower surfaces of the pillbox. Since $\mathbf{n}_2 = -\mathbf{n}_1$, and since either of these normals may serve as the

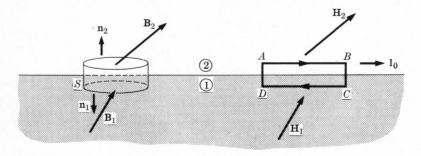

Figure 9–10 Boundary conditions on the field vectors at the interface between two media may be obtained by applying Gauss's law to S, and integrating $\mathbf{H} \cdot d\mathbf{l}$ around the path $ABCDA$.

normal to the interface,

$$(\mathbf{B}_2 - \mathbf{B}_1) \cdot \mathbf{n}_2 = 0, \tag{9-41a}$$

or

$$B_{2n} - B_{1n} = 0. \tag{9-41b}$$

Thus the normal component of **B** is continuous across an interface.

A boundary condition on the **H**-field may be obtained by applying Ampère's circuital law, Eq. (9–31), to the rectangular path $ABCD$ in Fig. 9–10. On this path the lengths AB and CD will be taken equal to Δl and the segments AD and BC will be assumed negligibly small. The current through the rectangle is negligible unless there is a true surface current. Therefore

$$(\mathbf{H}_2 - \mathbf{H}_1) \cdot \mathbf{l}_0 = \mathbf{j} \cdot (\mathbf{n}_2 \times \mathbf{l}_0) = \mathbf{j} \times \mathbf{n}_2 \cdot \mathbf{l}_0,$$

or

$$(\mathbf{H}_2 - \mathbf{H}_1)_t = \mathbf{j} \times \mathbf{n}_2, \tag{9-42a}$$

where $\mathbf{j}$ is the *surface* current density (transport current per unit length in the surface layer) and $\mathbf{l}_0$ is a unit vector in the direction of $\Delta \mathbf{l}$. Thus the tangential component of the magnetic intensity is continuous across an interface unless there is a true surface current. Finally, by taking the cross product of Eq. (9–42a) with $\mathbf{n}_2$, the equation may be written as

$$\mathbf{n}_2 \times (\mathbf{H}_2 - \mathbf{H}_1) = \mathbf{j}. \tag{9-42b}$$

This form is convenient for determining $\mathbf{j}$ if $\mathbf{H}_2$ and $\mathbf{H}_1$ are known.

Before completing this section, we shall prove one other important property of the magnetic induction **B**, namely, that its flux is everywhere continuous. Let us focus our attention on a region of space, and construct *magnetic field lines*, which are imaginary lines drawn in such a way that the direction of a line at any point is the direction of **B** at that point. Next we imagine a tube of flux, a volume bounded on the sides by lines of **B** but not cut by them (Fig. 9–11). The tube is terminated

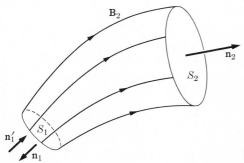

Figure 9–11 A tube of magnetic induction.

on the ends by the surfaces S_1 and S_2. Applying the divergence theorem, we obtain

$$\int_V \mathbf{\nabla} \cdot \mathbf{B} \, dv = 0$$

$$= \int_{S_2} \mathbf{B} \cdot \mathbf{n} \, da - \int_{S_1} \mathbf{B} \cdot \mathbf{n}' \, da$$

$$= \Phi(S_2) - \Phi(S_1). \qquad (9\text{–}43)$$

Thus the same magnetic flux enters the tube through S_1 as leaves through S_2. The flux lines can never terminate, but must eventually join back onto themselves, forming closed loops.

The previous statements apply, of course, to the **B**-field; it is perhaps worth while noting that they do not apply to the **H**-field, since $\mathbf{\nabla} \cdot \mathbf{H} = -\mathbf{\nabla} \cdot \mathbf{M}$, which is not everywhere zero. Thus, from the divergence theorem applied to a tube of magnetic intensity, we find

$$\int_{S_2} \mathbf{H} \cdot \mathbf{n} \, da - \int_{S_1} \mathbf{H} \cdot \mathbf{n}' \, da = \int_V \rho_M \, dv. \qquad (9\text{–}44)$$

The discontinuity in the magnetic intensity flux is determined by the total magnetic pole strength intercepted by the flux tube.

9–8 BOUNDARY-VALUE PROBLEMS INVOLVING MAGNETIC MATERIALS

Since **B** and **H** obey boundary conditions similar to those for **D** and **E**, the problems of linear media or specified magnetization are similar to the dielectric problems discussed in Chapter 4. In this section we shall be concerned with a particular class of problems, namely, calculation of magnetic fields inside magnetic material in which no transport current exists. This is formally identical to the dielectric with no external charge density.

When $\mathbf{J} = 0$, the fundamental magnetic equations (9–28) and (9–30) reduce to

$$\mathbf{\nabla} \cdot \mathbf{B} = 0, \qquad (9\text{–}28)$$

$$\mathbf{\nabla} \times \mathbf{H} = 0. \qquad (9\text{–}45)$$

Equation (9–45) implies that **H** can be derived as the gradient of a scalar field. This should not surprise us because, according to the source equation, (9–27), the contribution to **H** from magnetic material is already expressed in this form, and in Section 8–8 we showed that the field (actually the proof presented there must be generalized to the **H**-field) produced by transport currents can be so derived when the local current density is zero. In accordance with Eq. (9–45), we write

$$\mathbf{H} = -\mathbf{\nabla}\varphi^*, \qquad (9\text{–}46)$$

where φ^* is now the magnetic scalar potential due to all sources.

There are two types of magnetic material for which the magnetic field calculation reduces to a simple boundary-value problem: (1) *linear* or approximately linear magnetic material for which $\mathbf{B} = \mu\mathbf{H}$, and (2) a *uniformly* magnetized piece of material for which $\nabla \cdot \mathbf{M} = 0$. In both cases Eq. (9–28) reduces to

$$\nabla \cdot \mathbf{H} = 0. \tag{9–47}$$

Combining this result with (9–46), we obtain

$$\nabla^2 \varphi^* = 0, \tag{9–48}$$

which is Laplace's equation. Thus the magnetic problem reduces to finding a solution to Laplace's equation which satisfies the boundary conditions. $\mathbf{H}$ may then be calculated as minus the gradient of the magnetic potential, and $\mathbf{B}$ obtained from

$$\mathbf{B} = \mu\mathbf{H}$$

or

$$\mathbf{B} = \mu_0(\mathbf{H} + \mathbf{M}),$$

whichever is appropriate.

Two magnetic problems serve to illustrate the usefulness of the method just described; additional exercises of this type will be found in the problems at the end of the chapter. Our first example deals with a sphere of linear magnetic material of radius a and permeability μ placed in a region of space containing an *initially* uniform magnetic field, $\mathbf{B}_0$. We should like to determine how the magnetic field is modified by the presence of the sphere and, in particular, to determine the magnetic field in the sphere itself. The problem is closely analogous to the case of a dielectric sphere in a uniform electric field, which was solved in Section 4–9. Thus, choosing the origin of our coordinate system at the center of the sphere and the direction of $\mathbf{B}_0$ as the polar direction (z-direction), we may express the potential as a sum of zonal harmonics. Again, all boundary conditions can be satisfied by means of the $\cos \theta$ harmonics:

$$\varphi_1^*(r, \theta) = A_1 r \cos \theta + C_1 r^{-2} \cos \theta \tag{9–49}$$

for the vacuum region (1) outside the sphere, and

$$\varphi_2^*(r, \theta) = A_2 r \cos \theta + C_2 r^{-2} \cos \theta \tag{9–50}$$

for the magnetic material region (2). The constants A_1, A_2, C_1, and C_2 must be determined from the boundary conditions.

At distances far away from the sphere, the magnetic field retains its uniform character: $\mathbf{B} = B_0\mathbf{k}$, and $\varphi_1^* \rightarrow -(B_0/\mu_0)r \cos \theta$. Hence $A_1 = -(B_0/\mu_0)$. Since φ_2^* and its associated magnetic field cannot become infinite at any point, the coefficient C_2 must be set equal to zero. Having applied the boundary conditions at $r = \infty$ and at $r = 0$, we next turn our attention to the interface at $r = a$:

$$H_{1\theta} = H_{2\theta},$$

$$B_{1r} = B_{2r},$$

or

$$-\left(\frac{B_0}{\mu_0}\right)\sin\theta + \frac{C_1}{a^3}\sin\theta = A_2\sin\theta, \tag{9-51}$$

$$B_0\cos\theta + 2\mu_0\frac{C_1}{a^3}\cos\theta = -\mu A_2\cos\theta. \tag{9-52}$$

Solving these two equations simultaneously yields

$$A_2 = -\frac{3B_0}{(\mu + 2\mu_0)},$$

and

$$C_1 = [(\mu/\mu_0) - 1]\frac{B_0 a^3}{(\mu + 2\mu_0)},$$

whence the magnetic fields inside and outside the sphere are given by

$$\mathbf{B}_2 = \frac{3B_0\mathbf{k}}{1 + 2(\mu_0/\mu)} \tag{9-53}$$

and

$$\mathbf{B}_1 = B_0\mathbf{k} + \left[\frac{(\mu/\mu_0) - 1}{(\mu/\mu_0) + 2}\right]\left(\frac{a}{r}\right)^3 B_0(2\mathbf{a}_r\cos\theta + \mathbf{a}_\theta\sin\theta). \tag{9-54}$$

The second problem we wish to solve deals with a permanent magnet. We should like to determine the magnetic field produced by a uniformly magnetized sphere of magnetization $\mathbf{M}$ and radius a when no other magnetic fields are present. Taking the magnetization along the z-axis and the origin of our coordinate system at the center of the sphere, we may expand the potential in zonal harmonics:

$$\varphi_1^*(r,\theta) = \sum_{n=0}^{\infty} C_{1,n}r^{-(n+1)}P_n(\theta) \tag{9-55}$$

for the vacuum region (1) outside the sphere, and

$$\varphi_2^*(r,\theta) = \sum_{n=0}^{\infty} A_{2,n}r^n P_n(\theta) \tag{9-56}$$

for the permanent magnet region (2). Here we have purposely left out the harmonics with positive powers of r from expansion (9-55) since these would become large at large distances, and we have left out the negative powers of r in (9-56) since these would become infinite at the origin. From the boundary conditions at $r = a$:

$$H_{1\theta} = H_{2\theta},$$

$$B_{1r} = B_{2r},$$

we obtain

$$\sum_{n=0}^{\infty} (C_{1,n}a^{-(n+1)} - A_{2,n}a^n)a^{-1} \frac{d}{d\theta} P_n(\theta) = 0 \qquad (9\text{-}57)$$

and

$$\mu_0 C_{1,0} a^{-2} + \mu_0 \sum_{n=1}^{\infty} P_n(\theta)[C_{1,n}(n+1)a^{-(n+2)} + A_{2,n}na^{n-1}] - \mu_0 M \cos \theta = 0. \qquad (9\text{-}58)$$

Since each $P_n(\theta)$ is a distinct function of θ, none of them can be constructed from a linear combination of other P_n's. Hence, in order for Eqs. (9–57) and (9–58) to hold, each of the terms involving a P_n or a $dP_n/d\theta$ must vanish individually. From the $n = 0$ terms,

$$\frac{dP_0}{d\theta} = 0,$$

and

$$\mu_0 C_{1,0} a^{-2} = 0.$$

Therefore $C_{1,0} = 0$, and $A_{2,0}$ is undetermined. But $A_{2,0}$ is just the constant term in the potential; this may be set equal to zero without affecting $\mathbf{H}$ or $\mathbf{B}$.

From the $n = 1$ terms,

$$C_{1,1} a^{-3} - A_{2,1} = 0$$

and

$$2C_{1,1} a^{-3} + A_{2,1} - M = 0,$$

which may be solved simultaneously to yield

$$C_{1,1} = \tfrac{1}{3}Ma^3$$

and

$$A_{2,1} = \tfrac{1}{3}M.$$

For all $n \geq 2$, the only $C_{1,n}$ and $A_{2,n}$ compatible with the two equations are $C_{1,n} = 0$ and $A_{2,n} = 0$.

Putting these results back into Eqs. (9–55) and (9–56), we obtain

$$\varphi_1^*(r, \theta) = \tfrac{1}{3}M(a^3/r^2) \cos \theta \qquad (9\text{-}59)$$

and

$$\varphi_2^*(r, \theta) = \tfrac{1}{3}Mr \cos \theta. \qquad (9\text{-}60)$$

The magnetic intensity $\mathbf{H}$ may be calculated from the gradient operation, with the result:

$$\mathbf{H}_1 = \tfrac{1}{3}M(a^3/r^3)[2\mathbf{a}_r \cos \theta + \mathbf{a}_\theta \sin \theta], \qquad (9\text{-}61)$$

$$\mathbf{H}_2 = -\tfrac{1}{3}M\mathbf{k}. \qquad (9\text{-}62)$$

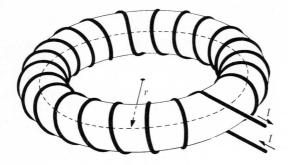

Figure 9–12 A toroidal winding.

Thus the external field of the uniformly magnetized sphere is exactly a dipole field, arising from the dipole moment $\frac{4}{3}\pi a^3 \mathbf{M}$. The magnetic intensity *inside* the sphere is a demagnetizing field, a result that is in accord with the E-field inside a uniformly polarized dielectric. We see, therefore, that the magnetized sphere is subjected to its own demagnetizing field. The factor $\frac{1}{3} = (1/4\pi)(4\pi/3)$ in Eq. (9–62) depends explicitly on the spherical geometry; the quantity $4\pi/3$ is known as the *demagnetization factor* of a sphere. The demagnetization factors for other geometrical shapes have been calculated and tabulated.*

The external magnetic field $\mathbf{B}_1$ is just μ_0 times Eq. (9–61). The magnetic induction in the sphere is

$$\mathbf{B}_2 = \tfrac{2}{3}\mu_0 M\mathbf{k} = \tfrac{2}{3}\mu_0\mathbf{M}. \tag{9–63}$$

9–9 CURRENT CIRCUITS CONTAINING MAGNETIC MEDIA

In Chapter 8 we dealt with magnetic fields produced by current circuits *in vacuum*. One of the examples taken up in the problems (Problem 8–19) was that of a uniformly wound toroid of N turns carrying current I (Fig. 9–12). Let us solve the toroid problem again, but this time with the region inside the windings filled with a ferromagnetic material, which we shall assume to be homogeneous, isotropic, and originally unmagnetized. The field vector obtained most easily is the magnetic intensity, since this is related to the current in the windings by means of Ampere's circuital law, Eq. (9–31). If we apply Eq. (9–31) to a circular path that is coaxial with the hole in the toroid, such as the dotted path shown in the figure, symmetry arguments tell us that $\mathbf{H}$ is the same at all points on the path:

$$H_t l = NI,$$

or

$$H_t = \frac{NI}{l}. \tag{9–64}$$

---

* See, for example, *American Institute of Physics Handbook*, Third Edition (New York: McGraw-Hill, 1972), p. 5–247.

Here the subscript stands for the component tangent to the path, and $l = 2\pi r$ is the total path length. From Eq. (9–26),

$$B_t = \frac{\mu_0 NI}{l} + \mu_0 M_t. \qquad (9\text{–}65)$$

Thus the magnetic field differs from that in the vacuum case by the additive term $\mu_0 M_t$.

Only the tangential component of **B** (and of **H**) is obtained by the above procedure; nevertheless, this is the only component we expect to be present. According to Eq. (9–27) there are two kinds of sources for the magnetic intensity: transport currents and magnetized material. It is easy to show that the current in the toroidal winding produces only a tangential field. The winding is equivalent to N circular loops of current; if we combine the loops in pairs (Fig. 9–13), it is evident that each pair of loops produces a tangential field at the point in question.

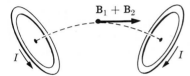

Figure 9–13 Axial nature of the field in a toroidal winding is shown by combining the magnetic field of the current loops in pairs.

The second source of **H**, the magnetized material itself, may possibly provide a contribution through the pole densities: $\rho_M = -\nabla \cdot \mathbf{M}$ and $\sigma_M = \mathbf{M} \cdot \mathbf{n}$. Since the ferromagnetic material in the toroid is isotropic, **M** will have the same direction as **H**. But **M** arose in response to currents in the toroidal windings, and this field is a tangential one. Thus we expect an M_t only and we can drop the subscript t. On this basis, there are no surfaces in the toroidal specimen that are normal to **M**, and hence no σ_M. Finally, ρ_M must equal zero; although M may be a function of r (the distance from the axis of the toroid), the term $\partial M/\partial r$ does not contribute to $\nabla \cdot \mathbf{M}$. The interesting result is that the magnetized material provides no contribution to **H** in this case, and Eq. (9–65) gives the entire magnetic field.

Another problem, somewhat more complicated than the preceding one, is that of a toroidal winding of N turns surrounding a ferromagnetic specimen in which a narrow air gap of width d has been cut (Fig. 9–14). We shall not distinguish between an air gap and a vacuum gap, since it is evident from Table 9–1 that the permeability of air differs only slightly from μ_0. In this problem Ampere's circuital law does not suffice to determine **H**, because symmetry arguments cannot be invoked to state that **H** is the same at all points on a circular path. Thus we first go to the source equation, Eq. (9–27).

Again we note that there are two contributions to the magnetic intensity, one from transport currents and one from the magnetization. Since the toroidal winding is identical to that of the preceding problem, the contribution to **H** from the

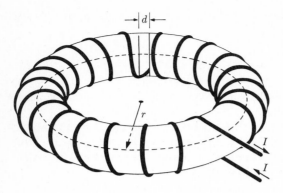

Figure 9-14 A toroidal winding surrounding a ring of magnetic material with an air gap.

transport currents must be the same as before. Denoting this contribution by the subscript 1, we may write

$$H_1 = \frac{NI}{l}. \tag{9-66}$$

Our problem then, is to evaluate H_2, or the $\nabla\varphi^*$ term. To keep the problem simple, we make the plausible assumption of uniform tangential magnetization M throughout the ferromagnetic material; this will provide us with all the essential physics without complicating the algebra. Then ρ_M equals zero, but $\sigma_M = \pm M$ on the pole faces bordering the air gap. The situation here is strongly reminiscent of the electrostatic problem involving a charged parallel-plate capacitor. In fact, the mathematical formulation of the potential is identical in the two cases. If the air gap is exceedingly narrow, then, approximately,

$$H_2 = M \quad \text{(in the gap)},$$

$$H_2 = 0 \quad \text{(elsewhere)}. \tag{9-67}$$

However, this result is not consistent with Ampere's circuital law, since

$$\oint H \, dl = \oint (H_1 + H_2) \, dl = NI + Md \neq NI$$

unless d is negligibly small. For a narrow, but not negligibly small, air gap, a better approximation is

$$H_2 = M\left(1 - \frac{d}{l}\right) \quad \text{(in the gap)},$$

$$H_2 = -M\frac{d}{l} \quad \text{(in the material)}, \tag{9-68}$$

which not only satisfies Ampere's circuital law, but also provides for the continuity of the normal component of **B** across the pole faces.

Combining Eqs. (9–66) and (9–68), and substituting the result in Eq. (9–26):

$$\mathbf{B} = \mu_0(\mathbf{H} + \mathbf{M}),$$

we find

$$B = \frac{\mu_0 NI}{l} + \mu_0 M \left(1 - \frac{d}{l}\right) \tag{9–69}$$

both in the gap and in the magnetic material. In order to solve the problem completely, we have only to know the relation

$$M = \chi_m(H)H.$$

For "soft iron" χ_m can be taken as a constant.

*9–10 MAGNETIC CIRCUITS

The magnetic flux lines, as we have seen, form closed loops. If all the magnetic flux (or substantially all of it) associated with a particular distribution of currents is confined to a rather well-defined path, then we may speak of a *magnetic circuit*. Thus the examples discussed in Section 9–9 are magnetic circuits, since the magnetic flux is confined to the region inside the toroidal winding. In the first example, the circuit consisted of just one material, a ferromagnetic ring; in the second case, however, we encountered a series circuit of two materials: a ferromagnetic material and an air gap.

Let us consider a more general series circuit of several materials surrounded by a toroidal winding of N turns carrying current I, such as that shown in Fig. 9–15. From the application of Ampere's circuital law to a path following the circuit (the dotted path in the figure),

$$\oint H \, dl = NI.$$

It is convenient to express H at each point along the path in terms of the magnetic flux Φ; using $B = \mu H$ and $\Phi = BA$, where A is the cross-sectional area of the circuit at the point under consideration, we find

$$\oint \frac{\Phi \, dl}{\mu A} = NI.$$

Figure 9–15 A magnetic circuit.

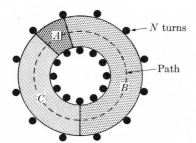

N turns

Path

Since we are dealing with a magnetic circuit, we expect Φ to be essentially constant at all points in the circuit; hence we may take Φ outside the integral:

$$\Phi \oint \frac{dl}{\mu A} = NI. \tag{9-70}$$

This is the basic magnetic circuit equation, which enables us to solve for the flux Φ in terms of the circuit parameters.

Equation (9-70) reminds us of the similar equation for a series current circuit: $IR = \mathscr{V}$. By analogy, we define the magnetomotive force (mmf):

$$\text{mmf} = NI, \tag{9-71}$$

and the reluctance $\mathscr{R}$,

$$\mathscr{R} = \oint \frac{dl}{\mu A}. \tag{9-72}$$

Using these definitions, we may rewrite Eq. (9-70), as

$$\Phi = \frac{\text{mmf}}{\mathscr{R}}. \tag{9-70a}$$

If the circuit is made from several homogeneous pieces, each of uniform cross section, the reluctance may be approximated:

$$\mathscr{R} = \sum_j \frac{l_j}{\mu_j A_j} = \sum_j \mathscr{R}_j. \tag{9-72a}$$

Hence the total reluctance of the series circuit is just the sum of the reluctances of the individual elements. The analogy between magnetic and current circuits is even closer than has been indicated, since the resistance of a current circuit is given by

$$R = \oint \frac{dl}{gA}, $$

which differs from Eq. (9-72) only through the substitution of g for μ. Because of this analogy, it is apparent that series and parallel reluctance combinations may be combined in the same manner as series and parallel resistance combinations.

The magnetic circuit concept is of most use when applied to circuits containing ferromagnetic materials, but it is for just these materials that we experience a certain amount of difficulty. For a ferromagnetic material, $\mu = \mu(H)$, and we do not know H in the material until the circuit problem is completely solved and Φ determined. The situation is not hopeless, however; in fact, the problem can be solved rather easily by an iterative procedure: (1) As a first guess, we might take $H = NI/l_{\text{total}}$, where l_{total} is the total length of the circuit. (2) The permeability of each material in the circuit is obtained for this value of H from the appropriate magnetization curve. (3) The total reluctance of the circuit is computed, and (4) the flux Φ is calculated from Eq. (9–70a). (5) From Φ, the magnetic intensities in the various elements may be found and the permeabilities redetermined. (6) The procedure is repeated, starting again with item (3). One or two iterations are usually sufficient to determine Φ to within a few percent.

The reluctance $\mathscr{R}_j$ is inversely proportional to the permeability μ_j. Since the permeability of ferromagnetic material may be 100 times μ_0, $10^3\mu_0$, or even $10^5\mu_0$ in certain circumstances, it is apparent that ferromagnetic material forms a low-reluctance path for the magnetic flux. If the magnetic flux encounters two parallel paths, one high reluctance $\mathscr{R}_h$ and the other low reluctance $\mathscr{R}_l$, then most of the flux will pass through the low reluctance path, and the equivalent reluctance of the combination is given by $\mathscr{R} = \mathscr{R}_h\mathscr{R}_l/(\mathscr{R}_h + \mathscr{R}_l)$. Looking now at Fig. 9–16, we note that if materials A, B, and C are ferromagnetic, most of the flux will follow the ferromagnetic ring, because the air path between the ends of the solenoid is of relatively high reluctance. Thus the magnetic circuits of Figs. 9–15 and 9–16 are essentially equivalent.

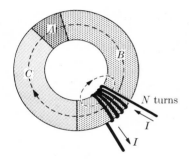

Figure 9–16 This magnetic circuit is equivalent to the magnetic circuit of Fig. 9–15 if the permeabilities of A, B, and C are large.

If materials B and C are ferromagnetic, but A represents an air gap, the two circuits are no longer equivalent because there is some *leakage* of flux from the ends of the solenoid in Fig. 9–16. How much flux leaks out of the circuit depends on the reluctance ratio of magnetic circuit to leakage path. When the air gap A is small compared with the length of the solenoid, the leakage flux is small and in approximate calculations may be neglected. Reluctance of the leakage path has been worked out for many common geometries, and is given in a number of standard reference books.* The circuit concept is certainly a cruder approximation in the magnetic case than in the electrical one because (1) the ratio of circuit reluctance to leakage reluctance is not as small as the corresponding resistance ratio of the electrical case, and (2) the lateral dimensions of the magnetic circuit are usually not negligible in comparison with its length; nevertheless, the magnetic circuit concept has proved to be extremely useful.

---

* See, for example, S. A. Nasar and L. E. Unnewehr, *Electromechanics and Rotating Electric Machines* (New York: Wiley, 1978); and F. N. Bradley, *Materials for Magnetic Functions* (New York: Hayden Book Co., 1971), p. 162.

*9–11 MAGNETIC CIRCUITS CONTAINING PERMANENT MAGNETS

The magnetic circuit concept is useful also when applied to permanent-magnet circuits, i.e., to flux circuits in which Φ has its origin in permanently magnetized material. We shall find it convenient to use the abbreviation P-M for permanent magnet. Because of the complicated *B-H* relationship in P-M material, the procedure outlined in the preceding section is not well suited to the problem at hand. Instead, we start again with Ampere's circuital law, applied now to the flux path of the P-M circuit:

$$\oint H \, dl = 0,$$

or

$$\int_a^b H \, dl = -\int_{b \, (\text{P-M})}^a H \, dl. \tag{9–73}$$

In writing Eq. (9–73) we assume explicitly that the P-M material lies between the points b and a of the flux path, whereas from a to b the flux path encounters no P-M material. The use of $B = \mu H$ and $\Phi = BA$ in the left side of Eq. (9–73) yields

$$\Phi \int_a^b \frac{dl}{\mu A} = -\int_{b \, (\text{P-M})}^a H \, dl. \tag{9–74a}$$

The magnetic flux Φ is continuous throughout the entire circuit, so $\Phi = B_m A_m$, where B_m is the magnetic field in the permanent magnet and A_m is its cross-sectional area. The right side of Eq. (9–74) may be written $-H_m l_m$, where H_m is the average magnetic intensity in the magnet and l_m is the length of the magnet. Thus

$$B_m A_m \mathscr{R}_{ab} = -H_m l_m \tag{9–74b}$$

is the equation that links the unknown quantities B_m and H_m. This equation can be solved simultaneously with the hysteresis curve of the magnet to yield both B_m and H_m.

As an example of a P-M circuit, consider the circuit composed of a magnet, an air gap, and soft iron (Fig. 9–17). It is important to realize that soft iron is *not* a P-M material; its hysteresis is actually negligible compared with that of the

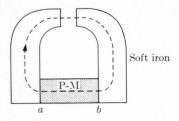
Soft iron
P-M
a   b

Figure 9–17 A permanent-magnet circuit. For the circuit shown, the magnet has a rather large demagnetizing field acting upon it; the demagnetizing field can be reduced by increasing the length of P-M material, for example, by placing additional magnets in the side arms of the circuit.

magnet, and $\mu_i = B_i/H_i$ is a positive quantity. The reluctance $\mathscr{R}_{ab}$ is given by

$$\mathscr{R}_{ab} = \frac{l_i}{\mu_i A_i} + \frac{l_g}{\mu_0 A_g}, \tag{9-75}$$

where the subscripts i and g refer to the soft iron and air gap, respectively. If the air gap is not too narrow, Eq. (9-75) may usually be approximated by

$$\mathscr{R}_{ab} \approx \frac{l_g}{\mu_0 A_g},$$

which, when combined with (9-74b), yields

$$B_m = -\frac{l_m A_g}{l_g A_m} \mu_0 H_m, \tag{9-76}$$

a linear relationship between B_m and H_m. This equation is plotted along with the hysteresis curve of the magnet in Fig. 9-18. The intersection of the two curves gives the operating point of the magnet. The problem is now essentially solved: from a knowledge of B_m, the flux Φ and the flux density B_g are easily determined.

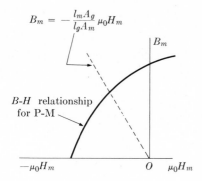

Figure 9-18 Demagnetizing line for a magnetic circuit. (The subscript m stands for magnet.) Since $\mu_0 H_m$ is plotted instead of H_m, the slope of the demagnetizing line is just $-(l_m A_g/l_g A_m)$, in other words, a pure number.

There are, however, two points which deserve to be mentioned. The first is: What does one use for the effective area A_g of the gap? As a first approximation, we might take A_g equal to the pole-face area of the soft iron, and if the air gap is not too large, this approximation is adequate. We shall not enter into a detailed discussion of this point, but instead refer the interested reader to the references cited in the previous section. Secondly, the problem of leakage flux is just as important in P-M circuits as it is in other types of magnetic circuits. For the problems presented in this book, however, we shall usually make the assumption that leakage flux may be neglected.

Finally, we note that H_m as determined from Fig. 9-18 is negative, i.e., the magnetic intensity in the magnet is a *demagnetizing* effect. This is a general result; when the magnetic flux has its origin in a permanent magnet, then the magnet itself is subjected to a demagnetizing field.

9–12 SUMMARY

In Chapter 4 we saw that the response of a (dielectric) medium to an E-field was a polarization charge density $\rho_P = -\nabla \cdot \mathbf{P}$ ($\sigma_P = \mathbf{n} \cdot \mathbf{P}$), where $\mathbf{P} = \chi \mathbf{E}$; in Chapter 7, a (conducting) medium was seen to respond to an E-field with a transport current density $\mathbf{J} = g\mathbf{E}$. We now find that a (magnetic) medium responds to a B-field with another sort of current density, the atomic magnetization current,

$$\mathbf{J}_M = \nabla \times \mathbf{M} \quad (\mathbf{j}_M = -\mathbf{n} \times \mathbf{M}),$$

where $\mathbf{M} = d\mathbf{m}/dv$ is the magnetic moment per unit volume of the material. The vector potential due to the magnetization is

$$\mathbf{A}(\mathbf{r}) = \frac{\mu_0}{4\pi} \int \frac{\mathbf{J}_M(\mathbf{r}')\, dv'}{|\mathbf{r} - \mathbf{r}'|}.$$

The total B-field due to steady transport currents plus magnetization currents satisfies

$$\nabla \times \mathbf{B} = \mu_0(\mathbf{J} + \mathbf{J}_M).$$

Note that $\nabla \cdot \mathbf{J}_M = 0$ identically. It is convenient to define the vector field

$$\mathbf{H} = \frac{1}{\mu_0}\, \mathbf{B} - \mathbf{M},$$

such that $\nabla \times \mathbf{H} = \mathbf{J}$ with only the conventional transport currents as sources. For a particular medium, the constitutive equation

$$\mathbf{M} = \chi_m(\mathbf{H})\mathbf{H}$$

must be known. Combined with the definition of $\mathbf{H}$, this gives

$$\mathbf{B} = \mu(\mathbf{H})\mathbf{H},$$

where $\mu = \mu_0(1 + \chi_m)$. Then this relation, together with the differential equations

$$\nabla \cdot \mathbf{B} = 0,$$

$$\nabla \times \mathbf{H} = \mathbf{J}$$

determines $\mathbf{B}$ and $\mathbf{H}$, subject to the boundary conditions

$$\mathbf{B}_{2n} - \mathbf{B}_{1n} = 0,$$

$$\mathbf{H}_{2t} - \mathbf{H}_{1t} = \mathbf{j} \times \mathbf{n}_2.$$

1. Most materials are either diamagnetic ($\chi_m < 0$) or paramagnetic ($\chi_m > 0$); in either case, $|\chi_m| \ll 1$. The magnetic materials of practical importance are ferromagnetic. For these $|\chi_m|$ may be more than 1000, but $\mathbf{B} = \mathbf{B}(\mathbf{H})$ is not linear and not even single-valued (hysteresis).

2. In problems with no transport currents, since $\nabla \times \mathbf{H} = 0$, it is convenient to use the scalar potential,

$$\mathbf{H} = -\nabla \varphi^*.$$

Since $\mathbf{V} \cdot \mathbf{B} = 0$, $\mathbf{V} \cdot \mathbf{H} = -\mathbf{V} \cdot \mathbf{M}$, and so φ^* satisfies Poisson's equation

$$\nabla^2 \varphi^* = \mathbf{V} \cdot \mathbf{M}.$$

A solution is

$$\varphi^*(\mathbf{r}) = \frac{1}{4\pi} \int_{V_0} \frac{\rho_M(\mathbf{r}') \, dv'}{|\mathbf{r} - \mathbf{r}'|} + \frac{1}{4\pi} \int_{S_0} \frac{\sigma_M(\mathbf{r}') \, da'}{|\mathbf{r} - \mathbf{r}'|},$$

where $\rho_M = -\mathbf{V} \cdot \mathbf{M}$, $\sigma_M = \mathbf{n} \cdot \mathbf{M}$. (This is useful if $\mathbf{M}$ is a given function.)

3. In such problems with linear media or with uniform $\mathbf{M}$, $\mathbf{V} \cdot \mathbf{H} = 0$ and φ^* satisfies Laplace's equation. These problems are identical to the corresponding electrostatic problems with no external charge density.

4. With transport currents, the integral of $\mathbf{V} \times \mathbf{H} = \mathbf{J}$ is Ampere's law

$$\oint \mathbf{H} \cdot d\mathbf{l} = I.$$

5. The solution for $\mathbf{H}$ divides into one part due to transport currents and another part due to the magnetic materials:

$$\mathbf{H}(\mathbf{r}) = \frac{1}{4\pi} \int \frac{\mathbf{J}(\mathbf{r}') \times (\mathbf{r} - \mathbf{r}')}{|\mathbf{r} - \mathbf{r}'|^3} \, dv' - \nabla\varphi^*(\mathbf{r}).$$

(Evaluation of the second term depends on knowing $\mathbf{M}(\mathbf{H})$.)

6. In the presence of ferromagnetic materials with large μ, it may be a useful approximation to assume that all of the flux Φ is confined to a known volume. Then

$$NI = \Phi\mathscr{R}$$

where the reluctance

$$\mathscr{R} = \frac{l}{\mu A}$$

can be calculated for each element of the magnetic circuit.

PROBLEMS

9–1 A permanent magnet has the shape of a right circular cylinder of length L. If the magnetization $\mathbf{M}$ is uniform and has the direction of the cylinder axis, find the magnetization current densities, J_M and j_M. Compare the current distribution with that of a solenoid.

9–2 Find the distribution of magnetization currents corresponding to a uniformly magnetized sphere with magnetization $\mathbf{M}$. According to Eq. (9–63) the magnetic induction $\mathbf{B}$ is uniform inside such a sphere. Can you use this information to design a current winding that will produce a uniform magnetic field in a spherical region of space?

9–3 (a) The magnetic moment of a macroscopic body is defined as $\int_V \mathbf{M} \, dv$. Prove the relationship

$$\int_V \mathbf{M} \, dv = \int_V \mathbf{r}\rho_M \, dv + \oint_S \mathbf{r}\sigma_M \, da,$$

where S is the surface bounding V. (*Hint:* Refer to the similar problem involving **P** in Chapter 4.) (b) A permanent magnet in the shape of a sphere of radius R has uniform magnetization $\mathbf{M}_0$ in the direction of the polar axis. Determine the magnetic moment of the magnet from both the right and left sides of the equation in part (a).

9–4 (a) Given a magnet with magnetization specified: $\mathbf{M}(x, y, z)$. Each volume element dv may be treated as a small magnetic dipole, $\mathbf{M}\, dv$. If the magnet is placed in a uniform magnetic field $\mathbf{B}_0$, find the torque on the magnet in terms of its magnetic moment (defined in Problem 9–3). (b) A magnet in the shape of a right circular cylinder of length L, and cross section A is uniformly magnetized parallel to the cylinder axis with magnetization $\mathbf{M}_0$. The magnet is placed in a uniform magnetic field $\mathbf{B}_0$. Find the torque on the magnet in terms of its pole densities.

9–5 An ellipsoid with principal axes of lengths $2a$, $2a$, and $2b$ is magnetized uniformly in a direction parallel to the $2b$-axis. The magnetization of the ellipsoid is $\mathbf{M}_0$. Find the magnetic pole densities for this geometry.

9–6 Given a spherical shell, inside radius R_1 and outside radius R_2, which is uniformly magnetized in the direction of the z-axis. The magnetization in the shell is $\mathbf{M}_0 = M_0\mathbf{k}$. Find the scalar potential φ^* for points on the z-axis, both inside and outside the shell.

9–7 A permanent magnet in the shape of a right circular cylinder of length L and radius R is oriented so that its symmetry axis coincides with the z-axis. The origin of coordinates is at the center of the magnet. If the cylinder has uniform axial magnetization M, (a) determine $\varphi^*(z)$ at points on the symmetry axis, both inside and outside the magnet, and (b) use the results of part (a) to find the magnetic induction B_z at points on the symmetry axis, both inside and outside the magnet.

9–8 A sphere of magnetic material of radius R is placed at the origin of coordinates. The magnetization is given by $\mathbf{M} = (ax^2 + b)\mathbf{i}$, where a and b are constants. Determine all pole densities and magnetization currents.

9–9 An annealed iron ring of mean length 15 cm is wound with a toroidal coil of 100 turns. Determine the magnetic induction in the ring when the current in the winding is (a) 0.1 A, (b) 0.2 A, and (c) 1.0 A.

9–10 A soft-iron ring with a 1.0 cm air gap is wound with a toroidal winding such as is shown in Fig. 9–14. The mean length of the iron ring is 20 cm, its cross section is 4 cm^2, and its permeability, assumed constant, is 3000 μ_0. The 200-turn winding carries a current of 10 A. Find B and H inside the iron ring and in the air gap.

9–11 A long cylinder of radius a and permeability μ is placed in a uniform magnetic field $\mathbf{B}_0$ such that the cylinder axis is at right angles to $\mathbf{B}_0$. Calculate the magnetic induction inside the cylinder. Make a semiquantitative sketch showing typical lines of induction through the cylinder. (Assume from the beginning that φ^* can be completely specified in terms of the $\cos\theta$ cylindrical harmonics. This assumption is justified, since all boundary conditions can be satisfied in terms of the $\cos\theta$ harmonics.)

9–12 A long straight copper wire and a long straight iron wire each carry the same current I in a uniform B-field, $\mathbf{B}_0$. Show that the force on the iron wire is nearly twice the force on the copper wire. (*Hint:* Use the result of Problem 9–11.)

9–13 A wire carrying a current I is in a cylindrical iron conduit. The conduit has inner and outer radii a and b, has constant susceptibility χ, and is coaxial with the wire. Find the magnetization current density and the total magnetization current.

9-14 Two magnetic media are separated by a plane interface. Show that the angles between the normal to the boundary and the B-fields on either side satisfy

$$\mu_2 \tan \theta_1 = \mu_1 \tan \theta_2.$$

9-15 A straight wire carrying a current I is parallel to an infinite plane, at a distance d from the plane. The wire is in air, and the medium on the far side of the plane boundary has constant permeability μ. Find an image current that gives the correct B-field in the air if (a) $\mu = \infty$, (b) $\mu = 0$. (Case (a) approximates a ferromagnetic material and (b) describes a superconductor.)

*9-16 Calculate the demagnetizing factor of a long cylinder that is permanently magnetized at right angles to the cylinder axis. The magnetization is uniform.

*9-17 A long cylindrical shell (outside radius b, inside radius a, relative permeability K_m) is oriented normal to a uniform magnetic induction field $\mathbf{B}_0$. (a) Show that the magnetic induction $\mathbf{B}_i$ in the vacuum region inside the shell is parallel to $\mathbf{B}_0$. (b) Show that the magnetic shielding factor h_m is given by

$$h_m \equiv \frac{B_0}{B_i} = 1 + \frac{(K_m - 1)^2}{4K_m}\left(1 - \frac{a^2}{b^2}\right).$$

9-18 A magnetic circuit in the shape of

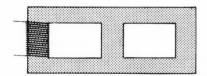

is wound with 100 turns of wire carrying a current of 1 A. The winding is located on the extreme left-hand leg of the circuit. The height of the circuit is 10 cm, its length is 20 cm, the cross section of each leg is 6 cm^2, and its permeability, assumed constant, is 5000 μ_0. Neglecting leakage, calculate the magnetic flux through the central leg and also through the extreme right-hand leg of the circuit.

9-19 A magnetic circuit in the form shown in Fig. 9-17 has a permanent magnet of length 8 cm, a soft-iron path length of 16 cm, and an air gap of 0.8 cm. The cross section of the iron and of the magnet is 4 cm^2 on the average, whereas the effective cross-sectional area of the air gap is 3 cm^2. The relative permeability of the iron is 5000. (a) Calculate the magnetic flux density in the gap for two different magnet materials: sintered oxide and 35 percent Co steel. Neglect leakage. (b) The dimensions of the magnetic circuit are altered in *one* respect: the air gap is decreased to 0.8 mm. Repeat the calculation called for in part (a).

9-20 Find the magnetic induction in a uniformly magnetized sphere for each of the materials shown in Fig. 9-9.

9-21 A magnetic circuit in the form shown in Fig. 9-17 has an Alnico V magnet of length 10 cm, a soft-iron path of 16 cm, and an air gap of 1 cm. It is also wound with 800 ampere-turns of wire (in a direction to aid the flux produced by the magnet). Find the magnetic flux density in the air gap. (Neglect leakage, take $K_m = 5000$ for the soft iron, and assume that the cross sections of the magnet, soft iron, and air gap are the same.)

*CHAPTER 10 Microscopic Theory of Magnetism

In the preceding chapter we were concerned with the macroscopic aspects of magnetization. The magnetic properties of matter were introduced explicitly through the function $\mathbf{M}$, and this was related to the magnetic induction by means of experimentally determined parameters. We now look at matter from the microscopic point of view (i.e., as an assembly of atoms or molecules) and see how the individual molecules respond to an imposed magnetic field. If this procedure were carried through completely, we should end up with theoretical expressions for susceptibility, and *B-H* relationships for all types of materials. Such a procedure is certainly beyond the scope of this book; nevertheless, we can show rather simply how the various kinds of magnetic behavior come about and, in addition, derive expressions that predict the correct order of magnitude for susceptibility in certain cases. A more thorough discussion of the topics presented here is to be found in books on solid state physics.†

In the macroscopic formulation we dealt with two field vectors, $\mathbf{B}$ and $\mathbf{H}$, which we related through the equation $\mathbf{B} = \mu_0(\mathbf{H} + \mathbf{M})$. From the microscopic viewpoint the distinction between $\mathbf{B}$ and $\mathbf{H}$ largely disappears because we deal with an assembly of molecules (i.e., with an assembly of magnetic dipoles or dipole groups) in vacuum. We are concerned with the magnetic field near a molecule in vacuum or at the position of a molecule when that molecule is removed from the system. Thus $\mathbf{B}_m = \mu_0\mathbf{H}_m$. Here the subscript m stands for microscopic, but in the following sections of this chapter the symbol $\mathbf{B}_m$ (and $\mathbf{H}_m$) will denote a particular value of the microscopic field, namely, the field at the position of a molecule.

It is customary when discussing the microscopic field inside matter to relate $\mathbf{H}_m$ to the macroscopic $\mathbf{H}$ field, instead of $\mathbf{B}_m$ to the $\mathbf{B}$ field, because both $\mathbf{H}$ and $\mathbf{H}_m$ can be written simply in terms of integrals over the current and dipole distributions. It makes very little difference, however, whether we calculate $\mathbf{H}_m$ or $\mathbf{B}_m$, since they differ from each other only by the scale factor μ_0.

---

* This chapter may be omitted without loss of continuity.

† See, for example, C. Kittel, *Introduction to Solid State Physics*, Fifth Edition (New York: Wiley, 1976), Chapters 14 and 15.

10–1 MOLECULAR FIELD INSIDE MATTER

The magnetic field that is effective in its interaction with atomic currents in an atom or molecule is called the molecular field $\mathbf{B}_m = \mu_0\mathbf{H}_m$. In some textbooks it is called the *local field*. This is the magnetic field at a molecular (or atomic) position in the material; it is produced by all external sources and by all molecular dipoles in the material *with the exception* of the one molecule (or atom) at the point under consideration. It is evident that $\mathbf{B}_m$ need not be the same as the macroscopic magnetic induction field, since the latter quantity is related to the force on a current element whose dimensions are large compared with molecular dimensions.

The molecular field may be calculated by a procedure similar to that of Section 5–1 for the molecular electric field in a dielectric. We consider a material object of arbitrary shape, which for convenience we take to be uniformly magnetized with magnetization $\mathbf{M}$. Let us cut out a small piece of the object, leaving a spherical cavity surrounding the point at which the molecular field is to be computed. The material which is left is to be treated as a continuum, i.e., from the macroscopic point of view. Next we put the material back into the cavity, molecule by molecule, except for the molecule at the center of the cavity, where we wish to compute the molecular field. The molecules that have just been replaced are to be treated, not as a continuum, but as individual dipoles or dipole groups.

The *macroscopic* field $\mathbf{H}$, the magnetic intensity in the specimen, can be expressed, according to Eq. (9–27), as

$$\mathbf{H} = \frac{1}{4\pi}\int \frac{\mathbf{J}\times(\mathbf{r}-\mathbf{r}')}{|\mathbf{r}-\mathbf{r}'|^3}\,dv' + \frac{1}{4\pi}\int \frac{\rho_M(\mathbf{r}-\mathbf{r}')}{|\mathbf{r}-\mathbf{r}'|^3}\,dv' + \frac{1}{4\pi}\int_S \frac{\sigma_M(\mathbf{r}-\mathbf{r}')}{|\mathbf{r}-\mathbf{r}'|^3}\,da',$$

where the integrals extend over all sources: $\mathbf{J}$, ρ_M, and σ_M. The molecular field $\mathbf{H}_m$ may be expressed in a similar way, except that now there are additional contributions from the surface of the cavity and from the individual dipoles in the cavity. The integral of $\rho_M(\mathbf{r}-\mathbf{r}')/|\mathbf{r}-\mathbf{r}'|^3$ over the cavity volume need not be excluded specifically, since $\rho_M = -\nabla\cdot\mathbf{M} = 0$ in the uniformly magnetized specimen. Thus

$$\mathbf{H}_m = \mathbf{H} + \mathbf{H}_s + \mathbf{H}', \tag{10–1}$$

where $\mathbf{H}$ is the macroscopic magnetic intensity in the specimen, $\mathbf{H}_s$ is the contribution from the surface pole density $\sigma_M = M_n$ on the cavity surface, and $\mathbf{H}'$ is the contribution of the various dipoles inside the cavity.

From the corresponding derivation in Section 5–1, $\mathbf{H}_s$ is seen to be

$$\mathbf{H}_s = \tfrac{1}{3}\mathbf{M}. \tag{10–2}$$

Furthermore, the dipole contribution,

$$\mathbf{H}' = \frac{1}{4\pi}\sum_i\left[\frac{3(\mathbf{m}_i\cdot\mathbf{r}_i)\mathbf{r}_i}{r_i^5} - \frac{\mathbf{m}_i}{r_i^3}\right], \tag{10–3}$$

where $\mathbf{r}_i$ is the distance from the ith dipole to the center of the cavity, is of the same form as the corresponding electric dipole term $\mathbf{E}'$ in Section 5–1. Thus if we restrict our interest to the rather large class of materials for which Eq. (10–3) vanishes, Eq. (10–1) reduces to

$$\mathbf{H}_m = \mathbf{H} + \tfrac{1}{3}\mathbf{M}, \qquad (10\text{–}4)$$

and

$$\mathbf{B}_m = \mu_0 \mathbf{H}_m. \qquad (10\text{–}5)$$

Equations (10–4) and (10–5) give the molecular field in terms of the macroscopic magnetic intensity and the magnetization in the sample. For most diamagnetic and paramagnetic materials the term $\tfrac{1}{3}\mathbf{M} = \tfrac{1}{3}\chi_m\mathbf{H}$ is negligibly small, but for ferromagnetic materials the correction is quite important.

10–2 ORIGIN OF DIAMAGNETISM

In order to calculate the diamagnetic susceptibility of an assembly of atoms we must know something about the electronic motion in the atom itself. We shall assume that each electron circulates around the atomic nucleus in some kind of orbit, and for simplicity we choose a circular orbit of radius R in a plane at right angles to the applied magnetic field. Quantum mechanics tells us that although this picture is approximately correct, the electrons do not circulate in well-defined orbits. To solve the problem properly, we would have to solve the Schroedinger equation for an atomic electron in a magnetic field; nevertheless, our rather naive "classical" calculation will give the correct order of magnitude for the diamagnetic susceptibility.

Before the magnetic induction field is applied, the electron is in equilibrium in its orbit:

$$F_q = m_e \omega_0^2 R, \qquad (10\text{–}6)$$

where F_q is the electric force holding the electron to its atom, ω_0 is the angular frequency of the electron in its orbit, and m_e is the electron mass. Application of a magnetic field exerts an additional force $-e\mathbf{v} \times \mathbf{B}_m$ on the electron; assuming that the electron stays in the same orbit, we find

$$F_q \pm e\omega R B_m = m_e \omega^2 R,$$

which, when combined with Eq. (10–6), yields

$$\pm e\omega B_m = m_e(\omega - \omega_0)(\omega + \omega_0). \qquad (10\text{–}7)$$

The quantity $\Delta\omega = \omega - \omega_0$ is the change in angular frequency of the electron. Thus the electron either speeds up or slows down in its orbit, depending on the detailed geometry (i.e., on the direction of $\mathbf{v} \times \mathbf{B}_m$ relative to $\mathbf{F}_q$), but in either case the *change* in orbital magnetic moment is in a direction *opposite* to the applied field. This statement may be easily verified by the reader.

Even for the largest fields that can be obtained in the laboratory (~ 100 T), $\Delta\omega$ is small compared with ω_0, so that (10–7) may always be approximated by

$$\Delta\omega = \pm\frac{e}{2m_e}\, B_m. \qquad (10\text{–}8)$$

The quantity $(e/2m_e)B_m$ is known as the Larmor frequency.

Up to this point we have merely *assumed* that the electron stays in the same orbit. We have used this assumption together with the equilibrium of forces to derive Eq. (10–8). For the electron to stay in its orbit, the change in its kinetic energy as determined from Faraday's law of induction must be consistent with Eq. (10–8). When the magnetic field is switched on, there is a change in flux through the orbit given by $\pi R^2\, \Delta B_m$. This flux is linked by Δn electron loops, where Δn is the number of revolutions made by the electron during the time in which the field changes. The changing flux produces an emf according to Faraday's law:

$$\mathscr{E} = \pi R^2\, \frac{dB_m}{dt}\, \Delta n = \pi R^2\, \frac{dn}{dt}\, \Delta B_m. \qquad (10\text{–}9)$$

The energy given to the electron in this process is $\mathscr{E}e$, and this appears as a change in kinetic energy:

$$\tfrac{1}{2}m_e R^2(\omega^2 - \omega_0^2) = e\pi R^2\, \frac{dn}{dt}\, \Delta B_m. \qquad (10\text{–}10)$$

But ΔB_m is just the final value of the field B_m, and the average value of $dn/dt = (\omega + \omega_0)/4\pi$. Thus

$$\Delta\omega = \frac{e}{2m_e}\, B_m,$$

in agreement with Eq. (10–8). Thus the assumption of a constant orbit leads to no contradiction between Eq. (10–9) and the force equation. Diamagnetism is the result of Lenz's law operating on an atomic scale. Upon the application of a magnetic field, the electronic currents in each atom are modified in such a way that they tend to weaken the effect of this field.

The change in angular velocity predicted by Eq. (10–8) produces a change in magnetic moment given by

$$\Delta\mathbf{m} = -\frac{e}{2\pi}\, \pi R^2\, \frac{e}{2m_e}\, \mathbf{B}_m$$

$$= -\frac{e^2}{4m_e}\, R^2 \mu_0 \mathbf{H}_m. \qquad (10\text{–}11)$$

In order to find the magnetization, this result must be summed over all electrons in a unit volume. For a substance containing N molecules per unit volume, all of the same molecular species,

$$\mathbf{M} = -\frac{Ne^2 \mu_0}{4m_e} \mathbf{H}_m \sum_i R_i^2, \tag{10-12}$$

where the summation is over the electrons in one molecule. For diamagnetic materials, $\mathbf{H}_m$ differs very little from $\mathbf{H}$, so the diamagnetic susceptibility

$$\chi_m = -\frac{Ne^2 \mu_0}{4m_e} \sum_i R_i^2. \tag{10-13a}$$

This result has been obtained on the assumption that all electrons circulate in planes perpendicular to the field $\mathbf{H}_m$. When the orbit is inclined, so that a normal to the orbit makes an angle θ_i with the field, only the component of $\mathbf{H}_m$ along this normal $(H_m \cos \theta_i)$ is effective in altering the angular velocity of the electron. Furthermore, the component of $\Delta\mathbf{m}$ parallel to the field is smaller by the factor $\cos \theta_i$. Hence a better approximation to the diamagnetic susceptibility is

$$\chi_m = -\frac{Ne^2 \mu_0}{4m_e} \sum_i R_i^2 \cos^2 \theta_i. \tag{10-13b}$$

Diamagnetism is presumably present in all types of matter, but its effect is frequently masked by stronger paramagnetic or ferromagnetic behavior that can occur in the material simultaneously. Diamagnetism is particularly prominent in materials that consist entirely of atoms or ions with "closed electron shells," since in these cases all paramagnetic contributions cancel out.

10–3 ORIGIN OF PARAMAGNETISM

The orbital motion of each electron in an atom or molecule can be described in terms of a magnetic moment; this follows directly from Eq. (8–22). In addition, it is known that the electron has an intrinsic property called *spin*, and an intrinsic magnetic moment associated with this spinning charge. Each molecule, then, has a magnetic moment $\mathbf{m}_i$ which is the vector sum of orbital and spin moments from the various electrons in the molecule. Briefly, paramagnetism results from the tendency of these molecular moments to align themselves with the applied field, just as the current circuit of Eq. (8–19) tends to align itself with the field.

The situation is not quite so straightforward as that for a current circuit, however. There are, in fact, two complications: (1) in the presence of a magnetic field the electronic motions are quantized such that each orbital and spin moment has only a discrete set of orientations relative to the field direction. Furthermore, no two electrons in the molecule can occupy the same quantum state, so that if there are just enough electrons per molecule to fill "electron shells," then all possible orientations must be used and $\mathbf{m}_i$ is zero. Of course, paramagnetism can occur only when $\mathbf{m}_i \neq 0$. (2) The electronic motion inside an atom that gives rise to $\mathbf{m}_i$ also produces an angular momentum about the atomic nucleus; in fact, $\mathbf{m}_i$ is linearly related to this angular momentum. Under these conditions the magnetic torque does not directly align the dipole moment $\mathbf{m}_i$ with the field, but causes it to precess around the field at constant inclination.* The atoms (or molecules) in our material system are in thermal contact with each other. In a gas or liquid the atoms are continually making collisions with one another; in a solid the atoms are undergoing thermal oscillation. Under these conditions the various $\mathbf{m}_i$ can interchange magnetic energy with the thermal energy of their environment and make transitions from one precessional state to another of a different inclination. The thermal energy of the system tries to act in such a way as to produce a completely random orientation of the $\mathbf{m}_i$, but orientations along or near the field direction have a lower magnetic energy and thus are favored. The situation is quite similar to that of polar molecules in an electric field, which was discussed in Section 5–3.

For a material composed entirely of one molecular species, each molecule having magnetic moment m_0, the fractional orientation is given approximately by the Langevin function, Eq. (5–21), with

$$y = \frac{m_0 \mu_0 H_m}{kT}. \tag{10–14}$$

The magnetization is given by

$$|\mathbf{M}| = Nm_0 \left[\coth y - \frac{1}{y} \right], \tag{10–15a}$$

where N is the number of molecules per unit volume. Except for temperatures near absolute zero, the Langevin function can be approximated by the first term in its power series:

$$\mathbf{M} = \frac{Nm_0^2}{3kT} \mu_0 \mathbf{H}_m, \tag{10–15b}$$

which yields the paramagnetic susceptibility

$$\chi_m = \frac{Nm_0^2 \mu_0}{3kT}. \tag{10–16}$$

According to atomic theory, m_0 is in the range of a few Bohr magnetons (1 Bohr magneton $= eh/4\pi m_e$, where h is Planck's constant). Equations (10–16) and (10–13b) account for the order of magnitude of the χ_m's in Table 9–1.

We may summarize the results of this section briefly as follows: In order to exhibit paramagnetic behavior, the atoms (or molecules) of the system must have permanent magnetic moments, and these tend to orient in the applied field. The various molecular moments are decoupled, i.e., they precess around the magnetic field as individuals (not in unison), but they are able to exchange energy because of thermal contact with their environment. Except for temperatures near absolute zero and simultaneous large fields, the magnetization is far below the saturation value which would obtain when all the dipole moments are aligned.

---

* A discussion of the precession of $\mathbf{m}_i$ in a uniform magnetic field is given in many textbooks. See, for example, H. Goldstein, *Classical Mechanics* (Reading, Mass.: Addison-Wesley, 1950), pp. 176–7.

10–4 THEORY OF FERROMAGNETISM

In ferromagnetic materials the atomic (or molecular) magnetic moments are very nearly aligned even in the absence of an applied field. The cause of this alignment is the molecular field $\mathbf{H}_m$ which, according to Eq. (10–4), does not vanish when $\mathbf{H} = 0$ unless $\mathbf{M}$ vanishes simultaneously. A magnetization $\mathbf{M}$ does give rise to a molecular field, but unless this molecular field *produces* the same magnetization $\mathbf{M}$ that is presumed to exist in the material, the solution is inconsistent. Our problem is to determine in what set of circumstances the magnetization can maintain itself via the molecular field.

It will prove necessary to generalize Eq. (10–4) to a certain extent. For the molecular field, let us write $\mathbf{H}_m = \mathbf{H} + \gamma \mathbf{M}$, which, for $\mathbf{H} = 0$, reduces to

$$\mathbf{H}_m = \gamma \mathbf{M}. \tag{10–4a}$$

According to the simple theory of Section 10–1, $\gamma = \frac{1}{3}$. If the terms in Eq. (10–3) do not sum to zero, γ may be different from $\frac{1}{3}$; nevertheless, we expect γ to be of this order of magnitude.

Let us restrict our attention to a material composed entirely of one atomic species, each atom having magnetic moment m_0. There are N atoms per unit volume. If the atomic moments are to be very nearly aligned, M must be a substantial fraction of Nm_0; for the sake of definiteness, however, let us say

$$M > 0.7Nm_0. \tag{10–17}$$

According to Eq. (10–15), this implies that $[\coth y - (1/y)] > 0.7$, or y [which is defined by Eq. (10–14)] > 3. Thus

$$y = \frac{m_0 \mu_0 H_m}{kT} > 3,$$

which, when combined with Eqs. (10–4a) and (10–17), yields

$$0.7 \frac{\gamma N \mu_0 m_0^2}{kT} > 3. \tag{10–18}$$

This (approximately) is the condition for the occurrence of ferromagnetism.

In the previous section it was stated that atomic theory predicts m_0 to be in the range of a few Bohr magnetons. On this basis, Eq. (10–18) requires a γ of about 10^3, which is orders of magnitude larger than can be accounted for in the derivation presented in Section 10–1. It would thus appear that the origin of ferromagnetism is considerably more complex than the corresponding situation in ferroelectrics (discussed in Section 5–4).

In 1907 Pierre Weiss* formulated his theory of ferromagnetism. Weiss appreciated the essential role played by the molecular field; he could not explain the large value of γ, but he accepted it as a fact and proceeded to develop his theory

---

* P. Weiss, *Journal de Physique*, vol. 6, p. 667 (1907).

from this point. The predictions of his theory were found to be in close accord with experiment. For this reason the molecular field of Eq. (10–4a) is often called the *Weiss molecular field*.

It was left to Heisenberg,* some twenty years later, to explain the origin of the large value of γ. Heisenberg showed, first, that it is only the spin magnetic moments that contribute to the molecular field, and second, that the field is produced basically by electrostatic forces. On the basis of quantum mechanics he showed that when the spins on neighboring atoms change from parallel alignment to antiparallel alignment, there must be a simultaneous change in the electron charge distribution in the atoms. The change in charge distribution alters the electrostatic energy of the system and, in certain cases, favors parallel alignment (i.e., ferromagnetism). A spin-dependent energy, i.e., an energy that depends on the spin configuration of the system, can be viewed in terms of the force (or torque) that is produced on one of the atoms when the configuration is altered. The equivalent field turns out to be proportional to **M**, but with a coefficient that depends in detail upon the charge distribution in the atom under consideration.

The Weiss-Heisenberg theory can be used to predict the way in which the magnetization of a ferromagnet changes with temperature. It is evident that the theory depicts ferromagnetism as the limiting case of paramagnetism in an extremely large magnetic field, but with this field coming from the magnetization itself. Combining Eq. (10–4a) with Eqs. (10–14) and (10–15) yields

$$M = Nm_0 \left[\coth y - \frac{1}{y} \right], \tag{10–19}$$

and

$$M = \frac{kTy}{\gamma \mu_0 m_0}. \tag{10–20}$$

The *spontaneous magnetization*, i.e., the magnetization at zero external field, for a given temperature is obtained from the simultaneous solution of Eqs. (10–19) and (10–20). This is easily done by a graphical procedure: We plot M versus y for *both* (10–19) and (10–20), as shown in Fig. 10–1. The intersection of the two curves

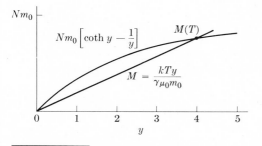

Figure 10–1 Determination of the spontaneous magnetization $M(T)$ with the aid of the Langevin function.

---

* W. Heisenberg, *Zeitschrift für Physik*, vol. 49, p. 619 (1928).

gives a magnetization $M(T)$ that is consistent with both equations. As the temperature is increased, the linear curve, Eq. (10–20), becomes steeper, but Eq. (10–19) is unchanged. Thus the intersection point moves to the left in the figure, and a lower value for the spontaneous magnetization obtains. Finally, a temperature is reached at which Eq. (10–20) is tangent to (10–19) at the origin; at this and higher temperatures the spontaneous magnetization is zero. This temperature is the *Curie temperature*, T_c, above which the spontaneous magnetization vanishes and ordinary paramagnetic behavior results.

A plot of $M(T)$ versus temperature, obtained according to the above procedure, is displayed in Fig. 10–2. It is in approximate agreement* with experimentally determined values of the spontaneous magnetization for a ferromagnetic material.

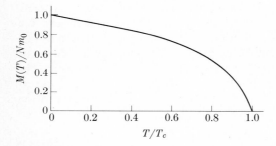

Figure 10–2 The magnetization of a ferromagnetic material as a function of temperature. T_c is called the Curie temperature. (The curve shown has been calculated with the aid of the classical Langevin function; quantum-mechanical corrections change the shape of the curve somewhat, bringing it into agreement with experimental data.)

10–5 FERROMAGNETIC DOMAINS

According to the preceding section, a ferromagnetic specimen should be magnetized very nearly to saturation (regardless of its previous history) at temperatures below the Curie temperature. This statement appears to be contrary to observation. We know, for example, that a piece of iron can exist in either a magnetized or unmagnetized condition. The answer to this apparent paradox is that a ferromagnetic material breaks up into *domains*; each domain is fully magnetized in accord with the results of the preceding section, but the various domains can be randomly oriented (Fig. 10–3) and thus present an unmagnetized appearance from the macroscopic point of view. The presence of domains was first postulated by Weiss in 1907.

---

* Detailed quantum corrections to the theory presented here bring the theoretical curve into good agreement with experiment.

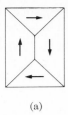

(a)

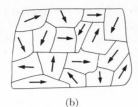

(b)

Figure 10–3 Ferromagnetic domain structures: (a) single crystal, (b) polycrystalline specimen. Arrows represent the direction of magnetization.

In passing from one domain to an adjacent one, the atomic moment vector m_0 gradually rotates from its original to its new direction in the course of about 100 atoms (Fig. 10–4). This region between the two domains is called a *domain wall*. It would appear that an atomic spin moment in the wall region is subjected to a slightly lower molecular field than is an atomic spin moment inside the domain proper. This observation by itself would favor a single domain configuration. On the other hand, a specimen consisting of a single domain must maintain a large external magnetic field, whereas a multidomain specimen has a lower "magnetic energy" associated with its field structure. Thus the multidomain structure is usually energetically favored.

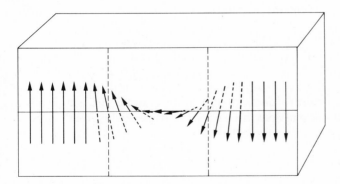

Figure 10–4 Structure of transition region, or "Bloch wall," between domains in a ferromagnetic material.

The macroscopic aspects of magnetization in ferromagnetic materials are concerned with changes in domain configuration. The increase in magnetization resulting from the action of an applied magnetic field takes place by two independent processes: by an increase in the volume of domains that are favorably oriented relative to the field at the expense of domains that are unfavorably oriented (domain wall motion), or by rotation of the domain magnetization toward the field direction. The two processes are illustrated schematically in Fig. 10–5.

In weak applied fields the magnetization usually changes by means of domain wall motion. In pure materials consisting of a single phase, the wall motion is to a

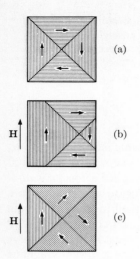

(a)

(b)

H

(c)

H

Figure 10–5 Magnetization of a ferromagnetic material: (a) unmagnetized, (b) magnetization by domain wall motion, (c) magnetization by domain rotation.

large extent reversible in the weak-field region. In stronger fields the magnetization proceeds by irreversible wall motion, and finally by domain rotation; in these circumstances the substance remains magnetized when the external magnetic field is removed.

The experimental study of domains was made possible by a technique first developed by F. H. Bitter.* A finely divided magnetic powder is spread over the surface of the specimen, and the powder particles, which collect along the domain boundaries, may be viewed under a microscope. By means of this technique, it has even proved possible to observe domain wall motion under the action of an applied magnetic field. The size of domains varies widely, depending on the type of material, its previous history, etc.; typical values are in the range from 10^{-6} to 10^{-2} cm^3.

10–6 FERRITES

According to the Heisenberg theory of ferromagnetism, there is a change in electrostatic energy associated with the change from parallel to antiparallel spin alignment of neighboring atoms. If this energy change favors parallel alignment and is at the same time of sufficient magnitude, the material composed of these atoms is ferromagnetic. If the energy change favors antiparallel alignment, it is still possible to obtain an *ordered* spin structure, but with spins alternating from atom to atom as the crystal is traversed.

An ordered spin structure with zero net magnetic moment is called an *antiferromagnet* (Fig. 10–6b). The most general ordered spin structure contains both

---

* F. H. Bitter, *Physical Review*, vol. 41, p. 507 (1932). For a brief discussion of the technique, see B. D. Cullity, *Introduction to Magnetic Materials* (Reading, Mass.: Addison-Wesley, 1972), p. 293.

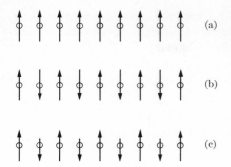

Figure 10-6 Schematic representation of atomic spins in ordered spin structures: (a) ferromagnetic, (b) antiferromagnetic, (c) ferrimagnetic.

"spin-up" and "spin-down" components but has a net, nonzero magnetic moment in one of these directions; such a material is called a *ferrimagnet* or simply a *ferrite*. The simplest ferrites of magnetic interest are oxides represented by the chemical formula $MOFe_2O_3$, where M is a divalent metal ion such as Co, Ni, Mn, Cu, Mg, Zn, Cd, or divalent iron. These ferrites crystallize in a rather complicated crystal structure known as the *spinel structure*. The classic example of a ferrite is the mineral magnetite (Fe_3O_4), which has been known since ancient times.

Ferrites are of considerable technical importance because, in addition to their relatively large saturation magnetization, they are poor conductors of electricity. Thus they can be used for high-frequency applications where the eddy-current losses in conducting materials pose serious problems. Typical resistivities of ferrites fall in the range from 1 to $10^4 \, \Omega \cdot m$; for comparison, the electrical resistivity of iron is approximately $10^{-7} \, \Omega \cdot m$.

10-7 SUMMARY

The macroscopic magnetization M of a magnetic material results from the molecular magnetic dipole moment (or its component), which arises in response to the local field at the molecule—the molecular field H_m. The molecular field depends on the applied field H and also on the magnetization itself. The latter contribution that stems from the dipole magnetic fields of all the other molecules, to give

$$H_m = H + \tfrac{1}{3}M$$

in analogy to the dielectric case, is negligibly small for most linear materials because of the smallness of the magnetic susceptibility in

$$M = \chi_m H.$$

Nevertheless spontaneous magnetization occurs in ferromagnetic materials because the contribution of the magnetization to the effective molecular field has a coefficient much greater than $\tfrac{1}{3}$.

1. All molecules exhibit an induced magnetic dipole moment in a magnetic field due to the deformation of the electronic current distribution. The response is always such as to weaken the applied field—i.e., the (diamagnetic) contribution to the susceptibility is always negative. A linear approximation leads to the constant diamagnetic susceptibility,

$$\chi_m = -\frac{Ne^2\mu_0}{4m_e} \sum_i R_i^2.$$

2. Molecules which have a permanent magnetic dipole moment m_0 show an additional orientational response. This is described approximately by the Langevin function, as for polar molecules in an electric field. Except near absolute zero, the resulting paramagnetic susceptibility is

$$\chi_m = \frac{Nm_0^2\mu_0}{3kT}.$$

3. In order to understand ferromagnetism, one assumes that

$$H_m = H + \gamma M,$$

with $\gamma \gg \frac{1}{3}$. (This contribution derives from a quantum mechanical energy that depends on the relative orientation of spin magnetic moments; it adds to the magnetic energy $\mathbf{m}_0 \cdot \mathbf{H}$ and can therefore be expressed in terms of an effective magnetic field, even though it is electrostatic in origin.) Then this equation and the Langevin equation admit a solution with $H = 0$, $M \neq 0$, so long as T is below the Curie temperature.

4. Even below the Curie temperature, a macroscopic piece of ferromagnetic material may show no net magnetic moment because of its domain structure.

PROBLEMS

10–1 A Bohr magneton is defined as the magnetic moment of an electron circulating in the classic "Bohr orbit" of the hydrogen atom. This orbit is a circular orbit of exactly one *de Broglie wavelength*, for which the coulomb attraction provides the centripetal acceleration. Show that 1 Bohr magneton = $eh/4\pi m_e$, where m_e is the mass of the electron and h is Planck's constant.

10–2 The Bohr magneton is a natural unit for measuring the magnetic moment of an atom. Calculate the magnetic moment per atom, in Bohr magneton units, for iron, nickel, and cobalt, under conditions of saturation magnetization. Use the data in Table 9–2.

10–3 Calculate the relative strength of the interaction between two typical magnetic dipoles, compared with the interaction between two typical electric dipoles. To be explicit: calculate the torque exerted on one dipole by the other when they are oriented perpendicularly to each other at a distance of one angstrom unit; take each magnetic dipole = 1 Bohr magneton, each electric dipole = $e \times 0.1$ angstrom. This calculation shows that the basic magnetic interaction is several orders of magnitude smaller than the electrical interaction in matter.

10–4 Calculate the diamagnetic susceptibility of neon at standard temperature and pressure (0°C, 1 atm) on the assumption that only the eight outer electrons in each atom contribute, and that their mean radius is $R = 4.0 \times 10^{-9}$ cm.

10–5 The magnetization of a ferromagnetic material drops essentially to zero at the Curie temperature. In Fig. 10–1 the Curie temperature is represented by the straight line which is tangent to the Langevin function at the origin. Use the experimental value of the Curie temperature of iron to determine γ for iron.

10–6 The gyromagnetic ratio of a current distribution is defined as the ratio of magnetic moment to angular momentum. Calculate the gyromagnetic ratio of a sphere of mass M and charge Q which is rotating with angular velocity ω about an axis through its center, on the assumption that the mass is distributed uniformly throughout and the charge is distributed uniformly on the surface of the sphere.

CHAPTER 11 Electromagnetic Induction

The induction of electromotive force by changing magnetic flux was first observed by Faraday and by Henry in the early nineteenth century. From their pioneering experiments have developed modern generators, transformers, etc. This chapter is primarily concerned with the mathematical formulation of the law of electromagnetic induction and its exploitation in simple cases.

The equation that characterized electrostatics was

$$\mathbf{\nabla} \times \mathbf{E} = 0,$$

or, in integral form,

$$\oint \mathbf{E} \cdot d\mathbf{l} = 0.$$

These equations follow immediately from Coulomb's law, and they are not spoiled by the magnetic force due to a steady current. They do not hold for more general time-dependent fields, however, and these cases are what we will now consider. We define the *electromotive force*, or *emf*, around a circuit by

$$\oint_C \mathbf{E} \cdot d\mathbf{l} = \mathscr{E}. \tag{11–1}$$

With static E- and B-fields, this emf was always zero. We now take up cases where it is not zero. Since this E-field cannot be defined from Coulomb's law, it is legitimate to ask what does define it. It is defined so that the Lorentz force

$$\mathbf{F} = q(\mathbf{E} + \mathbf{v} \times \mathbf{B})$$

is *always* the electromagnetic force on a test charge q.

11–1 ELECTROMAGNETIC INDUCTION

The results of a large number of experiments can be summarized by associating an emf

$$\mathscr{E} = -\frac{d\Phi}{dt} \tag{11–2}$$

234

with a change in magnetic flux through a circuit. This result, which is known as Faraday's law of electromagnetic induction, is found to be independent of the way in which the flux is changed—the value of **B** at various points inside the circuit may be changed in any way. It is extremely important to realize that Eq. (11–2) represents an independent experimental law—it cannot be derived from other experimental laws and it certainly is not, as is sometimes stated, a consequence of conservation of energy applied to the energy balance of currents in magnetic fields.

Since by definition

$$\mathcal{E} = \oint_C \mathbf{E} \cdot d\mathbf{l} \qquad (11\text{–}1)$$

and

$$\Phi = \int_S \mathbf{B} \cdot \mathbf{n} \, da, \qquad (11\text{–}3)$$

Eq. (11–2) can be written

$$\oint_C \mathbf{E} \cdot d\mathbf{l} = -\frac{d}{dt} \int_S \mathbf{B} \cdot \mathbf{n} \, da. \qquad (11\text{–}4)$$

If the circuit is a rigid stationary circuit, the time derivative can be taken inside the integral, where it becomes a partial time derivative. Furthermore, Stokes's theorem can be used to transform the line integral of **E** into the surface integral of $\nabla \times \mathbf{E}$. The result of these transformations is

$$\int_S \nabla \times \mathbf{E} \cdot \mathbf{n} \, da = -\int_S \frac{\partial \mathbf{B}}{\partial t} \cdot \mathbf{n} \, da. \qquad (11\text{–}5)$$

If this must be true for all fixed surfaces S, it follows that

$$\nabla \times \mathbf{E} = -\frac{\partial \mathbf{B}}{\partial t}, \qquad (11\text{–}6)$$

which is the differential form of Faraday's law. It is the required generalization of $\nabla \times \mathbf{E} = 0$, which held for static fields. (Moving media and other subtleties require a more careful treatment, beyond the scope of this text.)

The negative sign in Faraday's law indicates, as can be easily demonstrated, that the direction of the induced emf is such as to oppose the change that produces it. Thus if we attempt to increase the flux through a circuit, the induced emf tends to cause currents in such a direction as to decrease the flux. If we attempt to thrust one pole of a magnet into a coil, the currents caused by the induced emf set up a magnetic field which tends to repel the pole. All these phenomena are covered by Lenz's law, which may be stated as:

In case of a change in a magnetic system, that thing happens which tends to oppose the change.

The utility of Lenz's law should not be underestimated. In many cases it represents the quickest way of obtaining this information about electromagnetic reactions. Even if other methods are available, it affords a valuable check.

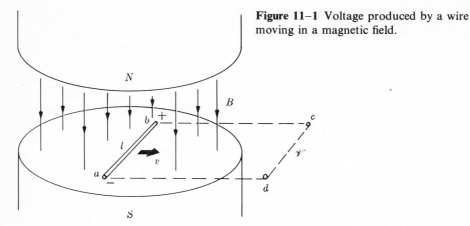

Figure 11–1 Voltage produced by a wire moving in a magnetic field.

In order to get some insight into Faraday's law, it may be useful to consider an example that is usually regarded as an instance of the law but which *can* be analyzed completely according to the electrostatic theory developed in the preceding chapters. Suppose a straight metal wire of length l moves in a direction perpendicular to its length, with velocity $\mathbf{v}$. Let there be a magnetic field $\mathbf{B}$ perpendicular to the plane in which the wire moves, as shown in Fig. 11–1. The free charges in the wire will experience the Lorentz force

$$\mathbf{F} = q(\mathbf{E} + \mathbf{v} \times \mathbf{B}), \tag{11-7}$$

which drives positive and negative charges to opposite ends of the wire because of the $q\mathbf{v} \times \mathbf{B}$ term. In the steady state, when the free charges are not moving with respect to the wire, the total force on a charge must equal zero; that is, the magnetic force must be balanced at each point in the wire by an equal and opposite electric force due to the charge separation,

$$E = vB. \tag{11-8}$$

If the B-field is uniform, then E is constant along the wire, and the potential difference between the ends is

$$\Delta\varphi = -\int_a^b \mathbf{E} \cdot d\mathbf{l} = El. \tag{11-9}$$

If we call this potential difference $\mathscr{V}$, then combining (11–8) with (11–9) gives

$$\mathscr{V} = Blv. \tag{11-10}$$

In this example the B-field is independent of time, and so $\nabla \times \mathbf{E} = 0$ and $\oint \mathbf{E} \cdot d\mathbf{l} = 0$ as in electrostatics. The integral $\int \mathbf{E} \cdot d\mathbf{l}$ is independent of path; in particular, if we imagine a circuit *abcda* extending outside of the magnetic field, $\mathscr{V}$ is also the potential difference along the path *bcda*. In fact, if *bc* and *da* are connected by perfectly conducting wires, $\mathscr{V}$ will be the voltage between the terminals c and d outside the magnetic field. The right side of Eq. (11–10) can be expressed in another way by noting that the flux Φ through the circuit $abcda$ is changing according to $d\Phi/dt = B\ dA/dt = Bl\ dx/dt = -Blv$. Thus

$$\mathscr{V} = -\frac{d\Phi}{dt}, \tag{11–11}$$

which has just the form of Faraday's law, Eq. (11–2); except that $\mathscr{V}$ is *not* an emf in the sense defined by (11–1) because $\oint \mathbf{E} \cdot d\mathbf{l} = 0$ around every closed path in this problem. Equation (11–10) can be generalized by writing it in vector notation. If $\mathbf{v}$ is arbitrarily oriented with respect to l, then only the component of $\mathbf{v}$ which is perpendicular to l contributes to $\mathscr{V}$. Thus $\mathscr{V}$ is proportional to $l \times \mathbf{v}$. For arbitrary $\mathbf{B}$, only the component perpendicular to the plane of l and $\mathbf{v}$ contributes to $\mathscr{V}$. Since $l \times \mathbf{v}$ is perpendicular to the $l,\mathbf{v}$-plane, $\mathscr{V}$ may be written as

$$\mathscr{V} = \mathbf{B} \cdot l \times \mathbf{v} \tag{11–12}$$

The voltage in Eq. (11–12) is sometimes called a motional emf.

Let us now look at the same problem from the standpoint of the wire—i.e., let us imagine that we are in a coordinate system moving with the wire, so that in this system the wire is at rest and the magnet is moving toward the left in Fig. 11–1 with velocity v. One can readily believe that, moving with the wire, one would still observe the same charge separation and the same potential difference between the ends as before. The *explanation* is completely different, however; in this coordinate system, there can be no magnetic force since the wire is at rest. On the other hand the magnetic field is no longer constant in time—at any point it changes from B to approximately zero as the edge of the moving magnet passes by that point. We shall see that the modified curl E equation (11–6) is just sufficient to give the same result $\mathscr{V}$ for the potential difference in this coordinate system also. In the steady state the force on a free charge inside the wire must still be zero,

$$\mathbf{F} = q\mathbf{E} = 0,$$

but there is no magnetic force since $\mathbf{v} = 0$. Thus the electric force must vanish inside the wire,

$$\mathbf{E} = 0 = \mathbf{E}_1 + \mathbf{E}_2. \tag{11–13}$$

There is still a field $\mathbf{E}_1$ due to the charge separation, which is the same as in the previous case; this field is canceled inside the wire by a field $\mathbf{E}_2$ associated with the changing magnetic field,

$$\nabla \times \mathbf{E}_2 = -\frac{\partial \mathbf{B}}{\partial t}.$$

If we again consider the closed curve $abcda$,

$$\mathscr{E} = \oint \mathbf{E} \cdot d\mathbf{l} = \int_a^b \mathbf{E} \cdot d\mathbf{l} + \int_b^a \mathbf{E} \cdot d\mathbf{l}$$

$$= 0 + \mathscr{V}.$$

The first term on the right is zero because **E** vanishes inside the wire, and the second integral along the path *bcda* is what we called $\mathscr{V}$ in the previous case. Thus from this and Eq. (11–2) we again find that

$$\mathscr{V} = -\frac{d\Phi}{dt}. \tag{11–11}$$

It is the generalized curl *E* Eq. (11–6) which, in conjunction with the Lorentz force Eq. (11–7), gives the same result Eq. (11–11) in *either* inertial coordinate system. Equation (11–6) is therefore generally valid.* Since the integral result Eq. (11–11) holds in both coordinate systems, it is not illogical to refer to it as Faraday's law in both cases, even though strictly speaking there is an emf, as defined by Eq. (11–1), only in the second case. In some situations it may not be immediately obvious what circuit should be used to calculate Φ in Eq. (11–11), as for example in Problem 11–4. The equations that always apply to the *E*- and *B*-fields in any inertial coordinate system are Eqs. (11–6) and (11–7). No ambiguity arises in determining the emf or "motional emf" using them.

This example holds some further interest as a prototype of practical electrical generators. If a resistance were connected between terminals *c* and *d*, a current *I* would flow through the circuit.† In this case a mechanical force applied to the wire (or to the magnet in the second case) would be required in order to maintain a constant velocity *v*, so that the sum of the applied force and the magnetic force *BIl* on the wire would be zero. The power input of the applied mechanical force compensates for the I^2R power dissipation in the resistor. As far as the terminal voltage between *c* and *d* is concerned, it makes no difference whether the wire moves or the magnet moves in the generator (commonly the wire moves). In either case, $\oint \mathbf{E} \cdot d\mathbf{l} = 0$ around any path that does not encircle the magnetic field of the generator (in particular, any path confined to the laboratory).

In our two examples, the generalized equation

$$\nabla \times \mathbf{E} = -\frac{\partial \mathbf{B}}{\partial t} \tag{11–6}$$

applies to both; it happened that in the magnet's coordinate system $\partial \mathbf{B}/\partial t = 0$, and so an electrostatic analysis was possible. It would be wrong to conclude, however, that one can always find a coordinate system where $\partial \mathbf{B}/\partial t$ vanishes. A third example based on Fig. 11–1 will illustrate this point. Suppose that *neither* the wire nor the magnet is moving; but suppose the magnet is an electromagnet, whose field can be increased or decreased in magnitude by increasing or decreasing the current in its windings. Now there is no coordinate system in which $\partial \mathbf{B}/\partial t$

---

* From another point of view, we shall find in Chapter 22 that the field $E_2 = vB$ which justifies Eqs. (11–8) and (11–13) arises in the "moving" coordinate system from the relativistic Lorentz transformation of the *E*- and *B*-fields. This replaces the magnetic force that vanishes in the "moving" system.

† This could not be a practical direct-current generator because of the finite extent of the magnet, of course; but if the wire moves back and forth, an alternating current is generated. (See Problem 11–5.)

vanishes. Still Eq. (11–6) applies, however, and Faraday's law Eq. (11–2) gives the emf around any circuit (e.g., *abcda*). This is the situation in transformers and other practical devices with no mechanically moving parts, which are the subject of the rest of this chapter.

11–2 SELF-INDUCTANCE

In this section, the relationship between the flux and current associated with an isolated circuit will be considered and exploited for the purpose of introducing the practical circuit parameter: self-inductance. The magnetic flux linking an isolated circuit depends on the geometry of the circuit and, according to Eq. (8–26), is linearly dependent on the current in the circuit. Thus for a rigid stationary circuit the only changes in flux result from changes in the current. That is,

$$\frac{d\Phi}{dt} = \frac{d\Phi}{dI} \frac{dI}{dt},$$

(11–14)

which is valid even when Eq. (8–26) is not; the only requirement is that Φ depend only on the current. If, however, Eq. (8–26) is valid or, more generally, if Φ is directly proportional to the current, then $d\Phi/dI$ is a constant, equal to Φ/I. In any case, the inductance, L, is defined as

$$L = \frac{d\Phi}{dI}$$

(11–15)

When it is essential to distinguish between this and Φ/I, $d\Phi/dI$ is called the *incremental* inductance; unless otherwise specified, it is safest to associate the word inductance with Eq. (11–15). From Eqs. (11–14), (11–15), and (11–2) it follows that the expression for the induced emf,

$$\mathscr{E} = -L \frac{dI}{dt},$$

(11–16)

is an equation of considerable practical importance.

As an illustration of the use of Eq. (11–15) for the calculation of inductance, the self-inductance of a toroidal coil will be calculated. Such a coil is shown in Fig. 11–2. Equation (11–15) applies to an entire circuit, that is, not only to the toroidal coil of Fig. 11–2, but also to the external circuit connected to terminals 1 and 2. By using twisted leads or a coaxial cable, which produce essentially no external magnetic field, the field-producing portion of the external circuit can be removed to a sufficiently great distance that it does not contribute to the flux in the toroid. If this is done and if by emf we understand the voltage between terminals 1 and 2, then Eq. (11–15) can be used to obtain the inductance of the toroidal coil. From Ampere's circuital law, the magnetic induction inside the toroidal coil.

$$B = \frac{\mu_0 NI}{l},$$

(11–17)

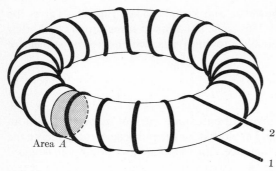

Figure 11-2 A toroidal winding.

where N is the number of turns, l the mean length, and I the current in the winding. (Equations (11–17) and (11–18) involve the approximation of neglecting the variation of the magnetic induction over the cross-sectional area. In Problem 11–8 the details of this approximation are considered.) The flux linking each turn is then

$$\Phi_1 = \frac{\mu_0 N I A}{l},\qquad (11\text{--}18)$$

and the total flux linking the N turns is

$$\Phi = \frac{\mu_0 N^2 A}{l} I.\qquad (11\text{--}19)$$

The inductance is then simply

$$L = \frac{d\Phi}{dI} = \frac{\mu_0 N^2 A}{l}.\qquad (11\text{--}20)$$

The mks unit of inductance is the henry (H), which, from Eq. (11–16), is equal to one volt second/ampere since the unit of emf is the volt. Equation (11–20) indicates that the dimensions of μ_0, which have been previously given as webers/ampere-meter or tesla-meters/ampere, can alternatively be given as henries/meter.

11-3 MUTUAL INDUCTANCE

In the preceding section only isolated circuits were considered, so that all of the flux linking the circuit was due to the current in the circuit itself. This restriction can be lifted by assuming that there are n circuits, labeled 1, 2, ... The flux linking one of these circuits, say the one labeled i, can be written as

$$\Phi_i = \Phi_{i1} + \Phi_{i2} + \cdots + \Phi_{ii} + \cdots + \Phi_{in} = \sum_{j=1}^{n} \Phi_{ij}.\qquad (11\text{--}21)$$

That is, it may be written as a sum of fluxes due to each of the n circuits, Φ_{i1} being the flux through the ith circuit due to circuit 1, etc. The emf induced in the ith circuit, $\mathscr{E}_i$, can then be written as

$$\mathscr{E}_i = -\frac{d\Phi_i}{dt} = -\left\{\frac{d\Phi_{i1}}{dt} + \cdots + \frac{d\Phi_{ii}}{dt} + \cdots + \frac{d\Phi_{in}}{dt}\right\} = -\sum_{j=1}^{n} \frac{d\Phi_{ij}}{dt}. \qquad (11\text{--}22)$$

If each of the circuits is a rigid stationary circuit, the only changes in the Φ_{ij}'s are those that result from changes in the currents. Thus

$$\frac{d\Phi_{ij}}{dt} = \frac{d\Phi_{ij}}{dI_j}\frac{dI_j}{dt}. \qquad (11\text{--}23)$$

The coefficients $d\Phi_{ij}/dI_j$ are constants, independent of the current, if Eq. (8–26) is appropriate. If they are not constants, they may depend on the current because of the nonlinearity of magnetic media associated with the circuit configuration. In any case,

$$M_{ij} = \frac{d\Phi_{ij}}{dI_j}, \qquad i \neq j \qquad (11\text{--}24)$$

is defined as the mutual inductance between circuit i and circuit j. It will be seen later that $M_{ij} = M_{ji}$ and hence there is no possibility of ambiguity in the subscripts. Of course $d\Phi_{ii}/dI_i$ is just the self-inductance of the ith circuit, for which L_i or M_{ii} is written. The units of mutual inductance are the same as those of self-inductance, namely, henries.

As an example of the calculation of mutual inductance, consider the configuration of Fig. 11–2 with a second toroidal winding of N_2 turns added. For this situation, a current I_1 in the first winding produces a magnetic induction

$$B = \frac{\mu_0 N_1 I_1}{l},$$

and consequently fluxes

$$\Phi_{11} = \frac{\mu_0 N_1^2 A I_1}{l}$$

and

$$\Phi_{21} = \frac{\mu_0 N_1 N_2 A I_1}{l}.$$

From these fluxes it follows that

$$L_1 = \frac{\mu_0 N_1^2 A}{l} \qquad (11\text{--}25)$$

as before, and

$$M_{21} = \frac{\mu_0 N_1 N_2 A}{l}. \qquad (11\text{--}26)$$

Reversing the procedure and considering a current I_2 gives

$$L_2 = \frac{\mu_0 N_2^2 A}{l},$$

(11–27)

and

$$M_{12} = \frac{\mu_0 N_1 N_2 A}{l},$$

(11–28)

thus demonstrating that for this case $M_{12} = M_{21}$. Furthermore, Eqs. (11–25), (11–26), and (11–27) may be combined to yield

$$M_{12} = \sqrt{L_1 L_2}.$$

(11–29)

Equation (11–29) represents a limit that is imposed on the mutual inductance between two circuits, namely, it is always less than or equal to the square root of the product of the self-inductances of the two circuits. In view of this limit, a coupling coefficient k is often introduced and defined by

$$M = k\sqrt{L_1 L_2}, \qquad |k| \le 1.$$

(11–30)

11–4 THE NEUMANN FORMULA

For two rigid stationary circuits in a linear medium (vacuum for the present) the mutual inductance is just

$$M_{21} = \frac{\Phi_{21}}{I_1}.$$

(11–31)

This is valid simply because Φ_{21} is proportional to I_1, making Φ_{21}/I_1 and $d\Phi_{21}/dI_1$ equal. In this case, Eq. (8–26) can be used to calculate M_{21}. The flux is given by

$$\Phi_{21} = \frac{\mu_0}{4\pi} I_1 \int_{S_2} \left\{ \oint_{C_1} \frac{d\mathbf{l}_1 \times (\mathbf{r}_2 - \mathbf{r}_1)}{|\mathbf{r}_2 - \mathbf{r}_1|^3} \right\} \cdot \mathbf{n} \, da_2.$$

(11–32)

However,

$$\oint_{C_1} \frac{d\mathbf{l}_1 \times (\mathbf{r}_2 - \mathbf{r}_1)}{|\mathbf{r}_2 - \mathbf{r}_1|^3} = \nabla_2 \times \oint_{C_1} \frac{d\mathbf{l}_1}{|\mathbf{r}_2 - \mathbf{r}_1|} ;$$

(11–33)

hence

$$M_{21} = \frac{\Phi_{21}}{I_1} = \frac{\mu_0}{4\pi} \int_{S_2} \nabla_2 \times \left\{ \oint_{C_1} \frac{d\mathbf{l}_1}{|\mathbf{r}_2 - \mathbf{r}_1|} \right\} \cdot \mathbf{n} \, da_2.$$

(11–34)

Using Stokes's theorem to transform the surface integral gives

$$M_{21} = \frac{\mu_0}{4\pi} \oint_{C_2} \oint_{C_1} \frac{d\mathbf{l}_1 \cdot d\mathbf{l}_2}{|\mathbf{r}_2 - \mathbf{r}_1|},$$

(11–35)

which is known as *Neumann's formula* for the mutual inductance. The symmetry alluded to earlier is apparent in Eq. (11–35).

Neumann's formula is equally applicable to self-inductance, in which case it is written as

$$L = \frac{\mu_0}{4\pi} \oint_{C_1} \oint_{C_1} \frac{d\mathbf{l}_1 \cdot d\mathbf{l}_1'}{|\mathbf{r}_1 - \mathbf{r}_1'|}. \tag{11–36}$$

Some care must be used in the application of Eq. (11–36) because of the singularity at $\mathbf{r}_1 = \mathbf{r}_1'$; however, if care is taken, Eq. (11–36) is sometimes useful.

Equations (11–35) and (11–36) are usually difficult to apply to the calculation of inductance except for circuits in which the geometry is simple. But Eq. (11–35) in particular is very important in the study of forces and torques exerted by one circuit on another. This application will be exploited in Chapter 12.

11–5 INDUCTANCES IN SERIES AND IN PARALLEL

Inductances are often connected in series and in parallel, and, as with resistors and capacitors, it is important to know the result of such connections. We could proceed with a derivation based simply on $\mathscr{E} = -L(dI/dt)$ and obtain formulas for the effective inductance of two inductances in series or in parallel; however, to do so would be to ignore the practical fact that an inductor always has a certain internal resistance. A perfect inductance is much more difficult to approximate practically than a perfect capacitance or a perfect resistance. For this reason, the series and parallel combinations of this section will always involve resistances as well as inductances.

For two inductors in series, the circuit of Fig. 11–3 is appropriate. In adding the voltage drops along the circuit it is important to note that M can be either positive or negative (changing the direction in which either C_1 or C_2 is described reverses the sign of M in Eq. (11–35)). Bearing this in mind, the sum of the voltage drops for the circuit of Fig. 11–3 is found to be

$$V + \mathscr{E}_1 + \mathscr{E}_2 = R_1 I + R_2 I,$$

or

$$V = R_1 I + L_1 \frac{dI}{dt} + M \frac{dI}{dt} + R_2 I + L_2 \frac{dI}{dt} + M \frac{dI}{dt}. \tag{11–37}$$

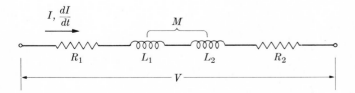

Figure 11–3 Series connection of two inductors.

This is equivalent to

$$V = (R_1 + R_2)I + (L_1 + L_2 + 2M)\frac{dI}{dt}. \tag{11-38}$$

The circuit thus resembles a resistor of resistance $R_1 + R_2$ in series with an inductance $L_1 + L_2 + 2M$. The magnitude of the inductance is $L_1 + L_2 + 2|M|$ for positive coupling (i.e., for fluxes due to I_1 and I_2 in the same direction in each coil), and is $L_1 + L_2 - 2|M|$ for negative coupling. An alternative description of the mutual inductance is

$$M = k\sqrt{L_1 L_2}, \qquad -1 \le k \le 1. \tag{11-39}$$

The effective inductance of the series circuit is then

$$L_{\text{eff}} = L_1 + 2k\sqrt{L_1 L_2} + L_2. \tag{11-40}$$

If k can be varied, then a variable inductance can be constructed. (In the early days of radio this was a popular way of tuning resonant circuits; see Chapter 13.)

The parallel connection shown in Fig. 11–4 is not as simple as the series circuit. In fact, the circuit shown does not behave like a simple series L-R circuit. Thus it is not possible to say that the effective inductance and effective resistance are certain functions of L_1, L_2, R_1, and R_2. If, however, R_1 and R_2 are negligible, then

$$V = L_1 \frac{dI_1}{dt} + M \frac{dI_2}{dt}$$

$$V = L_2 \frac{dI_2}{dt} + M \frac{dI_1}{dt}. \tag{11-41}$$

If first dI_1/dt and then dI_2/dt are eliminated from between Eqs. (11–41), there results

$$V(L_2 - M) = (L_1 L_2 - M^2)\frac{dI_1}{dt},$$

$$V(L_1 - M) = (L_1 L_2 - M^2)\frac{dI_2}{dt}. \tag{11-42}$$

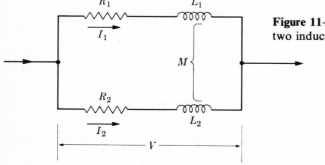

Figure 11–4 Parallel connection of two inductors.

Adding these gives

$$V = \frac{L_1 L_2 - M^2}{L_1 + L_2 - 2M} \frac{dI}{dt}. \tag{11-43}$$

Thus the effective inductance of two inductors in parallel is

$$L_{\text{eff}} = \frac{L_1 L_2 - M^2}{L_1 + L_2 - 2M}, \tag{11-44}$$

where again the sign of M depends on the way in which the inductors are connected.

The most important use of inductances is in alternating current circuits. For a circuit operating at a single frequency, an equivalent series circuit for Fig. 11-4 can be obtained; however, both the equivalent resistance and equivalent inductance are frequency dependent. This frequency dependency is the root of the difficulty encountered above.

11-6 SUMMARY

In this chapter we go a step beyond static fields to the so-called slowly varying case. The new generalization of the field equations is

$$\nabla \times \mathbf{E} = -\frac{\partial \mathbf{B}}{\partial t}.$$

This is the third of Maxwell's four equations, which always holds, along with the two divergence equations and the Lorentz force

$$\mathbf{F} = q(\mathbf{E} + \mathbf{v} \times \mathbf{B}).$$

(At this point in our development of the fundamental equations of electricity and magnetism, we have three of the four Maxwell equations in final form. Only the curl H equation still needs to be generalized.) The equation is known as the differential form of Faraday's law; in integral form.

$$\mathscr{E} = -\frac{d\Phi}{dt},$$

where the emf $\mathscr{E}$ around a fixed circuit C is defined by

$$\mathscr{E} = \oint_C \mathbf{E} \cdot d\mathbf{l}.$$

(It may happen in certain problems that a moving coordinate system can be found in which $\partial \mathbf{B}/\partial t = 0$, and the problem can be analyzed electrostatically, but such is not necessarily the case.)

1. The easiest way to determine the correct polarity of an induced voltage is by means of Lenz's law.

2. The "motional emf" of a straight wire moving in a magnetic field is

$$\mathscr{V} = \mathbf{B} \cdot \mathbf{l} \times \mathbf{v}.$$

3. The self-inductance of a fixed circuit (or circuit element) is defined as

$$L = \frac{d\Phi}{dI},$$

so that

$$\mathscr{E} = -L\frac{dI}{dt}.$$

For a toroid (or long solenoid) L is easily found to be

$$L = \frac{\mu_0 N^2 A}{l}.$$

4. The mutual inductance of two circuits is defined as

$$M_{ij} = \frac{d\Phi_{ij}}{dI_j}.$$

It follows that

$$M_{ii} = L_i,$$

and

$$M_{12} = M_{21} = k\sqrt{L_1 L_2}, \qquad |k| \le 1.$$

5. Pure inductances connected in series or parallel add according to the same formulas as resistances, assuming their mutual inductance and inherent resistance can be neglected.

PROBLEMS

11–1 A metallic conductor in the shape of a wire segment of length l is moved in a magnetic field **B** with velocity **v**. From a detailed consideration of the Lorentz force on the electrons in the wire, show that the ends of the wire are at the potential difference: $\mathbf{B} \cdot \mathbf{l} \times \mathbf{v}$.

11–2 A metal rod one meter long rotates about an axis through one end and perpendicular to the rod, with an angular velocity of 12 rad/s. The plane of rotation of the rod is perpendicular to a uniform magnetic field of 0.3 T. What is the motional emf induced between the ends of the rod?

11–3 In a betatron accelerator, an ion of charge q and mass m traverses a circular orbit at a distance R from the symmetry axis of the machine. The magnetic field has cylindrical symmetry, i.e., the z-component is $B_z = B(r)$ in the plane of the orbit, where r is the distance from the symmetry axis. Show that the ion's velocity is $v = qB(R)R/m$. If the magnetic field is slowly increased in magnitude, show that the emf induced around the ion's orbit is such as to accelerate the ion. Show that in order for the ion to stay in its same orbit the radial variation of the B-field inside the orbit must satisfy the following condition: the spatial average of the increase in $B(r)$ (averaged over the area enclosed by the orbit) must equal twice the increase in $B(R)$ during the same time interval.

11–4 Faraday's homopolar generator consists of a metal disk that rotates in a uniform magnetic field perpendicular to the plane of the disk. Show that the potential difference produced between the center of the disk and its periphery is $\mathscr{V} = f\Phi$, where Φ is the flux through the disk and f is its frequency of rotation. What is the voltage if $f = 3000$ r/min and $\Phi = 0.1$ Wb?

11–5 An alternator consists of an N-turn coil of area A, which rotates in a field B about a diameter perpendicular to the field, with a frequency of rotation f. Find the emf in the coil. What is the amplitude of the alternating voltage if $N = 100$ turns, $A = 10^{-2}$ m^2, $B = 0.1$ T, and $f = 2000$ r/min?

11–6 A dielectric cylinder of permittivity ϵ rotates about its axis with angular velocity ω. If a uniform magnetic field **B** exists parallel to the cylinder axis, find the induced polarization charge in the dielectric.

11–7 Two coupled circuits, A and B, are situated as shown in Fig. 11–5. Use Lenz's law to determine the direction of the induced current in resistor ab when (a) coil B is brought closer to coil A, (b) the resistance of R is decreased, (c) switch S is opened.

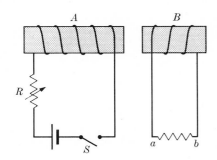

Figure 11–5

11–8 A 100-turn coil of circular cross section is wound compactly so that all loops lie in approximately the same plane. The average radius of the coil is 3 cm. The coil rotates about one of its diameters at 900 r/min. When the rotation axis is vertical, the root-mean-square induced motional emf in the coil is found to be 0.50 mV. What can be concluded about the earth's magnetic field at the location of the coil?

11–9 A circular disk rotates about its axis with angular velocity ω. The disk is made of metal with conductivity g, and its thickness is t. The rotating disk is placed between the pole faces of a magnet which produces a uniform magnetic field B over a small square area of size a^2 at the average distance r from the axis; B is perpendicular to the disk. Calculate the approximate torque on the disk. (Make a reasonable assumption about the resistance of the "eddy current circuit.")

11–10 A toroidal coil of N turns, such as is shown in Fig. 11–2, is wound on a nonmagnetic form. If the mean radius of the coil is b and the cross-sectional radius of the form is a, show that the self-inductance of the coil is given by $L = \mu_0 N^2(b - \sqrt{b^2 - a^2})$.

11–11 A circuit consists of two coaxial cylindrical shells of radii R_1 and R_2 ($R_2 > R_1$) and common length L, connected by flat end plates. The charge flows down one shell and back up the other. What is the self-inductance of this circuit?

11–12 The toroidal coil of Problem 11–10 has 150 turns, $b = 4$ cm, and $a = 1.5$ cm. What is the self-inductance of the coil, in henries?

11–13 Two small circular loops of wire (of radii a and b) lie in the same plane at distance r apart. What is the mutual inductance between the loops if the distance r is sufficiently large that the dipole approximation may be used?

11–14 Two circular current loops with *parallel* axes are located at a distance r from each other that is sufficiently large so that the dipole approximation may be used. Show how one of the loops should be placed relative to the other so that their mutual inductance is zero.

11–15 Given two circuits: a very long straight wire, and a rectangle of dimensions h and d. The rectangle lies in a plane through the wire, the sides of length h being parallel to the wire and at distances r and $r + d$ from it. Calculate the mutual inductance between the two circuits.

11–16 Given two circuits: a very long straight wire, and a circle of radius a. The circle lies in a plane through the wire, with its center at a distance r. Calculate the mutual inductance between the two circuits.

11–17 A transmission line consists of two very long wires of radius a separated by a distance d. Calculate the self-inductance per unit length, assuming $d \gg a$, so that the flux inside the wires themselves can be ignored.

11–18 Given two coaxial, circular loops of wire of radii a and b, separated by the axial distance x. Through the use of Neumann's formula, show that the mutual inductance of the loops is

$$M = \mu_0(ab)^{1/2} \left[\left(\frac{2}{k} - k \right) K(k) - \frac{2}{k} E(k) \right],$$

where

$$k^2 = \frac{4ab}{(a + b)^2 + x^2},$$

and $K(k)$ and $E(k)$ are complete elliptic integrals defined by

$$K(k) = \int_0^{\pi/2} \frac{d\phi}{(1 - k^2 \sin^2 \phi)^{1/2}},$$

and

$$E(k) = \int_0^{\pi/2} (1 - k^2 \sin^2 \phi)^{1/2} \, d\phi.$$

11–19 Consider again the preceding problem. By expanding $1/|\mathbf{r}_2 - \mathbf{r}_1|$ in Neumann's formula according to the binomial theorem, integrate term by term to obtain

$$M = \frac{\mu_0 \pi a^2 b^2}{2h^3} \left(1 + 3 \frac{ab}{h^2} + \frac{75}{8} \frac{a^2 b^2}{h^4} + \cdots \right),$$

where $h^2 = x^2 + (a + b)^2$.

11–20 Two circuits with inductances L_1 and L_2 and resistances R_1 and R_2 are located near each other. If the mutual inductance between the circuits is M, show that a quantity of charge $Q = \mathcal{V} M/R_1 R_2$ will circulate through one of them if an applied voltage $\mathcal{V}$ is suddenly connected in series with the other.

11–21 Given a nonmagnetic conducting medium of conductivity g, which is subjected to a time-dependent magnetic field $\mathbf{B}(\mathbf{r}, t)$. Starting with the differential form of Faraday's law, Eq. (11–6), show that on the assumption of no accumulation of charge (i.e., $\nabla \cdot \mathbf{J} = 0$) the induced eddy-current density in the medium satisfies the differential equation $\nabla^2 \mathbf{J} = g\mu_0(\partial \mathbf{J}/\partial t)$. Show that $\mathbf{E}$ and $\mathbf{B}$ satisfy the same equation.

11–22 Show that the emf in a fixed circuit C is given by

$$-\frac{d}{dt}\oint_C \mathbf{A} \cdot d\mathbf{l},$$

where $\mathbf{A}$ is the vector potential.

11–23 Suppose that the current in a very long solenoid is increasing linearly with time, so that $\partial B/\partial t = K$. Find the E-field inside and outside the solenoid.

11–24 The E-field induced by $\dot{\mathbf{B}} = \partial \mathbf{B}/\partial t$ can be written explicitly as

$$\mathbf{E}(\mathbf{r}) = \frac{1}{4\pi} \int \frac{(\mathbf{r} - \mathbf{r}') \times \dot{\mathbf{B}}(\mathbf{r}')}{|\mathbf{r} - \mathbf{r}'|^3}\, dv'.$$

Verify that $\nabla \times \mathbf{E} = -\partial \mathbf{B}/\partial t$ and $\nabla \cdot \mathbf{E} = 0$, by differentiating inside the integral. Show that the gradient of any solution of Laplace's equation can be added to $\mathbf{E}$.

11–25 A "force-free field" is one such that $\mathbf{J} \times \mathbf{B} = 0$. Show that such a field satisfies the equation

$$\nabla^2 \mathbf{B} = -\alpha^2 \mathbf{B},$$

where α is a constant. (*Hint:* In the force-free field $\nabla \times \mathbf{B}$ is parallel to $\mathbf{B}$.) Show that $\mathbf{J}$ satisfies the same equation. Use the result of Problem 11–21 to find the time dependence of the current density and the fields.

CHAPTER 12   Magnetic Energy

Establishing a magnetic field requires the expenditure of energy; this follows directly from Faraday's law of induction. If a source of voltage $\mathscr{V}$ is applied to a circuit, then, in general, the current through the circuit can be expressed by the equation

$$\mathscr{V} + \mathscr{E} = IR, \tag{12-1}$$

where $\mathscr{E}$ is the induced emf and R is the resistance of the current circuit. The work done by $\mathscr{V}$ in moving the charge increment $dq = I\,dt$ through the circuit is

$$\mathscr{V}\,dq = \mathscr{V}I\,dt = -\mathscr{E}I\,dt + I^2R\,dt$$

$$= I\,d\Phi + I^2R\,dt, \tag{12-2}$$

the last form of which is obtained with the aid of Faraday's law, Eq. (11–2). The term $I^2R\,dt$ represents the irreversible conversion of electrical energy into heat by the circuit, but this term absorbs the entire work input *only* in cases where the flux change is zero. The additional term, $I\,d\Phi$, is the work done against the induced emf in the circuit; it is that part of the work done by $\mathscr{V}$ which is effective in altering the magnetic field structure. Disregarding the $I^2R\,dt$ term, we write

$$dW_b = I\,d\Phi, \tag{12-3}$$

where the subscript b indicates that this is work performed by external electrical energy sources (e.g., by batteries). The work increment Eq. (12–3) may be either positive or negative. It is positive when the flux change $d\Phi$ through the circuit is in the same direction as the flux produced by the current I.

For a rigid stationary circuit showing no energy losses other than Joule heat loss (e.g., no hysteresis), the term dW_b is equal to the change in *magnetic energy* of the circuit. Hysteresis loss will be discussed in Section 12–4, but for the present we shall restrict our attention to reversible magnetic systems. The development closely parallels that of Chapter 6.

12–1 MAGNETIC ENERGY OF COUPLED CIRCUITS

In this section we shall derive an expression for the magnetic energy of a system of interacting current circuits. If there are n circuits, then, according to Eq. (12–3), the electrical work done against the induced emf's is given by

$$dW_b = \sum_{i=1}^{n} I_i \, d\Phi_i. \tag{12–4}$$

This expression is perfectly general; it is valid independently of how the flux increments $d\Phi_i$ are produced. We are particularly interested, however, in the case where the $d\Phi_i$ are produced by current changes in the n circuits themselves. In these circumstances the flux changes are directly correlated with changes in these currents:

$$d\Phi_i = \sum_{j=1}^{n} \frac{d\Phi_{ij}}{dI_j} \, dI_j = \sum_{j=1}^{n} M_{ij} \, dI_j. \tag{12–5}$$

If the circuits are rigid and stationary, then no mechanical work is associated with the flux changes $d\Phi_i$, and dW_b is just equal to the change in magnetic energy, dU, of the system. Note that here we restrict our attention to stationary circuits, so that the magnetic energy can be calculated as a work term. Later we shall let the various circuits move relative to one another, but then we will not be able to identify dU with dW_b.

The magnetic energy U of a system of n rigid stationary circuits is obtained by integrating Eq. (12–4) from the zero flux situation (corresponding to all $I_i = 0$) to the final set of flux values. For a group of *rigid circuits* containing, or located in, *linear magnetic media*, the Φ_i are linearly related to the currents in the circuits, and the magnetic energy is independent of the way in which these currents are brought to their final set of values. Since this situation is of considerable importance, let us restrict our attention to the rigid-circuit, linear case.

Because the final energy is independent of the order in which the currents are varied, we may choose a particular scheme for which W is easily calculated. This scheme is one in which all currents (and hence all fluxes) are brought to their final values in concert, i.e., at any instant of time all currents (and all fluxes) will be at the same fraction of their final values. Let us call this fraction α. If the final values of the current are given the symbols

$$I_1, I_2, \ldots, I_n,$$

then at any stage $I_i' = \alpha I_i$; furthermore, $d\Phi_i = \Phi_i \, d\alpha$. Integration of Eq. (12–4) yields

$$\int dW_b = \int_0^1 d\alpha \sum_{i=1}^{n} I_i' \Phi_i = \sum_{i=1}^{n} I_i \Phi_i \int_0^1 \alpha \, d\alpha$$

$$= \tfrac{1}{2} \sum_{i=1}^{n} I_i \Phi_i.$$

Thus the magnetic energy is

$$U = \tfrac{1}{2} \sum_{i=1}^{n} I_i \Phi_i \qquad \text{(rigid circuits, linear media).} \qquad (12\text{–}6)$$

With the aid of Eq. (12–5), which for a rigid-circuit, linear system may be integrated directly, the magnetic energy may be expressed in the following form:

$$
\begin{aligned}
U &= \tfrac{1}{2} \sum_{i=1}^{n} \sum_{j=1}^{n} M_{ij} I_i I_j \\
&= \tfrac{1}{2} L_1 I_1^2 + \tfrac{1}{2} L_2 I_2^2 + \cdots + \tfrac{1}{2} L_n I_n^2 \\
&\quad + M_{12} I_1 I_2 + M_{13} I_1 I_3 + \cdots + M_{1n} I_1 I_n \\
&\quad + M_{23} I_2 I_3 + \cdots + M_{n-1,n} I_{n-1} I_n \qquad (12\text{–}7)
\end{aligned}
$$

(rigid circuits, linear media).

Here we have used the results and notation of Sections 11–3 and 11–4: $M_{ij} = M_{ji}$; $M_{ii} \equiv L_i$.

For *two* coupled circuits, the last equation reduces to

$$U = \tfrac{1}{2} L_1 I_1^2 + M I_1 I_2 + \tfrac{1}{2} L_2 I_2^2, \qquad (12\text{–}8)$$

where, for simplicity, we have written M for M_{12}. The term $M I_1 I_2$ may be either positive or negative, but the total magnetic energy U must be positive (or zero) for any pair of current values: I_1 and I_2. Denoting the current ratio I_1/I_2 by x, we obtain

$$U = \tfrac{1}{2} I_2^2 (L_1 x^2 + 2Mx + L_2) \geq 0.$$

The value of x which makes U a minimum (or maximum) is found by differentiating U with respect to x and setting the result equal to zero:

$$x = -\frac{M}{L_1}. \qquad (12\text{–}9)$$

The second derivative of U with respect to x is positive, which shows that Eq. (12–9) is the condition for a minimum. The magnetic energy $U \geq 0$ for any x; in particular, the minimum value of U (defined by $x = -M/L_1$) is greater than or equal to zero. Thus

$$\frac{M^2}{L_1} - \frac{2M^2}{L_1} + L_2 \geq 0$$

or

$$L_1 L_2 \geq M^2, \qquad (12\text{–}10)$$

a result that was stated, but not proved, in Section 11–3.

For a single circuit,

$$\Phi = LI,$$

$$U = \tfrac{1}{2}I\Phi = \tfrac{1}{2}LI^2 = \tfrac{1}{2}\frac{\Phi^2}{L}. \tag{12-11}$$

12–2 ENERGY DENSITY IN THE MAGNETIC FIELD

Equation (12–7) gives the magnetic energy of a current system in terms of circuit parameters: currents and inductances. Such a formulation is particularly useful because these parameters are capable of direct experimental measurement. On the other hand, an alternative formulation of the magnetic energy in terms of the field vectors **B** and **H** is of considerable interest because it provides a picture in which energy is stored in the magnetic field itself. This picture can be extended, as is done in Chapter 16, to show how energy moves through the electromagnetic field in nonstationary processes.

Consider a group of rigid current-carrying circuits, none of which extends to infinity, immersed in a medium with linear magnetic properties. The energy of this system is given by Eq. (12–6). For the present discussion it is convenient to assume that each circuit consists of only a single loop; then the flux Φ_i may be expressed as

$$\Phi_i = \int_{S_i} \mathbf{B} \cdot \mathbf{n}\, da = \oint_{C_i} \mathbf{A} \cdot d\mathbf{l}_i, \tag{12-12}$$

where **A** is the local vector potential. Substitution of this result into Eq. (12–6) yields

$$U = \tfrac{1}{2} \sum_i \oint_{C_i} I_i \mathbf{A} \cdot d\mathbf{l}_i. \tag{12-13a}$$

We should like to make Eq. (12–13a) somewhat more general. Suppose that we do not have current circuits defined by wires, but instead each "circuit" is a closed path in the medium (assumed to be conducting) that follows a line of current density. Equation (12–13a) may be made to approximate this situation very closely by choosing a large number of contiguous circuits (C_i), replacing $I_i\, d\mathbf{l}_i \rightarrow \mathbf{J}\, dv$, and, finally, by the substitution of

$$\int_V \quad \text{for} \quad \sum_i \oint_{C_i}.$$

Hence

$$U = \tfrac{1}{2} \int_V \mathbf{J} \cdot \mathbf{A}\, dv. \tag{12-13b}$$

The last equation may be further transformed by using the field equation $\nabla \times \mathbf{H} = \mathbf{J}$, and the vector identity (1–1–8):

$$\nabla \cdot (\mathbf{A} \times \mathbf{H}) = \mathbf{H} \cdot \nabla \times \mathbf{A} - \mathbf{A} \cdot \nabla \times \mathbf{H},$$

whence

$$U = \tfrac{1}{2} \int_V \mathbf{H} \cdot \nabla \times \mathbf{A} \; dv - \tfrac{1}{2} \int_S \mathbf{A} \times \mathbf{H} \cdot \mathbf{n} \; da, \qquad (12\text{–}14)$$

where S is the surface which bounds the volume V. Since, by assumption, none of the current "circuits" extends to infinity, it is convenient to move the surface S out to a very large distance so that all parts of this surface are far from the currents. Of course, the volume of the system must be increased accordingly. Now $\mathbf{H}$ falls off at least as fast as $1/r^2$, where r is the distance from an origin near the middle of the current distribution to a characteristic point on the surface S; $\mathbf{A}$ falls off at least as fast as $1/r$; and the surface area is proportional to r^2. Thus the contribution from the surface integral in Eq. (12–14) falls off as $1/r$ or faster, and if S is moved out to infinity, this contribution vanishes.

By dropping the surface integral in Eq. (12–14) and extending the volume term to include all space, we obtain

$$U = \tfrac{1}{2} \int_V \mathbf{H} \cdot \mathbf{B} \; dv, \qquad (12\text{–}15)$$

since $\mathbf{B} = \nabla \times \mathbf{A}$. This result is completely analogous to the expression for electrostatic energy, Eq. (6–17). Equation (12–15) is restricted to systems containing linear magnetic media, since it was derived from Eq. (12–6).

By reasoning similar to that of Section 6–3, we are led to the concept of energy density in a magnetic field:

$$u = \tfrac{1}{2}\mathbf{H} \cdot \mathbf{B}, \qquad (12\text{–}16a)$$

which, for the case of isotropic, linear, magnetic materials reduces to

$$u = \tfrac{1}{2}\mu H^2 = \tfrac{1}{2}\frac{B^2}{\mu}. \qquad (12\text{–}16b)$$

12–3 FORCES AND TORQUES ON RIGID CIRCUITS

Up to this point we have developed a number of alternative expressions for the magnetic energy of a system of current circuits. These are given by Eqs. (12–6) and (12–7), and (12–15). We shall now show how the force, or torque, on one component of such a system may be calculated from a knowledge of the magnetic energy.

Suppose we allow one of the parts of the system to make a rigid displacement $d\mathbf{r}$ under the influence of the magnetic forces acting upon it, all currents remaining constant. The mechanical work performed by the force $\mathbf{F}$ acting on the system is

$$dW = \mathbf{F} \cdot d\mathbf{r}, \qquad (12\text{–}17)$$

as in Eq. (6–32). The work contains two contributions in these circumstances, as in Eq. (6–37):

$$dW = dW_b - dU, \tag{12–18}$$

where dU is the change in magnetic energy of the system and dW_b is the work performed by external energy sources against the induced emf's to keep the currents constant.

Before we can proceed to an expression linking U and the force on a part of the system, it will be necessary to eliminate dW_b from Eq. (12–18). This is easily done for a system of rigid circuits in *linear* magnetic media. If the geometry of the system is changed but all currents remain unaltered, then, according to Eq. (12–6),

$$dU = \tfrac{1}{2} \sum_i I_i \, d\Phi_i.$$

But, from Eq. (12–4),

$$dW_b = \sum_i I_i \, d\Phi_i.$$

Thus

$$dW_b = 2 \, dU. \tag{12–19}$$

Using this equation to eliminate dW_b from Eq. (12–18) and combining the result with Eq. (12–17), we obtain

$$dU = \mathbf{F} \cdot d\mathbf{r},$$

or

$$\mathbf{F} = \nabla U,$$

$$F_x = \left(\frac{\partial U}{\partial x} \right)_I. \tag{12–20}$$

The force on the circuit is the gradient of the magnetic energy when I is maintained constant.

If the circuit under consideration is constrained to move in such a way that it rotates about an axis, then Eq. (12–17) may be replaced by

$$dW = \boldsymbol{\tau} \cdot d\boldsymbol{\theta} = \tau_1 \, d\theta_1 + \tau_2 \, d\theta_2 + \tau_3 \, d\theta_3,$$

where $\boldsymbol{\tau}$ is the magnetic torque on the circuit and $d\boldsymbol{\theta}$ is an angular displacement. Under these conditions,

$$\tau_1 = \left(\frac{\partial U}{\partial \theta_1} \right)_I, \tag{12–21}$$

and so on. The results Eqs. (12–20) and (12–21) for constant current are analogous to the electrostatic case of constant potential, where work by the battery is needed to keep the potentials constant.

In some other interesting cases, the fluxes through the circuits can be treated as constant instead. Then, according to Eq. (12–4), $dW_b = 0$, and so the system can be taken as isolated.* Consequently,

$$\mathbf{F} \cdot d\mathbf{r} = dW = -dU,$$

$$F_x = -\left(\frac{\partial U}{\partial x}\right)_\Phi, \tag{12–22}$$

$$\tau_1 = -\left(\frac{\partial U}{\partial \theta_1}\right)_\Phi. \tag{12–23}$$

Just as in the electrostatic case, in order to make use of the energy method it is necessary to express U in analytic form, i.e., the specific dependence of U on the variable coordinates $(x, y, z, \theta_1, \theta_2, \text{or } \theta_3)$ must be given. When this is done, however, the energy method becomes a powerful technique for calculating forces and torques.

We shall illustrate the method by considering two examples. Additional exercises of this type will be found in the problems at the end of the chapter. For our first example let us calculate the force between two rigid circuits carrying constant currents. The magnetic energy is given by Eq. (12–8), and the force on circuit 2 is

$$\mathbf{F}_2 = \nabla_2 U = I_1 I_2 \nabla_2 M,$$

where the mutual inductance M must be written so that it displays its dependence on $\mathbf{r}_2$. Neumann's formula, Eq. (11–35), shows this dependence explicitly, so we may write

$$\mathbf{F}_2 = \frac{\mu_0}{4\pi} I_1 I_2 \oint_{C_1} \oint_{C_2} (d\mathbf{l}_1 \cdot d\mathbf{l}_2) \nabla_2 \frac{1}{|\mathbf{r}_2 - \mathbf{r}_1|}$$

$$= -\frac{\mu_0}{4\pi} I_1 I_2 \oint_{C_1} \oint_{C_2} (d\mathbf{l}_1 \cdot d\mathbf{l}_2) \frac{(\mathbf{r}_2 - \mathbf{r}_1)}{|\mathbf{r}_2 - \mathbf{r}_1|^3}, \tag{12–24}$$

an expression that evidently shows the proper symmetry, i.e., $\mathbf{F}_2 = -\mathbf{F}_1$. However, we already have an expression for the force between two circuits, Eq. (8–25), and this appears to be at variance with the formula just derived. Actually, the two expressions are equivalent, as may be easily verified. Let us expand the triple product in the integrand of Eq. (8–25):

$$d\mathbf{l}_2 \times [d\mathbf{l}_1 \times (\mathbf{r}_2 - \mathbf{r}_1)] = d\mathbf{l}_1[d\mathbf{l}_2 \cdot (\mathbf{r}_2 - \mathbf{r}_1)] - (\mathbf{r}_2 - \mathbf{r}_1)(d\mathbf{l}_1 \cdot d\mathbf{l}_2).$$

The integral containing the last term on the right is identical with Eq. (12–24); that containing the first term may be written

$$\frac{\mu_0}{4\pi} I_1 I_2 \oint_{C_1} d\mathbf{l}_1 \oint_{C_2} \frac{d\mathbf{l}_2 \cdot (\mathbf{r}_2 - \mathbf{r}_1)}{|\mathbf{r}_2 - \mathbf{r}_1|^3}. \tag{12–25}$$

---

* In a normal circuit a battery would still be needed to supply the $I^2 R$ power dissipation, but we are disregarding this. If the wires were superconducting $(R = 0)$, the system could in fact be isolated.

Now $d\mathbf{l}_2 \cdot (\mathbf{r}_2 - \mathbf{r}_1)$ is $|\mathbf{r}_2 - \mathbf{r}_1|$ times the projection of $d\mathbf{l}_2$ on the vector $\mathbf{r}_2 - \mathbf{r}_1$. Let us denote $|\mathbf{r}_2 - \mathbf{r}_1|$ by r_{21}; then the projection of $d\mathbf{l}_2$ is just dr_{21}. The integral over C_2 may be carried out at fixed $d\mathbf{l}_1$:

$$\oint_{C_2} \frac{dr_{21}}{r_{21}^2} = -\frac{1}{r_{21}}\bigg|_a^a,$$

the upper and lower limit being identical because of the complete circuit. Thus the expression (12–25) vanishes, and Eq. (12–24) is equivalent to Eq. (8–25).

As a second example, consider a long solenoid of N turns, and length l carrying constant current I. An iron rod of constant permeability μ and cross-sectional area A is inserted along the solenoid axis. If the rod is withdrawn (Fig. 12–1a) until only one-half of its length remains in the solenoid, calculate approximately the force tending to pull it back into place.

Solution. The magnetic field structure associated with this problem is complicated if end-effects are included; fortunately, however, we do not have to calculate the entire magnetic energy of the system but merely the difference in energy between the two configurations shown in Fig. 12–1(a) and (b). The field structure is relatively uniform far from the ends of the rod and the solenoid. The essential difference between configurations (a) and (b) is that a length Δx from the extreme right-hand end of the rod (outside the field region) is effectively transferred to the uniform field region inside the solenoid, at a place beyond the demagnetizing influence of the magnet pole. Thus since $\mathbf{H}$ is nearly longitudinal in the region Δx,

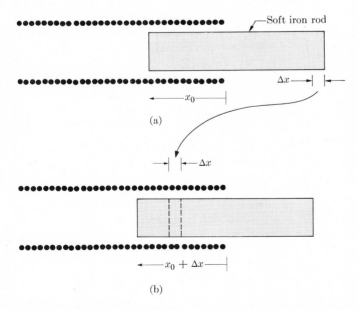

Figure 12–1 Force on soft-iron rod inserted into a solenoid (by the energy method).

and since the tangential component of **H** is continuous at the cylindrical boundary of the rod, let us use

$$U = \tfrac{1}{2} \int \mu H^2 \, dv,$$

where **H** is constant inside and outside the rod because I is constant. Consequently,

$$U(x_0 + \Delta x) \approx U(x_0) + \tfrac{1}{2} \int_{A \, \Delta x} (\mu - \mu_0) H^2 \, dv$$

$$= U(x_0) + \tfrac{1}{2}(\mu - \mu_0) \frac{N^2 I^2}{l^2} A \, \Delta x,$$

and from Eq. (12–20)

$$F_x \approx \tfrac{1}{2}(\mu - \mu_0) \frac{N^2 I^2 A}{l^2} = \tfrac{1}{2} \chi_m \mu_0 H^2 A. \tag{12–26}$$

in the direction of increasing x_0.

An example where Φ is constant is found in Problem 12–7.

*12–4 HYSTERESIS LOSS

In the preceding sections we have limited our discussion to reversible magnetic systems, and in most instances to linear systems. We shall now say something about energy changes in systems containing permanent magnet material, that is, systems in which hysteresis plays a prominent role. Let us consider an electrical circuit, in the form of a closely wound coil of N turns, which surrounds a piece of ferromagnetic material (Fig. 12–2). If the coil is connected to an external source of electrical energy, the work done against the induced emf in the coil is given by Eq. (12–3). In Eq. (12–3), however, the flux change $d\Phi$ is the total flux change through the circuit; for the present purpose it is convenient to let the symbol $d\Phi$ stand for the flux change through a single turn of the coil. Thus, on the assumption that the same flux links every turn,

$$\delta W_b = NI \, \delta\Phi. \tag{12–3a}$$

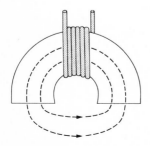

Figure 12–2 A ferromagnetic specimen forming part of a magnetic circuit.

Let us treat the ferromagnetic specimen as forming part of a magnetic circuit. Then NI may be replaced by $\oint \mathbf{H} \cdot d\mathbf{l}$ around a typical flux path, and Eq. (12–3a) becomes*

$$\delta W_b = \oint \delta \Phi \mathbf{H} \cdot d\mathbf{l} = \oint A \, \delta B \mathbf{H} \cdot d\mathbf{l},$$

where A is the cross section of the magnetic circuit appropriate to the length interval $d\mathbf{l}$. Since $d\mathbf{l}$ is always tangent to the flux path, the preceding equation may be written as

$$\delta W_b = \oint A \, \delta \mathbf{B} \cdot \mathbf{H} \, dl = \int_V \delta \mathbf{B} \cdot \mathbf{H} \, dv, \qquad (12-27)$$

where V is the volume of the magnetic circuit, i.e., the region of space in which the magnetic field is different from zero.

If the ferromagnetic material in the system shows reversible magnetic behavior, Eq. (12–27) may be integrated from $\mathbf{B} = 0$ to its final value, to yield the magnetic energy of the system. For linear material, the energy so obtained is identical with that expressed by Eq. (12–15). But Eq. (12–27) is much more general than this; it correctly predicts the work done on the magnetic system even for cases in which there is hysteresis.

According to Eq. (12–27), a change in the magnetic field structure implies a work input

$$dw_b = \mathbf{H} \cdot d\mathbf{B} \qquad (12-28)$$

associated with each unit volume of magnetic material (or vacuum) in the system. Of particular interest is the case where the material is cycled, as it would be when the coil surrounding the specimen is subjected to alternating current operation. In one cycle the magnetic intensity H (for a typical point in the specimen) starts at zero, increases to a maximum, H_{max}, decreases to $-H_{max}$, and then returns to zero. The magnetic induction B shows a similar variation, but for a typical ferromagnetic will lag behind H, thus tracing out a hysteresis curve (Fig. 12–3). The work input (per unit volume) required to change the magnetic induction from point a to b on the hysteresis curve,

$$(w_b)_{ab} = \int_a^b H \, dB,$$

---

* The analysis presented here may be put on a somewhat more rigorous basis by replacing the magnetic circuit with a large number of magnetic flux paths of various lengths (magnetic circuits in parallel). Equation (12–3a) then becomes

$$\delta W_b = NI \sum_j \delta \Phi_j = \sum_j \oint_j \delta \Phi_j \mathbf{H} \cdot d\mathbf{l}_j,$$

where $\delta \Phi_j$ is the flux change associated with one of these paths. The final result, Eq. (12–27), is unchanged.

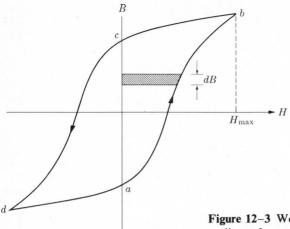

Figure 12–3 Work done per unit volume in cycling a ferromagnetic material.

is just the area between the hysteresis segment ab and the B-axis; it is positive because both H and dB are positive. The contribution $(w_b)_{bc}$ is also the area between the appropriate hysteresis segment (bc) and the B-axis, but it must be taken negative, since H and dB are of opposite sign. Similar arguments can be made about $(w_b)_{cd}$ and $(w_b)_{da}$. Thus, in cycling the material once around the hysteresis loop, the work required per unit volume is

$$w_b = \oint H \, dB, \qquad (12\text{–}29)$$

which is the area enclosed by the hysteresis loop.

At the end of one complete cycle, the magnetic state of the material is the same as it was at the start of the cycle; hence the "magnetic energy" of the material is the same. It is evident, then, that Eq. (12–29) represents an energy loss. This loss appears as heat; it comes about through the irreversible changes in domain structure of the material. Hysteresis loss is an important factor in circuits subjected to alternating current operation. Equation (12–29) represents the energy loss per unit volume *per cycle*; thus the energy loss per unit time is directly proportional to the frequency of the alternating current.

According to Eq. (12–28), the work required to change the magnetic induction in a unit volume of material is

$$dw_b = \mathbf{H} \cdot d\mathbf{B} = \mu_0 H \, dH + \mu_0 \mathbf{H} \cdot d\mathbf{M}. \qquad (12\text{–}28a)$$

It is sometimes convenient to regard the $\mu_0 H \, dH$ term (the work done on the vacuum) as taking place whether the material is present or not. From this point of view, then, the term $\mu_0 \mathbf{H} \cdot d\mathbf{M}$ is the specific work done on the material. This is the approach usually taken in thermodynamics textbooks; it forms the basis for discussion of such processes as "magnetic cooling."

Since the integral of $H\ dH$ vanishes for a complete cycle, Eq. (12–29) is equivalent to

$$w_b = \mu_0 \oint H\ dM. \qquad (12\text{–}29\text{a})$$

From $d(MH) \equiv H\ dM + M\ dH$, this can also be written

$$w_b = -\mu_0 \oint M\ dH. \qquad (12\text{–}29\text{b})$$

12–5 SUMMARY

The work that is done by an outside agency, e.g., a battery, in altering the magnetic field of a system of current circuits is

$$dW_b = \sum_{i=1}^{n} I_i\, d\Phi_i$$

(exclusive of the work that supplies the Joule heat loss of resistive circuits). The magnetostatic potential energy of a system of current circuits and linear magnetic media is

$$U = \tfrac{1}{2} \sum_{i=1}^{n} I_i\Phi_i,$$

where

$$\Phi_i = \sum_{j=1}^{n} M_{ij} I_j.$$

For a continuous current distribution in linear media, the magnetic energy becomes

$$U = \tfrac{1}{2} \int \mathbf{J} \cdot \mathbf{A}\ dv,$$

where the vector potential $\mathbf{A}$ is that produced by the current density $\mathbf{J}$. Integration by parts transforms the energy in linear magnetic materials into an integral,

$$U = \int u\ dv,$$

over the energy density of the magnetic field,

$$u = \tfrac{1}{2}\mathbf{H} \cdot \mathbf{B} = \tfrac{1}{2}\mu H^2 = \frac{1}{2}\frac{B^2}{\mu}.$$

1. For a single circuit,

$$U = \tfrac{1}{2}I\Phi,$$

with

$$\Phi = LI.$$

2. The magnetic force on part of an isolated system, with constant flux through each circuit, is the negative gradient of the magnetostatic energy,

$$F_x = -\left(\frac{\partial U}{\partial x}\right)_{\Phi}.$$

If the system is not isolated, but instead the current in each circuit is maintained constant by an outside agency (battery), the force is given by

$$F_x = +\left(\frac{\partial U}{\partial x}\right)_{I}.$$

3. In the presence of nonlinear material, including hysteresis,

$$dw_b = \mathbf{H} \cdot d\mathbf{B}.$$

In a complete cycle of a cyclic process,

$$w_b = \oint H\, dB = \mu_0 \oint H\, dM = -\mu_0 \oint M\, dH.$$

PROBLEMS

12–1 Given a current circuit (not necessarily rigid) in a *prescribed* magnetic field. The magnetic force on each circuit element $d\mathbf{l}$ is given by $I\, d\mathbf{l} \times \mathbf{B}$. If the circuit is allowed to move under the influences of the magnetic forces, such that an element is displaced $\delta\mathbf{r}$ and at the same time the current I is held constant, show by direct calculation that the mechanical work done by the force is $\delta W = I\, \delta\Phi$, where $\delta\Phi$ is the additional flux through the circuit.

12–2 Given a set of interacting current circuits in a linear magnetic medium. All circuits with the exception of circuit 1 are held stationary, but circuit 1 is allowed to move rigidly. The currents are all held constant by means of batteries. Show from the combination of Eqs. (12–4), (12–6), and (12–18), that the mechanical work done by the moving circuit is $dW = I_1\, d\Phi_1$, where $d\Phi_1$ is the change in flux through circuit 1.

12–3 Consider two interacting current circuits characterized by the inductances $L_1 = \beta I_1^s$, $M_{12} = M_{21} = \beta I_1^{s/2} I_2^{s/2}$, and $L_2 = \beta I_2^s$, where β and s are constants. This is a reversible magnetic system but not a linear one. Calculate the magnetic energy of the system in terms of the final currents I_1 and I_2. Do this in two ways: first, by bringing the currents to their final values in concert; second, by keeping $I_2' = 0$ while I_1' is brought to its final value, then changing I_2'.

12–4 A circuit in the form of a circular turn of wire of radius b is placed at the center of a larger turn of radius a, $b \ll a$. The small circuit is fixed so that it is free to rotate about one of its diameters, this diameter being located in the plane of the larger circuit. The circuits carry the steady currents I_b and I_a, respectively. If the angle between the normals to the two circuits is θ, calculate the torque on the movable circuit. In what direction is this torque when I_b and I_a circulate in the same sense?

*12–5 A U-shaped electromagnet of length l, pole separation d, and permeability μ has a square cross section of area A. It is wound with N turns of wire carrying a current I.

Calculate the force with which the magnet holds a bar of the same material, of same cross section, against its poles.

12–6 A permanent magnet with constant magnetization, and a circuit which is connected to a battery, form an isolated system. The circuit is allowed to move relative to the magnet, the current I in the circuit being maintained constant. The mechanical work done by the circuit is given in Problem 12–1. What conclusion can you draw about the change in magnetic energy of this system?

12–7 The magnetic induction field between the poles of an electromagnet is relatively uniform and is held at the constant value $\mathbf{B}_0$. A thin paramagnetic slab, which is constrained to move vertically, is placed in the field as shown in Fig. 12–4. The susceptibility of the slab is χ_m and its cross-sectional area is A. (a) Calculate the force on the slab. (b) Obtain a numerical value for the force if the slab material is titanium, $A = 1$ cm^2 and $B_0 = 0.25$ T.

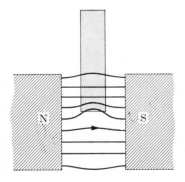

Figure 12–4 A paramagnetic slab inserted between the pole faces of a magnet.

*12–8 From the result of Problem 12–1, the force on a current circuit in a prescribed magnetic field is given by $\mathbf{F} = I\,\nabla\Phi$. If the circuit is very small, the magnetic field $\mathbf{B}$ may be treated as constant over the surface bounded by the circuit; furthermore, the circuit itself may be characterized by its magnetic dipole moment $\mathbf{m}$. Show that when the prescribed magnetic field has no sources (i.e., $\mathbf{J}$, $\mathbf{J}_M = 0$) at the position of the dipole, the force on the dipole is

$$\mathbf{F} = (\mathbf{m} \cdot \nabla)\mathbf{B}.$$

12–9 A rigid circuit consisting of a single loop of wire is located in a radial, inverse-square, magnetic induction field, $\mathbf{B} = K\mathbf{r}/r^3$. Show that the force on the circuit is $\mathbf{F} = KI\nabla\Omega$, where Ω is the solid angle the circuit subtends at the field center, and I is the current in the circuit.

12–10 The center of a plane circular circuit of radius R and consisting of one turn lies on the x-axis at distance x from the origin. The circuit carries the current I, and its positive normal points in the $-x$ direction. Find the force exerted on the circuit by a radial induction field diverging from the origin, $\mathbf{B} = K\mathbf{r}/r^3$.

12–11 Consider a very long solenoid of N/l turns per unit length and radius R, such that the field inside is approximately uniform and the field outside is zero. Find the radial force on one turn of the winding, per unit length of circumference, from the magnetic energy. (a) Assume that the current I is maintained constant by a battery. (b) Repeat assuming that the flux remains constant and the system is isolated (with superconducting windings).

12–12 Solve the example of Fig. 12–1 with the magnetic energy in the form $U = \frac{1}{2}LI^2$, where $L = L(x_0)$ is the inductance of the solenoid with the iron rod inserted a distance x_0. Assume that the rod's diameter is nearly as large as the solenoid's, and that both are so long that end effects are negligible.

12–13 For the toroid of Problem 11–10 find the radial force on the coil if it carries a current I. Does the force tend to expand the coil or collapse it?

12–14 Find the force between the straight wire and the rectangular circuit of Problem 11–15 if the currents are I_1 and I_2.

12–15 Two isolated superconducting circuits carry certain currents when they are positioned so that their mutual inductance is zero. Now they are moved so that their mutual inductance is M. If the circuits are identical and had the same initial currents I_0, find the final currents I.

12–16 Prove that $\frac{1}{2}\int_V \mathbf{H}\cdot\mathbf{B}\,dv = 0$, where V is all space, if the fields are produced solely by magnets (i.e., no transport currents). Is the magnetic energy zero?

12–17 Estimate the areas enclosed by the two hysteresis curves shown in Fig. 9–8, and calculate the power loss per unit volume due to hysteresis in these materials at 60 Hz operation.

12–18 The core of a generator armature is made of iron whose average hysteresis loop under operating conditions has an area of 2000 joules/m³. The core is cylindrical in shape, with a length of 0.4 m and a diameter of 0.15 m. If it rotates at 1800 r/min, calculate the rate at which heat is produced in the core.

12–19 A current circuit in a prescribed magnetic field moves under the influence of magnetic forces. The mechanical work done by the circuit is given in Problem 12–1. Suppose now that the circuit is an *atomic circuit*, and that the atomic current is held constant because of general quantum principles (note that we are neglecting a small change in current due to diamagnetism). What is the change in magnetic energy of the circuit? The result of this problem is the basis for the magnetic dipole energy in the calculation in Section 10–3.

CHAPTER 13 Slowly Varying Currents

13–1 INTRODUCTION

In Chapter 7 the idea of an electrical circuit was introduced, and an analysis was made of the currents in such circuits when they are excited by constant applied voltages. These ideas will now be expanded to include slowly varying voltages and the resulting slowly varying currents. To understand properly what is meant by "slowly varying," Maxwell's equations* must be used; however, the general ideas can be understood without recourse to the details of these equations.

For sinusoidal voltage variations in circuits containing linear elements, the basis for elementary circuit theory, the behavior of a circuit is characterized by a frequency ω.† An electromagnetic wave of this frequency in free space has a wavelength $\lambda = 2\pi c/\omega$, where c is the velocity of light. The principal restriction to be imposed in order that the current in the circuit may be called slowly varying is that the circuit should not radiate an appreciable amount of power. This restriction can be met by requiring that the maximum linear dimension of the system, L, be much smaller than the free space wavelength associated with the driving frequency, that is,

$$L \ll \frac{2\pi c}{\omega} \qquad \text{or} \qquad \omega \ll \frac{2\pi c}{L}. \tag{13–1}$$

If this condition is satisfied, then for every element $d\mathbf{l}$ of the circuit carrying a current I there is, much less than one wavelength away, a corresponding element

---

* Maxwell's equations are treated in detail in Chapter 16. For those who are particularly interested, it is worth correlating the material presented in Chapter 16 with that presented here. In this context, "slowly varying" simply means that we are neglecting the displacement current $\partial \mathbf{D}/\partial t$, which is discussed in Sections 16–1 and 16–2, so that $\nabla \times \mathbf{H} = \mathbf{J}$, as has been assumed heretofore.

† The quantity ω is 2π times the frequency and is sometimes called the angular frequency. The use of ω instead of $2\pi f$ is of considerable advantage in many branches of physics. In particular, for the present discussion it eliminates a multitude of 2π's from the circuit equations.

Table 13–1

f, Hz	ω, rad/s	λ, m	L, m
60	376	5×10^6	5×10^5 (300 miles)
10^6	6.28×10^6	300	30
10^8	6.28×10^8	3	0.3
10^{10}	6.28×10^{10}	0.03	0.003

$-d\mathbf{l}$ carrying the same current. This clearly ensures cancellation of the fields produced by these elements at distances of the order of a few wavelengths in all directions, and thus shows that the fields associated with the circuit are confined to the vicinity of the circuit. To see what practical restrictions are imposed by Eq. (13–1), $L \sim \lambda/10$ has been used as the maximum circuit dimension in constructing Table 13–1. The frequencies chosen are a power line frequency, a low radiofrequency (AM broadcast band), a high radiofrequency (FM and TV), and a microwave frequency. It is clear that for the first three frequencies ordinary circuits satisfy the criterion; however, for the last one the circuit must be built in a cube about 0.1 inch on a side, which limits the applicability to integrated circuits. It should also be noted that at 100 MHz the wavelength and circuit dimensions are of laboratory size, and hence that care must be used in applying ordinary circuit theory at this and higher frequencies. In the balance of this chapter it will be assumed that the slowly varying criterion is satisfied, without further explicit comment.

13–2 TRANSIENT AND STEADY-STATE BEHAVIOR

If a network of passive elements is suddenly connected to a source or sources of voltage, currents arise. Regardless of the nature of the applied voltages, the initial variation of the currents with time is nonperiodic. If, however, the voltages vary periodically with the time,* then a long time after the application of the voltages the currents will also be found to vary periodically with the time. (Actually, of course, they become strictly periodic only after infinite time; however, any desired approximation to periodicity can be attained by waiting a sufficiently long time.) It is convenient to discuss the behavior of circuits in two phases, according to whether the periodic or nonperiodic behavior is important. The periodic behavior is referred to as the *steady-state* behavior, while the nonperiodic behavior is known as the *transient* behavior. Both aspects are governed by the same basic integro-differential equations; however, the elementary techniques used in solving them are radically different in the two cases. The analysis presented here will be restricted to elementary transient analysis (primarily excitation by constant volt-

---

* A constant voltage should be understood as a special case of a periodic voltage, in which the period is infinite or the frequency is zero.

ages) and steady-state analysis for sinusoidal excitations. For further details the reader is referred to the classic books of Guillemin and of Bode,* and to other more recent engineering texts.†

13–3 KIRCHHOFF'S LAWS

In Chapter 7, Kirchhoff's laws were introduced for direct current (d-c) circuits; these must now be generalized to include slowly varying currents. The first generalization is to note that not only resistors but also capacitors and inductors must be included as circuit elements. Each such element has a potential difference between its terminals, which must be included in Kirchhoff's loop law. The name "IR-drop" is no longer appropriate for all of these, therefore the name *counter voltage* will be adopted to specify the difference in potential between the terminals of a passive element. The other generalization is to observe that both of Kirchhoff's laws must apply at each instant of time, that is, they must apply to the instantaneous values of the currents, applied voltages, and counter voltages. The laws may now be stated:

I. *The algebraic sum of the instantaneous currents flowing toward a junction is zero.*

II. *The algebraic sum of the instantaneous applied voltages in a closed loop equals the algebraic sum of the instantaneous counter voltages in the loop.*

The meaning of the first of these laws is clear: if currents directed toward a junction are called positive then those oppositely directed should be called negative, and the law says that as much current enters the junction as leaves it. Basically, the second law represents the integral of the electric field around the loop; however, it is necessary to establish the sign convention. The sign convention to which we will adhere is best explained in terms of a single simple mesh, as shown in Fig. 13–1. In this figure an applied voltage $\mathscr{V}(t)$ is shown connected in series with a resistance R, an inductance L, and a capacitance C. An arrow labeled $I(t)$ has been drawn to indicate the assumed (arbitrary) positive direction for the current. All signs are ultimately referred to this direction. The voltage $\mathscr{V}(t)$ is positive if it tends to cause the current to move in the assumed direction, i.e., if the top terminal in Fig. 13–1 is positive with respect to the bottom terminal. The resistive counter voltage is just IR, as in d-c circuits. If dI/dt is positive, an emf will be induced in the inductance that tends to cause a current in the opposite direction to that assumed for I, i.e., the upper terminal of L must be positive with

---

* E. A. Guillemin, *Communication Networks* (2 vols.) (New York: Wiley, 1931 and 1935); and H. W. Bode, *Network Analysis and Feedback Amplifier Design* (Princeton, N. J.: D. Van Nostrand, 1945), (Huntington, N. Y.: Krieger, 1975, reprint of 1945 edition).

† For example, N. Balabanian and T. Bickart, *Electrical Network Theory* (New York: Wiley, 1969); and J. B. Murdoch, *Network Theory* (New York: McGraw-Hill, 1970).

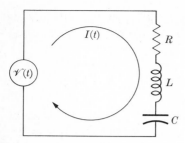

Figure 13–1 A series circuit of circuit elements.

respect to the lower terminal. Since this is the same sense as IR with respect to the direction of I, the counter voltage is just $L(dI/dt)$.* The capacitative counter voltage depends on the charge on the capacitor, which may be either positive or negative, depending on whether we consider the upper or the lower conductor. This difficulty is resolved by writing

$$Q = \int_{t_0}^{t} I(t) \, dt, \qquad (13\text{–}2)$$

where t_0 is chosen so that $Q(t_0)$ is zero. With this choice of Q a positive Q makes the upper terminal of the capacitor positive, and thus produces the capacitative counter voltage $+Q/C$. Kirchhoff's voltage law for the circuit of Fig. 13–1 is

$$\mathscr{V}(t) = RI + L\frac{dI}{dt} + \frac{1}{C}\int_{t_0}^{t} I \, dt, \qquad (13\text{–}3)$$

which is typical of the integrodifferential equations of circuit theory.

13–4 ELEMENTARY TRANSIENT BEHAVIOR

The only transient behavior to be considered here is that associated with the sudden application of a constant voltage $\mathscr{V}$ to a network of resistors, capacitors, and inductors, the first example being the simple R-L circuit shown in Fig. 13–2. For this circuit, Eq. (13–3) becomes

$$\mathscr{V} = RI + L\frac{dI}{dt} \qquad (13\text{–}4)$$

after the switch S is closed. Before the switch is closed the solution is trivial, being just $I = 0$. Equation (13–4) is a first-order linear differential equation with constant coefficients and hence can always be solved with one arbitrary constant in

---

* It is worth noting that the induced emf is written $-L(dI/dt)$; however, being an emf, it would normally be written on the other side of the equation from the counter voltages. Thus no inconsistency is introduced by writing $+L(dI/dt)$ for the counter voltage.

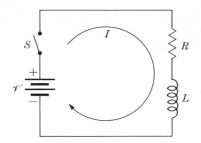

Figure 13–2 Transient response of an *R-L* circuit. Circuit diagram.

the solution. The solution is

$$I(t) = \frac{\mathscr{V}}{R} - Ke^{-tR/L}, \tag{13-5}$$

with K the arbitrary constant. Since the circuit contains an inductance that prevents an abrupt change in the current, the current just after the switch is closed must be the same as the current just before the switch is closed, i.e., zero. If the switch is closed at $t = t_0$, this requires that

$$\frac{\mathscr{V}}{R} - Ke^{-t_0 R/L} = 0 \tag{13-6}$$

or

$$K = \frac{\mathscr{V}}{R} e^{t_0 R/L}. \tag{13-7}$$

The complete solution is then

$$I(t) = \frac{\mathscr{V}}{R} [1 - e^{-R(t-t_0)/L}], \tag{13-8}$$

which is plotted in Fig. 13–3. There are several useful, easily obtained facts that can be found from Eq. (13–8) and Fig. 13–3. First, L/R has the dimensions of time and is called the *time constant*. Since $1/e \cong 0.368$, the time constant is the time required for the current to reach 0.632 times its final value, $\mathscr{V}/R$. In five time constants the current reaches 0.993 times its final value, which is conveniently

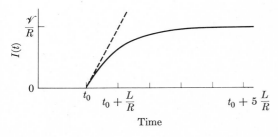

Figure 13–3 Transient response of an *R-L* circuit.

remembered as 99 percent. The initial slope dI/dt is just the final current $\mathscr{V}/R$ divided by one time constant L/R, i.e., a slope such that if the current continued to increase at this rate it would reach its final value in one time constant. The usefulness of these facts is that, by simply sketching a standard exponential curve, they enable evaluation of the exponential function involved in a simple transient problem to an accuracy of a few percent. Many other aspects of a resistance-inductance circuit can be explored, and a similar treatment can be applied to resistance-capacitance circuits. Several of the problems at the end of this chapter are devoted to accomplishing this end.

The second example to be considered is a series R-L-C circuit that is suddenly connected to a constant voltage $\mathscr{V}$. Such a circuit is shown in Fig. 13–4. The appropriate equation after the switch is closed is

$$\mathscr{V} = RI + L\frac{dI}{dt} + \frac{1}{C}\int_{t_0}^{t} I(t)\,dt, \tag{13–9}$$

where again t_0 is a time at which the charge on the capacitor is zero. In the interest of simplicity it will be assumed that the capacitor is initially uncharged and that the switch S is closed at $t_0 = 0$. Equation (13–9) may be unfamiliar; however, by simply differentiating it once with respect to the time it becomes

$$\frac{d\mathscr{V}}{dt} = R\frac{dI}{dt} + L\frac{d^2I}{dt^2} + \frac{I}{C}, \tag{13–10}$$

which is an ordinary second-order linear differential equation with constant coefficients (the harmonic oscillator equation). The technique for solving such equations is well known, and in fact for the case at hand, $d\mathscr{V}/dt = 0$, the solution is*

$$I = \{Ae^{i\omega_n t} + Be^{-i\omega_n t}\}e^{-Rt/2L}, \tag{13–11}$$

where

$$\omega_n = \sqrt{\frac{1}{LC} - \frac{R^2}{4L^2}},$$

so long as neither L nor C is zero. If either vanishes, an indeterminacy appears in Eq. (13–11); however, Eq. (13–10) can still be solved for $L = 0$; in fact, the solution

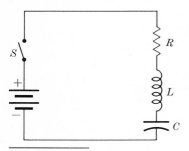

Figure 13–4 Transient response of an R-L-C circuit. Circuit diagram.

---

* Here i is the unit imaginary number, that is, $i \equiv \sqrt{-1}$.

is simpler than Eq. (13–11). Furthermore, the case $C = 0$ corresponds to the uninteresting case of an open circuit. To complete the discussion of this point, if $C = \infty$, which corresponds to short-circuiting the capacitor, Eq. (13–11) reduces to Eq. (13–5), with now two arbitrary constants to be obtained by fitting boundary conditions. This, of course, reflects the fact that all knowledge of $\mathscr{V}$ was lost in going from Eq. (13–9) to Eq. (13–10).

We return now to the solution Eq. (13–11), where it remains to evaluate the constants A and B. For the current to be real, B must be the complex conjugate of A. Since the switch is closed at $t = 0$, at $t = 0$ the current must be zero, which means that the two imaginary exponentials must combine to give a sine function. These observations lead to

$$I(t) = De^{-Rt/2L} \sin \omega_n t, \tag{13–12}$$

where D is a single real constant still to be evaluated. This evaluation is accomplished by noting that at $t = 0$, Q and I are both zero, and hence that

$$\mathscr{V} = L \left. \frac{dI}{dt} \right|_{t=0}. \tag{13–13}$$

Using this initial condition gives

$$D = \frac{\mathscr{V}}{\omega_n L} = \frac{\mathscr{V}}{\sqrt{\dfrac{L}{C} - \dfrac{R^2}{4}}}. \tag{13–14}$$

The solution is now complete. The current oscillates with the *natural frequency*

$$\omega_n = \sqrt{\frac{1}{LC} - \frac{R^2}{4L^2}},$$

with, however, an amplitude that decreases with time and is given by $De^{-Rt/2L}$. This behavior is shown in Fig. 13–5. If the time t_0 of closing the switch is not 0, it is only necessary to replace t with $t - t_0$.

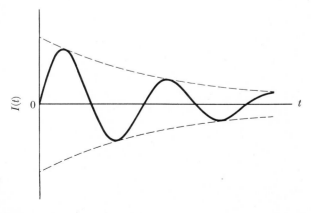

Figure 13–5 Transient response of an *R-L-C* circuit.

This completes the elementary transient analysis to be presented here. The balance of this chapter will be devoted to circuits excited by sinusoidal voltages in the steady state, i.e., sufficiently long after the excitation has been applied to ensure that the transients are negligible.

13-5 STEADY-STATE BEHAVIOR OF A SIMPLE SERIES CIRCUIT

The behavior of the circuit of Fig. 13-1, with the following excitation, will now be studied:

$$\mathscr{V}(t) = \mathscr{V}_0 \cos \omega t, \tag{13-15}$$

where ω is a given frequency, not necessarily equal to ω_n. This could simply be written in place of $\mathscr{V}(t)$ in Eq. (13-1) or Eq. (13-10) and the resulting equation solved; however, a more fruitful procedure is to note that $\mathscr{V}_0 \cos \omega t$ is the real part of $\mathscr{V}_0 e^{i\omega t}$. If a fictitious complex voltage $\mathscr{V}_1 + i\mathscr{V}_2$ were applied to the circuit the resulting current would most certainly also be complex, $I_1 + iI_2$ (it is implied here that $\mathscr{V}_1$, $\mathscr{V}_2$, I_1, and I_2 are all real). Putting these fictitious quantities into Eq. (13-10) gives us

$$\frac{d\mathscr{V}_1}{dt} + i\frac{d\mathscr{V}_2}{dt}$$

$$= \left(L\frac{d^2 I_1}{dt^2} + R\frac{dI_1}{dt} + \frac{I_1}{C}\right) + i\left(L\frac{d^2 I_2}{dt^2} + R\frac{dI_2}{dt} + \frac{I_2}{C}\right). \tag{13-16}$$

The only way this equation can be satisfied is if the real parts on the left and right are equal and the imaginary parts on the left and right are equal. Thus I_1 satisfies Eq. (13-10) with $d\mathscr{V}_1/dt$ on the left, and I_2 satisfies Eq. (13-10) with $d\mathscr{V}_2/dt$ on the left. This means that if $\mathscr{V}(t)$ is the real part of some complex function it is sufficient to solve Eq. (13-10) with the complex function for $\mathscr{V}(t)$, and then obtain the physical current by taking the real part of the complex solution. For the excitation $\mathscr{V}_0 \cos \omega t$ it is appropriate to use $\mathscr{V}_0 e^{i\omega t}$ and take the real part of the solution to be the physical current. In some instances it may be preferable to use $e^{i(\omega t + \phi)}$ in order to obtain the response to $\cos(\omega t + \phi)$, where ϕ is some given phase angle.

If $\mathscr{V}_0 e^{i\omega t}$ is used in Eq. (13-10), then the current will be $I_0 e^{i\omega t}$, with I_0 some complex constant. Direct substitution into the equation gives

$$i\omega\mathscr{V}_0 e^{i\omega t} = \left[-\omega^2 L + i\omega R + \frac{1}{C}\right] I_0 e^{i\omega t}. \tag{13-17}$$

Dividing by $i\omega$ changes this to

$$\mathscr{V}_0 e^{i\omega t} = \left[R + i\omega L + \frac{1}{i\omega C}\right] I_0 e^{i\omega t}, \tag{13-18}$$

which is in the form

$$\mathscr{V}_0 e^{i\omega t} = ZI_0 e^{i\omega t} \tag{13-19}$$

with

$$Z = R + i\omega L + \frac{1}{i\omega C},$$ (13–20a)

or

$$Z = R + i\left(\omega L - \frac{1}{\omega C}\right).$$ (13–20b)

The *impedance* Z of the circuit consists of two parts: the real part or *resistance* (R), and the imaginary part or *reactance* (X). The reactance is further divided into the *inductive reactance* $X_L = \omega L$ and the *capacitive reactance* $X_C = -1/\omega C$. The fact that the impedance is complex means that the current is not in phase with the applied voltage. It is sometimes convenient to write the impedance in polar form:

$$Z = |Z|e^{i\theta},$$ (13–21)

with

$$|Z| = [R^2 + (\omega L - 1/\omega C)^2]^{1/2}$$ (13–22)

and

$$\theta = \tan^{-1}\left(\frac{\omega L - 1/\omega C}{R}\right).$$ (13–23)

Using this form for the impedance, we may write the complex current as

$$I(t) = \frac{\mathscr{V}_0}{|Z|}e^{i(\omega t - \theta)},$$ (13–24a)

and the physical current as

$$\frac{\mathscr{V}_0}{|Z|}\cos(\omega t - \theta).$$ (13–24b)

If θ is greater than zero, the current reaches a specified phase later than the voltage and is said to lag the voltage. In the opposite case, the current leads the voltage. This formally completes the study of the simple series circuit, although later we shall examine the solution with care, to enhance our physical understanding of the situation.

13–6 SERIES AND PARALLEL CONNECTION OF IMPEDANCES

If two impedances are connected in series, then the same current flows through each of them. The voltages* across the two impedances are $V_1 = Z_1 I$ and $V_2 = Z_2 I$. The voltage across the combination is $V_1 + V_2 = (Z_1 + Z_2)I$. It is clear, then, that the connection of impedances in series adds the impedances, that is,

$$Z = Z_1 + Z_2 + Z_3 + \cdots \quad \text{(series connection)}.$$ (13–25)

---

* In this and the remaining sections of the chapter, we shall use the symbol V in place of $\Delta\varphi$ for the potential difference across an element, or group of elements.

Thus Eq. (13–20a) is the sum of the impedance of a resistance R,

$$Z_1 = R,$$

an inductance L,

$$Z_2 = i\omega L,$$

and a capacitance C,

$$Z_3 = \frac{1}{i\omega C},$$

all in series. It is important to note that the impedances add as complex numbers. If $Z_1 = R_1 + iX_1$ and $Z_2 = R_2 + iX_2$, then

$$Z = Z_1 + Z_2 = (R_1 + R_2) + i(X_1 + X_2). \tag{13–26}$$

In polar form,

$$Z = |Z|e^{i\theta}, \qquad |Z| = [(R_1 + R_2)^2 + (X_1 + X_2)^2]^{1/2},$$

$$\theta = \tan^{-1}\frac{X_1 + X_2}{R_1 + R_2}. \tag{13–27}$$

Note that the magnitude of Z is **not** the sum of the magnitudes of Z_1 and Z_2.

If impedances are connected in parallel, then the same voltage appears across each, and the currents are given by $I_1 = V/Z_1, I_2 = V/Z_2$, etc. The total current is

$$I = I_1 + I_2 + \cdots = \frac{V}{Z_1} + \frac{V}{Z_2} + \cdots = V\left(\frac{1}{Z_1} + \frac{1}{Z_2} + \cdots\right),$$

from which it is clear that

$$\frac{1}{Z} = \frac{1}{Z_1} + \frac{1}{Z_2} + \cdots \qquad \text{(parallel connection)}. \tag{13–28}$$

Here, too, the addition is the addition of complex numbers.

Equations (13–25) and (13–28) provide the basis for solving problems involving more complex configurations with a single applied voltage. As an example, we now consider the circuit of Fig. 13–6. The impedance consists of a resistor in series

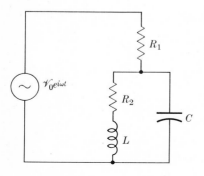

Figure 13–6 A typical a-c circuit.

with the parallel combination of a capacitor and an inductor. This is written as

$$Z = R_1 + \cfrac{1}{\cfrac{1}{R_2 + i\omega L} + \cfrac{1}{1/i\omega C}}. \tag{13-29}$$

Alternatively,

$$Z = R_1 + \frac{R_2 + i\omega L}{1 + i\omega C(R_2 + i\omega L)} \tag{13-30}$$

or

$$Z = R_1 + \frac{(R_2 + i\omega L)[(1 - \omega^2 LC) - i\omega R_2 C]}{(1 - \omega^2 LC)^2 + \omega^2 R_2^2 C^2}. \tag{13-31}$$

The only other worth-while manipulation at this time is the separation into real and imaginary parts:

$$Z = R_1 + \frac{R_2}{(1 - \omega^2 LC)^2 + \omega^2 R_2^2 C^2} + i\,\frac{\omega L(1 - \omega^2 LC) - \omega R_2^2 C}{(1 - \omega^2 LC)^2 + \omega^2 R_2^2 C^2}. \tag{13-32}$$

Having found Z, we now determine the current by dividing Z into $\mathscr{V}_0 e^{i\omega t}$. The study of this circuit will be continued later in connection with resonance phenomena.

13-7 POWER AND POWER FACTORS

The power delivered to a resistor may be determined by multiplying the voltage across the resistor by the current through the resistor. However, for the more general case, such as the impedance shown in Fig. 13-7, a more subtle approach is required. If $V(t)$ and $I(t)$ are the complex voltage and current as shown, then the instantaneous power is

$$P(t) = \operatorname{Re} I(t) \operatorname{Re} V(t). \tag{13-33}$$

The *average* power is a more important quantity, with the average being taken over either one full period or a very long time (many periods). If the phases are chosen so that V_0 is real and, as usual, $Z = |Z|e^{i\theta}$, then it is straightforward to show (Problem 13-11) that

$$\bar{P} = \overline{\operatorname{Re} I(t) \operatorname{Re} V(t)} = \tfrac{1}{2}|I_0|\,|V_0| \cos \theta. \tag{13-34}$$

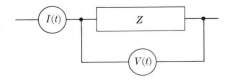

Figure 13-7 Measurement of power.

The factor one-half in Eq. (13–34) represents the fact that the average of $\sin^2 \omega t$ or $\cos^2 \omega t$ is one-half. The other interesting factor is cosine θ, which takes into account the fact that the current and voltage are not in phase. Cosine θ is frequently called the *power factor* of an alternating current (a-c) circuit. In Section 17–3 it is shown that

$$\overline{\text{Re } (I_0\, e^{i\omega t}) \text{ Re } (V_0\, e^{i\omega t})} = \tfrac{1}{2} \text{ Re } (I_0^*\, V_0), \qquad (13\text{–}35)$$

where I_0^* is the complex conjugate of I_0. This form is conveniently remembered and leads at once to Eq. (13–34).

As a final comment, we mention that the *effective values* of the voltage and current are often defined by

$$V_{\text{eff}} = \frac{\sqrt{2}}{2}\,|V_0|, \qquad I_{\text{eff}} = \frac{\sqrt{2}}{2}\,|I_0|. \qquad (13\text{–}36)$$

The virtue of these definitions is that a given V_{eff} applied to a resistance dissipates the same power as a constant voltage of the same magnitude. The specification of effective values is very common, e.g., 115-volt a-c lines are 115 effective volt lines.

13–8 RESONANCE

Equation (13–22) shows that a simple series *L-R-C* circuit has a frequency-dependent impedance that is a minimum at $\omega^2 = \omega_0^2 \equiv 1/LC$. At this frequency the impedance is just R, the phase angle is zero, and the current is a maximum of magnitude $\mathscr{V}_0/R$. This is a resonant phenomenon much like that observed in force-damped mechanical oscillators. If the current is plotted as a function of frequency, a curve of the form shown in Fig. 13–8 is obtained. Several curves are shown; all are based on the same values of L and C, but the series resistance varies from curve to curve. It is clear that the curves are sharper for small than for large values of the series resistance. The current falls to $\sqrt{2}/2$ times its maximum value at a frequency where the magnitude of the impedance is $\sqrt{2}$ times R, or where

$$\left| \omega L - \frac{1}{\omega C} \right| = R. \qquad (13\text{–}37)$$

Figure 13–8 Resonance curves for a series *R-L-C* circuit.

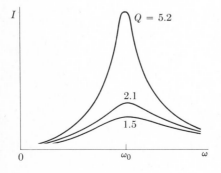

I

$Q = 5.2$

2.1

1.5

$0 \qquad \omega_0 \qquad \omega$

For relatively sharply peaked responses this obtains at values of ω not far removed from ω_0. We then write $\omega = \omega_0 + \Delta\omega$, and obtain

$$\left| \omega_0 L + \Delta\omega L - \frac{1}{\omega_0 C} \frac{1}{1 + \Delta\omega/\omega_0} \right| = R. \tag{13-38}$$

Using $\omega_0^2 = 1/LC$ and $(1 + \Delta\omega/\omega_0)^{-1} \cong 1 - \Delta\omega/\omega_0$ gives

$$2|\Delta\omega|L = R$$

or

$$\frac{2|\Delta\omega|}{\omega_0} = \frac{R}{\omega_0 L}. \tag{13-39}$$

The quantity

$$Q = \omega_0 L/R \qquad \text{or} \qquad Q = \frac{\omega_0}{2|\Delta\omega|} \tag{13-40}$$

characterizes the sharpness of the resonance and is known as the quality factor Q of the circuit.* For practical purposes, Q may be considered to be a property of the inductor only, since most of the unavoidable series resistance is associated with the wire with which the inductor is wound. However, a more refined treatment shows that the capacitor losses must also be included in computing Q's. The curves of Fig. 13–8 are labeled with the appropriate Q values.

As the driving frequency is varied, not only the magnitude but also the phase of the current varies. This variation is shown in Fig. 13–9 for the same Q values used in Fig. 13–8. Below resonance, the phase angle of the impedance function is negative; therefore the phase of the current is positive and it leads the voltage. Above resonance, the opposite is true and the current lags the voltage.

It is interesting to note that the usual radiofrequency resonant circuits found in communications equipment are series resonant circuits, in spite of their parallel-circuit appearance. In the most simple case this is because the driving power is inductively coupled into L and thus appears as an emf in series with L.

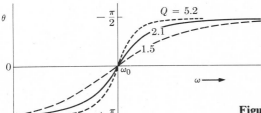

Figure 13–9 Phase angle of the impedance in a typical R-L-C series circuit.

* This Q has nothing to do with charge.

Resonance is not restricted to series circuits like those just discussed; parallel circuits may also exhibit resonant characteristics. The circuit of Fig. 13–6 exhibits such a resonance. Defining the resonant frequency for a parallel resonant circuit is not as simple as it is for a series circuit. Some of the possibilities are: (1) $\omega_0 = 1/\sqrt{LC}$, (2) the frequency at which the impedance [given by Eq. (13–31)] is a maximum, or (3) the frequency at which the power factor is unity. Each of these three choices gives a different frequency; however, for high Q circuits they are very nearly the same. The first choice is by far the most useful in practice because it makes many series resonance results directly applicable to the parallel resonant case. One very interesting result is obtained by using Eq. (13–31) to evaluate Z, with $R_1 = 0$ and $\omega_0 = 1/\sqrt{LC}$. The result is

$$ Z = \omega_0 L \left[\frac{\omega_0 L}{R} - i \right], \qquad (\omega = \omega_0). \tag{13–41} $$

For a high Q circuit the i can be neglected, with the result that the impedance at resonance is Q times the inductive reactance at resonance.

The subject of resonant circuits can be pursued at great length; however, to do so here is unwarranted. Some of the problems extend this section, and more comprehensive details are given elsewhere.*

*13–9 MUTUAL INDUCTANCES IN a-c CIRCUITS

Solving a-c circuit problems involving mutual inductances presents a minor difficulty in assigning the correct sign to the mutual inductance. This difficulty can be readily resolved by noting that the sign to be associated with the mutual inductance depends on the assumed direction of the current in the two circuits involved, and on the way in which the windings are connected. The notation M_{ij} will be used for the pure mutual inductance between two circuits.

It was shown in Chapter 11 that the emf in winding 2, due to a changing current in winding 1, is given in magnitude by

$$ \mathscr{E}_2 = M_{21} \frac{dI_1}{dt}. \tag{13–42} $$

For sinusoidal currents, using complex notation, we have

$$ \mathscr{E}_2 = i\omega M_{21} I_{10} e^{i\omega t} \tag{13–43} $$

or

$$ \mathscr{E}_2 = i\omega M_{21} I_1. \tag{13–44} $$

In what follows, the symbol M_{21} will be taken to be a positive quantity and the sign of $\mathscr{E}_2$ will be displayed explicitly; in other words, M_{21} in Eq. (13–44) will be replaced by $\pm M_{21}$, with M_{21} a positive quantity.

---

* K. Henney, *Radio Engineering Handbook*, Fifth Edition (New York: McGraw-Hill, 1959).

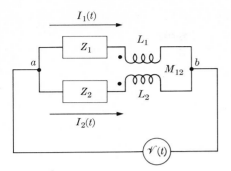

Figure 13–10 Circuit with mutual inductance.

To demonstrate the technique for assigning signs, we now consider the circuit shown in Fig. 13–10, in which two impedances Z_1 and Z_2 are combined with a mutual inductance and connected to an applied voltage $\mathscr{V}(t) = \mathscr{V}_0\, e^{i\omega t}$. The mutual inductance is labeled M_{12} and is taken to be a positive number. The black dots in the figure indicate the ends of the two windings that are simultaneously positive; that is, if the lower winding is excited by a sinusoidal current that makes the left-hand terminal positive at some time t_1, then the voltage induced in the upper winding makes the left-hand terminal of the upper winding positive at t_1. The equation for the upper branch, in accordance with Kirchhoff's law, is

$$Z_1 I_1 + i\omega L_1 I_1 + i\omega M_{12} I_2 = \mathscr{V}. \tag{13-45}$$

The plus sign is used with the mutual inductance because a positive I_2 gives a voltage in the upper branch that has the same sense as an $I_1 R$ drop. The second equation is

$$i\omega M_{12} I_1 + Z_2 I_2 + i\omega L_2 I_2 = \mathscr{V}, \tag{13-46}$$

where $M_{12} = M_{21}$ has been written in the interests of symmetry.

The assignment of the sign is on the same basis as before and may be checked by noting that M_{12} should appear in the branch-one equation with the same sign as M_{21} in the branch-two equation. Equations (13–45) and (13–46) may be solved simultaneously by standard techniques to yield

$$
\begin{aligned}
I_1 &= \mathscr{V}\, \frac{Z_2 + i\omega L_2 - i\omega M_{12}}{(Z_1 + i\omega L_1)(Z_2 + i\omega L_2) + \omega^2 M_{12}^2}\,; \\[2mm]
I_2 &= \mathscr{V}\, \frac{Z_1 + i\omega L_1 - i\omega M_{12}}{(Z_1 + i\omega L_1)(Z_2 + i\omega L_2) + \omega^2 M_{12}^2}\,.
\end{aligned}
\tag{13-47}
$$

Combining the two to obtain the total current $I_1 + I_2$ gives

$$I = I_1 + I_2 = \mathscr{V}\, \frac{Z_1 + i\omega L_1 + Z_2 + i\omega L_2 - 2i\omega M_{12}}{(Z_1 + i\omega L_1)(Z_2 + i\omega L_2) + \omega^2 M_{12}^2}. \tag{13-48}$$

The coefficient of $\mathscr{V}$ on the right side is the reciprocal of the impedance presented to the generator or the net impedance between points a and b. It is obvious that if

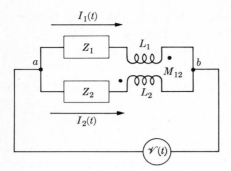

Figure 13–11 Circuit of Fig. 13–10 with the sign of the mutual inductance reversed.

M_{12} is zero, the impedance is the parallel combination of the two branch impedances. For the connection shown, as M_{12} increases so does the impedance.

The circuit obtained by interchanging the leads on one winding of the mutual inductance is shown in Fig. 13–11. Note that the only difference is that the black dot has been moved from the left end of the upper winding to the right end. The result is to change the sign of the M_{12} term in Eqs. (13–45) and (13–46), with the result that

$$(Z_1 + i\omega L_1)I_1 - i\omega M_{12}I_2 = \mathscr{V},$$

and

$$-i\omega M_{12}I_1 + (Z_2 + i\omega L_2)I_2 = \mathscr{V}.$$

$$(13\text{–}49)$$

The currents are easily found and combined to obtain the impedance:

$$Z_{ab} = \frac{(Z_1 + i\omega L_1)(Z_2 + i\omega L_2) + \omega^2 M_{12}^2}{Z_1 + i\omega L_1 + Z_2 + i\omega L_2 + 2i\omega M_{12}}, \qquad (13\text{–}50)$$

which is the same as in the previous case when the mutual inductance is zero. The relationship between Z_{ab} for finite M_{12} and Z_{ab} for $M_{12} = 0$ depends on the parameter in a rather complicated way. We will state here only that Z_{ab} may be larger or smaller than the Z_{ab} for $M_{12} = 0$.

The basic circuit for the most common mutual inductance device, the transformer, is shown in Fig. 13–12. R_1 and R_2 are the resistances of the primary (driving) and secondary (driven) windings, L_1 and L_2 are their self-inductances,

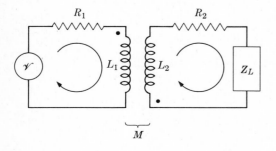

Figure 13–12 A transformer.

and M is the (positive) mutual inductance between them. Z_L is the impedance of the load connected to the secondary winding, and $\mathscr{V}(t) = \mathscr{V}_0 e^{i\omega t}$ is the voltage across the primary winding. If currents $I_1 e^{i\omega t}$ and $I_2 e^{i\omega t}$ are assumed to be in the directions indicated, then Kirchhoff's voltage law requires that the equations

$$\mathscr{V}_0 = I_1 R_1 + i\omega L_1 I_1 + i\omega M I_2,$$

and

$$0 = I_2 R_2 + i\omega L_2 I_2 + i\omega M I_1 + I_2 Z_L,$$

(13–51)

be satisfied. The solutions to these equations are

$$I_1 = \frac{Z_L + R_2 + i\omega L_2}{(R_1 + i\omega L_1)(Z_L + R_2 + i\omega L_2) + \omega^2 M^2} \mathscr{V}_0,$$

and

$$I_2 = \frac{-i\omega M}{(R_1 + i\omega L_1)(Z_L + R_2 + i\omega L_2) + \omega^2 M^2} \mathscr{V}_0.$$

(13–52)

These relatively complicated equations represent an exact solution for the circuit of Fig. 13–12.

For many purposes it is much more convenient to think in terms of an *ideal* transformer, i.e., one for which the relations

$$V_L = a\mathscr{V}_0, \qquad I_2 = -\frac{I_1}{a},$$

(13–53)

are satisfied, where the constant a is independent of frequency, V_L is the voltage across Z_L, and all other quantities are as shown in Fig. 13–12. Multiplying Eqs. (13–53) together shows that these relations require that the source power $\mathscr{V}_0 I_1$ is all delivered to the load, $V_L I_2$; in other words, there are no losses in the transformer. The condition that must be satisfied to ensure the second of these relations is

$$\frac{Z_L + R_2 + i\omega L_2}{i\omega M} = a,$$

(13–54)

which is satisfied if $\omega L_2 \gg |Z_L + R_2|$. Similar conditions can be found that will ensure that $V_L/\mathscr{V}_0 = a$.* The conditions are complicated and not easy to satisfy; however, practical transformers exist that approximate them over relatively wide frequency ranges. For such devices,

$$I_2 = -\frac{I_1}{a}, \qquad V_L = a\mathscr{V}_0,$$

and

$$\frac{\mathscr{V}_0}{I_1} = -\frac{V_L}{a^2 I_2} = \frac{Z_L}{a^2}.$$

(13–55)

---

* The details are given in Guillemin, *loc. cit.*, Chapter VIII.

The last of these relationships shows that the transformer acts also as an impedance transformer, with transformation ratio a^{-2}. It is left as an exercise to show that for very close coupling of the two windings $a = N_2/N_1$, that is, the turns ratio.

*13–10 MESH AND NODAL EQUATIONS

More complex a-c circuits may be approached in two ways: one based on Kirchhoff's voltage law and known as *mesh analysis*, and the other based on Kirchhoff's current law and known as *nodal analysis*. Each method has its advantages and disadvantages. Since choosing the expedient method can greatly simplify some problems, both methods will be considered in this section.

The first step in applying mesh analysis is the assignment of meshes. This is accomplished by assuming closed loop currents such that at least one current goes through each element. With such a choice of currents, Kirchhoff's law I is satisfied automatically. For example, in Fig. 13–13 three meshes are shown, labeled I_1, I_2, and I_3. This is, of course, not the only possible choice; several others are possible and useful. If Kirchhoff's voltage law II is applied to each of these meshes, we obtain

$$
\begin{aligned}
I_1(Z_3 + Z_4) \qquad & -I_2 Z_4 & -I_3 Z_3 & \quad = \mathscr{V}, \\
-I_1 Z_4 \qquad & +I_2(Z_1 + Z_2 + Z_4) & -I_3 Z_2 & \quad = 0, \qquad (13\text{–}56) \\
-I_1 Z_3 \qquad & -I_2 Z_2 & +I_3(Z_2 + Z_3 + Z_5) & \quad = 0.
\end{aligned}
$$

Note that the minus signs appear because in mesh one, for example, I_2 flows through Z_4 counter to the direction of I_1. Equations (13–56) can be solved most easily by matrix techniques, resulting in expressions for the set of mesh currents in the circuit. It is useful to note that the mesh equations can be written as

$$
\sum_{j=1}^{n} Z_{ij} I_j = \mathscr{V}_i \ (i = 1, 2, \ldots, n) \qquad (13\text{–}57)
$$

(with $n = 3$ in the circuit above). In this notation, $Z_{ij} = Z_{ji}$, which is a useful check on the mesh equations.

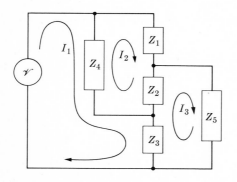

Figure 13–13 Illustration of the use of mesh analysis in a-c circuits.

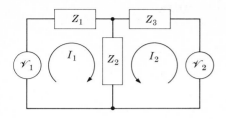

Figure 13–14 Further use of mesh equations.

As a second example, consider the circuit of Fig. 13–14. The appropriate equations for this circuit are written as

$$I_1(Z_1 + Z_2) + \quad I_2 Z_2 \quad = \mathscr{V}_1,$$
$$I_1 Z_2 \quad + I_2(Z_2 + Z_3) = \mathscr{V}_2. \tag{13-58}$$

There is no reason why $\mathscr{V}_1$ and $\mathscr{V}_2$ must be in phase; usually they will not be, but will be expressible as $\mathscr{V}_1 = |\mathscr{V}_{10}| e^{i\omega t}$, $\mathscr{V}_2 = |\mathscr{V}_{20}| e^{i(\omega t + \phi)}$. It is, however, very important to assign the phases correctly, and this is most conveniently accomplished by examining the relative phases at $t = 0$ and assigning directions (senses) with respect to the assigned mesh currents. It is also important to note that unless all of the generators have the same frequency the entire technique fails (more properly, the problem reduces to the superposition of two independent problems, each involving one generator and one frequency).

Before proceeding to discuss the alternative nodal equations, it is appropriate to discuss voltage and current generators. In the preceding sections, circuit problems have been phrased in terms of pure sources of applied voltage. Such idealized devices cannot be constructed, of course; practical devices always have a certain internal impedance. Thus a practical generator consists of a source of voltage, $\mathscr{V}(t)$, in series with an impedance Z_I, which is the internal impedance. Such a generator is shown in Fig. 13–15 connected to a load Z_L. Several observations may be made. First, for maximum power transfer to the external load, $Z_L = Z_I^*$; that is, Z_I and Z_L should have equal resistive parts, and reactive parts that are equal in magnitude but opposite in sign. The proof of this is left as an exercise. Secondly, a voltage generator is equivalent to a current generator delivering a current $\mathscr{I}(t) = \mathscr{V}(t)/Z_I$ shunted by the internal impedance. This equivalence for the circuit of Fig. 13–15 is shown in Fig. 13–16. It is easy to show this equivalence if it is noted that an ideal current generator delivers the current $\mathscr{I}(t)$ to *any* load

Figure 13–15 Practical generator connected to a load Z_L.

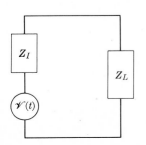

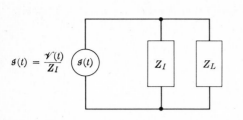

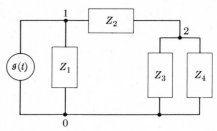

Figure 13–16 A "current generator" that is equivalent to the voltage generator of Fig. 13–15.

Figure 13–17 Illustrating the method of nodal analysis in a-c circuits.

connected to its terminals. The equivalence further means that in any circuit problem the generators may be taken either as voltage generators or as current generators, to suit the convenience of the situation.

The nodal equations for a circuit result from the application of Kirchhoff's current law I to each of the nodes, where a node is a point at which three or more elements join. In this procedure, Kirchhoff's voltage law II is satisfied automatically. As a simple example of the application of the nodal equations, we refer to the circuit of Fig. 13–17. The nodal equations are obtained by requiring that the algebraic sum of the currents to each node be zero. The nodes are numbered, starting with zero for the node whose potential is the reference for the circuit. If the potential at node 0 is taken to be zero, then at node 1

$$\mathscr{I}(t) = \frac{V_1}{Z_1} + \frac{V_1 - V_2}{Z_2}, \tag{13–59}$$

where V_1 and V_2 are the potentials of nodes 1 and 2 respectively. At node 2,

$$0 = \frac{V_2 - V_1}{Z_2} + \frac{V_2}{Z_3} + \frac{V_2}{Z_4}. \tag{13–60}$$

Before proceeding, we make the observation that a quantity that is the reciprocal of an impedance would be a great convenience. Such a quantity is the *admittance*, symbolized by Y. $Y = 1/Z$. Admittances in parallel add, while admittances in series combine by adding reciprocals. In terms of admittances, Eqs. (13–59) and (13–60) become

$$\mathscr{I}(t) = (Y_1 + Y_2)V_1 - Y_2 V_2,$$
$$0 = - Y_2 V_1 + (Y_2 + Y_3 + Y_4)V_2, \tag{13–61}$$

which are somewhat more convenient. The simultaneous solution of these equations yields the nodal voltages, V_1 and V_2.

We shall consider one more example of the use of nodal equations; namely, to the circuit shown in Fig. 13–18. The nodal equations are simply written down in

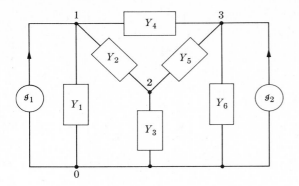

Figure 13-18 Another circuit illustrating nodal analysis.

the form

$$\mathscr{I}_1 = Y_1 V_1 + Y_2(V_1 - V_2) + Y_4(V_1 - V_3),$$
$$0 = Y_2(V_2 - V_1) + Y_3 V_2 + Y_5(V_2 - V_3), \qquad (13\text{--}62)$$
$$\mathscr{I}_2 = Y_6 V_3 + Y_5(V_3 - V_2) + Y_4(V_3 - V_1).$$

These equations may be solved by standard techniques to obtain the voltages at the nodes. The fact that voltages rather than currents are obtained when the equations are solved is a major advantage, particularly in communications circuits.

*13–11 DRIVING POINT AND TRANSFER IMPEDANCES

We shall now present simple definitions for the driving point and transfer impedance of a four-terminal network. These definitions are presented because these terms appear in the technical literature and because they are sometimes a serious stumbling block to the uninitiated. Consider a four-terminal network, and call terminals 1 and 2 the input and 3 and 4 the output. If a generator of voltage $\mathscr{V}$ and internal impedance Z_I is connected between terminals 1 and 2, and an impedance Z_L between terminals 3 and 4, as shown in Fig. 13–19, there will be a current I_I in Z_I and a current I_L in Z_L. The driving point impedance Z_D is

$$Z_D = \frac{\mathscr{V}}{I_I}, \qquad (13\text{--}63)$$

and the transfer impedance is

$$Z_T = \frac{\mathscr{V}}{I_L}. \qquad (13\text{--}64)$$

It should be noted that Z_D and Z_T both depend on Z_I and Z_L, as well as on the internal structure of the network.

A brief treatment such as the above cannot do justice to the subject of network theory; classics such as that of Guillemin, as well as the multitude of more recent books, should be consulted for the details of this complex subject.

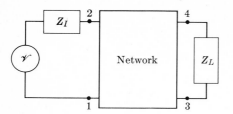

Figure 13–19 A four-terminal network.

13–12 SUMMARY

As for d-c circuits, the analysis of circuits carrying slowly varying currents depends on Kirchhoff's current and voltage laws, here applied at each instant of time. The current and voltage of a linear resistance are instantaneously related by Ohm's law

$$V_R = RI.$$

The analogous laws for a linear inductance and a linear capacitance are

$$V_L = L\,\frac{dI}{dt},$$

$$V_C = \frac{1}{C}\,Q,$$

where $I = dQ/dt$. For a single loop circuit containing an applied voltage $\mathscr{V}(t)$, Kirchhoff's laws result in the differential equation

$$L\,\frac{d^2Q}{dt^2} + R\,\frac{dQ}{dt} + \frac{1}{C}\,Q = \mathscr{V}(t).$$

(For a more complicated network, the result is a system of such linear second-order differential equations.) The general solution is a superposition of a particular solution (steady-state solution), plus the general solution of the corresponding homogeneous equation obtained by putting $\mathscr{V}(t) = 0$ (transient solution). The arbitrary constants in the latter are chosen so as to satisfy the initial conditions imposed.

1. The transient solution is exponential in t; if the exponent is complex, the imaginary part physically represents an oscillation of Q and I. The real part represents a decaying behavior (transient), with a decay time which is small for large R. For most practical circuits, the decay time is typically a fraction of a second.

2. The steady-state behavior is discussed only for a sinusoidal applied voltage (a-c). (Constant voltage is the special zero-frequency case; an arbitrary time dependence could be represented as a Fourier synthesis.) If the applied voltage $\mathscr{V}(t) = \mathscr{V}_0 \cos \omega t$ is represented in the complex form

$$\mathscr{V} = \mathscr{V}_0 e^{i\omega t},$$

the steady-state current has the same frequency ω,

$$I = I_0 e^{i\omega t}.$$

The complex impedance Z is defined by

$$\mathscr{V} = ZI.$$

When Z is expressed in polar form,

$$Z = |Z| e^{i\theta},$$

the modulus $|Z|$ gives the amplitude of the current,

$$|I_0| = \mathscr{V}_0/|Z|$$

and θ gives the phase relative to the applied voltage. Ohm's law and its extensions yield

$$Z_R = R,$$

$$Z_L = i\omega L,$$

$$Z_C = \frac{1}{i\omega C},$$

since $I = i\omega Q$.

3. The analysis of the steady-state behavior of linear a-c circuits is exactly parallel to that of d-c circuits, with the complex impedance serving as a generalization of the d-c resistance.

4. The instantaneous power dissipation is $P(t) = \operatorname{Re} I(t) \operatorname{Re} V(t)$. For a-c, this has the time-average value

$$\bar{P} = \tfrac{1}{2} |I_0| \, |V_0| \cos\theta;$$

the power factor $\cos\theta$ is 1 for a pure resistance and 0 for a pure inductance or capacitance. The "effective" voltage and current are $|V_0|/\sqrt{2}$ and $|I_0|/\sqrt{2}$. An equivalent expression for the time-average a-c power is

$$\bar{P} = \tfrac{1}{2} \operatorname{Re}(I^*V).$$

5. As a function of frequency, a series circuit exhibits a resonance near $\omega_0 = 1/\sqrt{LC}$, where $|Z|$ is a minimum and $|I_0|$ is a maximum. The sharpness of the resonance is given by

$$Q = \frac{\omega_0}{2|\Delta\omega|} = \frac{\omega_0 L}{R}.$$

A parallel circuit also has a resonance near $\omega_0 = 1/\sqrt{LC}$, where $|Z|$ is a maximum and $|I_0|$ is a minimum.

6. A transformer is the most common mutual inductance circuit element. An ideal transformer is one for which the ratio of secondary to primary current is the inverse of the ratio of secondary to primary voltage.

PROBLEMS

13–1 An inductance of 2 H and a resistance of 3 Ω are connected in series with a 5-V battery and a switch. Determine the current, and the rate of change of current (dI/dt) in the circuit at the following times after the switch is closed: (a) 0.3 s, (b) 1 s, (c) 4 s.

13–2 A circuit consisting of an inductance L_0, a resistance R_0, and a battery $\mathscr{V}_0$ has a steady current $I = \mathscr{V}_0/R_0$ through it. A switch in the circuit is opened at time $t = 0$, creating an arc across the switch. If the arc resistance is given by k/I, where the constant $k < \mathscr{V}_0$, determine the current through the arc as a function of time. What is the final steady value of current through the arc?

13–3 A capacitor C, a resistor R, and a battery $\mathscr{V}_0$ are connected in series with a switch. The switch is closed at time $t = 0$. Set up the differential equation governing the charge Q on the capacitor. Determine Q as a function of time.

13–4 A capacitor C with charge Q_0 is suddenly connected in series with a resistance R and inductance L. Determine the current as a function of time. Show that there are three different types of solution, depending upon whether $R^2 - 4L/C$ is less than, equal to, or greater than zero. The first of these conditions is called *underdamped*, the second *critically damped*, and the third *overdamped*.

13–5 A real capacitor C has a parallel leakage resistance R; it is connected in series with an ideal inductance L. Calculate $|Z|$; find the approximate values at high and low frequencies and at resonance, assuming that R is large. Sketch a graph of $|Z|$ versus ω.

13–6 Repeat Problem 5 if the leaky capacitor is in parallel with the perfect inductor.

13–7 The circuit of Fig. 13–1 has an additional capacitor C' shunting the entire R-L-C combination. $R = 100\ \Omega$, $L = 1$ H, $C = 100\ \mu$F, and $C' = 10\ \mu$F. Make a plot of the impedance $|Z|$ versus frequency from zero to $f = 10^4$ Hz.

13–8 The series combination of a resistance R and an inductance L is put in parallel with the series combination of resistance R and capacitance C. Show that if $R^2 = L/C$ the impedance is independent of frequency.

13–9 A wire-wound resistor has a d-c resistance of 90.00 Ω and an inductance of 8 μH. What is the phase angle of the impedance at 1000 Hz? A capacitor is placed in parallel with the resistor to reduce the phase angle to zero at 1000 Hz without changing the resistance appreciably. Over what range of frequency is the phase angle less than it was before the capacitor was added?

13–10 (a) A capacitance C in parallel with a resistance R has an impedance Z. Suppose a capacitance C' in series with a resistance R' has the same impedance Z; find the required values of C' and R' in terms of R and C for a given ω. (b) The *dissipation factor* is defined as $D = \omega R'C'$. Show that $D = 1/\omega RC$, and that the phase of the current is $\theta = \tan^{-1}(-1/D)$.

13–11 Prove Eq. (13–34) for the time-average power dissipation in a circuit which carries an a-c current $I(t) = I_0 e^{i\omega t}$, with $V(t) = ZI(t)$.

13–12 An a-c generator with internal impedance Z_I is connected in series with a variable load impedance Z_L. Prove that maximum power is transferred to the load when $Z_L = Z_I^*$.

13–13 Given the circuit of Fig. 13–6, with $L = 4$ mH, $C = 2\ \mu$F, $R_1 = 25\ \Omega$, $R_2 = 40\ \Omega$. Find the following set of frequencies: (a) where $\omega = 1/\sqrt{LC}$, (b) where the impedance is maximum, (c) where the current through R_1 is in phase with the generator voltage.

13–14 Show that the quality factor Q defined in the text can be expressed as 2π times the maximum energy stored in the circuit, divided by the energy dissipated in one cycle. This statement is sometimes used as the definition of Q and is independent of specific circuit parameters.

13–15 A crossover network for a hi-fi set is to be designed so that two loudspeakers (each of resistance R) are connected to the output stage of an amplifier. One speaker is to receive predominantly high frequencies, the other predominantly low frequencies. The network is as shown in Fig. 13–20. The two capacitors are each of capacitance C and the two inductors each of inductance L. (a) Find a relationship between L and C for a given R such that the network presents a purely resistive load ($= R$) to the amplifier at all frequencies. (b) The crossover frequency ω_c is defined as the frequency at which each speaker receives half of the power delivered by the amplifier. For a given R and ω_c determine L and C.

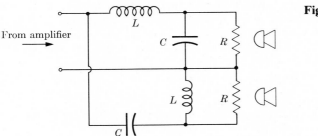

Figure 13–20

From amplifier

13–16 A 1-μF capacitor is first charged to 100 V by connecting it to a battery; it is then disconnected and immediately discharged through the 300-turn winding on a ring toroid. The toroid has a mean radius of 20 cm, a 4-cm² cross-sectional area, and an air gap of 2 mm (see Fig. 9–15). Neglecting copper losses, hysteresis, and fringing, calculate the maximum magnetic field subsequently produced in the air gap. Take the relative permeability of the toroid equal to 5000.

13–17 A potential difference of 1 V at a frequency $f = 10^6/\pi$ Hz is impressed across the circuit of Fig. 13–21. The mutual inductance of the coils is such that they are in opposition. Find the current in the upper branch.

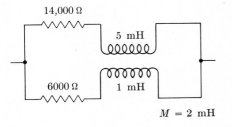

Figure 13–21

14,000 Ω

5 mH

6000 Ω   1 mH

$M = 2$ mH

13–18 A 60-Hz power transformer (turns ratio 2 : 1) has a primary inductance of 100 H and a d-c resistance of 20 Ω. The coupling coefficient between primary and secondary is close to unity. If 1000 V is placed across the primary, calculate the current in the primary winding (a) when the secondary is open-circuited, (b) when a load resistance of 20 Ω is in the secondary circuit.

*13–19 Three identical capacitors and three identical inductors are connected as shown in Fig. 13–22. Find the resonant frequencies of the system. (*Hint:* Use mesh analysis, with current of an assumed frequency ω, and show that the three equations obtained are compatible for certain ω only.)

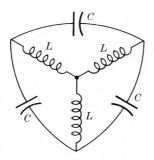

Figure 13–22

13–20 In the circuit shown in Fig. 13–14, $Z_1 = 2 + 5i$, $Z_2 = 8 - i$, $Z_3 = 4 + 3i$. The voltage generators are in phase with each other: $\mathscr{V}_1 = 10$ V, $\mathscr{V}_2 = 2$ V. Determine I_1 and I_2.

13–21 In the circuit shown in Fig. 13–17, $\mathscr{I}(t) = \mathscr{I}_1 e^{i\omega t}$ and Z_4 is *replaced* by a current generator $\mathscr{I}_2 e^{i\omega t}$. The two current generators are in phase with each other. Z_1 and Z_3 are capacitors of reactance 40 and 60 Ω respectively. Z_2 is a pure resistance of 20 Ω. $\mathscr{I}_1 = 5$ A, $\mathscr{I}_2 = 25$ A. Determine the node voltages at 1 and 2 relative to point 0.

*CHAPTER 14 Physics of Plasmas

Gases that are highly ionized are good conductors of electricity. The charged particles in such a gas interact with the local electromagnetic field; furthermore, the organized motion of these charge carriers (currents, fluctuations in charge density) can *produce* magnetic and electric fields. When subjected to a static electric field, the charge carriers in the gas rapidly redistribute themselves in such a way that most of the gas is shielded from the field. To the relatively field-free regions of the gas where positive and negative space charges are nearly balanced, Langmuir† gave the name *plasma*, while to the space-charge or strong-field regions on the boundary of the plasma he gave the name *sheaths*.

Equivalently, we may say: an ionized gas that has a sufficiently large number of charged particles to shield itself, electrostatically, in a distance small compared with other lengths of physical interest, is a plasma. A somewhat more precise definition in terms of the shielding distance will be given in Section 14-1. The earliest interest in plasmas was in connection with gaseous electronics (electrical discharges through gases, arcs, flames); recent interest has been directed toward problems in theoretical astrophysics, and the problem of ion containment in thermonuclear (fusion) reactors. A more thorough discussion can be found in books on plasma physics.‡

The general area of study embracing the interaction of ionized gases with time-dependent electromagnetic fields is called *plasma dynamics*. For many of the problems in this area, and these are the more important and interesting ones, it is impossible to treat a plasma adequately in terms of a purely macroscopic formulation. Instead, it is necessary to use what is known conventionally as kinetic theory. The motions of individual ions and electrons must be studied; their collisions with

---

* This chapter may be omitted without loss of continuity.

† I. Langmuir, *Physical Review*, vol. 33, p. 954 (1929).

‡ See, for example, T. J. M. Boyd and J. J. Sanderson, *Plasma Dynamics* (New York: Barnes and Noble, 1969); and F. F. Chen, *Introduction to Plasma Physics* (New York: Plenum Press, 1974).

other particles must be taken into account through solution of the Boltzmann transport equation. Thus a rigorous formulation for plasma problems exists, but their solution is extremely difficult in general, except for situations where it is permissible to neglect some of the terms in the Boltzmann equation. There are, however, three approximate formulations that provide considerable insight into what is happening inside the plasma.

The first of these methods is *equilibrium theory*, which rests on the premise that collisions between charged particles are sufficient to maintain the well-known Maxwell-Boltzmann velocity distribution for particles in the body of the plasma:

$$N_j(\mathbf{v}) \, dv_x \, dv_y \, dv_z = N_{0j} \left(\frac{m_p}{2\pi kT} \right)^{3/2} e^{-m_p v^2/2kT} \, dv_x \, dv_y \, dv_z,$$

where N_{0j} is the number of particles of type j per unit volume in the plasma, v_x (etc.) are the components of velocity, m_p is the mass of type j particles, and T is the absolute temperature. Kinetic and transport properties may then be calculated in terms of this velocity distribution.

The second approximate method is *orbit theory*, which treats the motion of charged particles (ions and electrons) in prescribed electric and magnetic fields. These fields may be functions both of position and of time. Orbit theory is a good approximation to particle motion in a plasma when collisions between particles do *not* play the dominant role, i.e., when the mean free path for collisions is large compared with characteristic dimensions of the orbit. Under these conditions the effect of collisions can be treated as a perturbation, and the primary problem centers around making the "prescribed" electromagnetic field self-consistent; in other words, the prescribed field must be the sum of the external field and the field produced by the orbiting particles.

The third approximate treatment is the *hydromagnetic formulation*. Here one uses the classical electromagnetic equations (Maxwell's equations) in conjunction with the classical equations of fluid motion. Evidently, the hydromagnetic treatment is just a macroscopic description of the plasma; it becomes a good approximation when the mean free path for collisions is very small compared with distances of physical interest in the plasma system. The hydromagnetic picture forms a good starting point for discussing the collective motion of particles in the plasma, e.g., plasma oscillations.

The rigorous kinetic theory approach to plasma problems is beyond the scope of this book. On the other hand, many important properties of plasmas can be discussed in terms of the approximations outlined above. For simplicity we shall assume that the plasma consists of electrons (charge, $-e$) and singly charged positive ions (charge, $+e$); neutral atoms may be present, but we shall ignore such complications as ionizing collisions and recombination of electrons and ions.

In Section 14–1, and again in Section 14–8, we encounter a plasma under stationary or steady-state conditions, for which equilibrium theory is well suited. In Sections 14–2 and 14–3, on the other hand, we shall be much concerned with individual particle motion, and here orbit theory is applicable. Finally, in Sections 14–4 through 14–7 we shall treat some dynamic aspects of the plasma, and we shall do this within the hydromagnetic framework.

14–1 ELECTRICAL NEUTRALITY IN A PLASMA

One of the most important properties of a plasma is its tendency to remain electrically neutral, i.e., its tendency to balance positive and negative space charge in each macroscopic volume element. A slight imbalance in the space-charge densities gives rise to strong electrostatic forces that act, wherever possible, in the direction of restoring neutrality. On the other hand, if a plasma is deliberately subjected to an external electric field, the space-charge densities will adjust themselves so that the major part of the plasma is shielded from the field.

Let us consider a rather simple example. Suppose a spherical charge $+Q$ is introduced into a plasma, thereby subjecting the plasma to an electric field. Actually, the charge $+Q$ would be gradually neutralized because of being continuously struck by charged particles from the plasma, but if the charged object is physically very small, this will take an appreciable period of time. Meanwhile, electrons find it energetically favorable to move closer to the charge, whereas positive ions tend to move away. Under equilibrium conditions (see Section 5–3), the probability of finding a charged particle in a particular region of potential energy U is proportional to the Boltzmann factor, $\exp\left(-U/kT\right)$. Thus the electron density N_e is given by

$$N_e = N_0 \exp\left(e\,\frac{\varphi - \varphi_0}{kT}\right), \qquad (14\text{–}1)$$

where φ is the local potential, φ_0 is the reference potential (plasma potential), T is the absolute temperature of the plasma, and k is Boltzmann's constant. N_0 is the electronic density in regions where $\varphi = \varphi_0$.

If N_0 is also the positive ion density in regions of potential φ_0, then the positive ion density N_i is given by

$$N_i = N_0 \exp\left(-e\,\frac{\varphi - \varphi_0}{kT}\right). \qquad (14\text{–}2)$$

The potential φ is obtained from the solution of Poisson's equation:

$$\frac{1}{r^2}\frac{d}{dr}\left(r^2\frac{d\varphi}{dr}\right) = -\frac{1}{\epsilon_0}(N_i e - N_e e) = \frac{2N_0 e}{\epsilon_0}\sinh\left(e\,\frac{\varphi - \varphi_0}{kT}\right). \qquad (14\text{–}3\text{a})$$

This differential equation is nonlinear, and hence must be integrated numerically. On the other hand, an approximate solution to (14–3a), which is rigorous at high temperature, is adequate for our purposes here. If $kT > e\varphi$, then $\sinh\left(e\varphi/kT\right) \approx e\varphi/kT$, and

$$\frac{1}{r^2}\frac{d}{dr}\left(r^2\frac{d\varphi}{dr}\right) = \frac{2N_0 e^2}{\epsilon_0 kT}(\varphi - \varphi_0), \qquad (14\text{–}3\text{b})$$

the solution of which is

$$\varphi = \varphi_0 + \frac{Q}{4\pi\epsilon_0 r} \exp\left(-\frac{r}{h}\right). \tag{14-4}$$

Here r is the distance from the spherical charge $+Q$, and h, the *Debye shielding distance*, is given by

$$h = \left(\frac{\epsilon_0 k T}{2N_0 e^2}\right)^{1/2}. \tag{14-5}$$

Thus the redistribution of electrons and ions in the gas is such as to screen out Q completely in a distance of a few h.

An ionized gas is called a plasma if the Debye length, h, is small compared with other physical dimensions of interest. This is not much of a restriction so long as ionization of the gas is appreciable; at $T = 2000$ K and $N_0 = 10^{18}$ electrons or ions/m^3, the Debye length is 2.2×10^{-6} m.

14-2 PARTICLE ORBITS AND DRIFT MOTION IN A PLASMA

The orbit of a charged particle q moving in a prescribed electric and magnetic field may be calculated directly from the force equation:

$$\mathbf{F} = q(\mathbf{E} + \mathbf{v} \times \mathbf{B}). \tag{14-6}$$

We shall find it convenient to start with relatively simple field configurations, and then to generalize to fields which are slowly varying in space.

A constant electric field applied to a plasma is not particularly interesting because the plasma adjusts itself by developing a thin sheath of space charge, which shields the main body of plasma from the field. On the other hand, a constant magnetic field causes the particles to gyrate about the field lines without altering the space-charge distribution.

Case 1 Uniform Magnetic Field. E = 0.

This is the same motion as that described in Problem 8–1, but because it forms the basis for more complicated orbital motion in plasmas, we discuss it here in some detail. It should be emphasized, however, that Case 1 is applicable to many other situations besides plasmas, e.g., it is fundamental to the operation of particle accelerators, such as the cyclotron and betatron.

The Lorentz force is always at right angles to the velocity $\mathbf{v}$ of the charged particle; hence its kinetic energy remains constant:

$$K = \tfrac{1}{2}m_p v^2 = \text{constant}, \tag{14-7}$$

where m_p is the mass of the particle. It is convenient to resolve the velocity $\mathbf{v}$ into two components: $\mathbf{v}_\|$, parallel to $\mathbf{B}$, and $\mathbf{v}_\perp$, in the plane perpendicular to $\mathbf{B}$. Since

$\mathbf{v}_{\parallel}$ is unaffected by the field, $K_{\parallel} = \frac{1}{2}m_p v_{\parallel}^2$ remains constant also. It follows that

$$K_{\perp} = \tfrac{1}{2}m_p v_{\perp}^2 = K - K_{\parallel} \tag{14-8}$$

is also a constant of the motion.

The Lorentz force provides a centripetal acceleration. Thus

$$qv_{\perp}B = \frac{m_p v_{\perp}^2}{R},$$

and R (the radius of the orbit) is given by

$$R = \frac{m_p v_{\perp}}{qB}. \tag{14-9}$$

The radius R is frequently called the *Larmor radius* of the particle. The complete motion of the charged particle is described as a gyration of the particle in an orbit (the Larmor orbit) superimposed on the uniform motion of the orbit center, or *guiding center*, along a magnetic field line. The resulting helical motion is shown in Fig. 14–1.

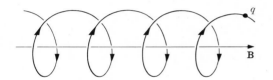

Figure 14–1 Particle motion in a uniform magnetic field.

The magnetic field acts to confine the plasma by bending the particles in circular orbits. Of course, no confinement is observed in the field direction. For ions and electrons of the same kinetic energy $K_{\perp}$, the electrons gyrate in much smaller orbits, the ratio of the two Larmor radii being equal to the square root of the mass ratio.

An interesting quantity, which we shall have occasion to use later, is the *magnetic moment* of the gyrating particle. By definition, the magnetic moment m is given by

$$m = \text{current} \times \text{area}$$

$$= \frac{qv_{\perp}}{2\pi R}\,\pi R^2 = \frac{K_{\perp}}{B}. \tag{14-10}$$

Inspection of Fig. 14–1 shows that m is directed opposite to the magnetic field and is thus a diamagnetic moment.

Case 2   Uniform Electric and Magnetic Fields. $\mathbf{E} \perp \mathbf{B}$.

If an electric and a magnetic field are simultaneously applied to a plasma, and $\mathbf{E}$ is perpendicular to $\mathbf{B}$, then there is no tendency to produce a sheath; in fact, we shall see that positive and negative space charge drift together in the same direction. For convenience, let the particle velocity **v** be written as

$$\mathbf{v} = \mathbf{u}_d + \mathbf{v}'; \tag{14-11}$$

then Eq. (14–6) may be written as

$$\mathbf{F} = q(\mathbf{E} + \mathbf{u}_d \times \mathbf{B} + \mathbf{v}' \times \mathbf{B}). \tag{14-12}$$

A particular choice for $\mathbf{u}_d$ causes the first two terms on the right of this equation to cancel each other:

$$\mathbf{u}_d = \frac{\mathbf{E} \times \mathbf{B}}{B^2}. \tag{14-13}$$

The remaining force, $q\mathbf{v}' \times \mathbf{B}$, is just what was studied under Case 1.

The total motion of the particle is thus made up of three terms: (a) constant velocity $\mathbf{v}'_{\parallel}$ parallel to **B**, (b) gyration about the magnetic field lines with angular frequency $v'_{\perp}/R = qB/m_p$, and (c) a constant drift velocity $u_d = E/B$ at right angles to both **E** and **B**. Some examples of this motion are shown in Fig. 14–2.

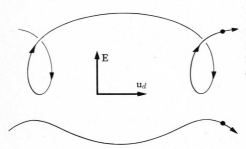

Figure 14–2 Crossed electric and magnetic fields. Particle motion in plane perpendicular to the magnetic field. The figure shows oppositely charged ions of different initial momenta.

The velocity $\mathbf{u}_d$ defined by Eq. (14–13) is called the *plasma drift velocity* or the *electric drift velocity*. It is important to note that $\mathbf{u}_d$ does not depend on the charge, mass, or velocity of the particle; thus all components of the plasma drift along together even though their individual gyrations may be vastly different.

Our derivation of Eq. (14–13) was obtained in a nonrelativistic fashion; if either $\mathbf{u}_d$ or **v** should approach c (the speed of light), then Eq. (14–11) must be replaced by an expression consistent with a Lorentz transformation. On the other hand, it turns out that Eq. (14–13) for the drift velocity is always correct* so long as $|\mathbf{E}| < c|\mathbf{B}|$. If $|\mathbf{E}| > c|\mathbf{B}|$, the magnetic field cannot prevent the particle from moving in the direction of **E**.

---

* The simplest way to treat the case where $|\mathbf{E}|$ is less than but not small compared with $c|\mathbf{B}|$ is to make a Lorentz transformation, transforming both the particle velocity and the fields. The velocity of the moving system is given by $\mathbf{u}_d$ (Eq. 14–13), and the force in the moving system is given by

$$\mathbf{F}' = q(\mathbf{v}' \times \mathbf{B})\left(\frac{c^2 - u_d^2}{c^2}\right)^{1/2}.$$

Case 3   Magnetic Field Constant in Time, but Space-Dependent. $E = 0$.

Let us suppose that a charged particle is moving in a nearly uniform magnetic field, one in which the field lines are slowly converging in space. The particle motion may be treated as a perturbation of the helical orbit in Fig. 14–1.

    The motion will be something like that shown in Fig. 14–3; the reader may easily verify that there is a force tending to push the particle into the weaker magnetic field region. To specify the problem precisely it will be assumed that the flux line through the guiding center coincides with the z-axis and that the magnetic field has azimuthal symmetry about the z-axis. Taking the z-component of Eq. (14–6), we obtain

$$F_z = m_p \frac{dv_z}{dt} = qv_\theta B_r \big|_{r=R}. \qquad (14\text{--}14)$$

But $\mathbf{V} \cdot \mathbf{B} = 0$ or, for the case in point,

$$\frac{1}{r} \frac{\partial}{\partial r}(rB_r) + \frac{\partial B_z}{\partial z} = 0.$$

Since the field lines are converging slowly, $\partial B_z / \partial z$ may be taken constant over the orbit cross section, yielding

$$B_r \big|_{r=R} = -\tfrac{1}{2} R \frac{\partial B_z}{\partial z}. \qquad (14\text{--}15)$$

Furthermore, v_θ is analogous to the $v_\perp$ of Case 1. Making these substitutions in Eq. (14–14) gives

$$m_p \frac{dv_\parallel}{dt} = -\tfrac{1}{2} qR v_\perp \frac{\partial B_z}{\partial z}$$

$$= -m \frac{\partial B_z}{\partial z}, \qquad (14\text{--}16)$$

the last form being obtained through the use of Eq. (14–10).

    The total kinetic energy K of the particle is unaltered in the magnetic field, since the Lorentz force, which is always at right angles to the velocity, can do no

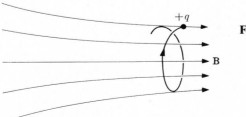

Figure 14–3

work. $K_\perp$, defined in (14–8), is not constant here; neither is $K_\parallel$, but we may write

$$\frac{d}{dt}\left(\tfrac{1}{2}m_p v_\parallel^2\right) = \frac{d}{dt}\left(K - K_\perp\right)$$

$$= -\frac{dK_\perp}{dt}$$

$$= -\frac{d}{dt}\left(mB_z\right), \tag{14–17}$$

the last form coming from Eq. (14–10). On the other hand, we may multiply Eq. (14–16) by $v_\parallel = \partial z/\partial t$ to obtain

$$\frac{d}{dt}\left(\tfrac{1}{2}m_p v_\parallel^2\right) = -m\,\frac{\partial B_z}{\partial z}\frac{\partial z}{\partial t}$$

$$= -m\,\frac{dB_z}{dt}, \tag{14–18}$$

where d/dt represents the time derivative taken along the dynamical path. By comparing Eqs. (14–17) and (14–18), we see that the magnetic moment m is a constant of the motion. It should be emphasized, however, that this is an approximate result which holds only so long as B_z varies slowly. If $\mathbf{B}$ were to change substantially in distances of the order of R, the approximations used in the derivation of (14–18) would break down.

Of further interest is the fact that the particle is constrained to move on the surface of a flux tube. This follows because the magnetic flux through the orbit is

$$\Phi = B_z \pi R^2 = \pi B_z\,\frac{m_p^2 v_\perp^2}{q^2 B_z^2}$$

$$= \frac{2\pi m_p}{q^2}\frac{K_\perp}{B_z} = \frac{2\pi m_p}{q^2}\,m, \tag{14–19}$$

and m is constant. The motion of the particle is depicted schematically in Fig. 14–4.

The z-component (parallel component) of the force, Eq. (14–16), is always in such a direction as to accelerate particles toward the weaker part of the field.

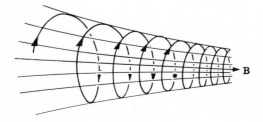

Figure 14–4 The particle winds in a tighter and faster helix until it is reflected.

Gyrating particles which are approaching regions of stronger magnetic field are thus slowed down, i.e., $v_\parallel$ is decreased. On the other hand, conservation of energy requires that simultaneously the orbital motion $v_\perp$ be speeded up. If the convergence of the magnetic field is sufficient, the particle will gyrate in an ever-tighter helical spiral until it is finally reflected back into the weaker field.

14–3 MAGNETIC MIRRORS

The results of the preceding section show that a slowly converging magnetic field can, in principle, confine a plasma. At right angles to the principal field direction the particles are bent into circular orbits; along the principal direction of the field the particles are slowed down and finally reflected by the converging field lines. Such a field configuration is called a *magnetic mirror*. At least two mirrors must be used in any confinement system; a system of this type is shown in Fig. 14–5.

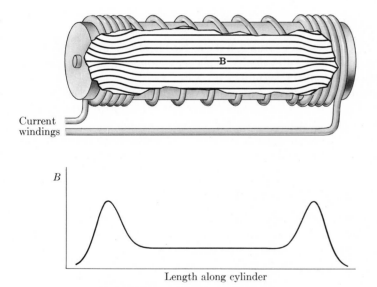

Figure 14–5 Magnetic mirror system.

Not all particles can be confined by the mirror system, however. The field lines cannot be made to converge to a point; thus there is a large but not infinite magnetic field B_m at the mirror. If the particle has too much "axial kinetic energy" it will not be turned back by the mirror field, and it will be able to escape.

Because the magnetic moment is a constant of the motion, we find, according to Eq. (14–10), that

$$\frac{K_{0\perp}}{B_0} = \frac{K_{1\perp}}{B_{1\perp}}.$$

Here the subscript 0 refers to the central region of Fig. 14–5, and the subscript 1 to the reflection point. At the reflection point, however, $K_\perp = K$. Furthermore, K, the total kinetic energy, is a constant of the motion. In order that the particle be reflected, the mirror field B_m must be greater than B_1; that is,

$$B_m > B_1 = \frac{K}{K_{0\perp}} B_0,$$

or

$$\frac{K_{0\perp}}{K} > \frac{B_0}{B_m}. \tag{14–20a}$$

If the initial velocity v_0 makes an angle θ_0 with the field direction, then $v_{0\parallel} = v_0 \cos \theta_0$ and $v_{0\perp} = v_0 \sin \theta_0$. Equation (14–20a) then reduces to

$$\sin^2 \theta_0 > \frac{B_0}{B_m}, \tag{14–20b}$$

as the criterion for reflection. For example, if the mirror field is one hundred times as intense as B_0, then particles with velocities making an angle of less than 6° with the field direction escape from the system.

The collisions between particles in the central region of the mirror system tend to produce an isotropic velocity distribution. Thus the net result of collisions is that particles are continually scattered into a region of velocity space such that they can escape from the system. As a result of collisions particles can also "diffuse" at right angles to the field direction, and so eventually escape.

14–4 THE HYDROMAGNETIC EQUATIONS

Collective motions of the particles in a plasma, such as the "pinch effect" and plasma oscillations, are handled best in the hydromagnetic formulation. According to this description, the plasma is regarded as a classical fluid which obeys the conventional equations of hydrodynamics. The fluid, however, is an electrical conductor, and thus electromagnetic forces must be taken into account explicitly.

The force on a unit volume of the plasma may be written as

$$\mathbf{F}_v = \mathbf{J} \times \mathbf{B} - \nabla p, \tag{14–21}$$

where $\mathbf{J}$ is the current density and p is the fluid pressure. Other forces, such as gravitational and viscous forces, may also be included, but are neglected here in the interest of simplicity. Because of the approximate electrical neutrality of the plasma, the term $\rho \mathbf{E}$ need not be included along with other force terms in (14–21). Deviations from neutrality must be considered, of course, in Poisson's equation, but they are usually ignored in the dynamical equations.

Momentum balance requires that

$$\zeta \frac{d\mathbf{v}}{dt} = \zeta \left[\frac{\partial \mathbf{v}}{\partial t} + (\mathbf{v} \cdot \nabla)\mathbf{v} \right]$$

$$= \mathbf{J} \times \mathbf{B} - \nabla p, \tag{14-22}$$

which is the equation of motion, or the *Euler equation*, of the fluid. Here ζ is the mass density of the plasma and $\mathbf{v}$ its fluid velocity. For problems in which the hydrodynamic motion is not particularly large, the term containing $(\mathbf{v} \cdot \nabla)\mathbf{v}$ can usually be neglected.*

It is sometimes convenient to interpret the $\mathbf{J} \times \mathbf{B}$ term of Eq. (14–21) as arising in part from a "magnetic pressure." This can be done with the aid of Ampere's circuital law, Eq. (9–29), which, specialized to the plasma case, is

$$\nabla \times \mathbf{B} = \mu_0 \mathbf{J}, \tag{14-23}$$

and the vector identity

$$\mathbf{B} \times \nabla \times \mathbf{B} = \nabla(\tfrac{1}{2}B^2) - (\mathbf{B} \cdot \nabla)\mathbf{B}. \tag{14-24}$$

Thus

$$\mathbf{J} \times \mathbf{B} = -\frac{1}{\mu_0} \mathbf{B} \times \nabla \times \mathbf{B}$$

$$= -\nabla \left(\frac{B^2}{2\mu_0} \right) + \frac{1}{\mu_0} (\mathbf{B} \cdot \nabla)\mathbf{B}. \tag{14-25}$$

The quantity $B^2/2\mu_0$, which is, of course, the magnetic energy density, thus plays the role of a magnetic pressure, p_m:

$$p_m = \frac{B^2}{2\mu_0}. \tag{14-26}$$

It should be emphasized, however, that $-\nabla p_m$ gives in most cases only part of the magnetic force; the remaining force comes from the $(1/\mu_0)(\mathbf{B} \cdot \nabla)\mathbf{B}$ term. When $\mathbf{J} = 0$, the two terms on the right of Eq. (14–25) cancel each other.

As an example of the utility of the magnetic pressure concept, consider a *unidirectional* magnetic field. The equation $\nabla \cdot \mathbf{B} = 0$ guarantees that $\mathbf{B}$ does not change along the field direction. Since space variations can occur only in directions at right angles to $\mathbf{B}$, it follows that $(\mathbf{B} \cdot \nabla)\mathbf{B} = 0$ for this case. Equation (14–21) reduces, therefore, to

$$\mathbf{F}_v = -\nabla(p + p_m),$$

---

* Although it may not be neglected in steady-flow problems for which the term $\partial \mathbf{v}/\partial t$ vanishes explicitly.

and the condition for static equilibrium of each volume element is

$$p + p_m = \text{constant}.$$

In other words, for this example the sum of the fluid pressure and the magnetic pressure must be space-independent.

In addition to Eq. (14–22) and the macroscopic equations governing electricity and magnetism,* we require two additional relationships to complete the hydromagnetic formulation. These are: (1) the equation of continuity for the plasma fluid:

$$\frac{\partial \zeta}{\partial t} + \nabla(\zeta \mathbf{v}) = 0, \tag{14–27}$$

and (2) an equation relating $\mathbf{J}$ to the field quantities. The latter relationship is simply a generalized form of Ohm's law that, under certain conditions, may be written as

$$\mathbf{J} = g(\mathbf{E} + \mathbf{v} \times \mathbf{B}) \tag{14–28a}$$

Here $\mathbf{v} \times \mathbf{B}$ is the "motional electric field" arising from hydrodynamic motion of the plasma in a magnetic field, and g is the conductivity of the plasma.

An approximation that is frequently made is that of infinite conductivity. The advantage of this approximation is that it permits a substantial simplification of the hydromagnetic equations, thus presenting a much clearer picture of the physical processes going on in the plasma. In some problems, particularly astrophysical ones, the approximation is quite good. For the case of infinite conductivity, Ohm's law reduces to

$$g \to \infty,$$

$$\mathbf{E} + \mathbf{v} \times \mathbf{B} = 0. \tag{14–28b}$$

Infinite conductivity (or, for practical purposes, high conductivity) has an important consequence, namely, that the magnetic flux is frozen into the plasma. If Eq. (14–28b) is combined with the differential form of Faraday's law of induction, we obtain

$$\frac{\partial \mathbf{B}}{\partial t} = \nabla \times (\mathbf{v} \times \mathbf{B}). \tag{14–29}$$

The normal component of this equation integrated over a fixed surface S yields

$$\frac{d}{dt} \int_S \mathbf{B} \cdot \mathbf{n} \, da = \int_S \nabla \times (\mathbf{v} \times \mathbf{B}) \cdot \mathbf{n} \, da,$$

---

* The Maxwell equations are summarized in Section 16–2. The reader will note that Eq. (16–10), the original Ampere's circuital law, has been modified through inclusion of the displacement current, $\partial \mathbf{D}/\partial t$. Actually, the displacement current does not play an important role in *most* hydromagnetic phenomena.

or

$$\frac{d\Phi}{dt} = \oint_C \mathbf{v} \times \mathbf{B} \cdot d\mathbf{l} = \oint_C \mathbf{B} \cdot (d\mathbf{l} \times \mathbf{v}), \tag{14-30}$$

where C is a fixed contour in space through which the plasma moves due to hydrodynamic motion. From Fig. 14–6 we see that $\oint_C d\mathbf{l} \times \mathbf{v}$ may be regarded as the increase in area, per unit time, of the cap surface that is bounded by C, and $\oint_C \mathbf{B} \cdot d\mathbf{l} \times \mathbf{v}$ is the magnetic flux associated with this increased area. Equation (14–30) simply states that the flux change per unit time through the contour C is just what we should calculate geometrically on the basis that all flux lines move along with the fluid. We conclude, therefore, that the lines of magnetic induction are "frozen" into the perfectly conducting material.

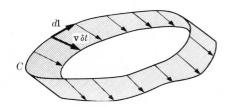

Figure 14–6

14-5 THE PINCH EFFECT

The tendency of a high-current discharge through a plasma to constrict itself laterally is known as the "pinch effect." The basic mechanism causing the pinch is the interaction of a current with its own magnetic field or, equivalently, the attraction between parallel current filaments. The pinch effect was first predicted by Bennett, and later independently by Tonks.* A somewhat different picture of the pinch, showing its inherent instability, was given by Rosenbluth.†

Let us consider a current discharge of cylindrical symmetry through the plasma. From Ampere's circuital law, the magnetic induction at distance r from the axis of the discharge is given by

$$B(r) = \frac{\mu_0}{r} \int_0^r J(r')r' \, dr'. \tag{14-31}$$

From this it follows that

$$\frac{\partial B}{\partial r} = -\frac{\mu_0}{r^2} \int_0^r J(r')r' \, dr' + \mu_0 J(r)$$

$$= -\frac{1}{r} B(r) + \mu_0 J(r). \tag{14-32}$$

---

* W. Bennett, *Physical Review*, vol. 45, p. 890 (1934); L. Tonks, *Physical Review*, vol. 56, p. 369 (1939).

† M. Rosenbluth, "Dynamics of a Pinched Gas," from *Magnetohydrodynamics*, edited by Rolf Landshoff (Stanford University Press, 1957).

The magnetic force per unit volume is

$$\mathbf{F}_v = \mathbf{J} \times \mathbf{B} = -J(r)B(r)\mathbf{a}_r, \tag{14-33}$$

where $\mathbf{a}_r$ is a unit vector in the r-direction. Eliminating $J(r)$ between Eqs. (14-32) and (14-33) yields

$$F_v = -\frac{1}{\mu_0} B \frac{\partial B}{\partial r} - \frac{1}{\mu_0 r} B^2. \tag{14-34}$$

This force can be converted to an equivalent pressure, p_{eq}, by writing $F_v = -\partial p_{eq}/\partial r$, and then integrating:

$$p_{eq} = \frac{1}{2\mu_0} B^2 + \frac{1}{\mu_0} \int_0^r \frac{B^2}{r} \, dr. \tag{14-35}$$

We are particularly interested in the pressure on the lateral boundaries of the discharge. Following Rosenbluth, we restrict our attention to the high-conductivity case where the magnetic field lines cannot penetrate appreciably into the conducting fluid.* Here the integral in Eq. (14-35) contains no contribution from the discharge region. At the boundary of the discharge, $r = R$, and the pressure is just what we have called the *magnetic pressure*, p_m:

$$p_m = \frac{1}{2\mu_0} B^2(R). \tag{14-36}$$

It is evident from Eq. (14-35) that the magnetic pressure is uniform in the outside region, but zero or very small inside the discharge. Thus the pinch effect can be viewed as coming about from the sudden buildup of magnetic pressure in the region external to the discharge.

The pinching of the discharge results in plasma compression. If the pinch could contract in a stable manner, it would proceed until the magnetic pressure in the external region was equal to the fluid pressure in the discharge. Let us treat the plasma as a perfect gas, whose fluid pressure $p = NkT$. Then, at the final radius R of the discharge,

$$\frac{1}{2\mu_0} B^2(R) = \frac{1}{2} \frac{\mu_0}{4\pi^2 R^2} I^2 = NkT,$$

where I is the current in the discharge. This expression may be solved for the current:

$$I^2 = 2 \left(\frac{\mu_0}{4\pi} \right)^{-1} \pi R^2 NkT$$

$$= 2 \left(\frac{\mu_0}{4\pi} \right)^{-1} A_0 N_0 kT,$$

---

* The nonpenetration of the field lines follows from the results of the preceding section and the fact that both the current and magnetic field are initially very small in the discharge.

since conservation of particles requires that $A_0 N_0 = \pi R^2 N$. Here A_0 is the initial cross section of the discharge, N_0 is its initial particle density, $\mu_0/4\pi = 10^{-7}$ T · m/A, and Boltzmann's constant $k = 1.38 \times 10^{-23}$ J/K. In order to achieve the temperature of 10^8 K required for a thermonuclear (fusion) reactor, with $A_0 = 0.04$ m² and $N_0 = 10^{21}$ particles/m³, a pinch current of approximately one million amperes is required.

It is easy to see that the pinch is an inherently unstable phenomenon. The magnetic pressure on the boundary of the discharge depends on its radius as well as on its detailed geometry. Small perturbations will grow if the pressure changes that result are such as to enhance these perturbations. Figure 14–7 shows that small ripples on the bounding surface of the discharge, as well as kinks, fall into this category, producing the so-called sausage and kink instabilities of the pinched plasma.

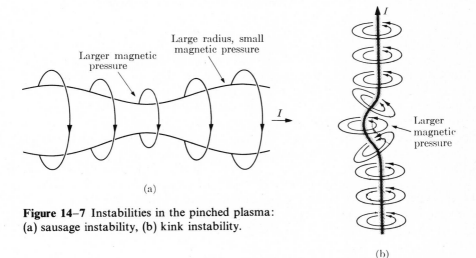

Figure 14–7 Instabilities in the pinched plasma: (a) sausage instability, (b) kink instability.

14–6 MAGNETIC CONFINEMENT SYSTEMS FOR CONTROLLED THERMONUCLEAR FUSION

Much of the current interest in plasma physics is motivated by our need to develop alternative sources of energy and the possibility of using a thermonuclear plasma of deuterium and tritium for an energy source. Both the pinch effect discussed in the preceding section and the magnetic mirror concept (Section 14–3) are being used in experimental fusion reactor systems in attempts to confine the thermonuclear plasma magnetically. Although the pinch effect is inherently unstable, its stability can be improved by supplying additional components to the magnetic field and by optimizing design parameters.

An important quantity for the design of fusion reactor systems is the ratio of kinetic pressure $p = NkT$ to the total pressure (kinetic plus magnetic p_m). This ratio is given the symbol β:

$$\beta = \frac{p}{[B^2/2\mu_0 + p]} = \frac{NkT}{[B^2/2\mu_0 + NkT]}, \tag{14-37}$$

where N is the sum of the ion and electron densities in the plasma. Fusion reactors are generally characterized by their β-value. Low β refers to values less than 0.01 and high β values lie between 0.1 and 1.0. The plasma is either deuterium or a deuterium-tritium plasma with temperature in excess of 10^8 K; its density is in the range 10^{19} m^{-3} to 10^{22} m^{-3}. Confinement does not have to be absolute, but must be for a long enough period τ so that more energy is produced in the thermonuclear reaction than is consumed in establishing the plasma conditions. Confinement is believed to be adequate when the *Lawson condition* is met:

$$N_i \tau > 10^{20} \text{ m}^{-3} \text{ s at } T > 10^8 \text{ K}, \tag{14-38}$$

where N_i is the ion density in the plasma.

The most active area of investigation relating to magnetic confinement involves a class of reactors designed to confine toroidal, or doughnut-shaped, plasmas. And within this class the *tokamaks* show considerable promise. The name derives from a series of Tokamak experiments carried out at the I. V. Kurchatov Institute of Atomic Energy in Moscow, but it now applies to a generic class of experimental reactors characterized by a *toroidal, diffuse pinch*. The tokamak is a low to moderate β device; the magnetic field in the plasma has a *poloidal** component due to current flowing in the plasma in response to a toroidal electric field, and a toroidal component produced by external coils, so that the resultant field traces out helical curves within the doughnut-shaped plasma. There are a number of tokamak devices in operation around the world; one of the largest U.S. machines is the Princeton Large Torus (PLT), which operates with a maximum plasma current of 1.6×10^6 amperes.

Examples of experimental devices with high β are the theta-pinch fusion reactor and the magnetic mirror reactor. The theta-pinch concept uses a toroidal-shaped plasma, but of substantially higher density than that of a tokamak. The magnetic mirror is not a true confinement system in that plasma leaks out the ends; however, it can amplify the power of an injected beam. The magnetic mirror is a high β device but uses a lower density plasma than the theta-pinch reactor.

For a more detailed discussion of experimental fusion reactor systems the reader is referred to one of the following revues:

"Fusion Reactor Systems" by F. L. Ribe, in *Reviews of Modern Physics*, vol. 47, p. 7 (1975).

---

* If the toroidal plasma were cut (for example, at the slot marked d in Fig. 9–15) and straightened into a cylinder, the poloidal component would become the azimuthal component and the toroidal component would become the axial component.

"The Tokamak Approach in Fusion Research" by B. Coppi and J. Rem, in *Scientific American*, vol. 227, no. 1 (July 1972).

"The Prospects of Fusion Power" by W. C. Gough and B. J. Eastland, in *Scientific American*, vol. 224, no. 2 (Feb. 1971).

"Fusion Energy in Context: Its Fitness for the Long Term" by J. P. Holdren, in *Science*, vol. 200, p. 168 (April 1978).

14–7 PLASMA OSCILLATIONS AND WAVE MOTION

One of the interesting properties of a plasma is its ability to sustain oscillations and propagate waves. Various types of oscillatory behavior are possible, and because of the nonlinear character of the hydrodynamic equations these oscillations can be quite complex. We find it expedient to restrict our attention to some rather simple cases, which, nevertheless, have been observed in controlled experiments.

Case 1   Electrostatic Plasma-Electron Oscillations

Electrostatic oscillations in a plasma were first discussed by Tonks and Langmuir.* Actually, there are two possible types of electrostatic oscillations: high-frequency oscillations that are too rapid for the heavy ions to follow, and oscillations of the ions that are so slow that the electrons are always distributed around the ions in a statistical manner. We discuss the first case only, the so-called electron oscillations.

Let us fix our attention on a region of plasma containing a uniform density of positive ions, N. There are no negative ions. Initially, the electrons also have uniform density N, but let us suppose that each electron is displaced in the x-direction by a distance ξ that is independent of the y- and z-coordinates and is zero on the plasma boundaries. The displacement of electrons disturbs the neutral plasma, producing a charge in each volume element $\Delta x\, \Delta y\, \Delta z$:

$$\delta\rho\, \Delta x\, \Delta y\, \Delta z = -Ne\, \Delta y\, \Delta z \left[\xi - \left(\xi + \frac{\partial \xi}{\partial x} \Delta x\right)\right]$$

$$= \Delta x\, \Delta y\, \Delta z\, Ne\, \frac{\partial \xi}{\partial x}.$$

The displacement of the electrons produces an electric field $\mathbf{E}(x, t)$ that, because of the symmetry of the problem, is in the x-direction. Thus

$$\mathbf{V} \cdot \mathbf{E} = \frac{1}{\epsilon_0} \delta\rho,$$

* L. Tonks and I. Langmuir, *Physical Review*, vol. 33, p. 195 (1929).

or

$$\frac{\partial E}{\partial x} = \frac{1}{\epsilon_0} Ne \frac{\partial \xi}{\partial x},$$

which, when integrated, yields

$$E = \frac{Ne}{\epsilon_0} \xi. \tag{14-39}$$

Here, the constant of integration has been taken equal to zero, since sheath formation will shield the plasma from a uniform electric field.

The force on each electron is $-eE$, which, according to Eq. (14–39), is proportional to the displacement ξ. It is also seen to be a restoring force. Thus each electron oscillates about its original position with simple harmonic motion. The equation of motion for each electron is

$$m_e \frac{d^2\xi}{dt^2} + \frac{Ne^2}{\epsilon_0} \xi = 0. \tag{14-40}$$

The "plasma frequency," $f_p = \omega_p/2\pi$, is defined, therefore, by

$$\omega_p = \left(\frac{Ne^2}{m_e \epsilon_0}\right)^{1/2}, \tag{14-41}$$

where m_e is the electron mass. As a numerical example, we have $f_p = 9.0 \times 10^9$ s^{-1} for a particle density $N = 10^{18}$ electrons/m^3.

Case 2 Hydromagnetic or Alfvén Waves

Hydromagnetic waves represent true wave propagation in a conducting medium that is subjected to a constant magnetic field. This behavior, which was first predicted by Alfvén* in 1942, is consistent with the hydromagnetic formulation of a plasma discussed in Section 14–4.

Before proceeding to the differential equations, let us look at the physical processes in the plasma from as elementary a viewpoint as possible. Consider an infinite plasma subjected to a constant, uniform magnetic field $\mathbf{B}_0$ that is directed along the z-axis. If a segment of the plasma, the rectangular section $ABCD$ in Fig. 14–8 that extends parallel to the y-axis, is given a velocity $\mathbf{v}$ directed parallel to the positive y-axis, then the charge carriers (ions and electrons) experience forces

$$q_i(\mathbf{v} \times \mathbf{B}_0)$$

that tend to separate the positive and negative carriers. The segment $ABCD$ thus becomes a seat of motional emf, its right-hand end tending to charge up positively, its left end negatively. But since we are dealing with a conducting medium, the plasma external to $ABCD$ completes the electrical circuit. A few of the current lines are shown in the figure.

* H. Alfvén, *Cosmical Electrodynamics* (New York: Oxford University Press, 1950; Second Edition, 1963).

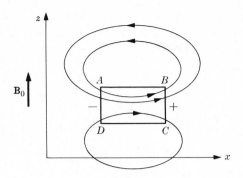

Figure 14–8 The segment, $ABCD$, of plasma moves in the positive y-direction. The currents that are generated are depicted schematically.

The induced current now interacts with the magnetic field $\mathbf{B}_0$. It is easy to verify that the force density $\mathbf{J} \times \mathbf{B}_0$ in the segment $ABCD$ is such as to oppose its motion, whereas the force on external parts of the plasma is such as to accelerate it in the positive y-direction. Eventually, $ABCD$ will have slowed down, and its motion will have been transferred to neighboring segments of the plasma. The mechanism is still operating, however, and the whole process is repeated, thus propagating the disturbance farther in the $\pm z$-direction.

We now turn to the differential equations. Let $\mathbf{B} = \mathbf{B}_0 + \mathbf{B}_1$, where $\mathbf{B}_0$ is the constant, uniform field parallel to the z-axis and $\mathbf{B}_1$ is the magnetic field set up by the induced currents. Using the results of the preceding paragraphs as a guide, we look for the simplest type of wave motion, characterized by v_y, E_x, J_x, and B_{1y}, other components vanishing. From Ampere's circuital law,

$$-\frac{\partial B_{1y}}{\partial z} = \mu_0 J_x, \tag{14–42}$$

and the Euler equation of the fluid, Eq. (14–22), gives the two relations

$$\zeta \frac{\partial v_y}{\partial t} = -J_x B_0, \tag{14–43a}$$

and

$$0 = J_x B_{1y} - \frac{\partial p}{\partial z}. \tag{14–43b}$$

Equations (14–43) may be combined with (14–42) to yield

$$\frac{\partial v_y}{\partial t} = \frac{B_0}{\mu_0 \zeta} \frac{\partial B_{1y}}{\partial z} \tag{14–44}$$

and

$$\frac{\partial p}{\partial z} = -\frac{1}{2\mu_0} \frac{\partial (B_{1y}^2)}{\partial z}. \tag{14–45}$$

The generalized Ohm's law may be written as

$$E_x = -v_y B_0 + \frac{1}{g} J_x$$

$$= -v_y B_0 - \frac{1}{g\mu_0} \frac{\partial B_{1y}}{\partial z}. \tag{14-46}$$

Finally, Faraday's law yields

$$\frac{\partial B_{1y}}{\partial t} = -\frac{\partial E_x}{\partial z}. \tag{14-47}$$

If v_y is eliminated between Eqs. (14-44) and (14-46), and E_x eliminated between the resulting equation and (14-47), we obtain, on the assumption of constant ζ,

$$\frac{\partial^2 B_{1y}}{\partial t^2} = \frac{B_0^2}{\mu_0 \zeta} \frac{\partial^2 B_{1y}}{\partial z^2} + \frac{1}{g\mu_0} \frac{\partial^3 B_{1y}}{\partial z^2 \partial t}, \tag{14-48}$$

which is the equation governing the propagation of Alfvén waves.

If the conductivity g of the plasma were infinite, then (14-48) would become identical with the wave equation whose solution is discussed in Sections 16-4 and 16-5. In these circumstances, Eq. (14-48) describes a plane, undamped wave moving parallel to the z-axis with phase velocity

$$v_p = \frac{B_0}{\sqrt{\mu_0 \zeta}}. \tag{14-49}$$

As a numerical example, take $B_0 = 0.01$ T, $\zeta = 10^{-5}$ kg/m^3 $= 10^{-8}$ g/cm^3; then $v_p = 2800$ m/s.

In order to see what results for finite conductivity, we try a solution to Eq. (14-48) of the form

$$B_{1y} = b_1 \exp [\alpha z + i\omega t].$$

This solution is satisfactory provided

$$\alpha^2 = \frac{-\omega^2}{v_p^2 + i\omega/g\mu_0}, \tag{14-50}$$

with v_p as defined in Eq. (14-49). For small damping,

$$\alpha \approx \pm \left(i\frac{\omega}{v_p} + \frac{\omega^2}{2g\mu_0 v_p^3} \right). \tag{14-51}$$

Thus the solution to Eq. (14-48) is a damped plane wave propagating in the $\pm z$-direction. The distance δ in which the amplitude of the wave is reduced to $1/e$ of its original value is

$$\delta = \frac{2g\mu_0 v_p^3}{\omega^2} = \frac{2gB_0^3}{\mu_0^{1/2}\zeta^{3/2}\omega^2}. \tag{14-52}$$

14-8 THE USE OF PROBES FOR PLASMA MEASUREMENTS

A plasma consists of electrons, ions, and perhaps neutral atoms. The electrons gain energy from electric fields at the boundary of the plasma as well as from the ionizing collisions in which they are produced, and the velocities of the electrons become random through collisions with ions. Thus we can speak of an electron temperature, T_e. In fact, for plasmas created in the laboratory (arcs, electrical discharges), the electrons are found to have a Maxwell-Boltzmann velocity distribution, which means, of course, that they may be characterized by a temperature. Electron temperatures in typical arc plasmas range from several thousand to 50,000 K.

To a certain extent, the preceding discussion also applies to the heavy ions; however, ions do not necessarily have the same temperatures as electrons. If a substantial difference between the mean kinetic energies of ions and electrons exists, then it takes several thousand collisions per particle to equalize the energy difference, and this may require a time longer than the mean life of an ion in the system.

Interesting quantities to be determined are the particle temperatures, particle densities, and random current densities in the plasma. Langmuir and Mott-Smith* showed that a small metal electrode or "probe" inserted into the plasma can be used to determine some of these quantities experimentally, by applying various potentials and measuring the corresponding collected currents. An electrode not at plasma potential will be enveloped by a sheath that shields the plasma from the disturbing field caused by the electrode. The sheath is, in most cases, quite thin, and if the probe is maintained negative, zero, or slightly positive with respect to plasma potential, it will barely disturb the bulk plasma.

The current-voltage relationship for a typical probe is shown in Fig. 14–9. When the probe is at plasma potential, it collects both the random electron and the random ion currents. But the random electron current is so much greater than the ion current that the former dominates, the reason for this being that the electrons have much larger average velocities than do the ions. As the probe is made negative it repels electrons and the electron current drops off; at point φ_F, the floating potential, the net current to the probe is zero; finally, if the probe is made negative enough, only the ion current density J_i is collected. If the probe is made slightly positive with respect to the plasma, the ions are repelled and the electron current density J_e is collected. If the probe is made even more positive, it will begin to act like a secondary anode and the current-voltage behavior will become complicated, depending in detail upon the nature of the plasma.

Let us consider a plasma consisting of positive ions (singly charged) and electrons. The ion density is equal to the electron density in the neutral region:

$$N_i = N_e = N_0. \qquad (14\text{--}53)$$

---

* I. Langmuir and H. Mott-Smith, *General Electric Review*, vol. 27, p. 449 (1924); *Physical Review*, vol. 28, p. 727 (1926). For a more recent discussion of electric probes, see F. F. Chen, *op. cit.*, or Chapter 4 of *Plasma Diagnostic Techniques*, edited by R. H. Huddlestone and S. L. Leonard (New York: Academic Press, 1965).

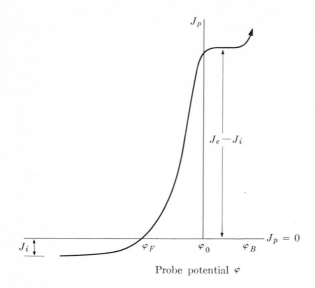

Figure 14–9 Current-voltage characteristic of a probe
inserted in a plasma. φ_0 is plasma potential.

If the electron distribution is characterized by the temperature T_e then, according
to kinetic theory, the random electron current density is

$$J_e = \tfrac{1}{4} N_0 e \bar{v} = N_0 e \left(\frac{kT_e}{2\pi m_e} \right)^{1/2}, \tag{14–54}$$

where $\bar{v}$ is the average thermal velocity of the electrons. This is the electron current
collected per unit area of the probe in the region $\varphi = \varphi_0$ to $\varphi = \varphi_B$. If the probe is
made negative, the electron current density falls off, because only a fraction of the
electrons has energy sufficient to penetrate the potential barrier:

$$J'_e = J_e \exp \left(e \, \frac{\varphi - \varphi_0}{kT_e} \right) = \tfrac{1}{4} N_e e \bar{v} \exp \left(e \, \frac{\varphi - \varphi_0}{kT_e} \right), \qquad \text{for } \varphi \le \varphi_0. \tag{14–55}$$

The ion current density, on the other hand, is constant in the negative potential
region, namely, J_i. The total probe current is thus

$$J_p = J_e \exp \left(e \, \frac{\varphi - \varphi_0}{kT_e} \right) - J_i,$$

and the electron temperature is found to be

$$T_e = \frac{e}{k} \left[\frac{d}{d\varphi} \ln (J_p + |J_i|) \right]^{-1}. \tag{14–56}$$

The particle density N_0 can now be determined from Eq. (14–54) by using the experimental value of J_p corresponding to the plateau region to the right of φ_0 in the figure. It should be noted that Eq. (14–56) and the shape of the J_p-φ characteristic are independent of the absolute value of φ; thus the potential of the probe can be measured with respect to any fixed potential (for example, an electrode potential) in the plasma.

Probe characteristics are well understood, but before the data obtained from probe measurements can be interpreted unambiguously it is necessary that certain conditions be satisfied: (1) the probe should be small compared with the mean free paths of electrons and ions, (2) the sheath should be small compared with the dimensions of the probe, (3) the ionization in the sheath must be negligible, (4) secondary emission from the probe must be negligible, and (5) there must be no plasma oscillations. In addition to these requirements, it is tacitly assumed that there is no magnetic field present; the use of probes in plasmas containing magnetic fields has been discussed by Bohm, Burhop, and Massey.*

We end this section with a discussion of the sheath surrounding the negatively charged probe. The equation governing the potential φ in the sheath region is Poisson's equation:

$$\nabla^2 \varphi = -\frac{1}{\epsilon_0} e(N_i - N_e), \tag{14–57}$$

where N_i and N_e are the local ion and electron densities. An approximate plot of φ versus distance from the probe is given in Fig. 14–10. It is convenient to make the substitution $\varphi = -V$, where V is a positive quantity, and, since the sheath thickness is small compared with the dimensions of the probe, we may use a one-dimensional version of Eq. (14–57):

$$\frac{d^2 V}{dx^2} = \frac{1}{\epsilon_0} e(N_i - N_e). \tag{14–58}$$

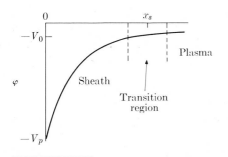

Figure 14–10 Plot of potential versus distance from the probe.

* Chapter 2 of *Characteristics of Electrical Discharges in Magnetic Fields*, edited by A. Guthrie and R. K. Wakerling (New York: McGraw-Hill, 1949). See also Chapter 4 of Huddlestone and Leonard, *op. cit.*

Electrons are distributed in the sheath in approximately a statistical fashion:

$$N_e = N_0 \exp\left[\frac{-e(V - V_0)}{kT_e}\right], \tag{14-59}$$

where N_0 is the electron density at plasma potential $-V_0$. The ion density is related to the ion current, J_i, in accordance with

$$J_i = N_i e v_i = N_i e \sqrt{\frac{2eV}{m_i}}. \tag{14-60a}$$

In the plasma, outside the sheath, the ion current is given by

$$J_i = N_0 e v_{io} = N_0 e \sqrt{\frac{2eV_0}{m_i}}, \tag{14-60b}$$

provided the plasma potential, $-V_0$, is measured relative to the point where the positive ions are formed. Thus

$$N_i = N_0 \sqrt{\frac{V_0}{V}}. \tag{14-61}$$

Substituting Eqs. (14–59) and (14–61) into Eq. (14–58) yields the so-called *plasma-sheath* equation:

$$\frac{d^2V}{dx^2} = \frac{1}{\epsilon_0} N_0 e \left[V_0^{1/2}V^{-1/2} - \exp\frac{-e(V - V_0)}{kT_e}\right]. \tag{14-62}$$

The last equation may be multiplied through by $(dV/dx)\, dx = dV$ and integrated to obtain

$$\frac{1}{2}\left(\frac{dV}{dx}\right)^2 = \frac{1}{\epsilon_0} N_0 e \left[2V_0^{1/2}V^{1/2} + \frac{kT_e}{e}\exp\frac{-e(V - V_0)}{kT_e}\right] + C, \tag{14-63}$$

where the constant C is determined from the condition that $dV/dx = 0$ at the sheath edge, i.e., where $V = V_0$. Thus

$$C = -\frac{1}{\epsilon_0} N_0[2eV_0 + kT_e]. \tag{14-64}$$

For all points in the sheath, $(dV/dx)^2 \geq 0$; examination of Eq. (14–63) shows that this condition is satisfied only if

$$V_0 \geq \frac{kT_e}{2e}, \tag{14-65}$$

a relation first pointed out by Bohm.* In other words, for a stable sheath to form, the ions that reach the sheath from the plasma must have a kinetic energy at least half as large as kT_e. Since stable sheaths always form under these circumstances, Eq. (14–65) effectively determines V_0; in fact, the inequality in Eq. (14–65) may usually be replaced by an equality sign.

---

* See Chapter 3 of the book edited by Guthrie and Wakerling, *op. cit.*

The sheath thickness may be found by integrating Eq. (14–63); we do this only for very negative probes, for which N_e may be neglected. Here

$$\left(\frac{dV}{dx}\right)^2 \approx \frac{4N_0 e V_0}{\epsilon_0} \left[\left(\frac{V}{V_0}\right)^{1/2} - 1\right] \approx \frac{4N_0 e V_0^{1/2} V^{1/2}}{\epsilon_0}$$

$$= 2\sqrt{\frac{2m_i}{e}} \frac{1}{\epsilon_0} J_i V^{1/2}, \tag{14–66}$$

which, when integrated, yields

$$x_s = \frac{4\epsilon_0^{1/2} V_p^{3/4}}{3(8m_i/e)^{1/4} J_i^{1/2}}. \tag{14–67}$$

14–9 SUMMARY

Highly ionized gases are good electrical conductors; a plasma is a region of highly ionized gas in which the static electric field and net charge density are nearly zero. Three different approaches to the analysis of plasma behavior are the equilibrium kinetic theory, orbit theory, and the macroscopic hydromagnetic theory.

1. The equilibrium theory, based on the statistical Boltzmann factor, shows that an external charge Q located in a plasma is shielded by the plasma in a distance called the *Debye length*. That is, the unshielded Coulomb potential $Q/4\pi\epsilon_0 r$ is replaced by

$$\varphi = \frac{Q}{4\pi\epsilon_0 r} e^{-r/h},$$

where the Debye length h is

$$h = \sqrt{\frac{\epsilon_0 kT}{2N_0 e^2}}.$$

2. Orbit theory is based on particle motions under the force

$$\mathbf{F} = q(\mathbf{E} + \mathbf{v} \times \mathbf{B}).$$

In a uniform magnetic field a particle of mass m_p moves freely along the field line but gyrates around it; the orbit is a helix with the Larmor radius

$$R = \frac{m_p v_\perp}{qB}.$$

The free particle has a diamagnetic moment. If the field is nonuniform, the particle gyrates in a tighter spiral as it moves into a stronger field along converging field lines; at the same time its axial motion slows down and is finally reversed. The result is a magnetic mirror.

3. The hydromagnetic approach is based on the macroscopic force law for unit volume

$$\mathbf{F}_v = \mathbf{J} \times \mathbf{B} - \nabla p,$$

where p is the fluid pressure. Sometimes the first term can be approximated by $-\nabla p_m$, where the "magnetic pressure" is equal to the energy density,

$$p_m = \frac{B^2}{2\mu_0}.$$

The magnetic pinch effect can be treated as a compression of the plasma by the magnetic pressure outside it.

4. The macroscopic approach also leads to plasma waves. The "electrostatic" waves are oscillations with a frequency, for the electrons, of

$$\omega_p = \sqrt{\frac{Ne^2}{m_e \epsilon_0}}$$

at infinite wavelength. The "Alfvén" waves propagate in a uniform magnetic field B_0 with a phase velocity

$$v_p = \frac{B_0}{\sqrt{\mu_0 \zeta}}.$$

PROBLEMS

14–1 The condition for orbit theory to be a good approximation to the motion of an electron in a plasma is that $\tau \gg 2\pi m_e/Be$, where τ is the mean collision time (see Chapter 7) and $2\pi m_e/Be$ is the cyclotron period in the magnetic field B. Show that this statement is equivalent to $\eta \ll \eta_H$, where $\eta_H \equiv B/N_0 e$ is the Hall resistivity.

14–2 Given a steady-flow hydromagnetic problem in which $\mathbf{v}$, $\mathbf{J}$, and $\mathbf{B}$ are mutually orthogonal. Assume that $\mathbf{v}$ is in the x-direction, and that $\mathbf{v}$, $\mathbf{J}$, and $\mathbf{B}$ are functions of x only. Assume also that the channel cross section (perpendicular to x) is independent of x. Show that

$$v = v_0 - \frac{1}{2\zeta_0 v_0} \left[2B_0 \int J \, dx + \mu_0 \left(\int J \, dx \right)^2 \right],$$

where v_0 is the velocity when $\zeta = \zeta_0$, $B = B_0$.

14–3 Derive Eq. (14–65) by examining Eq. (14–63) relative to the neighborhood of $V \approx V_0$.

14–4 The current-voltage characteristic is measured for a probe that is inserted into the plasma of a current-discharge tube. The probe has an area of 0.05 cm². All voltages are with respect to a fixed reference potential:

φ_p, V	I, mA	φ_p, V	I, mA
40.0	−20.5	35.0	−0.34
39.0	−20.4	34.0	−0.096
38.0	−7.5	33.0	−0.011
37.0	−2.7	31.0	+0.033
36.0	−0.98	29.0	+0.041

Determine the electron temperature in the plasma, the electron density, and the floating potential of the probe.

*14–5 A homogeneous sphere of radius a and electrical conductivity σ' moves with velocity $-\mathbf{v}_0$ in a nonviscous, incompressible fluid of conductivity σ in the presence of a uniform magnetic field $\mathbf{B}_0$. The velocity $\mathbf{v}_0$ is parallel to $\mathbf{B}_0$. Calculate the Joule loss resulting from induced currents in the system, and by equating this to the rate at which mechanical energy is dissipated by the sphere ($F_1 v_0$), calculate the drag force F_1. Assume potential flow in the fluid: in the coordinate system in which the sphere is at rest, the fluid velocity is given by

$$\mathbf{v} = \mathbf{v}_0 + \tfrac{1}{2}a^3 \, \mathbf{\nabla}(\mathbf{v}_0 \cdot \mathbf{r}/r^3),$$

relative to an origin at the center of the sphere. (For a discussion of this and related problems, see J. R. Reitz and L. L. Foldy, *Journal of Fluid Mechanics*, vol. 11, p. 133 (1961).)

CHAPTER 15 Electromagnetic Properties of Superconductors

15–1 THE HISTORY OF SUPERCONDUCTIVITY

Superconductivity was first recognized in 1911 by H. Kammerlingh Onnes at Leiden. He observed that as a sample of mercury was cooled, its resistance disappeared abruptly and apparently completely at 4.2 K. In a more sensitive experiment using a persistent current induced in a loop of superconducting wire, Onnes estimated the resistance in the superconducting state to be at most 10^{-12} the resistance in the normal state. More recently, at the Massachusetts Institute of Technology, it was found that an induced current of several hundred amperes in a superconducting lead ring showed no change in the magnitude of the current for a period of at least one year; this is strong evidence that the resistance in the superconducting state is indeed zero. The early experiments opened up a whole field of endeavor to characterize the new effect. It has been found that more than 20 elements and hundreds of alloys and intermetallic compounds are superconductors with transition temperatures ranging from substantially less than 1 K (e.g., 0.12 K for hafnium) to a high of about 20 K (e.g., 23 K for the compound Nb_3Ge). The transition, or critical, temperature is the temperature at which the transition from the normal state to the superconducting state takes place and is characteristic of the particular material being considered. The critical temperature depends to some extent on both the chemical purity and the metallurgical perfection of the sample being tested. Actually, inhomogeneities in the purity and strain of the sample generally tend to broaden the temperature range of the transition between the normal and superconducting states; a pure well-annealed sample may have a transition-temperature range as small as 0.001 K.

If a large enough magnetic field is applied parallel to a superconducting wire, it is found that the sample will become normal. The magnitude of the field that causes the transition is dependent on both the material and the temperature and is called the critical field. If a field is applied in some other direction the sample will start to become normal when the actual field at any point on the surface reaches the critical field. The field-temperature plot that can be made has essentially the same thermodynamic significance as the pressure-temperature diagram for ordinary phase transitions, and the curve itself may be considered as the phase boundary between the normal and superconducting thermodynamic states. The shape of the curve is generally parabolic and is given to a good approximation by the equation

$$H_c = H_0[1 - (T/T_c)^2],$$

where H_c is the critical field, T is the absolute (or Kelvin) temperature of observation, and T_c and H_0 represent the sample characteristics (critical temperature at zero field and critical field at zero absolute temperature). In addition to broadening the transition, inhomogeneities may also have a marked effect on H_0, sometimes increasing it by orders of magnitude. Such effects are of major importance in applications with high magnetic fields.

In the early history of superconductivity, the application of Maxwell's equations to a perfect conductor led to the conclusion that the time rate of change of the magnetic induction in the interior of the superconductor should be zero. Thus, depending on whether the sample was cooled to below the transition temperature in the presence or the absence of an applied magnetic field, magnetic flux should be trapped or excluded. This idea was so firmly believed that it was not until 1933 (22 years after the discovery of superconductivity) that W. Meissner and R. Ochsenfeld first tested it experimentally. The results of their experiments proved the hypothesis to be false and that, in all cases, irrespective of whether the sample is cooled in or out of a magnetic field, the magnetic induction of a superconductor is zero. This effect is called *flux exclusion* or, more commonly, the *Meissner effect*. An essentially equivalent statement is that a superconductor behaves as though it had zero permeability or a perfect diamagnetic susceptibility. This statement makes it easy to see that sample shape will have important effects, which are simple only when the specimen has the shape of a long cylinder whose axis is parallel to the applied magnetic field. The main significance of the Meissner effect is that it demonstrates that a superconductor is characterized by more complex electromagnetic properties than simple infinite conductivity. Any satisfactory explanation of superconductivity has to explain this effect in a natural way.

From a theoretical point of view, much has been done, starting with the application of thermodynamics to the transition by W. H. Keesom as early as 1924. Following this in 1934 was a phenomenological explanation of the second-order transition and other properties based on a two-fluid model developed by C. J. Gorter and H. B. G. Casimir. This was followed (1935) by F. and H. London's phenomenological theory of the electrodynamic properties of superconductors, in which Maxwell's equations were augmented by two additional equations to account for the Meissner effect. In this chapter, we will be mainly concerned with the London equations. From 1935, until the isotope effect was discovered in

1950,* little critical theoretical work was accomplished. However, in 1950, H. Fröhlich developed a theory based on the interaction of electrons with vibrating atoms in the crystal lattice, which explained the isotope effect but failed to predict other properties of the superconducting state. More recently (in 1957), J. Bardeen, L. N. Cooper, and J. R. Schrieffer developed a microscopic or quantum-mechanical theory of superconductivity that has been quite successful. This theory (BCS theory) accounts in a natural way for the second-order phase transition, the Meissner effect, and the other thermodynamic and electromagnetic properties of superconductors. As a result of their work, Bardeen, Cooper, and Schrieffer were awarded the Nobel Prize in Physics in 1972. According to BCS theory, superconductivity manifests itself as a phase transition arising from the *pairing* of electrons. This pairing results from the interaction of electrons with lattice vibrations in the material. In some ways, superconductivity is analogous to a Bose-Einstein condensation of bound pairs of electrons; both effects (superconductivity and the Bose-Einstein condensation) are essentially quantum mechanical and have no simple classical interpretation. BCS theory appears capable of predicting, at least qualitatively, all of the phenomenological results relating to superconductivity.

Technological applications of superconductivity require a material which remains superconducting in large magnetic fields, and for this one must use a *type II superconductor*. This is a more complicated form of superconductivity where, above a certain value of the magnetic field called H_{c1}, magnetic flux begins to penetrate the material even though it remains superconducting until a much higher magnetic field (the upper critical field H_{c2}) is reached. Between H_{c1} and H_{c2} the type II material does not exhibit a perfect Meissner effect, nor does it obey the London equations quantitatively; however, type II superconductivity is explained by BCS theory. H_{c2} is quite large in some materials; in Nb_3Sn, for example, $\mu_0 H_{c2}$ is larger than 10 tesla at 4.2 K. In this chapter we shall be concerned only with the simpler type of superconductivity (called *type I*).

The subject of superconductivity has developed into a very rich field of study. Nevertheless, the two complementary phenomenological theories—the Casimir-Gorter two-fluid theory and the London theory—together are adequate for the consideration of many problems involving superconductors. The Casimir-Gorter theory deals mainly with thermodynamic questions and is consequently only of peripheral interest here. The London theory, however, is, for the most part, an augmentation of Maxwell's equations for the purpose of constructing an electromagnetic theory that is competent to deal with situations involving superconductors. The balance of this chapter is concerned with the development of the London theory and its application to a few simple situations. This chapter attempts to

---

* Experiments on elemental superconductors of varying isotopic composition show that $T_c M^{1/2} \approx$ constant, with M the isotopic mass. The first experimental work was done by E. Maxwell and by C. A. Reynolds *et al*. The effect is now known as the *isotope effect* and is a clue that the interactions between the superconducting electrons and the ion cores of the crystal lattice play an important role in superconductivity.

provide a basis for the consideration of macroscopic electromagnetic problems involving superconductors, rather than to explore the contemporary microscopic theories of superconductivity.

15–2 PERFECT CONDUCTIVITY AND PERFECT DIAMAGNETISM OF SUPERCONDUCTORS

We noted in the preceding section that superconductors exhibit two unique properties. They have essentially infinite conductivity as revealed by the original Onnes experiments and subsequent extensions; they also completely exclude magnetic flux as demonstrated by the Meissner-Ochsenfeld experiment (so long as the magnetic field at the surface of the superconductor nowhere exceeds the critical field). These properties are independent in the sense that neither implies the other but, of course, both must and do emerge from satisfactory microscopic theories of superconductivity. To see more clearly what is meant by the independence of those two properties, we may cite the now classical consideration of a perfect conductor in a magnetic field.

Consider a sphere whose conductivity can be switched from a finite value to infinity in some way. For example, we can switch the conductivity of a superconductor by changing its temperature. When the conductivity is infinite the electric field is zero everywhere inside the superconductor, and consequently its curl and $\partial \mathbf{B}/\partial t$ are also zero. Thus, if the sphere is cooled (obtains perfect conductivity) in a uniform field $\mathbf{B}_0$, the flux density in the sphere remains $\mathbf{B}_0$ until the perfect conductivity is destroyed. On the other hand, if the sphere is cooled in zero field, the flux density remains zero until the perfect conductivity is destroyed in spite of, for example, its being placed in an initially uniform external field. Thus perfect conductivity does not imply flux exclusion and consequently $\mathbf{B} = 0$ is a postulate that must be introduced separately. Similarly $\mathbf{B} = 0$ does not imply perfect conductivity, for a material with susceptibility $\chi_m = -1$ would always have $\mathbf{B} = 0$, and this would not restrict the possible conductivity of the material.

In this chapter we are primarily concerned with the magnetic aspects of superconductivity (infinite conductivity will be further discussed but will not play an important role in the problems considered here), and appropriate formalisms for this will be developed. The first approach, which represents the smallest deviation from what has already been done (cf. Chapter 9), is to say that inside a superconductor $\mathbf{B} = \mu_0[\mathbf{H} + \mathbf{M}] = 0$ and at boundaries between superconductors and other media, the tangential component of $\mathbf{H}$ and the normal component of $\mathbf{B}$ are continuous. This approach views the superconductor as a magnetic material with susceptibility $\chi_m = -1$, that is, a medium which exhibits perfect diamagnetism. At the surface of the superconductor, *magnetization* currents flow with surface density (A/m) $\mathbf{j}_M = \mathbf{n} \times [\mathbf{M}_{\text{out}} - \mathbf{M}_{\text{in}}]$ with $\mathbf{n}$ the outward drawn normal to the surface (note that $\mathbf{M}_{\text{out}}$ is usually zero); inside the superconductor, volume magnetization currents flow with density $\mathbf{J}_M = \nabla \times \mathbf{M}$ (cf. Chapter 9, particularly Section 9–1).

An alternative description puts $\mathbf{B} = \mathbf{H} = \mathbf{M} = 0$ inside the superconductor and invokes a *real* surface current $\mathbf{j}_S = \mathbf{n} \times \mathbf{H}_{\text{out}}$ (since $\mathbf{H}_{\text{in}}$ is assumed to be zero). In this description there are no currents of any kind flowing in the interior of the superconductor. These two descriptions of a superconductor are so strikingly different that it is pertinent to ask how they are related. The usual statement is that they are equivalent when properly interpreted. However, it seems appropriate to consider the question in more detail. We first note that there are two differences between real transport currents and magnetization currents. The first of these is that transport currents are sources for $\mathbf{H}$ while both transport and magnetization currents are sources for $\mathbf{B}$. Since $\mathbf{B}$ is the magnetic field quantity which is accessible while $\mathbf{H}$ was introduced mainly to have a magnetic field quantity determined by the transport currents, this first distinction between the two kinds of currents is clearly convenient but somewhat artificial. The second distinction is that transport currents in normal materials are dissipative (i.e., give rise to joule heating) while magnetization currents are not. But for superconductors, even this distinction disappears. Furthermore, since it can be shown that the magnetization of superconductors is not due to spins (and consequently is associated with orbital motions of charge carriers), the two descriptions appear to be equivalent. An alternative succinct statement is that since only $\mathbf{B}$ is measurable, we can choose $\mathbf{M}$ and $\mathbf{H}$ according to relatively arbitrary rules so long as we apportion $\mathbf{J}$ and $\mathbf{J}_M$ accordingly and appreciate that they cannot be distinguished in a superconductor.

For most of what will be done the $\mathbf{H}, \mathbf{M} \neq 0$ description will be convenient. This is because it is a natural extension of what has been done earlier for normal materials and because this formulation leads to boundary value problems of a more conventional nature. In the next section, however, two problems will be considered, each in both formulations, in order to clarify the equivalence.

15–3 EXAMPLES INVOLVING PERFECT FLUX EXCLUSION

To reinforce the ideas presented in the preceding section, we will consider two elementary examples: a superconducting sphere in an asymptotically uniform field and an infinitely long, current-carrying, superconducting cylinder. Both formulations of Section 15–2 will be used to show explicitly that they are equivalent in these cases.

Consider first a superconducting sphere of radius a placed in a uniform external field $B_0\mathbf{k}$. In the first formulation, which treats the superconductor as a magnetic material, the boundary value problem takes the form

$$\text{Outside:} \quad \mathbf{B} \to B_0\mathbf{k} \text{ as } r \to \infty,$$

$$\nabla \cdot \mathbf{B} = 0,$$

$$\nabla \times \mathbf{H} = 0,$$

$$\mathbf{B} = \mu_0\mathbf{H}. \tag{15–1}$$

Inside: $\mathbf{B} = 0, \quad \mathbf{H} = -\mathbf{M},$

$$\nabla \times \mathbf{H} = 0, \tag{15-2}$$

$$\nabla \cdot \mathbf{M} = 0.$$

At $r = a$: B_r continuous,

$$H_\theta \text{ continuous.} \tag{15-3}$$

The only unusual equation is $\nabla \cdot \mathbf{M} = 0$, which is established on the basis that there are no magnetic poles inside the superconducting sphere. With these equations two magnetic scalar potentials, φ_1^* outside and φ_2^* inside, may be introduced. Both satisfy Laplace's equation, and from them the $\mathbf{H}$-field may be found by taking the negative gradient. Utilizing spherical coordinates and explicitly taking account of the first equation of Eq. (15-1), we have

$$\varphi_1^* = -\frac{B_0}{\mu_0} r \cos\theta + \sum_{\ell=0}^{\infty} c_\ell r^{-(\ell+1)} P_\ell(\cos\theta). \tag{15-4}$$

From this

$$B_r = B_0 \cos\theta + \mu_0 \sum_{\ell=0}^{\infty} (\ell+1) c_\ell r^{-(\ell+2)} P_\ell(\cos\theta) \quad \text{(outside).} \tag{15-5}$$

Since $\mathbf{B}$ is zero inside and B_r is continuous across $r = a$, each c_ℓ except c_1 must be zero and $c_1 = -B_0 a^3/2\mu_0$. This then completely solves the problem for $r > a$ *without* recourse to the boundary condition on the tangential component of $\mathbf{H}$; the only thing that has entered is $\mathbf{B} = 0$ inside and the continuity of the normal component of $\mathbf{B}$ at $r = a$. Inside the sphere, the potential φ_2^* must be regular at $r = 0$ and, in order to match the boundary conditions, can only involve $P_1(\cos\theta)$. Thus $\varphi_2^* = d_2 r \cos\theta$, with d_2 a constant to be determined, and by differentiation, $H_r = -d_2 \cos\theta$ and $H_\theta = d_2 \sin\theta$. Since outside $H_\theta = -\frac{3}{2}(B_0/\mu_0)\sin\theta$, it follows that $d_2 = -3B_0/2\mu_0$. There is no surface transport current, but a surface magnetization current $\mathbf{j}_M = -\frac{3}{2}(B_0/\mu_0)\sin\theta \, \mathbf{a}_\varphi$ exists because of the discontinuity of $\mathbf{M}$. These results may be summarized as follows:

Outside: $\mathbf{B} = \mu_0 \mathbf{H} = B_0 \mathbf{k} - B_0 \dfrac{a^3}{r^3} \cos\theta \, \mathbf{a}_r - \tfrac{1}{2} B_0 \dfrac{a^3}{r^3} \sin\theta \, \mathbf{a}_\theta.$

Inside: $\mathbf{B} = 0; \quad \mathbf{H} = \dfrac{3}{2} \dfrac{B_0}{\mu_0} \mathbf{k}; \quad \mathbf{M} = -\dfrac{3}{2} \dfrac{B_0}{\mu_0} \mathbf{k}. \tag{15-6}$

At $r = a$: $\mathbf{j}_M = -\dfrac{3}{2} \dfrac{B_0}{\mu_0} \sin\theta \, \mathbf{a}_\varphi.$

The second formulation is identical for the outside region but takes the form $\mathbf{B} = \mathbf{H} = \mathbf{M} = 0$ for the inside. There is also a real transport current $\mathbf{j}_S = \mathbf{n} \times$

$\mathbf{H}_{out} = -\frac{3}{2}(B_0/\mu_0) \sin \theta \, \mathbf{a}_\varphi$ at the surface. This description can be summarized as follows:

Outside: $\mathbf{B} = \mu_0 \mathbf{H} = B_0 \mathbf{k} - B_0 \dfrac{a^3}{r^3} \cos \theta \, \mathbf{a}_r - \frac{1}{2} B_0 \dfrac{a^3}{r^3} \sin \theta \, \mathbf{a}_\theta.$

Inside: $\mathbf{B} = \mathbf{H} = \mathbf{M} = 0.$ (15–7)

At $r = a$: $\mathbf{j}_S = -\dfrac{3}{2} \dfrac{B_0}{\mu_0} \sin \theta \, \mathbf{a}_\varphi.$

The relationship between the two descriptions is perhaps now clear. Outside, both are the same as they must be. Otherwise a simple experiment could be devised to select the correct description. Inside, both descriptions give $\mathbf{B} = 0$ but $\mathbf{H}$ and $\mathbf{M}$ are finite in one case and zero in the other. Neither $\mathbf{H}$ nor $\mathbf{M}$, however, is experimentally observable and, consequently, this distinction is unimportant. Identical surface currents exist in the two cases; in one, however, it is considered a transport current, while in the other it is called a magnetization current. What the current is called is important only in that it must be consistent with $\mathbf{H}$ and $\mathbf{M}$ inside the superconductor. For example, when we calculate the magnetic moment of the superconducting sphere, either $\mathbf{j}_M$ or $\mathbf{M}$ may be used, but not both; however, a real $\mathbf{j}_S$ always contributes to the magnetic moment.

A second example which further illustrates the indistinguishability of superconducting transport and magnetization currents is the case of an infinitely long, superconducting cylinder carrying a current. Before considering this problem in detail, however, we should note that inside a perfect superconductor the sum of $\mathbf{J}$ and $\mathbf{J}_M$ is always zero. This follows from $\mathbf{B} = 0$, which implies $\nabla \times \mathbf{B} = 0$ and thus, in turn, $\nabla \times \mathbf{H} + \nabla \times \mathbf{M} = \mathbf{J} + \mathbf{J}_M = 0$. At a surface of discontinuity this argument cannot be used, and a finite total surface current $\mathbf{j}_S + \mathbf{j}_M$ may exist. However, this argument shows clearly that the total current is always a surface current.

Returning now to the wire, which will be taken to have radius a and to be carrying a current I_0 (in the positive z-direction), we see from Ampere's law that outside the wire, $\mathbf{B} = \mu_0 \mathbf{H} = (\mu_0 I / 2\pi r) \mathbf{a}_\theta$ (cylindrical coordinates). If the first description, $\mathbf{M}, \mathbf{H} \neq 0$ inside, is used, we must make some assumption about the current density in the wire, and, because of the assumed continuity of the tangential component of $\mathbf{H}$, this assumption must not involve surface currents. The simplest possibility is uniform density: $\mathbf{J} = (I_0 / \pi a^2) \mathbf{k}$. Then inside,

$$\mathbf{H} = \frac{I_0}{2\pi} \frac{r}{a^2} \mathbf{a}_\theta \quad \text{and} \quad \mathbf{M} = -\frac{I_0}{2\pi} \frac{r}{a^2} \mathbf{a}_\theta.$$

The magnetization current density is $\mathbf{J}_M = -(I_0 / \pi a^2) \mathbf{k}$, and at the surface there is a surface magnetization current density

$$\mathbf{j}_M = +\mathbf{a}_r \times \left(\frac{I_0}{2\pi a} \mathbf{a}_\theta \right) = \frac{I_0}{2\pi a} \mathbf{k},$$

which is just sufficient to carry the total current I_0. The alternative description simply takes $\mathbf{B} = \mathbf{H} = \mathbf{M} = 0$ inside, and thus requires that the current exist entirely on the surface with real surface current density $\mathbf{j}_s = (I_0/2\pi a)\mathbf{k}$. These two descriptions are summarized in Table 15–1. Unless a method for separating transport currents from magnetization currents in superconductors or a way of directly measuring $\mathbf{H}$ or $\mathbf{M}$ inside a superconductor can be found, the descriptions are equivalent.

Table 15–1 Superconducting Current-Carrying Wire

Formulation 1 (Superconductor as magnetic material with $\chi_m = -1$)	Formulation 2 (Flux exclusion by surface transport currents)
$\mathbf{M} = -\mathbf{H} \neq 0$	$\mathbf{M} = \mathbf{H} = 0$
Outside: $\mathbf{B} = \mu_0\mathbf{H} = \dfrac{\mu_0 I_0}{2\pi r}\,\mathbf{a}_\theta$	$\mathbf{B} = \mu_0\mathbf{H} = \dfrac{\mu_0 I_0}{2\pi r}\,\mathbf{a}_\theta$
Inside: $\mathbf{B} = 0$	$\mathbf{B} = 0$
$\mathbf{H} = \dfrac{I_0 r}{2\pi a^2}\,\mathbf{a}_\theta$	$\mathbf{H} = 0$
$\mathbf{M} = -\dfrac{I_0 r}{2\pi a^2}\,\mathbf{a}_\theta$	$\mathbf{M} = 0$
$\mathbf{J} = \dfrac{I_0}{\pi a^2}\,\mathbf{k}$	$\mathbf{J} = 0$
$\mathbf{J}_M = -\dfrac{I_0}{\pi a^2}\,\mathbf{k}$	$\mathbf{J}_M = 0$
At $r = a$: $\mathbf{j}_M = (I_0/2\pi a)\mathbf{k}$	$\mathbf{j}_M = 0$
$\mathbf{j}_s = 0$	$\mathbf{j}_s = (I_0/2\pi a)\mathbf{k}$

In the two problems just considered the $\mathbf{M} = \mathbf{H} = 0$ formulation has an apparent advantage in simplicity. However, for more complicated problems, particularly those involving large demagnetizing factors, the distributed magnetization formulation is advantageous. Either method may be used and the results will be equivalent, but they must not be mixed in a single problem.

15–4 THE LONDON EQUATIONS

In the preceding section, flux exclusion was discussed on the basis of a highly idealized representation of a superconductor. This representation reproduces many of the observed features of superconductivity but fails to account adequately for some of the readily observable details. A more sophisticated theory can be developed by starting from the concept of perfect conductivity and making an appropriate modification to include the Meissner effect.

In a perfect conductor (*not* a superconductor) charge carriers would experience no retarding forces; consequently in an electric field $\mathbf{E}$ they would move according to

$$m_p \dot{\mathbf{v}} = q\mathbf{E}, \tag{15-8}$$

where m_p is the mass of the charge carrier and $\dot{\mathbf{v}}$ is its acceleration. But if $\mathbf{v}$ is the average velocity of the charge carriers and there are n of them per unit volume, then the current density is $\mathbf{J} = nq\mathbf{v}$. An alternative to Eq. (15-8) is then

$$\dot{\mathbf{J}} = (nq^2/m_p)\mathbf{E}, \tag{15-9}$$

where $\dot{\mathbf{J}} = d\mathbf{J}/dt$. Taking the curl of this equation and using $\nabla \times \mathbf{E} = -\partial \mathbf{B}/\partial t$, we have

$$\nabla \times \dot{\mathbf{J}} = -(nq^2/m_p)\dot{\mathbf{B}}. \tag{15-10}$$

Assuming that the fields are slowly varying* and using $\nabla \times \mathbf{H} = \mathbf{J}$ to eliminate $\dot{\mathbf{J}}$, we have

$$\nabla \times \nabla \times \dot{\mathbf{H}} = -(nq^2/m_p)\dot{\mathbf{B}}. \tag{15-11}$$

Assuming that $\mathbf{B} = \mu_0\mathbf{H}$ and using the definition of the Laplacian of a vector (with $\nabla \cdot \mathbf{B} = 0$) yields

$$\nabla^2\dot{\mathbf{B}} = (\mu_0 nq^2/m_p)\dot{\mathbf{B}}. \tag{15-12}$$

The significance of this equation may be seen best by considering a semi-infinite perfect conductor bounded by the plane $z = 0$ and extending in the positive z-direction. Assume that at the surface $\dot{B}_y = \dot{B}_z = 0$, $\dot{B}_x = \dot{B}_{x0}$, and that $\dot{B}_{x0}$ does not depend on x or y. The equation determining $\dot{B}_x$ is then

$$\frac{d^2\dot{B}_x}{dz^2} = \frac{\mu_0 nq^2}{m_p}\dot{B}_x, \tag{15-13}$$

which has as a general solution

$$\dot{B}_x = Ae^{-\sqrt{\mu_0 nq^2/m_p}\, z} + Be^{\sqrt{\mu_0 nq^2/m_p}\, z}.$$

The exponentially increasing solution may be rejected as being without any physical interpretation, and A may be chosen to give $\dot{B}_x$ correctly at $z = 0$; then

$$\dot{B}_x = \dot{B}_{x0}e^{-\sqrt{\mu_0 nq^2/m_p}\, z}. \tag{15-14}$$

It is easy to verify that $(m_p/\mu_0 nq^2)^{1/2}$ has the dimensions of length and that for q and m_p appropriate to an electron and n corresponding to one electron per atom this length is about 10^{-8} m. Equation (15-12) then indicates that, in the interior of the perfect conductor, the time derivative of $\mathbf{B}$ goes to zero exponentially with

---

* This assumption entails neglecting the displacement current, $\partial \mathbf{D}/\partial t$, which is discussed in Section 16-1 and 16-2.

distance from the surface. Thus, in the interior of the perfect conductor, $\dot{\mathbf{B}}$ is very small except in a thin surface layer. This is a reasonable refinement of the earlier conclusion that $\dot{\mathbf{B}} = 0$ everywhere in the interior of a perfect conductor.

The development just sketched shows again that perfect conductivity does not lead to flux exclusion. However, it also indicates how flux exclusion might be incorporated into a theory. If Eq. (15–12) described the behavior of $\mathbf{B}$ instead of $\dot{\mathbf{B}}$, then $\mathbf{B}$ itself would decrease exponentially from its value at the surface to zero in the interior of a superconductor. This was the motivation for the development of a theory of the electromagnetic behavior of superconductors by F. and H. London.*

In this theory it is assumed that the total current can be broken up into a supercurrent $\mathbf{J}_S$, a dissipative current $\mathbf{J}_{\text{diss}}$, and a displacement current $\mathbf{J}_{\text{disp}}$:

$$\mathbf{J} = \mathbf{J}_S + \mathbf{J}_{\text{diss}} + \mathbf{J}_{\text{disp}}. \qquad (15\text{–}15)$$

The dissipative and displacement currents are governed by the equations $\mathbf{J}_{\text{diss}} = g\mathbf{E}$ and $\mathbf{J}_{\text{disp}} = \partial \mathbf{D}/\partial t$. It remains to relate $\mathbf{J}_S$ to the electromagnetic field. This can be done, starting with Eq. (15–15), Maxwell's equations, and London's constitutive equation (similar in form to Eq. (15–10) but involving $\mathbf{B}$ and $\mathbf{J}$ rather than their derivatives). If this procedure is carefully followed, it can be shown that for frequencies less than about 10^{11} Hz, both $\mathbf{J}_{\text{diss}}$ and $\mathbf{J}_{\text{disp}}$ are negligible compared to $\mathbf{J}_S$. We should like to assume this result, namely, $\mathbf{J}_{\text{diss}} \approx 0$ and $\mathbf{J}_{\text{disp}} \approx 0$, without going through the detailed arguments; such an assumption is at least a reasonable one for steady-current problems of the type considered in this chapter. The remaining current $\mathbf{J}_S$ includes both transport and magnetization currents and consequently, from Maxwell's equation (8–51),

$$\mathbf{J}_S = (1/\mu_0)\nabla \times \mathbf{B}. \qquad (15\text{–}16)$$

To obtain an equation involving the magnetic field variables rather than their derivatives, London *postulated* that

$$\mu_0 \nabla \times \mathbf{J}_S = -(1/\lambda^2)\mathbf{B}. \qquad (15\text{–}17)$$

This equation differs from Eq. (15–10) in that it involves $\mathbf{J}_S$ and $\mathbf{B}$ rather than $\dot{\mathbf{J}}$ and $\dot{\mathbf{B}}$. Therefore it will lead to an equation analogous to Eq. (15–12) for the fields rather than for their derivatives. Also, a phenomenological penetration depth λ has been introduced as a specific parameter characteristic of the superconducting material (μ_0 occurs in order to make the dimensions of λ a length). Equation (15–17) will lead to the Meissner effect but, in order to include infinite conductivity, we must assume separately that

$$\mu_0 \dot{\mathbf{J}}_S = (1/\lambda^2)\mathbf{E}. \qquad (15\text{–}18)$$

The latter equation will, however, play no further role in the problems considered here.

---

* F. London and H. London, *Proc. Roy. Soc.*, vol. A149, p. 71 (1935).

Equations (15–16) and (15–17) may be combined to give

$$\mathbf{\nabla} \times \mathbf{\nabla} \times \mathbf{B} = -(1/\lambda^2)\mathbf{B}. \tag{15–19}$$

Since $\mathbf{\nabla} \cdot \mathbf{B} = 0$ this may be written

$$\nabla^2\mathbf{B} = (1/\lambda^2)\mathbf{B}. \tag{15–20}$$

Equation (15–20) may be solved in the case of a semi-infinite slab exactly as Eq. (15–13) was. The solution

$$B_x(z) = B_{x_0} e^{-z/\lambda} \tag{15–21}$$

now indicates that $\mathbf{B}$ rather than $\dot{\mathbf{B}}$ falls exponentially as the slab is penetrated. This is the desired generalization of $\mathbf{B} = 0$ inside a superconductor.

The penetration depth λ has been introduced here as a phenomenological parameter; various theories have, however, been constructed that attempt to compute its size. We are more interested in the experimental determination of λ. An obvious approach would be to construct a solenoid with a superconducting core. The inductance of such a solenoid would be very small if the superconductor were perfect and completely filled the volume enclosed by the solenoid. If, on the other hand, there were a finite penetration depth, then the inductance would be somewhat larger. If the penetration depth were a significant fraction of the radius of the solenoid, the penetration depth could be inferred from measurements of the inductance. The feasibility of such a determination depends on the ratio of the volume into which the field penetrates to the total volume of the sample. As it turns out, λ is typically a few millionths of a centimeter and, consequently, the simple experiment proposed above will not produce significant results. This difficulty can, however, be circumvented by using a sample with a large surface-to-volume ratio. The first successful experiments of this type, done with a mercury colloid, were performed by D. Shoenberg in 1939. These experiments demonstrated conclusively that the magnetic field penetrated into the small superconducting mercury spheres and that the penetration depth depended on temperature. The original Shoenberg experiments, which have been extended and supplemented, demonstrated that the concept of a penetration depth was valid and important.

Equations (15–15), (15–17), and (15–18), together with the four Maxwell equations, are often collectively called the *Maxwell–London equations* and are very useful for treating electromagnetic problems involving superconductors.

As is evident from the preceding discussion, the concept of perfect flux exclusion is an idealization. Instead, magnetic flux penetrates a thin layer at the surface of the superconductor and, according to the London theory, decreases exponentially toward the interior. The surface current density, $\mathbf{j}_M$ (or as the case may be, $\mathbf{j}_S$) is also an idealization. Here again, the supercurrent density $\mathbf{J}_S$ is spread out in a thin surface layer and decreases exponentially toward the interior. Thus there is no $\mathbf{j}_M$ in the London theory but only a total supercurrent density $\mathbf{J}_S$. In the next section the two problems considered earlier will be solved using the Maxwell–London equations.

*15–5 EXAMPLES INVOLVING THE LONDON EQUATIONS

In order to understand the Maxwell–London equations better, they will now be used to obtain more refined solutions to the problems considered in Section 15–3. The first problem is that of a superconducting sphere of radius a in an external field that, at large distances, is uniform and equal to $B_0\mathbf{k}$. The equations satisfied by the fields are

$$\text{Outside:}\quad \nabla \cdot \mathbf{B} = 0, \quad \nabla \times \mathbf{H} = 0, \quad \mathbf{B} = \mu_0\mathbf{H},$$

$$\text{Inside:}\quad \nabla^2\mathbf{B} = (1/\lambda^2)\mathbf{B}, \quad \nabla \cdot \mathbf{B} = 0, \tag{15-22}$$

where λ, the penetration depth, is considered as a phenomenological parameter. The boundary conditions that must be satisfied are

$$\text{At } r = \infty: \quad \mathbf{B} = B_0\mathbf{k},$$

$$\text{At } r = a: \quad B_r \text{ and } B_\theta \text{ are continuous.} \tag{15-23}$$

The only one of these boundary conditions that requires further comment is the continuity of B_θ at $r = a$. This follows from the assumption, in agreement with the discussion at the end of the preceding section, that the supercurrents (both transport and magnetization) are never infinite, i.e., there are no surface current densities $\mathbf{j}_M$ or $\mathbf{j}_S$. In this case the tangential components of both $\mathbf{H}$ and $\mathbf{M}$ are continuous, and hence the tangential component of $\mathbf{B}$ is also continuous.

The solution of the equations for the field outside the sphere poses no difficulties. A magnetic scalar potential satisfying Laplace's equation may be introduced exactly as was done in Section 15–3 and a general solution obtained. For the interior region, however, the equation $\nabla^2\mathbf{B} = (1/\lambda^2)\mathbf{B}$ must be solved. If in spherical coordinates the Laplacian of a vector could be obtained simply by taking the Laplacian of each of its coordinates, then the solutions to this equation could be found readily. This, however, is not the case, but rather the curl curl of the vector must be calculated. As a result, even in this simple problem, the r- and θ-components of $\nabla^2\mathbf{B} = (1/\lambda^2)\mathbf{B}$ involve both B_r and B_θ. This complexity is relatively well-known, and extensive techniques for solving the vector Helmholtz equation have been developed.* The development and application of these methods is, however, somewhat outside the scope of this book. Consequently, the results of Section 15–3 will be used to make a guess (*Ansatz*) at the form of the solution. The resulting final solution will be justified because it satisfies the equations and the boundary conditions and because these equations and boundary conditions have a unique solution. The uniqueness can, of course, be proved, but here it will be assumed.

In Section 15–3 it was found that only the $P_1(\cos \theta)$-term in φ^* survived in the solution for the region outside the sphere. It will be assumed that this is also true

---

* Cf. Morse and Feshbach, *Methods of Theoretical Physics* (New York: McGraw-Hill, 1953), Chapter 13.

for the Maxwell–London theory and that

$$\mathbf{B}(r, \theta) = B_0\mathbf{k} - b\left(\frac{a}{r}\right)^3 [\cos \theta \, \mathbf{a}_r + \tfrac{1}{2} \sin \theta \, \mathbf{a}_\theta] \text{ (outside).} \qquad (15\text{--}24)$$

This is very similar to the first of equations (15–7), the only difference being that B_0 has been replaced by b in that part of the field that is due to the magnetization of the sphere. The value of b will be determined by fitting the boundary conditions. For the interior of the sphere, Section 15–3 provides little by way of clues; however, from the form of $\mathbf{M}$ found there and the fact that in Eq. (15–24) B_r depends on θ through $\cos \theta$ while B_θ depends on θ through $\sin \theta$, a reasonable assumption might be that

$$B_r = u(r) \cos \theta \text{ (inside),} \qquad (15\text{--}25a)$$

$$B_\theta = v(r) \sin \theta \text{ (inside).} \qquad (15\text{--}25b)$$

The two functions $u(r)$ and $v(r)$ must be determined so that $\nabla^2\mathbf{B} = (1/\lambda^2)\mathbf{B}$ and also so that the boundary conditions at $r = a$ are satisfied. These boundary conditions are

$$u(a) = B_0 - b, \qquad (15\text{--}26a)$$

$$v(a) = -B_0 - b/2. \qquad (15\text{--}26b)$$

By expanding $\nabla \times \nabla \times \mathbf{B}$ and using the assumed forms (15–25), we find the equations

$$r\frac{dv}{dr} + v + u = -\frac{r^2}{2\lambda^2} u \qquad (15\text{--}27a)$$

and

$$r^2\frac{d^2v}{dr^2} + 2r\frac{dv}{dr} + r\frac{du}{dr} = \frac{r^2}{\lambda^2} v \qquad (15\text{--}27b)$$

for u and v. Differentiating Eq. (15–27a) with respect to r and subtracting from Eq. (15–27b) gives

$$v = -u - \tfrac{1}{2}ru'. \qquad (15\text{--}28)$$

Using this result to eliminate v and dv/dr from Eq. (15–27a) gives an equation for u:

$$r^2\frac{d^2u}{dr^2} + 4r\frac{du}{dr} = \frac{r^2}{\lambda^2} u. \qquad (15\text{--}29)$$

Introducing ξ by $\xi = ru$ and changing the independent variable to $\rho = r/i\lambda$ leads to the equation for spherical Bessel functions of order one [Eq. (17–84) with $l = 1$]. Using the solution $j_1(r/i\lambda)$ from Table 17–2 (p. 377) gives

$$u(r) = c(\lambda/r)^3[\sinh (r/\lambda) - (r/\lambda) \cosh (r/\lambda)] \qquad (15\text{--}30)$$

as the solution which is regular at the origin. From Eqs. (15–28) and (15–29) we find that

$$v = \frac{c}{2} \left(\frac{\lambda}{r}\right)^3 \left[\left(1 + \frac{r^2}{\lambda^2}\right) \sinh\left(\frac{r}{\lambda}\right) - \left(\frac{r}{\lambda}\right) \cosh\left(\frac{r}{\lambda}\right)\right]. \tag{15–31}$$

This completes the formal solution except for using Eqs. (15–26), (15–30), and (15–31) to determine b and c. These are

$$c = -3B_0 \left(\frac{a}{\lambda}\right) \sinh\left(\frac{a}{\lambda}\right), \tag{15–32}$$

$$b = B_0 \left[1 + 3\left(\frac{\lambda}{a}\right)^2 - 3\left(\frac{\lambda}{a}\right) \coth\left(\frac{a}{\lambda}\right)\right]. \tag{15–33}$$

It might be expected that for very small values of λ/a the fields would not be greatly different from those found in Section 15–3 for the perfect superconducting sphere. We can verify that this is the case by using the fact that coth x approaches unity exponentially for large values of x. Thus

$$b \simeq B_0 \left(1 - 3\frac{\lambda}{a} + 3\frac{\lambda^2}{a^2} + \cdots\right), \qquad \frac{\lambda}{a} \ll 1, \tag{15–34}$$

and the first correction to the field outside the sphere is of order λ/a.

The second example of the solution of the London equations is the long current-carrying wire. The radius of the wire will be taken to be a, the penetration depth λ, and the total external (real) current I_0. Outside the wire, **H** is given by Ampere's law and $\mathbf{B} = \mu_0 \mathbf{H}$. Therefore,

$$H_r = H_z = B_r = B_z = 0; \qquad B_\theta = \mu_0 H_\theta = \mu_0 \frac{I_0}{2\pi r} \text{ (outside).} \tag{15–35}$$

Inside, **B** satisfies

$$\nabla^2 \mathbf{B} = \frac{1}{\lambda^2} \mathbf{B}. \tag{15–36}$$

By symmetry, **B** has only a θ-component and this depends only on r. Then Eq. (15–36) becomes

$$r^2 \frac{d^2}{dr^2} B_\theta + r \frac{d}{dr} B_\theta - \left(1 + \frac{r^2}{\lambda^2}\right) B_\theta = 0. \tag{15–37}$$

This is just Bessel's equation for index one and argument ir/λ. The solution that is not infinite at the origin is

$$B_\theta = AJ_1(ir/\lambda). \tag{15–38}$$

The coefficient A is determined by matching B_θ inside and B_θ outside at $r = a$. The result is

$$B_\theta = \frac{\mu_0 I_0}{2\pi a} \frac{J_1(ir/\lambda)}{J_1(ia/\lambda)} \text{ (inside).} \tag{15–39}$$

Since $J_1(ir/\lambda) = iI_1(r/\lambda)$, where I_1 is a modified Bessel function, Eq. (15–39) may be written in terms of standard tabulated functions. From this result we can calculate the other fields and the current distribution and can show that the field and total current density fall off exponentially with distance from the surface of the wire. The details, however, are left to the exercises.

This discussion of the electromagnetic properties of superconductors has necessarily been fragmentary. In particular, problems involving time-dependent fields and the microscopic theory of superconductivity have been ignored. Many of these are discussed in recent books on superconductivity.* Two of the earlier books that supplement this chapter are those by F. London† and D. Shoenberg.‡

15–6 SUMMARY

Superconductors constitute a fairly large class of materials that have a phase transition to the superconducting state at low temperatures, usually below about 20 K. The transition depends on the magnetic field as well as the temperature, with the material transforming back to the normal state in fields greater than the critical field H_c. Ideally,

$$H_c = H_0[1 - (T/T_c)^2].$$

The macroscopic electric and magnetic behavior in the superconducting state is most simply described by the constitutive equations $\mathbf{J} = g\mathbf{E}$ and $\mathbf{M} = \chi_m\mathbf{H}$ with extreme values of the material parameters,

$$g = 0, \qquad \chi_m = -1 \qquad (\mu = 0),$$

representing perfect conductivity and perfect diamagnetism. The latter requires that the flux density B and the current density J are zero inside a superconductor, and any superconducting currents are surface currents. A more refined description is given by the London equations, which replace the linear constitutive equations with the differential equations

$$\mu_0 \dot{\mathbf{J}} = \left(\frac{1}{\lambda}\right)^2 \mathbf{E},$$

$$\mu_0 \nabla \times \mathbf{J} = -\left(\frac{1}{\lambda}\right)^2 \mathbf{B}.$$

---

* See, for example, A. C. Rose-Innes and E. H. Rhoderick, *Introduction to Superconductivity*, Second Edition (New York: Pergamon, 1977); and M. Tinkham, *Introduction to Superconductivity* (New York: McGraw-Hill, 1975). See also *Superconductivity in Science and Technology*, M. H. Cohen, editor (Chicago: University of Chicago Press, 1968).

† F. London, *Superfluids. The Macroscopic Theory of Superconductivity*, Vol. I (New York: Wiley, 1950; New York: Dover Publications, 1961).

‡ D. Shoenberg, *Superconductivity*, Second Edition (London: Cambridge University Press, 1965).

These, coupled with the Maxwell equations in vacuum, predict that the flux density and the current density fall off exponentially from the surface of a super-conductor, with the penetration depth λ, instead of falling discontinuously to zero.

1. The simple formulation with $\chi_m = -1, \mu = 0$, is most convenient for problems in which a superconductor is located in a magnetic field. The results are the same as those obtained by the techniques of Chapter 9, with the mere substitution of $\mu = 0$.

2. An alternative formulation, which may be more convenient for problems in which the superconductor carries an external current, assumes that inside the superconductor $\chi_m = 0, \mu = \mu_0$. The condition $\mathbf{B} = 0$ is met with $\mathbf{H} = 0 = \mathbf{M}$ inside, and the boundary conditions are satisfied by assuming suitable superconducting surface transport currents. The two formulations are equivalent, but the choice of one *or* the other must be adhered to in a given problem.

PROBLEMS

15–1 Consider an infinitely long superconducting circular cylinder of radius a in a transverse magnetic field. At large distances from the cylinder the field is uniform and of magnitude B_0. Compute the fields inside and outside the cylinder and the current density in the cylinder and on its surface. Assume that the superconducting properties are represented by perfect diamagnetism and perfect conductivity. Compare the two equivalent formulations.

15–2 Consider an infinite superconducting slab of thickness d, bounded by the planes $z = 0$ and $z = d$. Outside the slab the magnetic field is uniform and parallel to the surfaces, $B_x = B_0$. Find the field and the current density in the slab, using the London equations and the phenomenological penetration depth λ.

15–3 Do the computations of Problem 15–1, using the London equations and the phenomenological penetration depth λ.

15–4 Complete the discussion of the fields produced by an infinitely long current-carrying wire, using the London equations and starting from the results developed in Eqs. (15–35) through (15–39). (a) Compute $\mathbf{J}$ inside the cylinder. (b) Discuss the exponential fall-off of $\mathbf{B}$ in the region close to the surface of the cylinder.

15–5 Consider a superconducting sphere of radius a in a magnetic field that, at large distances from the sphere, is uniform and of magnitude B_0. Using the formulation of Section 15–5 as a basis, provide the following: (a) An expansion of $\nabla \times \nabla \times \mathbf{B}$ and from it the equations satisfied by the components of $\mathbf{B}$ inside the sphere. (b) A verification of Eq. (15–27). (c) A quantitative discussion of the exponential fall-off of $\mathbf{B}$ in the region close to the surface of the sphere.

CHAPTER 16 Maxwell's Equations

16–1 THE GENERALIZATION OF AMPERE'S LAW. DISPLACEMENT CURRENT

In Chapter 9 we found that the magnetic field due to a current distribution satisfied Ampere's circuital law,

$$\oint \mathbf{H} \cdot d\mathbf{l} = \int_S \mathbf{J} \cdot \mathbf{n} \, da. \tag{16-1}$$

We shall now examine this law, show that it sometimes fails, and find a generalization that is always valid.

Consider the circuit shown in Fig. 16–1, which consists of a small parallel-plate capacitor being charged by a constant current I (we need not worry about what causes the current). If Ampere's law is applied to the contour C and the surface S_1, we find

$$\oint_{C} \mathbf{H} \cdot d\mathbf{l} = \int_{S_1} \mathbf{J} \cdot \mathbf{n} \, da = I. \tag{16-2}$$

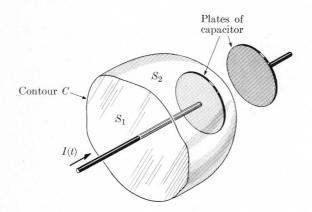

Plates of capacitor

Contour C

S_2

S_1

$I(t)$

Figure 16–1 The contour C and two surfaces, S_1 and S_2, for testing Ampere's circuital law.

If, on the other hand, Ampere's law is applied to the contour C and surface S_2, then J is zero at all points on S_2 and

$$\oint_C \mathbf{H} \cdot d\mathbf{l} = \int_{S_2} \mathbf{J} \cdot \mathbf{n} \, da = 0. \tag{16-3}$$

Equations (16–2) and (16–3) contradict each other and thus cannot both be correct. If C is imagined to be a great distance from the capacitor, it is clear that the situation is not substantially different from the standard Ampere law cases considered in Chapter 8. One is thus led to think that (16–2) is correct, since it is not dependent on the new feature, namely, the capacitor. Equation (16–3), on the other hand, requires consideration of the capacitor for its deduction. It would appear, then, that (16–3) requires modification. Since these equations arose by integrating Eq. (9–30),

$$\nabla \times \mathbf{H} = \mathbf{J}, \tag{16-4}$$

this also requires modification.

The proper modification can be made by noting that Eqs. (16–2) and (16–3) give different results because the integrals on the right-hand sides are different. Phrased mathematically,

$$\int_{S_2} \mathbf{J} \cdot \mathbf{n}_2 \, da - \int_{S_1} \mathbf{J} \cdot \mathbf{n}_1 \, da \neq 0. \tag{16-5}$$

S_1 and S_2 together form a closed surface (they join at C); however, $\mathbf{n}_2$ is outward drawn and $\mathbf{n}_1$ inward drawn. If this fact is taken into account, Eq. (16–5) may be written

$$\oint_{S_1+S_2} \mathbf{J} \cdot \mathbf{n} \, da \neq 0. \tag{16-6}$$

Phrased physically, the net transport current through the closed surface $S_1 + S_2$ does not vanish because charge is piling up on the plate of the condenser enclosed by the surface. Charge conservation requires, according to Eqs. (7–6) and (7–7),

$$\oint_{S_1+S_2} \mathbf{J} \cdot \mathbf{n} \, da = -\int_V \frac{\partial \rho}{\partial t} \, dv, \tag{16-7}$$

because inside the volume V enclosed by $S_1 + S_2$ the charge density ρ is changing with time on the condenser plate. In differential form, Eq. (16–7) is expressed by the equation of continuity,

$$\nabla \cdot \mathbf{J} + \frac{\partial \rho}{\partial t} = 0. \tag{16-8}$$

It is now clear what is wrong with (16–4): Taking its divergence we have

$$\nabla \cdot \nabla \times \mathbf{H} = 0 = \nabla \cdot \mathbf{J},$$

since the divergence of a curl is identically zero. Thus the relation $\mathbf{V} \cdot \mathbf{J} = 0$, as implied by Eq. (16–4), is inconsistent with charge conservation in the present situation, and so something must be added to the right side of Eq. (16–4) that will give $\partial \rho / \partial t$ in Eq. (16–8). What this could be is seen from the relation of ρ to electric displacement:

$$\mathbf{V} \cdot \mathbf{D} = \rho. \tag{16–9}$$

Inserting ρ from Eq. (16–9) into Eq. (16–8), we get

$$\mathbf{V} \cdot \mathbf{J} + \frac{\partial}{\partial t} \mathbf{V} \cdot \mathbf{D} = \mathbf{V} \cdot \left(\mathbf{J} + \frac{\partial \mathbf{D}}{\partial t} \right) = 0.$$

If $\partial \mathbf{D} / \partial t$ is added to Eq. (16–4), then its divergence will correctly give Eq. (16–8). (Of course, up until Chapter 11, we were assuming that the fields were independent of time, and so Eq. (16–4) itself was applicable. In Chapters 11 to 15, the fields were supposed to be "slowly varying," meaning that $\partial \mathbf{D} / \partial t$ was neglected in comparison to $\mathbf{J}$.) Inclusion of $\partial \mathbf{D} / \partial t$ gives the generalized Ampere's law:

$$\mathbf{V} \times \mathbf{H} = \mathbf{J} + \frac{\partial \mathbf{D}}{\partial t}. \tag{16–10}$$

The introduction of the second term on the right, which is known as the *displacement current*, represents one of Maxwell's major contributions to electromagnetic theory.

16–2 MAXWELL'S EQUATIONS AND THEIR EMPIRICAL BASIS

Equation (16–10) is one of the set of equations known as Maxwell's equations. The entire set consists of Eq. (16–10) plus three equations with which we are already familiar, namely:

$$\mathbf{V} \times \mathbf{H} = \mathbf{J} + \frac{\partial \mathbf{D}}{\partial t}, \tag{16–10}$$

$$\mathbf{V} \times \mathbf{E} = -\frac{\partial \mathbf{B}}{\partial t}, \tag{11–6) (16–11}$$

$$\mathbf{V} \cdot \mathbf{D} = \rho, \tag{4–29) (16–12}$$

$$\mathbf{V} \cdot \mathbf{B} = 0. \tag{8–30) (16–13}$$

Each of these equations represents a generalization of certain experimental observations: Eq. (16–10) represents an extension of Ampere's law; Eq. (16–11) is the differential form of Faraday's law of electromagnetic induction; Eq. (16–12) is Gauss's law, which in turn derives from Coulomb's law; Eq. (16–13) is usually said to represent the fact that single magnetic poles have never been observed.

It is clear that Maxwell's equations represent mathematical expressions of certain experimental results. In this light it is apparent that they cannot be proved; however, the applicability to any situation can be verified. As a result of extensive experimental work, Maxwell's equations are now believed to apply to all macroscopic situations and they are used, much like conservation of momentum, as guiding principles. They are the fundamental equations of the electromagnetic fields produced by the source charge and current densities ρ and $\mathbf{J}$. If material bodies are present, in order to use the Maxwell equations, one must also know the applicable constitutive equations—either experimentally or from a microscopic theory of the particular kind of matter: $\mathbf{D} = \mathbf{D}(\mathbf{E})$ and $\mathbf{H} = \mathbf{H}(\mathbf{B})$. The current density $\mathbf{J}$ in a material includes a contribution given by a third constitutive equation, $\mathbf{J} = \mathbf{J}(\mathbf{E})$, which must likewise be known experimentally or theoretically. Coupled with the Lorentz force equation, $\mathbf{F} = q(\mathbf{E} + \mathbf{v} \times \mathbf{B})$, which describes the action of the fields on charged particles, this set of laws gives the complete classical description of electromagnetically interacting particles.

We have just seen that the displacement current introduced in the preceding section is necessary in order to have charge conservation and that, when it is included in the Maxwell equations, they already imply the continuity equation, so that the latter need not be added to the set of fundamental equations. The Maxwell equations have two further interesting consequences, which are developed in the following sections. These will also be seen to depend crucially on the displacement current.

16–3 ELECTROMAGNETIC ENERGY

It was shown in Chapter 6 that the quantity

$$U_E = \tfrac{1}{2} \int_V \mathbf{E} \cdot \mathbf{D} \, dv \qquad (16\text{–}14)$$

can be identified with the electrostatic potential energy of the system of charges producing the electric field. This was derived by computing the work done in establishing the field. In a similar way

$$U_M = \tfrac{1}{2} \int_V \mathbf{H} \cdot \mathbf{B} \, dv \qquad (16\text{–}15)$$

was identified, in Chapter 12, with the energy stored in the magnetic field. The question of the applicability of these expressions to nonstatic situations now arises.

If the scalar product of Eq. (16–10) with $\mathbf{E}$ is taken, and the resulting equation is subtracted from the scalar product of Eq. (16–11) with $\mathbf{H}$, the resulting equation is

$$\mathbf{H} \cdot \nabla \times \mathbf{E} - \mathbf{E} \cdot \nabla \times \mathbf{H} = -\mathbf{H} \cdot \frac{\partial \mathbf{B}}{\partial t} - \mathbf{E} \cdot \frac{\partial \mathbf{D}}{\partial t} - \mathbf{E} \cdot \mathbf{J}. \qquad (16\text{–}16)$$

The left side of this expression can be converted into a divergence by using the identity

$$\nabla \cdot (\mathbf{F} \times \mathbf{G}) = \mathbf{G} \cdot \nabla \times \mathbf{F} - \mathbf{F} \cdot \nabla \times \mathbf{G}$$

to obtain

$$\nabla \cdot (\mathbf{E} \times \mathbf{H}) = -\mathbf{H} \cdot \frac{\partial \mathbf{B}}{\partial t} - \mathbf{E} \cdot \frac{\partial \mathbf{D}}{\partial t} - \mathbf{E} \cdot \mathbf{J}. \qquad (16\text{--}17)$$

If the medium to which (16–17) is applied is linear and nondispersive, i.e., if $\mathbf{D}$ is proportional to $\mathbf{E}$ and $\mathbf{B}$ is proportional to $\mathbf{H}$,* then the time derivatives on the right can be written as

$$\mathbf{E} \cdot \frac{\partial \mathbf{D}}{\partial t} = \mathbf{E} \cdot \frac{\partial}{\partial t} \epsilon \mathbf{E} = \tfrac{1}{2}\epsilon \frac{\partial}{\partial t} \mathbf{E}^2 = \frac{\partial}{\partial t} \tfrac{1}{2}\mathbf{E} \cdot \mathbf{D}$$

and

$$\mathbf{H} \cdot \frac{\partial \mathbf{B}}{\partial t} = \mathbf{H} \cdot \frac{\partial}{\partial t} \mu \mathbf{H} = \tfrac{1}{2}\mu \frac{\partial}{\partial t} \mathbf{H}^2 = \frac{\partial}{\partial t} \tfrac{1}{2}\mathbf{H} \cdot \mathbf{B}.$$

Using this relationship, Eq. (16–17) takes the form

$$\nabla \cdot (\mathbf{E} \times \mathbf{H}) = -\frac{\partial}{\partial t} \tfrac{1}{2}(\mathbf{E} \cdot \mathbf{D} + \mathbf{B} \cdot \mathbf{H}) - \mathbf{J} \cdot \mathbf{E}. \qquad (16\text{--}18)$$

---

* A medium is linear and nondispersive if $\mathbf{B} = \mu\mathbf{H}$ and $\mathbf{D} = \epsilon\mathbf{E}$, with μ and ϵ quantities that are independent of the field variables and that do not depend explicitly on the time. A notable exception to linearity occurs in the case of ferromagnetism, where the relationship between the magnetic induction and the magnetic intensity depends not only on the magnetic intensity but also on the past history of the specimen.

It should, however, be noted that anisotropy alone does not invalidate the expressions

$$\mathbf{E} \cdot \frac{\partial \mathbf{D}}{\partial t} = \frac{1}{2} \frac{\partial}{\partial t} (\mathbf{E} \cdot \mathbf{D}) \qquad \text{and} \qquad \mathbf{H} \cdot \frac{\partial \mathbf{B}}{\partial t} = \frac{1}{2} \frac{\partial}{\partial t} (\mathbf{H} \cdot \mathbf{B}).$$

In the case of anisotropic media, the relationship between $\mathbf{E}$ and $\mathbf{D}$ can be written as

$$D_i = \sum_{j=1}^{3} \epsilon_{ij} E_j.$$

Consequently,

$$\frac{1}{2} \frac{\partial}{\partial t} (\mathbf{E} \cdot \mathbf{D}) = \frac{1}{2} \sum_{i=1}^{3} \sum_{j=1}^{3} \epsilon_{ij} \left(E_i \frac{\partial E_j}{\partial t} + \frac{\partial E_i}{\partial t} E_j \right).$$

A simple argument based on the conservation of energy (Wooster, *Crystal Physics*, Cambridge University Press, 1938, p. 277) shows that $\epsilon_{ij} = \epsilon_{ji}$. Using this result to interchange i and j in the last term, we have

$$\frac{1}{2} \frac{\partial}{\partial t} (\mathbf{E} \cdot \mathbf{D}) = \sum_{i=1}^{3} \sum_{j=1}^{3} E_i \epsilon_{ij} \frac{\partial E_j}{\partial t}.$$

If $[\epsilon_{ij}]$ is a set of constants independent of $\mathbf{E}$ and of t, then

$$\frac{1}{2} \frac{\partial}{\partial t} (\mathbf{E} \cdot \mathbf{D}) = \sum_{i=1}^{3} E_i \frac{\partial}{\partial t} \sum_{j=1}^{3} \epsilon_{ij} E_j = \sum_{i=1}^{3} E_i \frac{\partial D_i}{\partial t} = \mathbf{E} \cdot \frac{\partial \mathbf{D}}{\partial t}.$$

Thus it is seen that anisotropy alone does not restrict the derivation.

The first term on the right is the time derivative of the sum of the electric and magnetic energy densities; the second term is in many cases, in particular if $\mathbf{J} = g\mathbf{E}$, just the negative of the Joule heating rate per unit volume.

Integrating over a fixed volume V bounded by the surface S gives

$$\int_V \boldsymbol{\nabla} \cdot (\mathbf{E} \times \mathbf{H}) \, dv = -\frac{d}{dt} \int_V \tfrac{1}{2}(\mathbf{E} \cdot \mathbf{D} + \mathbf{B} \cdot \mathbf{H}) \, dv - \int_V \mathbf{J} \cdot \mathbf{E} \, dv. \qquad (16\text{--}19)$$

Applying the divergence theorem to the left side, we obtain

$$\oint_S \mathbf{E} \times \mathbf{H} \cdot \mathbf{n} \, da = -\frac{d}{dt} \int_V \tfrac{1}{2}(\mathbf{E} \cdot \mathbf{D} + \mathbf{B} \cdot \mathbf{H}) \, dv - \int_V \mathbf{J} \cdot \mathbf{E} \, dv.$$

Rewriting this equation:

$$-\int_V \mathbf{J} \cdot \mathbf{E} \, dv = \frac{d}{dt} \int_V \tfrac{1}{2}(\mathbf{E} \cdot \mathbf{D} + \mathbf{B} \cdot \mathbf{H}) \, dv + \oint_S \mathbf{E} \times \mathbf{H} \cdot \mathbf{n} \, da, \qquad (16\text{--}20)$$

makes it clear that the $\mathbf{J} \cdot \mathbf{E}$ term is comprised of two parts: the rate of change of electromagnetic energy stored in V, and a surface integral. The left side of Eq. (16–20) is the power transferred *into* the electromagnetic field through the motion of free charge in volume V. If there are no sources of emf in V, then the left side of Eq. (16–20) is negative and equal to minus the Joule heat production per unit time. In certain circumstances, however, the left side of Eq. (16–20) may be positive. Suppose that a charged particle q moves with constant velocity $\mathbf{v}$ under the combined influence of mechanical, electric, and magnetic forces; the rate at which the mechanical force does work on the particle is

$$\mathbf{F}_m \cdot \mathbf{v} = -q(\mathbf{E} + \mathbf{v} \times \mathbf{B}) \cdot \mathbf{v} = -q\mathbf{E} \cdot \mathbf{v}.$$

But according to Eq. (7–4) the current density is defined by

$$\mathbf{J} = \sum_i N_i q_i \mathbf{v}_i;$$

thus the rate at which mechanical work is done (per unit volume) is

$$\sum_i N_i \mathbf{F}_m \cdot \mathbf{v}_i = -\mathbf{E} \cdot \mathbf{J},$$

and this power density is transferred into the electromagnetic field.

Since the surface integral in (16–20) involves only the electric and magnetic fields, it is feasible to interpret this term as the rate of energy flow across the surface. Equation (16–20) thus expresses the conservation of energy in a fixed volume V.

Let us return to the corresponding differential equation (16–18), which will express the local conservation of energy at a point. If we make the abbreviations

$$\mathbf{S} = \mathbf{E} \times \mathbf{H}, \qquad\qquad\qquad (16\text{--}21)$$

$$u = \tfrac{1}{2}(\mathbf{E} \cdot \mathbf{D} + \mathbf{B} \cdot \mathbf{H}), \qquad\qquad\qquad (16\text{--}22)$$

then Eq. (16–18) implies that at any point

$$\mathbf{V} \cdot \mathbf{S} + \frac{\partial u}{\partial t} = -\mathbf{J} \cdot \mathbf{E}. \tag{16–23}$$

There is no doubt that $\mathbf{J} \cdot \mathbf{E}$ is the work done by the local field on charged particles per unit volume. Previously u was interpreted as the energy density of the electric and magnetic fields. If $\mathbf{V} \cdot \mathbf{S} = 0$, Eq. (16–23) would therefore express local conservation of energy: The rate of change of field energy equals power dissipation per unit volume at each point. If, on the other hand, $\mathbf{V} \cdot \mathbf{S} \neq 0$ but $\mathbf{J} \cdot \mathbf{E} = 0$ (as for example in a nonconducting medium), then

$$\mathbf{V} \cdot \mathbf{S} + \frac{\partial u}{\partial t} = 0. \tag{16–24}$$

This has exactly the mathematical form of the equation of continuity (16–8) for charge, except that the energy density u takes the place of the charge density ρ. If (16–24) is still to describe conservation of energy, $\mathbf{V} \cdot \mathbf{S}$ must represent the divergence of an energy current density or, in other words, of a rate of energy flow per unit area. One usually treats $\mathbf{S} = \mathbf{E} \times \mathbf{H}$ itself, known as the *Poynting vector*, as the local energy flow per unit time per unit area.* We shall use these interpretations of u and $\mathbf{S}$, while recognizing that it is only their time derivative and divergence, respectively, whose interpretations are directly required by Maxwell's equations. It is usually only the latter that are physically measurable anyway. In any case, Eq. (16–23) expresses energy conservation locally, as does Eq. (16–20) in integral form.

16–4 THE WAVE EQUATION

One of the most important consequences of Maxwell's equations is the equations for electromagnetic wave propagation in a linear medium. The wave equation for $\mathbf{H}$ is derived by taking the curl of (16–10):

$$\mathbf{V} \times \mathbf{V} \times \mathbf{H} = \mathbf{V} \times \mathbf{J} + \mathbf{V} \times \frac{\partial \mathbf{D}}{\partial t}.$$

Putting $\mathbf{D} = \epsilon \mathbf{E}$ and $\mathbf{J} = g\mathbf{E}$ and assuming g and ϵ to be constants, we obtain

$$\mathbf{V} \times \mathbf{V} \times \mathbf{H} = g\mathbf{V} \times \mathbf{E} + \epsilon \frac{\partial}{\partial t} \mathbf{V} \times \mathbf{E}.$$

The order of time and space differentiation can be interchanged if $\mathbf{E}$ is a sufficiently well-behaved function, as we assume to be the case. Equation (16–11) can now be used to eliminate $\mathbf{V} \times \mathbf{E}$, yielding

$$\mathbf{V} \times \mathbf{V} \times \mathbf{H} = -g\mu \frac{\partial \mathbf{H}}{\partial t} - \epsilon\mu \frac{\partial^2 \mathbf{H}}{\partial t^2}, \tag{16–25}$$

---

* There is a continuing controversy over this point. For a recent discussion, see W. H. Furry, *Am. J. Phys.*, vol. 37, p. 621 (1969).

where $\mathbf{B} = \mu\mathbf{H}$, with μ a constant, has been used. The vector identity

$$\boldsymbol{\nabla} \times \boldsymbol{\nabla} \times = \boldsymbol{\nabla}\,\boldsymbol{\nabla}\cdot - \nabla^2 \tag{16-26}$$

is now used to obtain

$$\boldsymbol{\nabla}\,\boldsymbol{\nabla}\cdot\mathbf{H} - \nabla^2\mathbf{H} = -g\mu\,\frac{\partial\mathbf{H}}{\partial t} - \epsilon\mu\,\frac{\partial^2\mathbf{H}}{\partial t^2}. \tag{16-27}$$

Since μ is a constant,

$$\boldsymbol{\nabla}\cdot\mathbf{H} = \frac{1}{\mu}\,\boldsymbol{\nabla}\cdot\mathbf{B} = 0;$$

consequently, the first term on the left side of Eq. (16–27) vanishes. The final wave equation is

$$\nabla^2\mathbf{H} - \epsilon\mu\,\frac{\partial^2\mathbf{H}}{\partial t^2} - g\mu\,\frac{\partial\mathbf{H}}{\partial t} = 0. \tag{16-28}$$

The vector $\mathbf{E}$ satisfies the same wave equation, as is readily seen by first taking the curl of Eq. (16–11):

$$\boldsymbol{\nabla} \times \boldsymbol{\nabla} \times \mathbf{E} = -\boldsymbol{\nabla} \times \frac{\partial\mathbf{B}}{\partial t}.$$

Using Eq. (16–10) to eliminate the magnetic field and treating g, μ, and ϵ as constants yields

$$\boldsymbol{\nabla} \times \boldsymbol{\nabla} \times \mathbf{E} = -g\mu\,\frac{\partial\mathbf{E}}{\partial t} - \epsilon\mu\,\frac{\partial^2\mathbf{E}}{\partial t^2}.$$

Applying the vector identity (16–26) and restricting the application of the equation to a charge-free medium so that $\boldsymbol{\nabla}\cdot\mathbf{D} = 0$ gives

$$\nabla^2\mathbf{E} - \epsilon\mu\,\frac{\partial^2\mathbf{E}}{\partial t^2} - g\mu\,\frac{\partial\mathbf{E}}{\partial t} = 0. \tag{16-29}$$

The wave equations derived above govern the electromagnetic field in a homogeneous, linear medium in which the charge density is zero, whether this medium is conducting or nonconducting. However, it is not enough that these equations be satisfied; Maxwell's equations must also be satisfied. It is clear that Eqs. (16–28) and (16–29) are a necessary consequence of Maxwell's equation, but the converse is not true. In solving the wave equations, special care must be used to obtain solutions to Maxwell's equations. One method that works very well for monochromatic waves is to obtain a solution for $\mathbf{E}$. The curl of $\mathbf{E}$ then gives the time derivative of $\mathbf{B}$, which for monochromatic waves is sufficiently simply related to $\mathbf{B}$ so that $\mathbf{B}$ can be easily found.

Monochromatic waves may be described as waves that are characterized by a single frequency. The methods of complex variable analysis afford a convenient way of treating such waves. The time dependence of the field (for definiteness we take the vector $\mathbf{E}$) is taken to be as $e^{-i\omega t}$, so that

$$\mathbf{E}(\mathbf{r}, t) = \mathbf{E}(\mathbf{r})e^{-i\omega t}. \tag{16-30}$$

It must be remembered that the physical electric field is obtained by taking the real part* of Eq. (16–30); furthermore $\mathbf{E}(\mathbf{r})$ is in general complex, so that the actual electric field is proportional to $\cos(\omega t + \phi)$ where ϕ is the phase of $\mathbf{E}(\mathbf{r})$. Using Eq. (16–30) in Eq. (16–29) gives

$$e^{-i\omega t}\{\nabla^2\mathbf{E} + \omega^2\epsilon\mu\mathbf{E} + i\omega g\mu\mathbf{E}\} = 0 \tag{16-31}$$

for the equation governing the spatial variation of the electric field (the common factor $e^{-i\omega t}$ can, of course, be dropped). The next task is to solve Eq. (16–31) in various special cases of interest to determine the spatial variation of the electro-magnetic field. This will be treated in the following chapter; here let us merely look at some of the simplest possible cases.

First, suppose that the "medium" is empty space, so that $g = 0$, $\epsilon = \epsilon_0$, $\mu = \mu_0$. Further, suppose $\mathbf{E}(\mathbf{r})$ varies in only one dimension, say the z-direction, and is independent of x and y. Then Eq. (16–31) becomes

$$\frac{d^2\mathbf{E}(z)}{dz^2} + (\omega/c)^2\mathbf{E} = 0, \tag{16-32}$$

where we have written

$$\epsilon_0\mu_0 = 1/c^2,$$

as was suggested in Chapter 8 for dimensional reasons; c has the dimensions of a velocity. This equation (Helmholtz equation) is mathematically the same as the harmonic oscillator equation and has solutions

$$\mathbf{E}(z) = \mathbf{E}_0\,e^{\pm i\kappa z},$$

where $\mathbf{E}_0$ is a constant vector, provided that

$$\kappa = \omega/c. \tag{16-33}$$

Putting this $\mathbf{E}(\mathbf{r})$ into Eq. (16–30), we get the full solution

$$\mathbf{E}(\mathbf{r}, t) = \mathbf{E}_0\,e^{-i(\omega t \mp \kappa z)} \tag{16-34}$$

or, taking the real part,

$$\mathbf{E}(\mathbf{r}, t) = \mathbf{E}_0\,\cos(\omega t \mp \kappa z). \tag{16-35a}$$

With (16–33), an equivalent form is

$$\mathbf{E}(\mathbf{r}, t) = \mathbf{E}_0\,\cos\omega(t \mp z/c). \tag{16-35b}$$

---

* As discussed in Chapter 13, one goes from the convenient mathematical description in terms of complex variables to the physical quantities by taking either the real or imaginary part of the complex quantity. The choice of real or imaginary part is quite arbitrary. The two choices differ only by a phase shift of $\pi/2$; however, one must always make the same choice in a given problem. In this and the following chapters the real part of complex quantities will represent the physical quantities unless otherwise explicitly noted.

This represents a sinusoidal wave traveling to the right or left in the z-direction (depending on whether the minus or plus sign is used). The velocity of propagation of the wave is c. If light is a form of electromagnetic radiation, then the Maxwell equations *predict* that $c = 1/\sqrt{\epsilon_0 \mu_0} = 2.9979 \times 10^8$ m/s is the velocity of light in vacuum. Although this result is what we anticipated, when Maxwell first announced the result it was considered a great triumph of his theory, since up to that time the electromagnetic nature of light was only a speculation. The form (16–35a) shows that the wave frequency is $f = \omega/2\pi$ and the wavelength is $\lambda = 2\pi/\kappa$. Thus Eq. (16–33) is the familiar result for a wave,

$$\lambda f = c.$$

In a nonconducting, nonmagnetic dielectric, we still have $g = 0$, $\mu = \mu_0$, but now $\epsilon = K\epsilon_0$. The preceding derivation will carry through just the same, except that now Eq. (16–33) becomes

$$\kappa = \sqrt{K}\,\omega/c. \tag{16–33a}$$

Defining $n = \sqrt{K}$, we see that the results are the same as in vacuum, except that the velocity of wave propagation is now c/n instead of c. The quantity n is called the *index of refraction* of the dielectric medium; for vacuum $n = 1$. This accounts for refractive effects in transparent materials, as will be seen below.

If the medium is conducting, $g > 0$, the third term in Eq. (16–31) must be retained. When g is small the result will be merely that the wave is damped, as we shall see in the next chapter. By small g, we mean that the third term of Eq. (16–31) is small compared with the second term, which led to the wave solution, or

$$\omega g \mu \ll \omega^2 \epsilon \mu,$$

$$g \ll \omega \epsilon.$$

In the other extreme when $g \gg \omega\epsilon$, we may neglect the second term of Eq. (16–31). Again restricting attention to the one-dimensional case, we get

$$\frac{d^2\mathbf{E}(z)}{d^2 z} + i\omega g \mu \mathbf{E} = 0.$$

We can make the coefficient of $\mathbf{E}$ real if we assume that $\alpha = i\omega$ is real or, in other words, that the frequency is imaginary. Then, if

$$\kappa = \sqrt{\alpha g \mu},$$

the spatial dependence $\mathbf{E}(\mathbf{r})$ of the solution is just the same as before. The difference is that the time dependence (16–30) becomes

$$\mathbf{E}(\mathbf{r}, t) = \mathbf{E}(\mathbf{r})e^{-\alpha t}.$$

That is, the field simply decays exponentially with time, instead of oscillating in a wave-like manner. The transition between the decaying and the wave behavior occurs when

$$|\omega| = |\alpha| \cong \left| \frac{g}{\epsilon} \right| = 1/t_c,$$

where t_c is the relaxation time of the material that was discussed in Chapter 7. (We repeat that caution is needed when this condition is applied to a metal, since g/ϵ is itself strongly dependent on ω.)

Finally, by tracing the derivation of Eq. (16–31) back to Maxwell's equations, we notice that the second term, or $\partial^2 \mathbf{E}/\partial t^2$ in Eq. (16–29), derives from the displacement current $\partial \mathbf{D}/\partial t$ in Eq. (16–10), whereas the third term, or $\partial \mathbf{E}/\partial t$ in Eq. (16–29), derives from the transport current $\mathbf{J}$ in Eq. (16–10). Thus the very existence of electromagnetic wave propagation depends on Maxwell's introduction of the displacement current. Without it, only exponential decay of the fields could occur.

16–5 BOUNDARY CONDITIONS

The boundary conditions that must be satisfied by the electric and magnetic fields at an interface between two media are deduced from Maxwell's equations exactly as in the static case. The most straightforward and universal boundary condition applies to the magnetic induction $\mathbf{B}$, which satisfies the Maxwell equation

$$\nabla \cdot \mathbf{B} = 0. \tag{16–36}$$

At any interface between two media a pillbox-like surface may be constructed as shown in Fig. 16–2. The divergence theorem may be applied to the divergence of $\mathbf{B}$ over the volume enclosed by this surface, to obtain

$$\oint_S \mathbf{B} \cdot \mathbf{n} \, da = \int_{S_1} \mathbf{B} \cdot \mathbf{n}_1 \, da + \int_{S_2} \mathbf{B} \cdot \mathbf{n}_2 \, da + \int_{S_3} \mathbf{B} \cdot \mathbf{n}_3 \, da = 0. \tag{16–37}$$

If $\mathbf{B}$ is bounded, letting h approach zero causes the last term to vanish and S_1 to approach S_2 geometrically. Taking account of the opposite directions of $\mathbf{n}_1$ and $\mathbf{n}_2$, it is quickly concluded that

$$B_{1n} = B_{2n}, \tag{16–38}$$

exactly as in the static case.

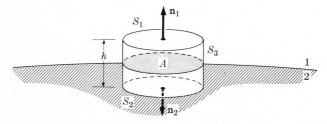

Figure 16–2 A pillbox-shaped surface at the interface between two media may be used to obtain boundary conditions on the field vectors.

The tangential component of the electric field can be treated in an equally simple way. The basic equation is again one of Maxwell's equations,

$$\mathbf{V} \times \mathbf{E} + \frac{\partial \mathbf{B}}{\partial t} = 0. \tag{16-39}$$

Integration of this equation over the surface bounded by a rectangular loop such as that shown in Fig. 16-3 yields

$$\int_S \mathbf{V} \times \mathbf{E} \cdot \mathbf{n}\, da = -\int_S \frac{\partial \mathbf{B}}{\partial t} \cdot \mathbf{n}\, da, \tag{16-40}$$

and applying Stokes's theorem to the left side gives

$$lE_{1t} - lE_{2t} + h_1 E_{1n} + h_2 E_{2n} - h_1 E'_{1n} - h_2 E'_{2n} = -\int_S \frac{\partial \mathbf{B}}{\partial t} \cdot \mathbf{n}\, da. \tag{16-41}$$

If the loop is now shrunk by letting h_1 and h_2 go to zero, the last four terms on the left vanish, as does the right-hand side, provided only that $\partial \mathbf{B}/\partial t$ is bounded. The resulting equation contains l as a common factor; dropping this gives

$$E_{1t} = E_{2t}. \tag{16-42}$$

Thus the tangential component of $\mathbf{E}$ must be continuous across the interface.

Figure 16-3 The rectangular path shown at the interface between two media may be used to obtain boundary conditions on the field vectors.

The boundary condition on the normal component of the electric displacement is more complex; however, it too is derived from one of Maxwell's equations. The appropriate equation in this case is

$$\mathbf{V} \cdot \mathbf{D} = \rho. \tag{16-43}$$

If we construct a pillbox-shaped volume, as shown in Fig. 16-2, and integrate (16-43) over this volume, we obtain

$$\int_V \mathbf{V} \cdot \mathbf{D}\, dv = \int_V \rho\, dv.$$

Applying the divergence theorem and letting h go to zero, we find

$$(D_{1n} - D_{2n})A = \sigma A, \tag{16-44}$$

where σ is the surface charge density on the interface. The fact that, in general, σ is not zero introduces some complexity in this boundary condition; however, noting that charge must be conserved, that is, that

$$\mathbf{V} \cdot \mathbf{J} = -\frac{\partial \rho}{\partial t}, \qquad (16\text{–}45)$$

makes possible certain simplifications. If we integrate this equation as we did Eq. (16–43) and shrink the pillbox in the same way, we obtain

$$J_{1n} - J_{2n} = -\frac{\partial \sigma}{\partial t}. \qquad (16\text{–}46)$$

If only monochromatic radiation is considered, the surface charge density must vary as $e^{-i\omega t}$; therefore the right side of Eq. (16–46) can be written as $i\omega\sigma$. Using the constitutive relations $\mathbf{D} = \epsilon\mathbf{E}$, $\mathbf{J} = g\mathbf{E}$ puts equations (16–44) and (16–46) in the form

$$\epsilon_1 E_{1n} - \epsilon_2 E_{2n} = \sigma, \qquad (16\text{–}47)$$

$$g_1 E_{1n} - g_2 E_{2n} = i\omega\sigma. \qquad (16\text{–}48)$$

Several cases of practical interest may be noted. If σ is zero, then

$$\frac{\epsilon_1}{g_1} = \frac{\epsilon_2}{g_2},$$

which can be true for appropriately chosen materials or, alternatively, if $g_1 = g_2 = 0$, or ∞. The case where both conductivities are infinite is not of great interest; however, the case where both conductivities vanish is approximately realized at the boundary between two good dielectrics. If σ is not zero, which is perhaps a more common case, then it may be eliminated from Eqs. (16–47) and (16–48). The result of this elimination is

$$\left(\epsilon_1 + i\frac{g_1}{\omega}\right)E_{1n} - \left(\epsilon_2 + i\frac{g_2}{\omega}\right)E_{2n} = 0. \qquad (16\text{–}49)$$

A final interesting case occurs when one conductivity, say g_2, is infinite. In this case E_{2n} must vanish and E_{1n} must equal σ/ϵ_1 in order for Eqs. (16–48) and (16–47) to be satisfied.

The final boundary condition is that imposed on the tangential component of the magnetic intensity $\mathbf{H}$. This boundary condition is obtained by integrating the Maxwell equation

$$\mathbf{V} \times \mathbf{H} = \frac{\partial \mathbf{D}}{\partial t} + \mathbf{J} \qquad (16\text{–}50)$$

over the area enclosed by a loop such as that shown in Fig. 16–3. If this is done and the loop is shrunk as before, the resulting boundary condition is

$$H_{1t} - H_{2t} = j_\perp, \qquad (16\text{–}51)$$

where $j_\perp$ is the component of the surface current density perpendicular to the direction of the **H**-component that is being matched. The idea of a surface current density is closely analogous to that of a surface charge density—it represents a finite current in an infinitesimal layer. The surface current density is zero unless the conductivity is infinite; hence, for finite conductivity,

$$H_{1t} = H_{2t}. \tag{16-52}$$

That is, unless one medium has infinite conductivity the tangential component of **H** is continuous. If the conductivity of medium 2 is infinite, then, as has already been shown, $E_{2n} = 0$. A more general result can be obtained by considering the Maxwell equation (16–50) as applied to medium 2:

$$\nabla \times \mathbf{H}_2 - \frac{\partial \mathbf{D}_2}{\partial t} = \mathbf{J}_2. \tag{16-53}$$

Using the constitutive relations and assuming that $\mathbf{E}_2$ varies with time as $e^{-i\omega t}$ yields

$$\mathbf{E}_2 = \frac{1}{g_2 - i\omega\epsilon_2} \nabla \times \mathbf{H}_2. \tag{16-54}$$

If the reasonable assumption that $\mathbf{H}_2$ is both bounded and differentiable is made, then Eq. (16–54) implies that $\mathbf{E}_2$ is zero in a medium of infinite conductivity. With the same assumptions as were made above,

$$\mathbf{H}_2 = \frac{1}{i\omega\mu_2} \nabla \times \mathbf{E}_2, \tag{16-55}$$

and the vanishing of $\mathbf{E}_2$ also implies the vanishing of $\mathbf{H}_2$. If $\mathbf{H}_2$ vanishes, then the boundary condition on the tangential component of **H** at an interface at which one medium has infinite conductivity is

$$H_{1t} = j_\perp. \tag{16-56}$$

The boundary conditions have now been obtained; for convenient reference they are tabulated in Table 16–1 for $g = 0$, $g = \infty$, and arbitrary g.

Table 16–1 Boundary Conditions

g	E_t	D_n	H_t	B_n
$g_1 = g_2 = 0$	$E_{1t} = E_{2t}$	$D_{1n} = D_{2n}$	$H_{1t} = H_{2t}$	$B_{1n} = B_{2n}$
$g_2 = \infty$	$E_{2t} = 0$	$D_{2n} = 0$	$H_{2t} = 0$	$B_{2n} = 0$
	$E_{1t} = 0$	$D_{1n} = \sigma$	$H_{1t} = j_\perp$	$B_{1n} = 0$
g_1, g_2 arb. $\neq \infty$	$E_{1t} = E_{2t}$	$\left(\epsilon_1 + i\dfrac{g_1}{\omega}\right)E_{1n}$ $= \left(\epsilon_2 + i\dfrac{g_2}{\omega}\right)E_{2n}$	$H_{1t} = H_{2t}$	$B_{1n} = B_{2n}$

16–6 THE WAVE EQUATION WITH SOURCES

In the preceding sections it was shown that Maxwell's equations predict the propagation of electromagnetic waves through a linear medium and also that the fields must match at an interface between two different media according to appropriate boundary conditions. It was assumed that the charge density ρ in the medium was zero and that the only current density $\mathbf{J}$ arose from the passive response of an ohmic medium to the electric field of the wave. We did not inquire how these waves were produced, but we shall ultimately find that they are fields produced by distant source charges that undergo accelerated motion.

The problem now is to consider prescribed charge and current distributions, $\rho(\mathbf{r}, t)$ and $\mathbf{J}(\mathbf{r}, t)$, and find the fields produced by them. There are several ways of approaching the problem, of which the most fruitful is the potential approach, which is developed analogously to the procedures used in electrostatics and magnetostatics. Since the magnetic induction has zero divergence it may always be represented as the curl of a vector potential, that is,

$$\mathbf{B} = \nabla \times \mathbf{A}. \tag{16–57}$$

Using this expression for $\mathbf{B}$ in Eq. (16–11) gives

$$\nabla \times \mathbf{E} + \frac{\partial}{\partial t} \nabla \times \mathbf{A} = 0. \tag{16–58}$$

Assuming sufficient continuity of the fields to interchange the spatial and temporal differentiations, this can be written

$$\nabla \times \left[\mathbf{E} + \frac{\partial \mathbf{A}}{\partial t} \right] = 0. \tag{16–59}$$

The vector $\mathbf{E} + \partial \mathbf{A}/\partial t$ thus has zero curl and can be written as the gradient of a scalar:

$$\mathbf{E} = -\nabla \varphi - \frac{\partial \mathbf{A}}{\partial t}. \tag{16–60}$$

Equations (16–57) and (16–60) give the electric and magnetic fields in terms of a vector potential $\mathbf{A}$ and a scalar potential φ. These potentials satisfy wave equations which are very similar to those satisfied by the fields. The wave equation for $\mathbf{A}$ is derived by substituting the expressions given in Eqs. (16–57) and (16–60) for $\mathbf{B}$ and $\mathbf{E}$ into Eq. (16–10), with the result

$$\frac{1}{\mu} \nabla \times \nabla \times \mathbf{A} + \epsilon \frac{\partial}{\partial t} \left[\nabla \varphi + \frac{\partial \mathbf{A}}{\partial t} \right] = \mathbf{J}. \tag{16–61}$$

Writing $\nabla \nabla \cdot - \nabla^2$ for $\nabla \times \nabla \times$ and multiplying by μ gives

$$-\nabla^2 \mathbf{A} + \epsilon\mu \frac{\partial^2 \mathbf{A}}{\partial t^2} + \nabla \nabla \cdot \mathbf{A} + \epsilon\mu \nabla \frac{\partial \varphi}{\partial t} = \mu \mathbf{J}. \tag{16–62}$$

Until now only the curl of **A** has been specified; the choice of the divergence of **A** is still arbitrary. It is clear from Eq. (16–62) that imposing the so-called *Lorentz condition*,

$$\mathbf{V} \cdot \mathbf{A} + \epsilon\mu \frac{\partial \varphi}{\partial t} = 0, \tag{16-63}$$

results in a considerable simplification. If this condition is satisfied, then **A** satisfies the wave equation

$$\nabla^2 \mathbf{A} - \epsilon\mu \frac{\partial^2 \mathbf{A}}{\partial t^2} = -\mu\mathbf{J}. \tag{16-64}$$

Furthermore, using Eq. (16–60) in (16–12) gives

$$-\epsilon \left[\mathbf{V} \cdot \mathbf{V}\varphi + \mathbf{V} \cdot \frac{\partial \mathbf{A}}{\partial t} \right] = \rho. \tag{16-65}$$

Interchanging the order of the divergence and the time derivative operating on **A** and using the Lorentz condition (Eq. 16–63) leads to

$$\nabla^2 \varphi - \epsilon\mu \frac{\partial^2 \varphi}{\partial t^2} = -\frac{1}{\epsilon}\rho. \tag{16-66}$$

Thus, by imposing the Lorentz condition, both the scalar and vector potentials are forced to satisfy inhomogeneous wave equations of similar forms.

The problem of finding the general solution of the inhomogeneous scalar wave equation is analogous to finding the general solution of Poisson's equation. In the latter case, it will be recalled, the general solution consists of a particular solution of the inhomogeneous equation plus a general solution of the homogeneous equation. The inclusion of the solutions of the homogeneous equation provides the means for satisfying arbitrary appropriate boundary conditions, while the particular solution ensures that the total function satisfies the inhomogeneous equation. Exactly the same considerations apply to the inhomogeneous wave equation—the general solution consists of a particular solution plus a general solution of the homogeneous equation. Methods for finding certain solutions of the homogeneous equation will be treated in Chapter 17. These methods may be extended and supplemented to yield solutions to almost any solvable problem. Approximate methods are available for problems that cannot be solved in terms of known functions. It remains, then, to find the needed particular solution of the inhomogeneous equation. The inhomogeneous scalar wave equation (16–66) reduces in the static case, $\partial\varphi/\partial t = 0$, to Poisson's equation, a particular solution of which we know from Eq. (3–1) (for vacuum):

$$\varphi(\mathbf{r}) = \frac{1}{4\pi\epsilon_0} \int_V \frac{\rho(\mathbf{r}')}{|\mathbf{r} - \mathbf{r}'|} \, dv' \tag{16-67}$$

The vector wave equation (16–64) has an analogous solution in the static (vacuum) case, given in Eq. (8–61). Unfortunately, we do not get solutions for the time dependent case simply by inserting $\rho(\mathbf{r}', t)$ and $\mathbf{J}(\mathbf{r}', t)$ into the static solutions, for reasons that we shall see.

Let us rewrite Eq. (16–66) for vacuum, using $1/\sqrt{\epsilon\mu} = c/n$, with the refractive index $n = 1$:

$$\nabla^2\varphi - \frac{1}{c^2}\frac{\partial^2\varphi}{\partial t^2} = -\frac{\rho}{\epsilon_0} \tag{16–68}$$

can be solved most readily by finding the solution for a point charge, and then later summing over all the charge elements $\rho\Delta v$ in the appropriate charge distribution. The most convenient location for the point charge is at the origin of coordinates. Thus the equation

$$\nabla^2\varphi - \frac{1}{c^2}\frac{\partial^2\varphi}{\partial t^2} = 0 \tag{16–69}$$

must be satisfied everywhere except at the origin, whereas in a small volume Δv surrounding the origin,

$$\int_{\Delta v} dv \left[\nabla^2\varphi - \frac{1}{c^2}\frac{\partial^2\varphi}{\partial t^2}\right] = -\frac{1}{\epsilon_0}q(t) \tag{16–70}$$

must be satisfied. The function $q(t)$ is supposed to represent an assumed point charge of magnitude q that is located at the origin at time t, simply as a mathematical device to solve the equation, without any assumption about where an actual conserved charge was at an earlier or later time. (It does not represent a physical moving point charge, and the resulting solution for φ is not the correct potential for a moving point charge. The latter is more complicated and will be treated in Chapter 21.) It is clear from the symmetry of the charge distribution that the spatial dependence of φ must be only on r. With this clue, an attempt to solve Eq. (16–69) may be made. Since φ does not depend on either the azimuthal angle or the colatitude, Eq. (16–69) becomes

$$\frac{1}{r^2}\frac{\partial}{\partial r}r^2\frac{\partial\varphi}{\partial r} - \frac{1}{c^2}\frac{\partial^2\varphi}{\partial t^2} = 0. \tag{16–71}$$

Now, by putting

$$\varphi(r, t) = \frac{\chi(r, t)}{r}, \tag{16–72}$$

Eq. (16–71) is converted to

$$\frac{\partial^2\chi}{\partial r^2} - \frac{1}{c^2}\frac{\partial^2\chi}{\partial t^2} = 0. \tag{16–73}$$

This equation is the one-dimensional wave equation that is satisfied by any function of $r - ct$, or $r + ct$. To verify this, let

$$u = r - ct$$

and let $f(u)$ be any function of u that can be twice differentiated; then

$$\frac{\partial f}{\partial r} = \frac{df}{du}\frac{\partial u}{\partial r} = \frac{df}{du}, \qquad \frac{\partial^2 f}{\partial r^2} = \frac{d^2 f}{du^2}\frac{\partial u}{\partial r} = \frac{d^2 f}{du^2} \qquad (16\text{--}74)$$

and

$$\frac{\partial f}{\partial t} = \frac{df}{du}\frac{\partial u}{\partial t} = -c\,\frac{df}{du}, \qquad \frac{\partial^2 f}{\partial t^2} = c^2\,\frac{d^2 f}{du^2} \qquad (16\text{--}75)$$

Substituting the results of Eqs. (16–74) and (16–75) into Eq. (16–73) verifies that any function of $(r - ct)$ that is twice differentiable is a solution of Eq. (16–73). A similar calculation verifies that a function of $(r + ct)$ is a solution. Thus

$$\chi = f(r - ct) + g(r + ct) \qquad (16\text{--}76)$$

is a very arbitrary solution of Eq. (16–73). It will be found that $g(r + ct)$ does not occur in our applications of the wave equation. For this reason it will be dropped, and only the first term of Eq. (16–76) will be carried, since this procedure simplifies the ensuing equations and causes no particular omissions. It may be noted that $f(r - ct)$ represents a wave propagating outward from the source charge q at the origin, while $g(r + ct)$ represents a wave propagating inward toward the source charge from infinity. We keep the former and drop the latter for somewhat the same reason we would keep a plane wave solution propagating to the right if we were to the right of the source and drop the one propagating leftwards.

A spherically symmetric solution of Eq. (16–69),

$$\varphi = \frac{f(r - ct)}{r}, \qquad (16\text{--}77)$$

is now available; furthermore, this solution contains an arbitrary function that may be chosen so that Eq. (16–70) is also satisfied. The proper choice is obtained by noting that for a static charge the potential compatible with Eqs. (16–69) and (16–70) is

$$\varphi = \frac{q}{4\pi\epsilon_0 r}. \qquad (16\text{--}78)$$

The functions (16–77) and (16–78) may be brought into concert by choosing

$$f(r - ct) = \frac{q(t - r/c)}{4\pi\epsilon_0}. \qquad (16\text{--}79)$$

The solution to Eqs. (16–69) and (16–70) is then

$$\varphi(r,\,t) = \frac{q(t - r/c)}{4\pi\epsilon_0 r}. \qquad (16\text{--}80)$$

With this result, we readily find that Eq. (16–66) is satisfied by

$$\varphi(\mathbf{r}, t) = \frac{1}{4\pi\epsilon_0} \int_V \frac{\rho(\mathbf{r}', t')}{|\mathbf{r} - \mathbf{r}'|} \, dv', \tag{16–81}$$

where $t' = t - |\mathbf{r} - \mathbf{r}'|/c$ is called the *retarded time*; φ is known as the *retarded scalar potential*.

The solution of Eq. (16–64) can be constructed in exactly the same way. The vectors $\mathbf{A}$ and $\mathbf{J}$ are first decomposed into rectangular components. The three resulting equations are closely analogous to Eq. (16–66), the x equation, for example, being

$$\nabla^2 A_x - \frac{1}{c^2} \frac{\partial^2 A_x}{\partial t^2} = -\mu_0 J_x. \tag{16–82}$$

Each of these equations may be solved exactly as was Eq. (16–66), giving, for example,

$$A_x(\mathbf{r}, t) = \frac{\mu_0}{4\pi} \int_V \frac{J_x(\mathbf{r}', t')}{|\mathbf{r} - \mathbf{r}'|} \, dv'. \tag{16–83}$$

These components are then combined to give

$$\mathbf{A}(\mathbf{r}, t) = \frac{\mu_0}{4\pi} \int_V \frac{\mathbf{J}(\mathbf{r}', t')}{|\mathbf{r} - \mathbf{r}'|} \, dv', \tag{16–84}$$

which is the *retarded vector potential*.

The physical interpretation of the retarded potentials is interesting. Eqs. (16–81) and (16–84) indicate that at a given point $\mathbf{r}$ and a given time t the potentials are determined by the charge and current that existed at other points in space $\mathbf{r}'$ at earlier times t'. The time appropriate to each source point is earlier than t by an amount equal to the time required to travel from source to field point $\mathbf{r}$ with velocity c. If, for example, a charge element located at the origin of coordinates were suddenly changed, then the effect of this change would not be felt at a distance r until a time r/c after the change was accomplished. The effect of the change propagates outward roughly as a spherical wavefront. (The actual situation is more complicated for a point charge because the charge density and current density are intimately related through $\nabla \cdot \mathbf{J} + \partial\rho/\partial t = 0$.)

Having found the scalar and vector potentials, we find the fields by applying the gradient to φ, and the time derivative and curl to $\mathbf{A}$. These operations are in principle straightforward; however, it will be seen that they are relatively complicated in practice.

In the above procedure it was essential to impose the Lorentz condition (16–63) on the potentials; otherwise it would not be the simple wave equations that they would have to satisfy. To see that we are always free to impose this condition, suppose that $\mathbf{A}$ and φ are a particular choice of potential functions that give the correct **E**- and **B**-fields according to Eqs. (16–57) and (16–60). Then, if we were to choose new potentials

$$\mathbf{A}' = \mathbf{A} + \nabla \xi, \qquad \varphi' = \varphi - \frac{\partial \xi}{\partial t}, \tag{16-85}$$

these would give exactly the same **E**- and **B**-fields when substituted into Eqs. (16–57) and (16–60), whatever we use for the function ξ, which is completely arbitrary. This change to new but physically equivalent potentials is called a *gauge transformation*. Now substituting $\mathbf{A}'$ and φ' into Eq. (16–63), we get, after rearranging terms, a scalar wave equation for ξ,

$$\nabla^2 \xi - \epsilon \mu \frac{\partial^2 \xi}{\partial t^2} = -\left(\nabla \cdot \mathbf{A} + \epsilon \mu \frac{\partial \varphi}{\partial t} \right),$$

as the condition that $\mathbf{A}'$, φ' should satisfy the Lorentz condition. Thus, if the original potentials satisfied the Lorentz condition, the new ones will also, provided that ξ satisfies the homogeneous scalar wave equation. If the original $\mathbf{A}$, φ did not, we can still find new potentials that will, by choosing ξ as a solution of the inhomogeneous scalar wave equation with

$$\nabla \cdot \mathbf{A} + \epsilon \mu \frac{\partial \varphi}{\partial t}$$

as the source term. We have just seen how to find such a solution. A choice of potentials that satisfies the Lorentz condition is called the *Lorentz gauge*. Other choices of gauge (i.e., other choices of $\nabla \cdot \mathbf{A}$) are useful in other circumstances.

With the development of the retarded potentials the basic work on radiation is completed. It remains to apply this material to the solution of practical problems. These are the concern of the next five chapters.

16–7 SUMMARY

This chapter contains the foundations of all classical electromagnetic theory. Maxwell's equations are the differential equations that determine (together with boundary conditions for a particular situation) the fields produced by sources of charge and current:

$$\nabla \cdot \mathbf{B} = 0, \qquad\qquad \nabla \cdot \mathbf{D} = \rho,$$

$$\nabla \times \mathbf{E} + \frac{\partial \mathbf{B}}{\partial t} = 0, \qquad \nabla \times \mathbf{H} - \frac{\partial \mathbf{D}}{\partial t} = \mathbf{J}.$$

The **E** and **B** fields are operationally defined by the Lorentz force

$$\mathbf{F} = q(\mathbf{E} + \mathbf{v} \times \mathbf{B}),$$

and the **D** and **H** fields are related to them by the constitutive equations of the medium, $\mathbf{D} = \mathbf{D}(\mathbf{E})$, $\mathbf{H} = \mathbf{H}(\mathbf{B})$.

The Maxwell equations have the following important consequences:

1. Electric charge is conserved, according to the equation of continuity

$$\mathbf{V} \cdot \mathbf{J} + \frac{\partial \rho}{\partial t} = 0.$$

2. Energy is conserved, according to

$$\mathbf{V} \cdot \mathbf{S} + \frac{\partial u}{\partial t} = -\mathbf{J} \cdot \mathbf{E},$$

where the field energy density is (in a linear medium)

$$u = \tfrac{1}{2}(\mathbf{E} \cdot \mathbf{D} + \mathbf{B} \cdot \mathbf{H})$$

and the energy flux per unit area is the Poynting vector

$$\mathbf{S} = \mathbf{E} \times \mathbf{H}.$$

3. Electromagnetic wave propagation can occur, with the velocity $c = 1/\sqrt{\epsilon_0 \mu_0}$ in vacuum, equal to the velocity of light.

4. Boundary conditions on the fields are determined at an interface between different media, the most important ones being that the tangential components of **E** and **H** are continuous.

5. The fields **E** and **B** are derivable from potential functions:

$$\mathbf{B} = \mathbf{V} \times \mathbf{A}, \qquad \mathbf{E} = -\mathbf{V}\varphi - \frac{\partial \mathbf{A}}{\partial t}.$$

6. The potentials satisfy the inhomogeneous wave equations

$$\nabla^2 \varphi - \epsilon\mu \frac{\partial^2 \varphi}{\partial t^2} = -\frac{1}{\epsilon} \rho,$$

$$\nabla^2 \mathbf{A} - \epsilon\mu \frac{\partial^2 \mathbf{A}}{\partial t^2} = -\mu\mathbf{J},$$

if the Lorentz condition

$$\mathbf{V} \cdot \mathbf{A} = -\epsilon\mu \frac{\partial \varphi}{\partial t}$$

is imposed. These will determine the generation of electromagnetic waves by prescribed charge and current distributions. Particular solutions (in vacuum) are

$$\varphi(\mathbf{r}, t) = \frac{1}{4\pi\epsilon_0} \int \frac{\rho(\mathbf{r}', t')}{|\mathbf{r} - \mathbf{r}'|} \, dv',$$

$$\mathbf{A}(\mathbf{r}, t) = \frac{\mu_0}{4\pi} \int \frac{\mathbf{J}(\mathbf{r}', t')}{|\mathbf{r} - \mathbf{r}'|} \, dv',$$

where

$$t' = t - \frac{|\mathbf{r} - \mathbf{r}'|}{c}$$

is the retarded time. These are called the *retarded potentials*.

PROBLEMS

16–1 A parallel-plate capacitor with plates having the shape of circular disks has the region between its plates filled with a dielectric of permittivity ϵ. The dielectric is imperfect, having a conductivity g. The capacitance of the capacitor is C. The capacitor is charged to a potential difference $\Delta\varphi$ and isolated. (a) Find the charge on the capacitor as a function of time. (b) Find the displacement current in the dielectric. (c) Find the magnetic field in the dielectric.

16–2 The Q of a dielectric medium is defined as the ratio of displacement current density to conduction current density. For monochromatic wave propagation, this reduces to $Q = \omega\epsilon/g$. Determine Q for quartz and for sulfur, at the following frequencies: $f = 1$, 10^6, 10^9 Hz.

16–3 Two circular plates of radius a separated by distance d form an ideal capacitor: Assume that the dielectric is a perfect insulator with uniform D-field (i.e., neglect the fringing field at the edge of the plates). The capacitor is being charged by a constant current I. (a) Find the H-field at a point P on the cylindrical surface of the dielectric. (b) Find the magnitude and direction of the Poynting vector $\mathbf{S}$ at P. (c) Integrate $\mathbf{S} \cdot \mathbf{n}$ over the cylindrical surface of the dielectric, and show that the result is equal to the time rate of change of the stored electrostatic energy.

16–4 A straight metal wire of conductivity g and cross-sectional area A carries a steady current I. Determine the direction and magnitude of the Poynting vector at the surface of the wire. Integrate the normal component of the Poynting vector over the surface of the wire for a segment of length L, and compare your result with the Joule heat produced in this segment.

16–5 Suppose that in a certain region there is an electrostatic field and also a magneto-static field. Show that although the Poynting vector may be nonzero, the surface integral of $\mathbf{S} \cdot \mathbf{n}$ vanishes over an arbitrary closed surface inside the region.

16–6 Given the one-dimensional wave equation

$$\frac{\partial^2 E}{\partial z^2} = \epsilon\mu \frac{\partial^2 E}{\partial t^2},$$

where E is the magnitude of the electric field vector. Assume that $\mathbf{E}$ has a constant direction, namely, the y-direction. By introducing the change of variables

$$\xi = t + \sqrt{\epsilon\mu}\, z,$$

$$\eta = t - \sqrt{\epsilon\mu}\, z,$$

show that the wave equation assumes a form which is easily integrated. Integrate the equation to obtain

$$E(z, t) = E_1(\xi) + E_2(\eta),$$

where E_1 and E_2 are arbitrary functions.

16–7 Given the electromagnetic wave

$$\mathbf{E} = \mathbf{i}E_0 \cos \omega(\sqrt{\epsilon\mu}\, z - t) + \mathbf{j}E_0 \sin \omega(\sqrt{\epsilon\mu}\, z - t),$$

where E_0 is a constant. Find the corresponding magnetic field $\mathbf{B}$ and the Poynting vector.

*16–8 Starting with an expression for the force per unit volume on a region of *free space* containing charges and currents:

$$\mathbf{F}_v = \rho\mathbf{E} + \mathbf{J} \times \mathbf{B},$$

and using the Maxwell equations and the vector identity of Eq. (14–24), show that

$$\mathbf{F}_v = -\epsilon_0 \frac{\partial}{\partial t}(\mathbf{E} \times \mathbf{B}) + \epsilon_0\mathbf{E}\,\boldsymbol{\nabla} \cdot \mathbf{E} - \frac{1}{2}\,\epsilon_0\,\boldsymbol{\nabla}\,(E^2) + \epsilon_0(\mathbf{E} \cdot \boldsymbol{\nabla})\mathbf{E}$$

$$+ \frac{1}{\mu_0}\,\mathbf{B}\,\boldsymbol{\nabla} \cdot \mathbf{B} - \frac{1}{2\mu_0}\,\boldsymbol{\nabla}\,(B^2) + \frac{1}{\mu_0}\,(\mathbf{B} \cdot \boldsymbol{\nabla})\mathbf{B}.$$

(The quantity $\epsilon_0\mathbf{E} \times \mathbf{B}$ is sometimes referred to as the momentum density of the electromagnetic field.)

16–9 Given a plane wave characterized by an E_x, B_y propagating in the positive z-direction,

$$\mathbf{E} = \mathbf{i}E_0 \sin \frac{2\pi}{\lambda}\,(z - ct).$$

Show that it is possible to take the scalar potential $\varphi = 0$, and find a possible vector potential $\mathbf{A}$ for which the Lorentz condition is satisfied.

16–10 Show that in free space with $\rho = 0$, $\mathbf{J} = 0$, the Maxwell equations are correctly obtained from a single vector function $\mathbf{A}$ satisfying

$$\boldsymbol{\nabla} \cdot \mathbf{A} = 0, \qquad \nabla^2\mathbf{A} - \frac{1}{c^2}\frac{\partial^2\mathbf{A}}{\partial t^2} = 0.$$

The gauge in which $\boldsymbol{\nabla} \cdot \mathbf{A} = 0$ is called the *Coulomb gauge*.

16–11 Show that in a linear conducting medium a suitable gauge can be chosen so that $\mathbf{A}$ and φ each satisfy the damped wave equation, Eq. (16–29). Assume that $\rho = 0$.

16–12 Given a medium in which $\rho = 0$, $\mathbf{J} = 0$, $\mu = \mu_0$, but where the polarization $\mathbf{P}$ is a given function of position and time: $\mathbf{P} = \mathbf{P}(x, y, z, t)$. Show that the Maxwell equations are correctly obtained from a single vector function $\mathbf{Z}$ (the Hertz vector), where $\mathbf{Z}$ satisfies the equation

$$\nabla^2\mathbf{Z} - \frac{1}{c^2}\frac{\partial^2\mathbf{Z}}{\partial t^2} = -\frac{\mathbf{P}}{\epsilon_0},$$

and

$$\mathbf{E} = \boldsymbol{\nabla} \times \boldsymbol{\nabla} \times \mathbf{Z} - \frac{1}{\epsilon_0}\,\mathbf{P}, \qquad \mathbf{B} = \frac{1}{c^2}\,\boldsymbol{\nabla} \times \frac{\partial\mathbf{Z}}{\partial t}.$$

16–13 Given a medium in which $\rho = 0$, $\mathbf{J} = 0$, $\epsilon = \epsilon_0$, but where the magnetization $\mathbf{M}(x, y, z, t)$ is a given function. Show that the Maxwell equations are correctly obtained from a single vector function $\mathbf{Y}$, where $\mathbf{Y}$ satisfies the equation

$$\nabla^2 \mathbf{Y} - \frac{1}{c^2} \frac{\partial^2 \mathbf{Y}}{\partial t^2} = -\mu_0 \mathbf{M}$$

and where

$$\mathbf{B} = \nabla \times \nabla \times \mathbf{Y}, \qquad \mathbf{E} = -\nabla \times \frac{\partial \mathbf{Y}}{\partial t}.$$

16–14 Show that Maxwell's equations for an isotropic, homogeneous, nonconducting, charge-free medium can be satisfied by taking either

1. $\mathbf{E} = $ real part of $\nabla \times \nabla \times (F\mathbf{a})$,

   $\mathbf{B} = $ real part of $\epsilon\mu \dfrac{\partial}{\partial t} \nabla \times (F\mathbf{a})$,

or

2. $\mathbf{B} = $ real part of $\nabla \times \nabla \times (F\mathbf{a})$,

   $\mathbf{E} = $ real part of $-\dfrac{\partial}{\partial t} \nabla \times (F\mathbf{a})$,

where $\mathbf{a}$ is a constant unit vector and F satisfies the scalar wave equation.

CHAPTER 17 Propagation of Electromagnetic Waves

Maxwell's equations have some special solutions that describe electromagnetic waves, as we saw by deriving the wave equation from the Maxwell equations. We will now consider these solutions in detail. We begin by considering the propagation of the waves through a linear medium, which we idealize to be infinite in extent. We leave until later chapters the question of how the waves are generated and how they enter the medium in the first place. The results will be applicable to radio waves, microwaves, thermal radiation, light, x-rays, etc. (although quantum mechanical effects may also be important at higher frequencies).

17–1 PLANE MONOCHROMATIC WAVES IN NONCONDUCTING MEDIA

The most easily treated solutions of Eq. (16–31) are those known as plane wave solutions. A plane wave is defined as a wave whose phase is the same, at a given instant, at all points in each *plane* perpendicular to some specified direction. If for example the specified direction is the z-direction, then **E** must have the same phase at all points that have the same z-value, i.e., all points on a plane parallel to the xy-plane. Thus the solution (16–34) that we already discussed is a plane wave solution, since $(\omega t - \kappa z)$ is a constant for a given t and z, no matter what the values of x and y. Plane waves traveling in the z-direction are adequate for problems in which the choice of z-direction is arbitrary; however, in many problems a system of axes is chosen for other reasons, for example because of boundary conditions. In such cases it is necessary to construct plane waves with arbitrary directions of propagation. Suppose a plane wave solution with direction of propagation **u** is to be constructed, where **u** is a unit vector. Then the variable z in the exponent must be replaced by $\mathbf{u} \cdot \mathbf{r}$, the projection of **r** in the **u** direction. Thus a plane wave with direction of propagation **u** is described by

$$e^{-i(\omega t - \kappa \mathbf{u} \cdot \mathbf{r})}.$$

We define a vector, called the *propagation vector*,

$$\boldsymbol{\kappa} = \kappa \mathbf{u}$$

and write the exponential space and time dependence of the plane wave as

$$e^{-i(\omega t - \boldsymbol{\kappa} \cdot \mathbf{r})}.$$

If $\mathbf{u} = \mathbf{k}$, the unit vector in the z-direction, then $\mathbf{u} \cdot \mathbf{r} = z$ as in the special case; but in every case the wavelength $\lambda = 2\pi/\kappa$.

The velocity of propagation of a plane monochromatic wave is precisely the velocity with which planes of constant phase move. Constant phase means, of course, that

$$\boldsymbol{\kappa} \cdot \mathbf{r} - \omega t = \text{constant.} \tag{17-1}$$

If $\boldsymbol{\kappa} \cdot \mathbf{r}$ is written $\kappa \xi$, with κ the magnitude of $\boldsymbol{\kappa}$ and ξ the projection of $\mathbf{r}$ in the $\boldsymbol{\kappa}$ direction, then Eq. (17–1) becomes

$$\kappa \xi - \omega t = \text{constant.}$$

Differentiating with respect to the time yields

$$v_p = \frac{d\xi}{dt} = \frac{\omega}{\kappa} = \frac{c}{n}, \tag{17-2}$$

where we have used the result of Eq. (16–33a) $\kappa = n\omega/c$. In free space

$$v_p = c = \frac{1}{\sqrt{\epsilon_0 \mu_0}} = 2.9979 \times 10^8 \text{ m/s.}$$

Now to obtain the detailed plane wave solutions for $\mathbf{E}$ and $\mathbf{B}$, we could go back to Eq. (16–31), but it is actually better to go all the way back to the Maxwell equations themselves. There are no prescribed charge or current distributions in the medium and the conductivity $g = 0$, so that the equations are

$$\nabla \cdot \mathbf{D} = 0, \tag{17-3}$$

$$\nabla \cdot \mathbf{B} = 0, \tag{17-4}$$

$$\nabla \times \mathbf{E} = -\frac{\partial \mathbf{B}}{\partial t}, \tag{17-5}$$

$$\nabla \times \mathbf{H} = \frac{\partial \mathbf{D}}{\partial t}. \tag{17-6}$$

From our previous discussion of the wave equation we already know the space and time dependence to be expected in a plane wave, so let us assume that the fields have the form

$$\mathbf{E}(\mathbf{r}, t) = \hat{\mathbf{E}} e^{-i(\omega t - \boldsymbol{\kappa} \cdot \mathbf{r})}, \text{ etc.,} \tag{17-7}$$

where $\hat{\mathbf{E}}$ is a complex constant vector amplitude of the plane wave, and let us substitute the assumed solutions into the Maxwell equations (17–3)–(17–6). This substitution will impose conditions that the assumed constants κ, $\hat{\mathbf{E}}$, etc., will have to satisfy in order that the plane wave functions will actually be solutions of the Maxwell equations.

By differentiating a function of the form $\hat{\mathbf{E}}e^{-i\omega t}$ with respect to t, one sees that the operator $\partial/\partial t$ is

$$\frac{\partial}{\partial t} = -i\omega$$

for a function of this particular form. Likewise one finds (Problem 17–1) that for a function of the form $\hat{\mathbf{E}}e^{i\kappa \cdot \mathbf{r}}$, the operator ∇ is

$$\nabla = i\kappa.$$

Thus the Maxwell equations become for plane waves (after i and the exponential are canceled)

$$\kappa \cdot \hat{\mathbf{D}} = 0, \qquad (17\text{–}8)$$

$$\kappa \cdot \hat{\mathbf{B}} = 0, \qquad (17\text{–}9)$$

$$\kappa \times \hat{\mathbf{E}} = \omega\hat{\mathbf{B}}, \qquad (17\text{–}10)$$

$$\kappa \times \hat{\mathbf{H}} = -\omega\hat{\mathbf{D}}. \qquad (17\text{–}11)$$

If we assume that the medium is linear the constitutive equations are

$$\hat{\mathbf{D}} = \epsilon\hat{\mathbf{E}}, \qquad (17\text{–}12)$$

$$\hat{\mathbf{H}} = \frac{1}{\mu}\,\hat{\mathbf{B}}. \qquad (17\text{–}13)$$

We also assume that the medium is homogeneous and isotropic, so that ϵ and μ are constant scalars. All of our applications will be to nonmagnetic media,* so for simplicity let us assume that $\mu = \mu_0$. Recalling that $\epsilon = K\epsilon_0$ and $\epsilon_0 \mu_0 = 1/c^2$, we get the Maxwell equations in the form

$$K\kappa \cdot \hat{\mathbf{E}} = 0, \qquad (17\text{–}14)$$

$$\kappa \cdot \hat{\mathbf{B}} = 0, \qquad (17\text{–}15)$$

$$\kappa \times \hat{\mathbf{E}} = \omega\hat{\mathbf{B}}, \qquad (17\text{–}16)$$

$$\kappa \times \hat{\mathbf{B}} = -\frac{\omega}{c^2}\,K\hat{\mathbf{E}}. \qquad (17\text{–}17)$$

---

* The only media for which μ differs appreciably from μ_0 at low frequency are the ferromagnetic ones, which are not linear anyway. For optical frequencies, $\mu \cong \mu_0$ for all materials. We exclude from consideration paramagnetic resonance, which is observable at radio and microwave frequencies under special circumstances.

If we take ω to be a given frequency and K to be a given material constant, we must try to satisfy this set of algebraic vector equations by suitable choices of $\boldsymbol{\kappa}$, $\hat{\mathbf{E}}$, and $\hat{\mathbf{B}}$. First, if we assume that $K \neq 0$, we see that $\boldsymbol{\kappa} \cdot \hat{\mathbf{E}} = 0$; always $\boldsymbol{\kappa} \cdot \hat{\mathbf{B}} = 0$. That is, both $\mathbf{E}$ and $\mathbf{B}$ must be perpendicular to $\boldsymbol{\kappa}$. Such a wave is called *transverse*. (The case $K = 0$ is actually possible and nontrivial, but we will defer its discussion.) Further, since $\hat{\mathbf{B}}$ is proportional to $\boldsymbol{\kappa} \times \hat{\mathbf{E}}$, $\mathbf{E}$ and $\mathbf{B}$ are also perpendicular to each other. The vectors $\boldsymbol{\kappa}$, $\mathbf{E}$, $\mathbf{B}$ (in that order) form a right-handed orthogonal set. The relative magnitude of $\hat{\mathbf{E}}$ and $\hat{\mathbf{B}}$ is also determined by Eq. (17–16), $\hat{B} = (\kappa/\omega)\hat{E}$. Finally, we find the magnitude of $\boldsymbol{\kappa}$ by taking the vector product of $\boldsymbol{\kappa}$ with Eq. (17–16) and using $\boldsymbol{\kappa} \times \hat{\mathbf{B}}$ from Eq. (17–17):

$$\boldsymbol{\kappa} \times (\boldsymbol{\kappa} \times \hat{\mathbf{E}}) = \omega\boldsymbol{\kappa} \times \hat{\mathbf{B}} = -K(\omega/c)^2\hat{\mathbf{E}}.$$

With the vector identity

$$\boldsymbol{\kappa} \times (\boldsymbol{\kappa} \times \hat{\mathbf{E}}) = (\boldsymbol{\kappa} \cdot \hat{\mathbf{E}})\boldsymbol{\kappa} - \kappa^2\hat{\mathbf{E}},$$

since $\boldsymbol{\kappa} \cdot \hat{\mathbf{E}} = 0$ for the transverse wave,

$$-K(\omega/c)^2\hat{\mathbf{E}} = -\kappa^2\hat{\mathbf{E}},$$

or

$$\kappa = \sqrt{K}\,\omega/c. \tag{17–18}$$

This relation, called the *transverse dispersion relation*, determines the magnitude of the wave vector $\boldsymbol{\kappa}$ in terms of the assumed ω and K.

To recapitulate: a plane monochromatic transverse wave propagated in the plus $\mathbf{u}$ direction is described by

$$\mathbf{E}(\mathbf{r}, t) = \hat{\mathbf{E}}e^{-i(\omega t - \boldsymbol{\kappa} \cdot \mathbf{r})},$$
$$\mathbf{B}(\mathbf{r}, t) = \hat{\mathbf{B}}e^{-i(\omega t - \boldsymbol{\kappa} \cdot \mathbf{r})}, \tag{17–7}$$

where $\boldsymbol{\kappa} = \kappa\mathbf{u}$. The direction $\mathbf{u}$ and the frequency ω are completely arbitrary. The amplitude $\hat{\mathbf{E}}$ is arbitrary except that it must be perpendicular to $\mathbf{u}$:

$$\mathbf{u} \cdot \hat{\mathbf{E}} = 0. \tag{17–19}$$

The magnitude of $\boldsymbol{\kappa}$ is determined, for the given frequency ω, by the refractive index of the material:

$$\kappa = n\omega/c, \tag{17–20}$$

where n is defined as

$$n = \sqrt{K}. \tag{17–21}$$

Then $\hat{\mathbf{B}}$ is completely determined in magnitude and direction:

$$\hat{\mathbf{B}} = \frac{n}{c}\mathbf{u} \times \hat{\mathbf{E}}. \tag{17–22}$$

Note that in vacuum $(n = 1)$, $c\hat{B} = \hat{E}$ in mks units.* The phase velocity of the wave is c/n. With these results it is possible to consider some extremely interesting and important optical problems. These, however, will be postponed until the next chapter.

Although the plane wave solutions are only a restricted class of solutions of Maxwell's equations, they are very important since they form the basis of a much wider class of solutions. Because the equations are linear, a linear combination of solutions (superposition of plane waves) is also a solution. Thus we could form other solutions by taking sums of plane waves

$$\mathbf{E}(\mathbf{r}, t) = \sum_i \hat{\mathbf{E}}(\mathbf{\kappa}_i, \omega_i) \exp\left[-i(\omega_i t - \mathbf{\kappa}_i \cdot \mathbf{r})\right], \qquad (17\text{--}23)$$

where each coefficient $\hat{\mathbf{E}}$ would depend on $\mathbf{\kappa}_i$ and ω_i. This superposition of plane waves has the form of a (complex) Fourier series, and, therefore, could represent any solution that was periodic—not necessarily sinusoidal. Each term of the series would have to satisfy the conditions of Eqs. (17–14)–(17–17) separately. For a solution that is not even periodic, the sum in (17–23) can be converted to an integral—the Fourier integral—with $\hat{\mathbf{E}}(\mathbf{\kappa}, \omega)$ a continuous function of $\mathbf{\kappa}$ and ω. The function $\hat{\mathbf{E}}(\mathbf{\kappa}, \omega)$ is called the *Fourier transform* of $\mathbf{E}(\mathbf{r}, t)$. In this case we would also have to consider the possibility that n depends on κ and ω as well,

$$n = n(\kappa, \omega).$$

This latter effect, known as *dispersion*, will be discussed in Chapter 19.

17–2 POLARIZATION

There is more to be said about the complex vector amplitudes $\hat{\mathbf{E}}$ and $\hat{\mathbf{B}}$. In fact, we have not yet stated explicitly what we mean by a complex vector. Two obvious meanings suggest themselves: a complex quantity whose real and imaginary parts are real vectors,

$$\hat{\mathbf{E}} = \mathbf{E}_r + i\mathbf{E}_i;$$

or a vector whose components (with respect to real basis vectors) are complex scalars,

$$\hat{\mathbf{E}} = \hat{E}_p\mathbf{p} + \hat{E}_s\mathbf{s} + \hat{E}_u\mathbf{u}.$$

We shall use the circumflex for quantities that are complex when it is necessary to distinguish them; in the second form, $\mathbf{p}$, $\mathbf{s}$, $\mathbf{u}$ are a right-handed set of real orthogonal unit vectors. By writing the first form in terms of components and the second in terms of real and imaginary parts, it is easily seen that the two formulations are equivalent, provided that

$$E_{p_r} = E_{r_p}, \qquad E_{p_i} = E_{i_p}, \qquad E_{s_r} = E_{r_s}, \text{ etc.}$$

---

* In Gaussian units, according to the discussion in Chapter 8, we replace B by B/c, so that $B = E$. That is, the E and B fields have equal magnitudes for a plane wave in vacuum.

For our present purpose the second form is more convenient. We take $\mathbf{u}$ to be the propagation direction of the plane wave, so that $\hat{E}_u = 0$ according to the result $\mathbf{u} \cdot \hat{\mathbf{E}} = 0$ of Eq. (17–19), but $\hat{E}_p$ and $\hat{E}_s$ are arbitrary,

$$\hat{\mathbf{E}} = \hat{E}_p \mathbf{p} + \hat{E}_s \mathbf{s}. \tag{17–24}$$

The unit vector $\mathbf{p}$ can be chosen in any direction perpendicular to $\mathbf{u}$; in the next chapter we will make a special choice that will account for the particular notation introduced here.

It is also more convenient to express the complex components in polar form instead of in terms of real and imaginary parts. Let

$$\hat{E}_p = E_p e^{i\phi_p}, \qquad \hat{E}_s = E_s e^{i\phi_s}. \tag{17–25}$$

Then, for example,

$$\hat{E}_s e^{-i(\omega t - \boldsymbol{\kappa} \cdot \mathbf{r})} = E_s e^{-i(\omega t - \boldsymbol{\kappa} \cdot \mathbf{r} - \phi_s)};$$

that is, ϕ_s is the phase of the E-field component in the s-direction. It is no restriction to let

$$\phi_p - \phi_s = \phi, \qquad \phi_s = 0,$$

since $\phi_s = 0$ merely dictates a certain choice of the origin of t. With this choice,

$$\hat{\mathbf{E}} = E_p e^{i\phi} \mathbf{p} + E_s \mathbf{s},$$

$$\mathbf{E}(\mathbf{r},\, t) = E_p \mathbf{p} e^{-i(\omega t - \boldsymbol{\kappa} \cdot \mathbf{r} - \phi)} + E_s \mathbf{s} e^{-i(\omega t - \boldsymbol{\kappa} \cdot \mathbf{r})},$$

or the real part is

$$\mathbf{E}(\mathbf{r},\, t) = E_p \mathbf{p} \cos\,(\omega t - \boldsymbol{\kappa} \cdot \mathbf{r} - \phi) + E_s \mathbf{s} \cos\,(\omega t - \boldsymbol{\kappa} \cdot \mathbf{r}). \tag{17–26}$$

The E-field is resolved into components in two directions, with real amplitudes E_p and E_s, which may have any values. In addition the two components may be oscillating out of phase by ϕ: That is, at any given point $\mathbf{r}$, the maximum of E in the $\mathbf{p}$ direction may be attained at a different time from the maximum of E in the $\mathbf{s}$ direction.

A detailed picture of the oscillating E-field at a certain point, say $\mathbf{r} = 0$, is best seen by considering some special cases. First, suppose that $\phi = 0$. Then

$$\mathbf{E}(0,\, t) = (E_p \mathbf{p} + E_s \mathbf{s}) \cos\,\omega t.$$

The E-field alternately decreases from $\sqrt{E_p^2 + E_s^2}$ through zero to $-\sqrt{E_p^2 + E_s^2}$ and back to the original value, always pointing along the direction $E_p \mathbf{p} + E_s \mathbf{s}$. This case is called *linear polarization*,* and is illustrated in Fig. 17–1. If $E_p = 0$ or $E_s = 0$, we also have linear polarization; then ϕ is undefined. For $\phi = \pi$,

$$\mathbf{E}(0,\, t) = (-E_p \mathbf{p} + E_s \mathbf{s}) \cos\,\omega t,$$

---

* This use of the term "polarization" has nothing to do with the use introduced in Chapter 4. It is unfortunate that the same word is conventionally used, but usually no confusion arises since one usage applies to a wave and the other to a medium.

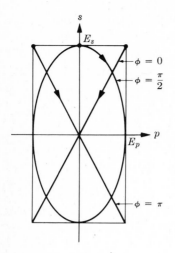

Figure 17–1 Trace of the tip of the E-vector at a given point in space as a function of time. The propagation direction **u** is pointing toward us. The traces for $\phi = 0$ and π are linearly polarized. The trace for $\phi = \pi/2$ is elliptically polarized, right-hand; for $-\pi/2$ (left-hand) it rotates in the opposite direction.

again linear polarization, as shown in Fig. 17–1. For $\phi = \pi/2$,

$$\mathbf{E}(0, t) = E_p\mathbf{p} \sin \omega t + E_s\mathbf{s} \cos \omega t.$$

The tip of the E-vector traces an elliptical path clockwise as shown. This case is called *right-hand* elliptical polarization*. For $\phi = -\pi/2$, the path is the same, but traced counterclockwise, called *left-hand elliptical polarization*. In the special case when $\phi = \pm\pi/2$ and $E_p = E_s$, we have (right- or left-hand) *circular polarization*. For other values of ϕ, we have elliptical polarization (even if $E_p = E_s$); the trace is still an ellipse inscribed in the box of Fig. 17–1, but the minor and major axes of the ellipse are at an angle to the p- and s-axes. With elliptical polarization the magnitude of the E-vector never goes to zero, although its component in any given direction does.

The complex amplitude of the B-vector is given by Eq. (17–22),

$$\hat{\mathbf{B}} = \frac{n}{c}\mathbf{u} \times \hat{\mathbf{E}}. \tag{17–22}$$

Taking the dot product of this with $\hat{\mathbf{E}}$, and interchanging dot and cross, we find that

$$\hat{\mathbf{B}} \cdot \hat{\mathbf{E}} = 0.$$

In general the vanishing of the dot product of two *complex* vectors does not mean that their real parts are perpendicular, but in this case they are. From Eq. (17–26), writing $\mathbf{E}(0, t) = \mathbf{E}$, we have for the real part

$$\mathbf{E} = E_p\mathbf{p} \cos (\omega t - \phi) + E_s\mathbf{s} \cos \omega t. \tag{17–27}$$

---

* It is too bad that the "right-hand rule" does not apply here, but this is the convention.

From the real part of $\mathbf{B}(0, t) = \mathbf{B} = \hat{\mathbf{B}}e^{-i\omega t}$ with Eq. (17–22),

$$\mathbf{B} = (n/c)[E_p\mathbf{s} \cos (\omega t - \phi) - E_s\mathbf{p} \cos \omega t]. \qquad (17\text{–}28)$$

Since the dot product of Eq. (17–28) with (17–27) is zero, the real E- and B-vectors are perpendicular at each instant. Also Re $\hat{\mathbf{E}} = \mathbf{E}(0, 0)$, Re $\hat{\mathbf{B}} = \mathbf{B}(0, 0)$, so that the real parts of $\hat{\mathbf{E}}$ and $\hat{\mathbf{B}}$ are perpendicular. The trace of the tip of the B-vector is the same as Fig. 17–1 rotated 90° counterclockwise.

Since the p- and s-axes were chosen arbitrarily in the plane perpendicular to $\mathbf{u}$, any other choice could be made. A new choice would rotate the coordinate axes in Fig. 17–1 and introduce new values of E_p, E_s, and ϕ; but the trace of the E-vector in Fig. 17–1 represents a physical reality, and this would not be changed merely by the coordinate transformation. At this point, however, even the physical state of polarization itself is part of the arbitrariness admitted by our assumed plane wave solution for an infinite medium. In the next chapter we shall see how a particular polarization can be produced and measured.

17–3 ENERGY DENSITY AND FLUX

We have freely used complex expressions for the E- and B-fields, with the understanding that the actual physical quantities are given by the *real parts* of the complex quantity. The mathematical justification of this procedure is that the Maxwell equations are *linear* equations, which are satisfied separately by the real and imaginary parts of a complex solution. The expressions

$$u = \tfrac{1}{2}(\mathbf{E} \cdot \mathbf{D} + \mathbf{B} \cdot \mathbf{H}), \qquad (17\text{–}29)$$

$$\mathbf{S} = \mathbf{E} \times \mathbf{H} \qquad (17\text{–}30)$$

for the energy density and flux per unit area, however, are nonlinear in the fields. Therefore, in these expressions it is essential to take the real parts of the fields before multiplying them together. (See Problem (17–6).) We may again calculate representative values at $\mathbf{r} = 0$, since the origin is arbitrary. Squaring Eqs. (17–27) and (17–28) gives

$$E^2 = E_p^2 \cos^2 (\omega t - \phi) + E_s^2 \cos^2 \omega t, \qquad (17\text{–}31)$$

$$B^2 = (n/c)^2 E^2 = \epsilon\mu_0 E^2. \qquad (17\text{–}32)$$

Since $D = \epsilon\mathbf{E}$, $B = \mu_0\mathbf{H}$, we find that the electric and magnetic fields each make equal contributions to the energy density:

$$\mathbf{B} \cdot \mathbf{H} = \mathbf{D} \cdot \mathbf{E},$$

$$u = \epsilon E^2 = \frac{1}{\mu_0} \left(\frac{n}{c}\right)^2 E^2. \qquad (17\text{–}33)$$

Further, $\mathbf{E} \times \mathbf{H} = EH\mathbf{u}$, so that the Poynting vector points in the propagation direction, with magnitude

$$S = \frac{1}{\mu_0} \frac{n}{c} E^2. \tag{17-34}$$

The expressions for energy density and energy flux per unit area have assumed especially simple forms for plane waves. Further, the two expressions can be combined to give an interesting result, which is independent of the particular value of the E-field:

$$S = \frac{c}{n} u. \tag{17-35}$$

If we write the phase velocity of the plane wave as a vector in the propagation direction with magnitude

$$v_p = \frac{c}{n},$$

then

$$\mathbf{S} = u\mathbf{v}_p.$$

This equation is analogous to the relation

$$\mathbf{J} = \rho \mathbf{v},$$

which defines the convective electric current density. This analogy reinforces the interpretation of $\mathbf{S}$ as an energy current density, i.e., an energy density u that is transported with the phase velocity $\mathbf{v}_p$ of the plane wave.

The time dependence of u and $\mathbf{S}$ is given by E^2 from Eq. (17–31) and depends on the polarization of the wave. For circular polarization ($\phi = \pm\pi/2$)

$$E^2 = E_p^2 \sin^2 \omega t + E_p^2 \cos^2 \omega t = E_p^2$$

is constant in time; for linear polarization ($\phi = 0, \pi$)

$$E^2 = (E_p^2 + E_s^2) \cos^2 \omega t$$

varies between zero and a maximum at twice the wave frequency. In every case E^2 is always positive of course. At high frequencies the time dependence is not measurable, however, and so what is of more interest is the time average of E^2. Since the time average over one period of $\cos^2 (\omega t - \phi)$ is $\frac{1}{2}$, for any polarization

$$\overline{E^2} = \tfrac{1}{2}(E_p^2 + E_s^2). \tag{17-36}$$

This and similar results can be obtained more quickly by using a theorem that was introduced in Chapter 13 but not proved; we now give the proof.

If $f = f_0 e^{i\omega t}$ and $g = g_0 e^{i\omega t}$, where f_0 and g_0 may depend on other variables but not on the time, then

$$\overline{\text{Re} f \, \text{Re} g} = \tfrac{1}{2} \, \text{Re} f^* g. \tag{17-37}$$

The bar indicates time averaging. To prove this relationship, let $f_0 = u + iv$ and $g_0 = \xi + i\eta$. Then

$$\text{Re } f \text{ Re } g = (u \cos \omega t - v \sin \omega t)(\xi \cos \omega t - \eta \sin \omega t), \qquad (17\text{--}38)$$

while

$$\text{Re } f^* g = u\xi + v\eta. \qquad (17\text{--}39)$$

The following integrals are easily verified:

$$\lim_{T \to \infty} \frac{1}{T} \int_0^T \sin^2 \omega t \, dt = \tfrac{1}{2},$$

$$\lim_{T \to \infty} \frac{1}{T} \int_0^T \cos^2 \omega t \, dt = \tfrac{1}{2},$$

$$\lim_{T \to \infty} \frac{1}{T} \int_0^T \sin \omega t \cos \omega t \, dt = 0.$$

By means of these integrals, it is easy to see that the time average of Eq. (17–38) is

$$\overline{\text{Re } f \text{ Re } g} = \tfrac{1}{2}(u\xi + v\eta). \qquad (17\text{--}40)$$

Comparison of (17–40) with (17–39) proves the theorem of Eq. (17–37). The theorem holds for the product of any two complex quantities that depend harmonically on t with the same frequency, but not necessarily having the same phase. If the quantities are vectors, the product can be either the dot- or cross-product. It is easy to check that the expression (17–36) is immediately given by putting the complex $\mathbf{E}$ of (17–26) into

$$\overline{E^2} = \tfrac{1}{2} \text{ Re } (\mathbf{E}^* \cdot \mathbf{E}).$$

17–4 PLANE MONOCHROMATIC WAVES IN CONDUCTING MEDIA

In a conducting medium we can obtain plane wave solutions that are formally very similar to those of Section 17–1, although physically their behavior is significantly more complicated. We still assume that there are no prescribed charge or current distributions, but now there can be an induced current density in response to the E-field of the wave, $\mathbf{J} = g\mathbf{E}$. Our starting point is the same as for a nonconducting medium, with the sole exception that instead of Eq. (17–6) we have

$$\nabla \times \mathbf{H} = \frac{\partial \mathbf{D}}{\partial t} + g\mathbf{E}.$$

Making the same assumptions and substitutions as before, we arrive at

$$\boldsymbol{\kappa} \times \hat{\mathbf{H}} = -\omega \hat{\mathbf{D}} - ig\hat{\mathbf{E}}$$

instead of Eq. (17–11). Still as before, this becomes

$$\kappa \times \hat{\mathbf{B}} = -\frac{\omega}{c^2}\left(K + i\,\frac{g}{\epsilon_0 \omega}\right)\hat{\mathbf{E}}.$$

Now to reduce the present case to the previous case, insofar as possible, let us define a *complex dielectric constant*

$$\hat{K} = K + i\,\frac{g}{\epsilon_0 \omega}, \qquad (17\text{–}41)$$

so that Eq. (17–17) becomes

$$\kappa \times \hat{\mathbf{B}} = -\frac{\omega}{c^2}\,\hat{K}\,\hat{\mathbf{E}}.$$

Then if we again assume that $\hat{K} \neq 0$ and $\kappa \cdot \hat{\mathbf{E}} = 0$, the transverse dispersion relation results,

$$\kappa = \sqrt{\hat{K}}\,\omega/c = \hat{n}\omega/c, \qquad (17\text{–}42)$$

where we define a *complex refractive index* $\hat{n}$ by

$$\hat{n}^2 = \hat{K}. \qquad (17\text{–}43)$$

To satisfy Eq. (17–42), we have to assume that either κ or ω is a complex quantity in a conducting medium. In the last chapter we discussed briefly the case of a complex ω. The solutions that will be of more interest in the following chapters result from assuming that ω is real and κ is complex, to be written $\hat{\kappa}$. The latter solutions will prove to be oscillating in time (not damped), but attenuated in space. All the results of Section 17–1 hold formally if K, n, κ, and $\mathbf{u}$ are replaced by $\hat{K}$, $\hat{n}$, $\hat{\kappa}$, and $\hat{\mathbf{u}}$. The only question is, what do they mean physically? To interpret the solutions with complex propagation vector $\hat{\kappa}$, it is useful to express it as

$$\hat{\kappa} = \kappa_r + i\kappa_i \qquad (17\text{–}44)$$

(in contrast to the representation adopted for the complex amplitude vector $\hat{\mathbf{E}}$). Inserting the complex $\hat{\kappa}$ into the solution (17–7), we find

$$\mathbf{E}(\mathbf{r},\,t) = (\hat{\mathbf{E}}e^{-\kappa_i \cdot \mathbf{r}})e^{-i(\omega t - \kappa_r \cdot \mathbf{r})},$$

$$\mathbf{B}(\mathbf{r},\,t) = (\hat{\mathbf{B}}e^{-\kappa_i \cdot \mathbf{r}})e^{-i(\omega t - \kappa_r \cdot \mathbf{r})}.$$

This is a plane wave propagating in the direction κ_r, with wavelength $\lambda = 2\pi/\kappa_r$; but instead of having a constant amplitude, it decreases in amplitude, most rapidly in the direction κ_i. The surfaces of constant phase are planes perpendicular to the propagation direction κ_r. There are also surfaces of constant amplitude that are planes perpendicular to κ_i. The scalar $\hat{\kappa}$ to be used in the dispersion relation $\hat{\kappa} = \hat{n}\omega/c$ is

$$\hat{\kappa} = \sqrt{\hat{\kappa} \cdot \hat{\kappa}} = \sqrt{\kappa_r^2 - \kappa_i^2 + 2i\kappa_r \cdot \kappa_i}.$$

We write

$$\hat{n} = n + ik, \tag{17-45}$$

where n and k are called the *optical constants*, in order to investigate the phase velocity and attenuation length of the wave.

Let us assume to start that κ_r and κ_i have the same direction. (This is a restrictive assumption, which fails to hold in some important cases to be discussed in the next chapter; but it may hold, namely in case the wave entered the conducting medium from outside at normal incidence to a plane boundary.) Then we can write

$$\hat{\kappa} = (\kappa_r + i\kappa_i)\mathbf{u} = \hat{\kappa}\mathbf{u},$$

where $\mathbf{u}$ is the real unit vector in the common direction of κ_r and κ_i. Since $\mathbf{u}$ is real, the equations

$$\mathbf{u} \cdot \hat{\mathbf{E}} = 0 = \mathbf{u} \cdot \hat{\mathbf{B}}$$

still mean that the wave has its E- and B-vectors perpendicular to the direction of propagation $\mathbf{u}$. Equation (17-22) becomes, however,

$$\hat{\mathbf{B}} = \frac{\hat{n}}{c} \mathbf{u} \times \hat{\mathbf{E}}. \tag{17-46}$$

The complex $\hat{n}$ means that $\mathbf{E}$ and $\mathbf{B}$ are not in phase with each other. It also means that the real E- and B-vectors are not perpendicular to each other, except for linear polarization (Problem 17-9). In terms of the complex refractive index

$$\hat{n} = n + ik,$$

we have

$$\kappa_r = n\omega/c, \qquad \kappa_i = k\omega/c. \tag{17-47}$$

Writing $\mathbf{u} \cdot \mathbf{r} = \xi$, we get for the E-field in this special case

$$\mathbf{E}(\mathbf{r}, t) = (\hat{\mathbf{E}}e^{-k\omega\xi/c})e^{-i\omega(t - n\xi/c)}.$$

The wave propagates with phase velocity c/n and attenuation constant $k\omega/c$. The latter quantity determines how rapidly the amplitude of the oscillating field falls off with distance. Its reciprocal

$$\delta = c/k\omega, \tag{17-48}$$

called the *skin depth*, measures the distance within which the field falls to $1/e$ of its value at a given point (say a surface where the wave enters the medium). For a nonconducting medium ($k = 0$) this distance was infinite. Since the wavelength in the medium is $\lambda = 2\pi c/n\omega$ ($= 1/n$ times the wavelength in vacuum at the same frequency), we can write the skin depth

$$\delta = \frac{n}{k} \frac{\lambda}{2\pi}. \tag{17-49}$$

In a material where k is comparable in magnitude with n, the "wave" decays in about one wavelength; but if $k \ll n$, as in an imperfect dielectric with a small conductivity, the wave propagates many wavelengths without appreciable loss. In the latter case the material is *transparent*.

When κ_r and κ_i have different directions, the unit vector $\hat{\mathbf{u}}$ must be taken as complex. Then the real E- and B-fields do not have to be perpendicular to κ_r (nor to κ_i), although the wave is still called "transverse" if $\hat{K} \neq 0$, $\hat{\mathbf{u}} \cdot \hat{\mathbf{E}} = 0 = \hat{\mathbf{u}} \cdot \hat{\mathbf{B}}$. The wavelength and attenuation still depend on the material constants n and k, but in a much more complicated way than Eq. (17–47). The polarization and energy relations can be generalized, with due care for the complex quantities. We will have no need for these general results, however.

Since the wave propagation is determined by the optical constants n and k, it is important to explore their relation to the quantities K and g, through which the properties of the material were originally expressed. The definitions

$$\hat{n} = n + ik, \tag{17–45}$$

$$\hat{K} = K + i\,\frac{g}{\epsilon_0 \omega}, \tag{17–41}$$

are related by

$$\hat{K} = \hat{n}^2. \tag{17–43}$$

This relationship looks deceptively simple; when it is expressed in terms of the real quantities n, k and K, g, it becomes fairly complicated. We write*

$$\hat{K} = K_r + iK_i,$$

where

$$K_r = K, \qquad K_i = g/\epsilon_0 \omega. \tag{17–50}$$

Squaring $(n + ik)$ and equating real and imaginary parts in Eq. (17–43) gives

$$K_r = n^2 - k^2,$$

$$K_i = 2nk. \tag{17–51}$$

These equations can be solved for n and k:

$$n = \sqrt{\tfrac{1}{2}[K_r + \sqrt{K_r^2 + K_i^2}]},$$

$$k = \sqrt{\tfrac{1}{2}[-K_r + \sqrt{K_r^2 + K_i^2}]}. \tag{17–52}$$

Positive square roots are chosen so that both n and k will always be real and positive, as is required by their meaning. It should be noted that $K_i = g/\epsilon_0 \omega$ is

---

* More common notations are $\hat{K} = K_1 + iK_2$ or $\hat{K} = K' + iK''$, but we want to reserve the numerical subscripts and primes for other purposes in the next chapter.

inherently positive, since g is positive: the energy dissipation $\mathbf{J} \cdot \mathbf{E} = gE^2$ in a passive medium must be positive; K_i always represents an energy loss. On the other hand, $K_r = K$ could be, and indeed can be, either positive or negative. For nonconducting media and static fields, it is positive and greater than 1, but for alternating fields in metals it may be less than 1, zero, or negative. Specific examples can be found in Chapter 19; for the present, we simply take K and g as given, recognizing that in any material they may depend on the frequency ω.

The equations above are exact but complicated; hence it is convenient to examine certain approximations. Most often one or another of the following approximations is valid:

1. $K_i \ll |K_r|, \; K_r > 0 \left(\omega \gg \dfrac{g}{\epsilon} \right)$:

$$n \cong \sqrt{K_r}, \qquad k = K_i/2n \ll n. \qquad (17\text{--}53)$$

2. $K_i \ll |K_r|, \; K_r < 0 \left(\omega \gg \dfrac{g}{-\epsilon} \right)$:

$$k \cong \sqrt{-K_r}, \qquad n = K_i/2k \ll k. \qquad (17\text{--}54)$$

3. $K_i \gg |K_r|, \left(\omega \ll \dfrac{g}{|\epsilon|} \right)$:

$$n \cong k \cong \sqrt{K_i/2}. \qquad (17\text{--}55)$$

For example, (1) holds for a good insulator down to very low frequencies, essentially d-c (strictly d-c only for a perfect insulator with $g = 0$), and gives $n = \sqrt{K}$ as for a nonconducting medium. Approximation (2) holds for metals in the upper infrared part of the frequency spectrum, while (3) holds for metals at microwave frequencies and below. The dividing frequency between the latter two cases is $\omega \approx 1/\tau$, where τ is the collision time for the free electrons. For pure metals at room temperature, $1/\tau \approx 10^{14} \text{ s}^{-1}$.

Where the skin depth is important in electrical problems, usually case (3) applies. Equation (17–49) shows that the "wave" is very strongly attenuated in terms of the wavelength. In terms of absolute distance, Eq. (17–48) becomes for this case

$$\delta \cong \frac{c}{\omega} \sqrt{\frac{2}{K_i}} = \sqrt{\frac{2}{\mu_0 \omega g}}. \qquad (17\text{--}56)$$

As ω approaches zero δ becomes infinite, consistently with our earlier result that the E-field and current density are uniform in a conductor (not a superconductor*) for d-c and also essentially for 60-Hz current. At higher frequencies δ becomes quite small within the range of validity of Eq. (17–56), however.

---

* The skin depth has no connection with the penetration depth discussed in Chapter 15.

Fine silver, for example, has a conductivity

$$g = 3 \times 10^7 \text{ S/m}†$$

at microwave frequencies. At a frequency of 10^{10} Hz, which is a common microwave region, the skin depth is

$$\delta = \sqrt{\frac{2}{(2\pi \times 10^{10})(3 \times 10^7)(4\pi \times 10^{-7})}} = 9.2 \times 10^{-5} \text{ cm}.$$

Thus at microwave frequencies the skin depth in silver is very small, and consequently the difference in performance between a pure silver component and a silver-plated brass component would be expected to be negligible. This is indeed the case, and the plating technique is used to reduce the material cost of high-quality waveguide components.

As a second example, we now calculate the frequency at which the skin depth in sea water is one meter. For sea water, $\mu = \mu_0$ and $g \approx 4.3$ S/m. The expression for the frequency corresponding to a given skin depth δ is

$$\omega = \frac{2}{g\mu_0 \, \delta^2} = \frac{2}{4.3 \times 4\pi \times 10^{-7} \, \delta^2} \text{ s}^{-1} = \frac{3.70 \times 10^5}{\delta^2} \text{ s}^{-1},$$

which yields

$$f = 58.6 \times 10^3 \text{ Hz},$$

or a frequency of 60 kHz for a skin depth of one meter. If a submarine is equipped with a very sensitive receiver and if a very powerful transmitter is used, it is possible to communicate with a submerged submarine. However, a very low radiofrequency must be used, and even then an extremely severe attenuation of the signal occurs. At five skin depths (5 m in the case calculated above), only 1 percent of the initial electric field remains, and only 0.01 percent of the incident power.

The anomalous case $\hat{K} = 0$, which we have so far excluded, allows the existence of a longitudinal, as opposed to a transverse, wave. (Problem 17–14.) In such a wave $\mathbf{V} \cdot \mathbf{E} \neq 0$ (even though $\mathbf{V} \cdot \mathbf{D} = 0$), so that the polarization charge density does not vanish as in the transverse wave case; only a charge density can give a longitudinal field. Such waves are of some importance in plasmas; the electrostatic oscillation discussed in Section 14–6 is an example.

*17–5 SPHERICAL WAVES

As an example of a more difficult wave problem, where in fact it is not easy to find even the elementary waves, we consider the wave equation in spherical coordinates. The wave equation for the electric field in a vacuum is

$$\nabla^2 \mathbf{E} - \frac{1}{c^2} \frac{\partial^2 \mathbf{E}}{\partial t^2} = 0. \tag{17–57}$$

---

† One siemens (S) is one reciprocal ohm; it is a unit of conductance or reciprocal resistance (see Chapter 7). A siemens was formerly called a mho.

For monochromatic waves, the equation for the spatial portion becomes

$$\nabla^2 \mathbf{E}(\mathbf{r}) + \left(\frac{\omega}{c}\right)^2 \mathbf{E}(\mathbf{r}) = 0. \tag{17-58}$$

The difficulty in using spherical coordinates is that one would like to express the vector $\mathbf{E}(\mathbf{r})$ in terms of radial, azimuthal, and meridional *components*, each expressed as functions of the radius, azimuth, and colatitude. If this is done, then it is not sufficient to use the expression for the Laplacian in spherical coordinates in Eq. (17–58); rather, it is necessary to define the Laplacian of a vector by

$$\nabla^2 \mathbf{E} = -\nabla \times \nabla \times \mathbf{E} + \nabla \nabla \cdot \mathbf{E} \tag{17-59}$$

The divergence of $\mathbf{E}$ is still zero; however, the radial component of $\nabla \times \nabla \times \mathbf{E}$ involves not only the radial component of $\mathbf{E}$, but also its azimuthal and meridional components. The θ and ϕ components are similarly complicated, and the final result is three simultaneous partial differential equations involving the three components of $\mathbf{E}$. The separation that occurs for the vector Laplace equation in rectangular coordinates does not occur in spherical coordinates; it is in fact peculiar to rectangular coordinates. It should be pointed out, however, that rectangular components of $\mathbf{E}$ may be used; in this instance they would be written: $E_x(r, \theta, \phi)$, $E_y(r, \theta, \phi)$, $E_z(r, \theta, \phi)$.

A simple procedure circumvents the difficulty discussed above. Consider the scalar Helmholtz equation:

$$\nabla^2 \psi + \left(\frac{\omega}{c}\right)^2 \psi = 0, \tag{17-60}$$

whose solutions are, as will be seen shortly, readily found. Suppose that ψ is any one of the solutions, then $\mathbf{E} = \mathbf{r} \times \nabla\psi$ satisfies the vector Helmholtz equation, Eq. (17–58):

$$-\nabla \times \nabla \times \mathbf{E} + \nabla \nabla \cdot \mathbf{E} + \left(\frac{\omega}{c}\right)^2 \mathbf{E} = 0. \tag{17-61}$$

To verify this, note the identity

$$\mathbf{E} = \mathbf{r} \times \nabla\psi = -\nabla \times (\mathbf{r}\psi), \tag{17-62}$$

which follows from the vector identity

$$\nabla \times (\mathbf{F}\varphi) = \varphi\nabla \times \mathbf{F} - \mathbf{F} \times \nabla\varphi \tag{17-63}$$

and

$$\nabla \times \mathbf{r} = 0. \tag{17-64}$$

Since the divergence of any curl is zero, it is necessary to consider only the curl curl term in Eq. (17–61). The curl of $\mathbf{E}$ can be found by using the vector identity

$$\nabla \times (\mathbf{F} \times \mathbf{G}) = \mathbf{F}\nabla \cdot \mathbf{G} - \mathbf{G}\nabla \cdot \mathbf{F} + (\mathbf{G} \cdot \nabla)\mathbf{F} - (\mathbf{F} \cdot \nabla)\mathbf{G} \tag{17-65}$$

to obtain

$$\mathbf{V} \times (\mathbf{r} \times \mathbf{V}\psi) = \mathbf{r}\mathbf{V}^2\psi - \mathbf{V}\psi\mathbf{V} \cdot \mathbf{r} + (\mathbf{V}\psi \cdot \mathbf{V})\mathbf{r} - (\mathbf{r} \cdot \mathbf{V})\mathbf{V}\psi. \qquad (17\text{--}66)$$

As was shown in Problem 1–13, $(\mathbf{A} \cdot \mathbf{V})\mathbf{r} = \mathbf{A}$ for any vector $\mathbf{A}$; also, the divergence of $\mathbf{r}$ is three (3). The first term of Eq. (17–66) can be reduced by using the fact that ψ satisfies the scalar Helmholtz equation, thus leaving only the last term as a possible source of complication. The vector identity

$$\mathbf{V}(\mathbf{F} \cdot \mathbf{G}) = (\mathbf{F} \cdot \mathbf{V})\mathbf{G} + (\mathbf{G} \cdot \mathbf{V})\mathbf{F} + \mathbf{F} \times \mathbf{V} \times \mathbf{G} + \mathbf{G} \times \mathbf{V} \times \mathbf{F}, \qquad (17\text{--}67)$$

with $\mathbf{F} = \mathbf{r}$ and $\mathbf{G} = \mathbf{V}\psi$, gives

$$\mathbf{V}(\mathbf{r} \cdot \mathbf{V}\psi) = (\mathbf{r} \cdot \mathbf{V})\mathbf{V}\psi + (\mathbf{V}\psi \cdot \mathbf{V})\mathbf{r}. \qquad (17\text{--}68)$$

The last two terms of Eq. (17–67) vanish because the curl of any gradient is zero, as is the curl of $\mathbf{r}$. Using these relationships in Eq. (17–66) leads to

$$\mathbf{V} \times (\mathbf{r} \times \mathbf{V}\psi) = -\left(\frac{\omega}{c}\right)^2 \mathbf{r}\psi - 3\mathbf{V}\psi + \mathbf{V}\psi - \mathbf{V}(\mathbf{r} \cdot \mathbf{V}\psi) + \mathbf{V}\psi. \qquad (17\text{--}69)$$

Finally, taking the curl of Eq. (17–69), we obtain

$$\mathbf{V} \times \mathbf{V} \times (\mathbf{r} \times \mathbf{V}\psi) = -\left(\frac{\omega}{c}\right)^2 \mathbf{V} \times \mathbf{r}\psi = \left(\frac{\omega}{c}\right)^2 \mathbf{r} \times \mathbf{V}\psi, \qquad (17\text{--}70)$$

which is just the vector Helmholtz equation. No explicit use of the spherical coordinate system has been made; however, since $\mathbf{r}$ is normal to a surface of constant radius in spherical coordinates, the solution $\mathbf{r} \times \mathbf{V}\psi$ would be expected to be particularly useful in this system. It is in fact not very useful in other coordinate systems.

Having found that $\mathbf{r} \times \mathbf{V}\psi$ is a solution of the vector Helmholtz equation, with ψ a solution of the scalar Helmholtz equation, it becomes pertinent to find out how such solutions can be used to construct electromagnetic waves. The procedure is very simple. The spatial variation of the electric field is taken as

$$\mathbf{E} = \mathbf{r} \times \mathbf{V}\psi. \qquad (17\text{--}62)$$

The magnetic field must be so chosen that it, together with $\mathbf{E}$, satisfies Maxwell's equations. To this end we write Eq. (17–5) as

$$\mathbf{V} \times \mathbf{E} = i\omega\mathbf{B}, \qquad (17\text{--}71)$$

where the standard $e^{-i\omega t}$ time dependence has been assumed. Equation (17–69) gives the curl of $\mathbf{E}$ explicitly or, in a shorter form,

$$\mathbf{B} = -i\,\frac{1}{\omega}\,\mathbf{V} \times (\mathbf{r} \times \mathbf{V}\psi). \qquad (17\text{--}72)$$

Since the divergence of any curl vanishes, Eq. (17–4) is satisfied. That Eq. (17–6) is satisfied is obvious from the fact that $\mathbf{E}$ and $\mathbf{B}$ are both solutions of the wave equation, which in turn represents a combination of Eqs. (17–5) and (17–6).

The solution represented by Eqs. (17–62) and (17–72) is not the most general solution that can be derived from a given ψ. Another solution is obtained by putting

$$\mathbf{B}' = \frac{1}{c}\, \mathbf{r} \times \nabla\psi \qquad (17\text{–}73)$$

and obtaining the electric field from Eq. (17–6),

$$\mathbf{E}' = \frac{ic}{\omega}\, \nabla \times (\mathbf{r} \times \nabla\psi). \qquad (17\text{–}74)$$

The considerations detailed above show that $\mathbf{E}'$, $\mathbf{B}'$ form a solution to Maxwell's equations, just as $\mathbf{E}$, $\mathbf{B}$ do. The solutions differ in that $\mathbf{E}$ at any point is tangent to a spherical surface through the point with center at the origin of coordinates; on the other hand, $\mathbf{B}'$ has the same property. These facts lead to the solution $\mathbf{E}$, $\mathbf{B}$ being sometimes called *transverse electric* (TE), and $\mathbf{E}'$, $\mathbf{B}'$ *transverse magnetic* (TM), transverse meaning perpendicular to the radial direction.

In the preceding paragraphs the problem of solving the vector Helmholtz equation has been reduced to that of solving the scalar Helmholtz equation. In spherical coordinates this is accomplished by the technique of separation of variables already familiar from potential problems (Chapter 3). In spherical coordinates, the scalar Helmholtz equation is

$$\frac{1}{r^2}\frac{\partial}{\partial r}\left(r^2\frac{\partial\psi}{\partial r}\right) + \frac{1}{r^2 \sin\theta}\frac{\partial}{\partial\theta}\left(\sin\theta\frac{\partial\psi}{\partial\theta}\right) + \frac{1}{r^2 \sin^2\theta}\frac{\partial^2\psi}{\partial\phi^2} + \kappa^2\psi = 0,$$
$$(17\text{–}75)$$

where $\kappa^2 = (\omega/c)^2$ and ψ is assumed to have the form

$$\psi = R(r)\Theta(\theta)\Phi(\phi). \qquad (17\text{–}76)$$

Substituting this assumed form for ψ in Eq. (17–75) and dividing by ψ gives

$$\frac{1}{R}\sin^2\theta\,\frac{d}{dr}\,r^2\,\frac{dR}{dr} + \frac{1}{\Theta}\sin\theta\,\frac{d}{d\theta}\sin\theta\,\frac{d\Theta}{d\theta} + \frac{1}{\Phi}\frac{d^2\Phi}{d\phi^2} + \kappa^2 r^2 \sin^2\theta = 0,$$
$$(17\text{–}77)$$

after multiplying by $r^2 \sin^2\theta$. The third term depends only on ϕ, and this is the only term that depends on ϕ. Consequently this term must be a constant, which is chosen to be $-m^2$. In other words,

$$\frac{d^2\Phi_m}{d\phi^2} + m^2\Phi_m = 0, \qquad (17\text{–}78)$$

where the subscript m serves to indicate that Φ depends on m. Rewriting Eq. (17–77) using (17–78) gives

$$\frac{1}{R}\frac{d}{dr}\,r^2\,\frac{dR}{dr} + \kappa^2 r^2 + \frac{1}{\Theta}\frac{1}{\sin\theta}\frac{d}{d\theta}\sin\theta\,\frac{d\Theta}{d\theta} - \frac{m^2}{\sin^2\theta} = 0. \qquad (17\text{–}79)$$

The first two terms depend only on r, while the last two depend only on θ. Thus the sum of the last two must be a constant, which is chosen as $-l(l+1)$. The sum of the first two terms must, of course, be $l(l+1)$. Thus there result two equations:

$$\frac{1}{\sin\theta}\frac{d}{d\theta}\sin\theta\frac{d\Theta_{lm}}{d\theta} + \left[l(l+1) - \frac{m^2}{\sin^2\theta}\right]\Theta_{lm} = 0 \qquad (17\text{–}80)$$

and

$$\frac{d}{dr}r^2\frac{dR_l}{dr} - [l(l+1) - \kappa^2 r^2]R_l = 0. \qquad (17\text{–}81)$$

The solutions of Eq. (17–78) are well known:

$$\Phi_m = e^{\mp im\phi}. \qquad (17\text{–}82)$$

The solutions of Eq. (17–80) are less well known, but some have already been met in Chapter 3, where solutions for $m = 0$ were discussed. These solutions* are the Legendre polynomials $P_l(\cos\theta)$. The solutions of (17–80) for arbitrary $m \lessgtr l$ are known as the associated Legendre polynomials. They may be defined by

$$P_l^m(u) = (1 - u^2)^{m/2}\frac{d^m}{du^m}P_l(u), \qquad (17\text{–}83)$$

with $u = \cos\theta$. It is clear that $P_l^0(u) = P_l(u)$, the ordinary Legendre polynomial. For $m \neq 0$ the functions are given in Table 17–1.

Finally, Eq. (17–81) must be considered. The change of variable from r to $\xi = \kappa r$ is readily accomplished; the resulting equation is

$$\frac{d}{d\xi}\xi^2\frac{d}{d\xi}R_l - [l(l+1) - \xi^2]R_l = 0. \qquad (17\text{–}84)$$

Table 17–1 Associated Legendre Polynomials,
$P_l^m(u)$, **where** $u = \cos\theta$

Designation	Function
$P_0(u)$	1
$P_1(u)$	$u = \cos\theta$
$P_1^1(u)$	$(1 - u^2)^{1/2} = \sin\theta$
$P_2(u)$	$\frac{1}{2}(3u^2 - 1) = \frac{1}{4}(3\cos 2\theta + 1)$
$P_2^1(u)$	$3u(1 - u^2)^{1/2} = \frac{3}{2}\sin 2\theta$
$P_2^2(u)$	$3(1 - u^2) = \frac{3}{2}(1 - \cos 2\theta)$
$P_3(u)$	$\frac{1}{2}(5u^3 - 3u)$
$P_3^1(u)$	$\frac{3}{2}(1 - u^2)^{1/2}(5u^2 - 1)$
$P_3^2(u)$	$15u(1 - u^2)$
$P_3^3(u)$	$15(1 - u^2)^{3/2}$

---

* In Chapter 3 these functions were written $P_l(\theta)$. Since, however, the Legendre polynomials are polynomials in $\cos\theta$, it is more common to write $P_l(\cos\theta)$; we follow this practice in the present chapter as well as in the succeeding ones.

The substitution $R_l = \xi^{-1/2}Z_l$ transforms this equation into

$$\xi^2 \frac{d^2Z_l}{d\xi^2} + \xi \frac{dZ_l}{d\xi} - [(l + \tfrac{1}{2})^2 - \xi^2]Z_l = 0. \tag{17-85}$$

This equation, which is very familiar to mathematical physicists, is known as Bessel's equation. The solutions of the equation are also well known, and have been extensively investigated and indeed tabulated. The common solutions are designated $J_{l+1/2}(\kappa r)$ and $N_{l+1/2}(\kappa r)$ and are known respectively as the Bessel function and the Neumann function, of order $l + \tfrac{1}{2}$. For purposes of the wave equation, it is extremely convenient to define spherical Bessel functions by

$$j_l(\kappa r) = \sqrt{\pi/2\kappa r}\, J_{l+1/2}(\kappa r), \qquad n_l(\kappa r) = \sqrt{\pi/2\kappa r}\, N_{l+1/2}(\kappa r); \tag{17-86}$$

and from these in turn we obtain

$$h_l^{(1)}(\kappa r) = j_l(\kappa r) + in_l(\kappa r), \qquad h_l^{(2)} = j_l(\kappa r) - in_l(\kappa r). \tag{17-87}$$

The functions $j_l(\kappa r)$, $n_l(\kappa r)$, $h_l^{(1)}(\kappa r)$, and $h_l^{(2)}(\kappa r)$ are all solutions of the radial equation, Eq. (17–81). These functions are tabulated for $l = 0$, 1, and 2 in Table 17–2. The h's are particularly convenient for radiation problems because for large values of r they behave as

$$h_l^{(1)}(\kappa r) \xrightarrow[\kappa r \to \infty]{} \frac{(-i)^{l+1}e^{i\kappa r}}{\kappa r},$$

$$h_l^{(2)}(\kappa r) \xrightarrow[\kappa r \to \infty]{} \frac{i^{l+1}e^{-i\kappa r}}{\kappa r},$$

and thus lead to outgoing and ingoing spherical waves.

Table 17–2 Spherical Bessel and Neumann Functions

Type	Function
$j_0(\rho)$	$(1/\rho)\sin\rho$
$n_0(\rho)$	$-(1/\rho)\cos\rho$
$h_0^{(1)}(\rho)$	$-(i/\rho)e^{i\rho}$
$h_0^{(2)}(\rho)$	$(i/\rho)e^{-i\rho}$
$j_1(\rho)$	$(1/\rho^2)\sin\rho - (1/\rho)\cos\rho$
$n_1(\rho)$	$-(1/\rho)\sin\rho - (1/\rho^2)\cos\rho$
$h_1^{(1)}(\rho)$	$-(1/\rho)e^{i\rho}(1 + i/\rho)$
$h_1^{(2)}(\rho)$	$-(1/\rho)e^{-i\rho}(1 - i/\rho)$
$j_2(\rho)$	$\left[\dfrac{3}{\rho^3} - \dfrac{1}{\rho}\right]\sin\rho - \dfrac{3}{\rho^2}\cos\rho$
$n_2(\rho)$	$-\dfrac{3}{\rho^2}\sin\rho - \left[\dfrac{3}{\rho^3} - \dfrac{1}{\rho}\right]\cos\rho$
$h_2^{(1)}(\rho)$	$(i/\rho)e^{i\rho}\left(1 + \dfrac{3i}{\rho} - \dfrac{3}{\rho^2}\right)$
$h_2^{(2)}(\rho)$	$-(i/\rho)e^{-i\rho}\left(1 - \dfrac{3i}{\rho} - \dfrac{3}{\rho^2}\right)$

A general form for ψ may be written as

$$\psi_{lm} = \sqrt{\pi/2\kappa r}\, Z_l(\kappa r) P_l^m (\cos \theta)\, e^{\mp im\phi}. \tag{17–88}$$

The corresponding vector fields are computed by using Eqs. (17–62) and (17–72) for the TE waves, and (17–73) and (17–74) for the TM waves. The simplest interesting choice of ψ is ψ_{10}, which is just

$$\psi_{10} = \frac{1}{\kappa r}\, e^{i\kappa r} \left[1 + \frac{i}{\kappa r} \right] \cos \theta. \tag{17–89}$$

The gradient of ψ_{10} is

$$\nabla \psi_{10} = \mathbf{a}_r e^{i\kappa r} \left[\frac{i}{r} - \frac{2}{\kappa r^2} - \frac{2i}{\kappa^2 r^3} \right] \cos \theta - \mathbf{a}_\theta e^{i\kappa r} \left[\frac{1}{\kappa r^2} + \frac{i}{\kappa^2 r^3} \right] \sin \theta. \tag{17–90}$$

The spatial portion of the electric field is

$$\mathbf{E} = \mathbf{r} \times \nabla \psi_{10} = -\mathbf{a}_\phi E_0\, e^{i\kappa r} \left[\frac{1}{\kappa r} + \frac{i}{\kappa^2 r^2} \right] \sin \theta, \tag{17–91}$$

where E_0 has been introduced to make the equation dimensionally correct. The surfaces of constant phase, $\kappa r = $ constant, are spheres. (The surfaces of constant amplitude are not, however.) The spatial dependence of the magnetic induction is given by

$$\mathbf{B} = -i\, \frac{1}{\omega}\, \nabla \times \mathbf{E}$$

$$= i\, \frac{1}{\omega}\, E_0\, e^{i\kappa r} \left[\frac{1}{\kappa r^2} + \frac{i}{\kappa^2 r^3} \right] 2 \cos \theta \mathbf{a}_r$$

$$-i\, \frac{1}{\omega}\, E_0\, e^{i\kappa r} \left[\frac{i}{r} - \frac{1}{\kappa r^2} - \frac{i}{\kappa^2 r^3} \right] \sin \theta \mathbf{a}_\theta. \tag{17–92}$$

As will be seen later, these are just the fields (TE) produced by a radiating magnetic dipole. It is interesting to note that only the portions of $\mathbf{E}$ and $\mathbf{B}$ which are proportional to $1/r$ contribute to the net radiation. All other terms give terms in the Poynting vector that fall off more rapidly than $1/r^2$ and consequently have integrals over spherical surfaces that vanish as the radii of these spherical surfaces go to infinity. The spherical wave solutions are applicable in considering the radiation from bounded sources, which will be treated in Chapter 20 from another point of view.

17–6 SUMMARY

The transverse wave solutions of Maxwell's equations are most simply expressed in terms of plane waves,

$$\mathbf{E}(\mathbf{r},\, t) = \hat{\mathbf{E}} e^{-i(\omega t - \boldsymbol{\kappa} \cdot \mathbf{r})},$$

$$\mathbf{B}(\mathbf{r},\, t) = \hat{\mathbf{B}} e^{-i(\omega t - \boldsymbol{\kappa} \cdot \mathbf{r})}.$$

These solutions exist for any frequency ω and amplitude $\hat{\mathbf{E}}$, so long as $\boldsymbol{\kappa} \cdot \hat{\mathbf{E}} = 0$. The magnitude of $\boldsymbol{\kappa}$ is determined by the transverse dispersion relation, although its direction is arbitrary. The vector $\boldsymbol{\kappa}$ (real part) specifies the direction of propagation and wavelength ($\lambda = 2\pi/\kappa$). The B-field of the wave is determined by the other quantities.

1. In vacuum, the dispersion relation is

$$\kappa = \omega/c,$$

where $c = 1/\sqrt{\epsilon_0 \mu_0}$ is the velocity of light. The vectors $\boldsymbol{\kappa}$, $\mathbf{E}$, $\mathbf{B}$ form a right-handed orthogonal set. The magnitude of $\mathbf{B}$ is

$$B = E/c$$

at every instant. If the two components of $\mathbf{E}$ (in the plane perpendicular to $\boldsymbol{\kappa}$) are in phase, the polarization is linear; if they are out of phase it is elliptical. The magnitude of the Poynting vector and the energy density u are related by

$$S = cu,$$

and the direction of $\mathbf{S}$ is along $\boldsymbol{\kappa}$.

2. In a linear dielectric medium (nonconducting, nonmagnetic), all the above results still hold if c is replaced by c/n, where the index of refraction n is related to the dielectric constant K by

$$n = \sqrt{K}.$$

(In a linear magnetic medium, $n = \sqrt{KK_m}$, but this generalization has little application.)

3. In a conducting medium all of the same formal results hold if the complex dielectric constant

$$\hat{K} = K + ig/\epsilon_0\omega = K_r + iK_i$$

is used instead of K. This has the consequence that the refractive index

$$\hat{n} = n + ik$$

and the propagation vector

$$\hat{\boldsymbol{\kappa}} = \boldsymbol{\kappa}_r + i\boldsymbol{\kappa}_i$$

are also complex. The wave is attenuated:

$$\mathbf{E}(\mathbf{r}, t) = (\hat{\mathbf{E}}e^{-\boldsymbol{\kappa}_i \cdot \mathbf{r}})e^{-i(\omega t - \boldsymbol{\kappa}_r \cdot \mathbf{r})}.$$

In the simplest case, $\kappa_r = n\omega/c$, $\kappa_i = k\omega/c$; but in general all of the physical relations in (1) are more complicated.

4. In general

$$\hat{K} = \hat{n}^2.$$

The relation between the real optical constants and dielectric constants is

$$K_r = n^2 - k^2, \qquad K_i = 2nk,$$

and these can be inverted explicitly. The result is complicated, unless certain extreme cases are applicable, as is usually true.

5. For conductors the attenuation length

$$\delta = c/\omega k = \frac{n}{k} \frac{\lambda}{2\pi},$$

also called the *skin depth*, is important. For at least moderately good conductors, below the infrared frequency range, the approximation

$$n \cong k \cong \sqrt{g/2\epsilon_0 \omega}$$

can be expected to hold, so that

$$\delta \cong \sqrt{2/\mu_0 \omega g}.$$

Here g can be taken as the d-c conductivity.

6. Aside from plane waves, the most interesting set of transverse vector wave solutions is the spherical waves. They are much more complicated, but applicable in radiation problems.

PROBLEMS

17–1 Show that if $\mathbf{F(r)} = \mathbf{A}e^{i\boldsymbol{\kappa} \cdot \mathbf{r}}$, where $\mathbf{A}$ is constant,

$$\nabla \cdot \mathbf{F} = i\boldsymbol{\kappa} \cdot \mathbf{F},$$

$$\nabla \times \mathbf{F} = i\boldsymbol{\kappa} \times \mathbf{F}.$$

17–2 Show that for a plane wave in vacuum

$$\frac{E}{H} = \sqrt{\frac{\mu_0}{\epsilon_0}} = 377 \ \Omega.$$

This resistance is called the *impedance of free space.*

17–3 Two plane waves have the same ω, $\boldsymbol{\kappa}$, and amplitude E, but opposite circular polarization (i.e., left and right). Show that the superposition of the two waves is linearly polarized, with amplitude $2E$.

17–4 Sketch a figure like Fig. 17–1 for $E_s = 2E_p$, with $\phi = 30°$ and with $\phi = 60°$.

17–5 The earth receives about 1300 W/m^2 radiant energy from the sun. Assuming the energy to be in the form of a plane polarized monochromatic wave and assuming normal incidence, compute the magnitude of the electric and magnetic field vectors in the sunlight.

17–6 Suppose $\mathbf{A}$ and $\mathbf{B}$ are complex vectors. Calculate Re $\mathbf{A} \cdot$ Re $\mathbf{B}$ and compare it with Re $(\mathbf{A} \cdot \mathbf{B})$. Show that $\mathbf{A} \cdot \mathbf{B} = 0$ does not imply that Re $\mathbf{A} \cdot$ Re $\mathbf{B} = 0$.

17–7 Consider two plane waves in vacuum with the same ω, $\boldsymbol{\kappa}$, and polarization direction $\mathbf{p}$, but with different amplitudes and phases: E_1, 0 and E_2, ϕ. Calculate the time-average Poynting vector $\bar{\mathbf{S}}$ of the superposition of the two waves. Note the interference effect due to the phase difference ϕ, which would not occur if the two waves had perpendicular polarization directions.

17–8 Consider a standing wave in vacuum which is the superposition of two plane waves of the same frequency, amplitude, and linear polarization, but with opposite κ. Calculate the Poynting vector $S(r, t)$ of the superposed waves. (Note that the origin is not arbitrary.) What is $\bar{S}$?

17–9 For a plane wave in a conducting medium

$$\mathbf{B} = \frac{\hat{n}}{c}\,\mathbf{u} \times \mathbf{E}.$$

Suppose $\mathbf{E}$ is elliptically polarized, with $\hat{\mathbf{E}} = E_p\,e^{i\phi}\mathbf{p} + E_s\mathbf{s}$. Prove that at each instant of time

$$\mathrm{Re}\,\mathbf{E} \cdot \mathrm{Re}\,\mathbf{B} = -\frac{k}{c}\,E_p E_s \sin\,\phi.$$

17–10 For metals in the infrared spectral region, it happens that $K_i = -K_r$, at a frequency $\omega = g/(-\epsilon) \approx 10^{14}\ \mathrm{s}^{-1}$. Calculate the optical constants n and k for this case, in terms of K_i.

17–11 For a dielectric which becomes absorbing at high frequencies, or for a semiconductor, it happens that $K_i = K_r$. Calculate n and k for this case, in terms of K_r. Find δ/λ, the ratio of the attenuation length to the wavelength.

17–12 Show that in a nearly transparent medium of refractive index n, the attenuation length δ is related to the conductivity g by

$$\delta = \frac{2n}{g\sqrt{\mu_0/\epsilon_0}},$$

where $\sqrt{\mu_0/\epsilon_0} = 377\ \Omega$.

17–13 A Poynting vector proportional to E^2 falls off as $e^{-\alpha z}$, where $\alpha = 2/\delta$ is called the *absorption coefficient*. The power loss is often expressed in decibels per meter (dB/m), where a *decibel* is defined as ten times the common logarithm of the ratio of the initial to the final energy flux per unit area.

 a) Show that

$$\text{power loss} = 4.34\ \alpha\ \mathrm{dB/m}.$$

 b) From the result of Problem 17–12, calculate the (optical, not d-c) conductivity a lightwave communications medium must have in order to achieve a loss as low as 1 dB/km, assuming a refractive index of $n = 1.5$.

17–14 Suppose the dielectric constant $K = 0$. Show that Maxwell's equations with no external charge or current densities have a *longitudinal* solution in which $\mathbf{H} = 0$ and $\kappa \times \mathbf{E} = 0$, but $\kappa \cdot \mathbf{E} \neq 0$. Show that there is a polarization charge density $\rho_P = i\kappa\epsilon_0 E$. (The equation $K(\kappa, \omega) = 0$ is called the *longitudinal dispersion relation*.)

CHAPTER 18 Waves in Bounded Regions

The solutions of Maxwell's equations found in the preceding chapters will now be used to solve problems of practical interest. Two general classes of problems will be considered: boundary-value problems, and radiation from prescribed charge-current distributions. In the first class of problems, solutions of the homogeneous wave equation are so combined as to satisfy the appropriate boundary conditions. In the second class, solutions of the inhomogeneous wave equation with specified sources are required and boundary conditions are largely ignored, except for such things as insistence on outgoing waves and that the fields fall off as $1/r$ at large distances. The first class of problems is the subject of this chapter.

18–1 REFLECTION AND REFRACTION AT THE BOUNDARY OF TWO NONCONDUCTING MEDIA. NORMAL INCIDENCE

Electromagnetic waves propagating in materials usually enter the material through a boundary between it and another medium, perhaps air or vacuum. Treating this problem requires the application of the boundary conditions derived in Chapter 16. We begin with the simplest possible case: a plane wave normally incident on a plane dielectric interface. Experience tells us that the incident wave will be accompanied by a reflected and transmitted wave; we will see that the boundary conditions can be satisfied only if they are present. The three waves are each constrained to satisfy the relations between κ, $\mathbf{E}$, and $\mathbf{B}$ developed in Section 17–1. The situation is described in Fig. 18–1.

In this figure $\mathbf{E}_1$, $\mathbf{H}_1$ describe the incident wave traveling in the plus z-direction, $\mathbf{E}'_1$, $\mathbf{H}'_1$ describe the reflected wave traveling in the minus z-direction, and $\mathbf{E}_2$, $\mathbf{H}_2$ describe the transmitted wave. The interface is taken as coincident with the xy-plane at $z = 0$, with medium 1 on the left and medium 2 on the right. The electric fields, which are at first assumed to be linearly polarized in the x-direction, are described by

$$\mathbf{E}_1 = \mathbf{i}E_{1x}e^{i(\kappa_1 z - \omega t)},$$
$$\mathbf{E}'_1 = -\mathbf{i}E'_{1x}e^{-i(\kappa_1 z + \omega t)}, \qquad (18\text{–}1)$$
$$\mathbf{E}_2 = \mathbf{i}E_{2x}e^{i(\kappa_2 z - \omega t)},$$

382

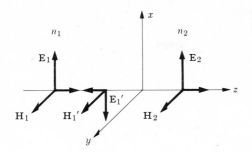

Figure 18–1 Reflection and transmission at normal incidence.

where

$$\kappa_1 = n_1 \frac{\omega}{c} \qquad \text{and} \qquad \kappa_2 = n_2 \frac{\omega}{c}. \tag{18-2}$$

From Eq. (17–22),

$$\hat{\mathbf{B}} = \frac{n}{c} \mathbf{u} \times \hat{\mathbf{E}},$$

where in this case $\mathbf{u} = \mathbf{k}$ for the incident and transmitted waves and $\mathbf{u} = -\mathbf{k}$ for the reflected wave, the magnetic fields associated with the electric fields of Eq. (18–1) are given by

$$c\mathbf{B}_1 = jn_1 E_{1x} e^{i(\kappa_1 z - \omega t)},$$
$$c\mathbf{B}'_1 = jn_1 E'_{1x} e^{-i(\kappa_1 z + \omega t)}, \tag{18-3}$$
$$c\mathbf{B}_2 = jn_2 E_{2x} e^{i(\kappa_2 z - \omega t)}.$$

Clearly the reflected and transmitted waves must have the same frequency ω as the incident wave if boundary conditions at $z = 0$ are to be satisfied for all t. Since the normal components of the fields vanish, only the boundary conditions on the tangential components need be considered. The E-field must be continuous at the boundary, so from Eq. (18–1) at $z = 0$

$$E_{1x} - E'_{1x} = E_{2x}. \tag{18-4}$$

The H-field must also be continuous, and for nonmagnetic media ($\mu_1 = \mu_2 = \mu_0$) so must the B-field:

$$n_1(E_{1x} + E'_{1x}) = n_2 E_{2x}. \tag{18-5}$$

Equations (18–4) and (18–5) can be solved simultaneously for the amplitudes E'_{1x} and E_{2x} in terms of the given amplitude E_{1x} of the incident wave:

$$E'_{1x} = \frac{n_2 - n_1}{n_2 + n_1} E_{1x}, \qquad E_{2x} = \frac{2n_1}{n_2 + n_1} E_{1x}.$$

The ratios of the reflected and transmitted wave amplitudes to the incident amplitude are determined entirely by the refractive indices of the two media. These amplitudes in turn determine the amplitudes of the magnetic fields through Eq. (18–3).

Since only the ratios are determined, it is convenient to introduce a special notation for them,

$$\frac{E'_{1x}}{E_{1x}} = r_{12}, \qquad \frac{E_{2x}}{E_{1x}} = t_{12}; \qquad (18\text{–}6)$$

r_{12} and t_{12} are called *Fresnel coefficients* for normal incidence reflection and transmission, respectively. The subscripts indicate that the wave is incident from medium 1 onto medium 2. Thus the solution is given as

$$r_{12} = \frac{n_2 - n_1}{n_2 + n_1}, \qquad t_{12} = \frac{2n_1}{n_1 + n_2}. \qquad (18\text{–}7)$$

What is usually measurable is not the reflected and transmitted *E*-fields, but the reflected and transmitted average energy fluxes per unit area. The latter are given by the Poynting vector and are called the *intensities* of the waves. According to Eqs. (17–34) and (17–36), in each medium

$$\bar{S} = \frac{1}{2} \frac{n}{\mu_0 c} (E_p^2 + E_s^2). \qquad (18\text{–}8)$$

Here we have chosen $E_p = E_x$, $E_s = 0$. We define the *reflectance* R_n and the *transmittance* T_n for normal incidence by the ratios of the intensities

$$\frac{\bar{S}'_1}{\bar{S}_1} = R_n, \qquad \frac{\bar{S}_2}{\bar{S}_1} = T_n. \qquad (18\text{–}9)$$

Then from Eqs. (18–8) and (18–6)

$$R_n = r_{12}^2, \qquad T_n = \frac{n_2}{n_1} t_{12}^2. \qquad (18\text{–}10)$$

With the Fresnel coefficients given by Eq. (18–7), the reflectance and transmittance of Eq. (18–10) satisfy

$$R_n + T_n = 1 \qquad (18\text{–}11)$$

for any pair of nonconducting media. This is an expression of energy conservation at the interface.

So far we have considered only linearly polarized radiation. If the incident wave is elliptically polarized, we must consider the perpendicular components $\hat{E}_y = \hat{E}_s$ in each medium, in addition to the *x*-components. They will also satisfy equations like (18–6), with the same Fresnel coefficients. The three *y*-components are all in phase with each other, although they are out of phase with the three *x*-components. From Eq. (18–8) we see that the intensities associated with the $x(p)$ and $y(s)$ components simply add,

$$\bar{S} = \bar{S}_p + \bar{S}_s, \qquad (18\text{–}12)$$

regardless of the phase difference between the p- and s-components, i.e., regardless of the degree of elliptical polarization. Thus Eqs. (18–9)–(18–11) hold separately for each polarization component and also for the total intensities.

For a typical air-glass interface, where $n_2 = 1.5$ and $n_1 = 1$, the reflection and transmission coefficients are

$$R_n = 0.04 \qquad \text{and} \qquad T_n = 0.96.$$

Thus, as is expected from Eq. (18–11), all of the incident energy is either reflected or transmitted—there is no place to store energy in the interface. A further interesting fact is obtained by examining Eq. (18–6); namely, if n_2 is greater than n_1 the first ratio is positive. This is precisely the familiar statement from optics that there is a phase change of π radians on reflection from a "more dense" medium but that there is no phase change on reflection from a "less dense" medium.

As a second example, in water $n_2 = 1.33$ for visible light, so that $R_n = 0.02$. Below $\omega \approx 10^{11}$ s^{-1}, however, $K_2 = 81$. For radio frequencies pure water approximates a nonconductor (Case 1 in Section 17–4) and so $n_2 = \sqrt{K_2} = 9$,

$$R_n = 0.64.$$

For sea water, a better conductor, the reflectance is much larger yet, as we will find in Section 18–4.

18–2 REFLECTION AND REFRACTION AT THE BOUNDARY OF TWO NONCONDUCTING MEDIA. OBLIQUE INCIDENCE

A more general case than that discussed in the preceding section is that of reflection of obliquely incident plane waves by a plane dielectric interface. Consideration of this case leads to three well-known optical laws: Snell's law, the law of reflection, and Brewster's law governing polarization by reflection.

The situation is described by Fig. 18–2, if the propagation vectors κ_1, κ_1', and κ_2 are coplanar and lie in the xz-plane, and the electric field vectors E_1, E_1', and E_2

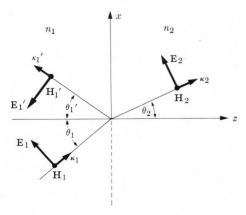

Figure 18–2 Reflection and refraction—oblique incidence. The xz-plane is the plane of incidence. The vectors H_1, H_1', and H_2 are directed out of the paper.

also lie in this plane.* The electric fields of the incident, reflected, and transmitted waves are given by

$$\mathbf{E}_1 = \hat{\mathbf{E}}_{1p}\, e^{i(\kappa_1 \cdot \mathbf{r} - \omega t)},$$

$$\mathbf{E}_1' = \hat{\mathbf{E}}_{1p}'\, e^{i(\kappa_1' \cdot \mathbf{r} - \omega t)}, \qquad (18\text{--}13)$$

$$\mathbf{E}_2 = \hat{\mathbf{E}}_{2p}\, e^{i(\kappa_2 \cdot \mathbf{r} - \omega t)},$$

where $\hat{\mathbf{E}}_{1p} = \hat{E}_{1p}\mathbf{p}_1$, $\hat{\mathbf{E}}_{1p}' = \hat{E}_{1p}'\mathbf{p}_1'$, $\hat{\mathbf{E}}_{2p} = \hat{E}_{2p}\mathbf{p}_2$. The propagation vectors are $\kappa_1 = \kappa_1\mathbf{u}_1$, etc., and the unit normal to the boundary is $\mathbf{n} = \mathbf{k}$. The plane defined by κ_1 and $\mathbf{n}$ is called the *plane of incidence*, and its normal is in the direction of $\kappa_1 \times \mathbf{n}$. The p-component of polarization has been chosen to lie parallel to the plane of incidence (p for "parallel"). In general there is also an s-component (not shown in Fig. 18–2) of the amplitude of each wave, $\hat{\mathbf{E}}_{1s} = \hat{E}_{1s}\mathbf{s}_1$, $\hat{\mathbf{E}}_{1s}' = \hat{E}_{1s}'\mathbf{s}_1'$, $\hat{\mathbf{E}}_{2s} = \hat{E}_{2s}\mathbf{s}_2$. For each of the three waves $\mathbf{s} = \mathbf{u} \times \mathbf{p}$ and $\mathbf{p} = \mathbf{s} \times \mathbf{u}$, so that

$$\mathbf{s}_1 = \mathbf{s}_1' = \mathbf{s}_2 = \mathbf{j}. \qquad (18\text{--}14)$$

The s-components are perpendicular to the plane of incidence (s for "senkrecht," the German word for perpendicular).

With oblique incidence it is clear from Eqs. (18–13) that not only must the reflected and transmitted waves have the same frequency as the incident wave, but also the phases must match everywhere on the boundary:

$$\kappa_1' \cdot \mathbf{r} = \kappa_1 \cdot \mathbf{r} = \kappa_2 \cdot \mathbf{r} \text{ on the boundary.} \qquad (18\text{--}15)$$

This condition has three interesting consequences, which we now derive algebraically. We need an algebraic expression of the fact that Eq. (18–15) holds only on the boundary $z = 0$ or $\mathbf{n} \cdot \mathbf{r} = 0$. Consider the vector identity

$$\mathbf{n} \times (\mathbf{n} \times \mathbf{r}) = (\mathbf{n} \cdot \mathbf{r})\mathbf{n} - \mathbf{r}.$$

Everywhere on the boundary, $\mathbf{n} \cdot \mathbf{r} = 0$,

$$\mathbf{r} = -\mathbf{n} \times (\mathbf{n} \times \mathbf{r}),$$

and so we substitute this into Eq. (18–15).

$$\kappa_1 \cdot \mathbf{r} = -\kappa_1 \cdot \mathbf{n} \times (\mathbf{n} \times \mathbf{r}) = -(\kappa_1 \times \mathbf{n}) \cdot (\mathbf{n} \times \mathbf{r})$$

and similarly for the other members of Eq. (18–15). Since $\mathbf{r}$ is an arbitrary vector on the boundary, Eq. (18–15) can hold if and only if

$$\kappa_1' \times \mathbf{n} = \kappa_1 \times \mathbf{n} = \kappa_2 \times \mathbf{n}. \qquad (18\text{--}16)$$

This implies first that κ_1' lies in the plane of incidence, since the normal to the plane defined by κ_1' and $\mathbf{n}$ is parallel to the normal to the plane of incidence; and

---

* It will be proved that the propagation vectors are always coplanar. The most general electric field vector can be resolved into a component in the xz-plane (plane of incidence) and a component perpendicular to this plane. The reflection and transmission of these two components are governed by different laws. The choice illustrated gives Brewster's law.

likewise κ_2 is in the plane of incidence. That is, $\mathbf{n}$, κ_1, κ_1', and κ_2 are *all coplanar*. Equation (18–14) is justified, and $\mathbf{s} = \mathbf{j}$ is the unit normal of the plane of incidence. Now the angle of incidence θ_1 is given by

$$\kappa_1 \cdot \mathbf{n} = \kappa_1 \cos\theta_1,$$

and also

$$\kappa_1' \cdot \mathbf{n} = -\kappa_1' \cos\theta_1',$$

$$\kappa_2 \cdot \mathbf{n} = \kappa_2 \cos\theta_2. \tag{18–17}$$

Therefore

$$\left|\kappa_1 \times \mathbf{n}\right| = \kappa_1 \sin\theta_1, \qquad \left|\kappa_1' \times \mathbf{n}\right| = \kappa_1' \sin\theta_1',$$

$$\left|\kappa_2 \times \mathbf{n}\right| = \kappa_2 \sin\theta_2,$$

and Eq. (18–16) implies that

$$\kappa_1' \sin\theta_1' = \kappa_1 \sin\theta_1 = \kappa_2 \sin\theta_2.$$

The magnitude κ_1' of the reflected wave equals κ_1 of the incident wave, since they are propagating with the same frequency in the same medium. Thus

$$\theta_1' = \theta_1, \tag{18–18}$$

the *angle of reflection equals the angle of incidence*. From the dispersion relations Eq. (18–2), the second equality above gives

$$n_1 \sin\theta_1 = n_2 \sin\theta_2. \tag{18–19}$$

This is *Snell's law*. Note that none of these three consequences depended on the boundary conditions for the electric and magnetic fields derived from Maxwell's equations. The first two must hold for any kind of wave; the third, Snell's law, depends on the particular dispersion relation for the waves.

To derive the Fresnel coefficients that will give the ratios of the field amplitudes for oblique incidence, we do need the boundary conditions on the tangential components of the fields. (The conditions on the normal components turn out to be satisfied automatically.) To express the tangential component of the E-field amplitude in vector form, consider the vector identity

$$\mathbf{n} \times (\mathbf{n} \times \hat{\mathbf{E}}) = (\mathbf{n} \cdot \hat{\mathbf{E}})\mathbf{n} - \hat{\mathbf{E}},$$

or

$$\hat{\mathbf{E}} = (\mathbf{n} \cdot \hat{\mathbf{E}})\mathbf{n} - \mathbf{n} \times (\mathbf{n} \times \hat{\mathbf{E}}).$$

Since $(\mathbf{n} \cdot \hat{\mathbf{E}})\mathbf{n}$ is just the normal component of $\hat{\mathbf{E}}$, the remainder has to be the tangential component, $-\mathbf{n} \times (\mathbf{n} \times \hat{\mathbf{E}})$, for any vector $\hat{\mathbf{E}}$. Then the boundary condition on the E-field becomes

$$\mathbf{n} \times (\hat{\mathbf{E}}_1 + \hat{\mathbf{E}}_1') = \mathbf{n} \times \hat{\mathbf{E}}_2 \tag{18–20}$$

after canceling the exponentials (because of Eq. (18–15)) and $\mathbf{n}\times$. Since we are assuming a nonmagnetic medium, the continuity of the tangential B-field is also required,

$$\mathbf{n} \times (\hat{\mathbf{B}}_1 + \hat{\mathbf{B}}_1') = \mathbf{n} \times \hat{\mathbf{B}}_2. \qquad (18\text{–}21)$$

The Maxwell curl equations relate $\hat{\mathbf{E}}$ and $\hat{\mathbf{B}}$ to each other in each medium. With the dispersion relations Eq. (18–2) these take the form of Eq. (17–22):

$$\hat{\mathbf{B}} = \frac{n}{c}\mathbf{u} \times \hat{\mathbf{E}} \qquad (18\text{–}22)$$

or its equivalent

$$\hat{\mathbf{E}} = -\frac{c}{n}\mathbf{u} \times \hat{\mathbf{B}}. \qquad (18\text{–}23)$$

If, for example, we substitute Eq. (18–22) into Eq. (18–21), it becomes

$$n_1\mathbf{n} \times (\mathbf{u}_1 \times \hat{\mathbf{E}}_1 + \mathbf{u}_1' \times \hat{\mathbf{E}}_1') = n_2\mathbf{n} \times (\mathbf{u}_2 \times \hat{\mathbf{E}}_2). \qquad (18\text{–}24)$$

Equations (18–20) and (18–24) are a pair of vector equations, which must be solved for $\hat{\mathbf{E}}_1'$ and $\hat{\mathbf{E}}_2$ in terms of $\hat{\mathbf{E}}_1$. We can expand the vector triple products:

$$\mathbf{n} \times (\mathbf{u}_1 \times \hat{\mathbf{E}}_1) = (\mathbf{n} \cdot \hat{\mathbf{E}}_1)\mathbf{u}_1 - (\mathbf{n} \cdot \mathbf{u}_1)\hat{\mathbf{E}}_1$$

and similarly for the others. For the s-component $\hat{\mathbf{E}}_{1s}$, for which $\mathbf{n} \cdot \hat{\mathbf{E}}_{1s} = 0$, this expression simplifies considerably:

$$\mathbf{n} \times (\mathbf{u}_1 \times \hat{\mathbf{E}}_{1s}) = -\cos\theta_1 \hat{\mathbf{E}}_{1s},$$

since $\mathbf{n} \cdot \mathbf{u}_1 = \cos\theta_1$ from Eq. (18–17). Thus for the s-component Eq. (18–24) becomes

$$n_1(\cos\theta_1 \hat{\mathbf{E}}_{1s} - \cos\theta_1' \hat{\mathbf{E}}_{1s}') = n_2\cos\theta_2 \hat{\mathbf{E}}_{2s},$$

and since $\theta_1' = \theta_1$,

$$n_1\cos\theta_1(\hat{\mathbf{E}}_{1s} - \hat{\mathbf{E}}_{1s}') = n_2\cos\theta_2 \hat{\mathbf{E}}_{2s}. \qquad (18\text{–}25)$$

Taking the cross product of Eq. (18–20) with $\mathbf{n}$ gives for the s-component

$$\hat{\mathbf{E}}_{1s} + \hat{\mathbf{E}}_{1s}' = \hat{\mathbf{E}}_{2s}. \qquad (18\text{–}26)$$

No such simplification occurs for the p-component, and so we are led to consider the two polarization components separately. Substitution of Eq. (18–23) into Eq. (18–20) will make a corresponding simplification for the p-polarized case,

1. s-polarization. Equations (18–25) and (18–26) are easily solved simultaneously. We get

$$\hat{\mathbf{E}}_{1s}' = r_{12s}\hat{\mathbf{E}}_{1s}, \qquad \hat{\mathbf{E}}_{2s} = t_{12s}\hat{\mathbf{E}}_{1s}, \qquad (18\text{–}27)$$

where

$$r_{12s} = \frac{n_1 \cos \theta_1 - n_2 \cos \theta_2}{n_1 \cos \theta_1 + n_2 \cos \theta_2}, \qquad (18\text{-}28)$$

$$t_{12s} = \frac{2n_1 \cos \theta_1}{n_1 \cos \theta_1 + n_2 \cos \theta_2}. \qquad (18\text{-}29)$$

2. p-polarization. When the E-vectors all lie in the plane of incidence, Eq. (18–22) shows that the corresponding B-vectors all lie along the s-direction. The choice of E-vectors in Fig. 18–2 was made so that all the corresponding B-vectors point in the $+\mathbf{j}$ direction. This time we substitute Eq. (18–23) into Eq. (18–20). Since $\mathbf{n} \cdot \hat{\mathbf{B}}_{1s} = 0 = \mathbf{n} \cdot \hat{\mathbf{B}}'_{1s} = \mathbf{n} \cdot \hat{\mathbf{B}}_{2s}$, the result simplifies to

$$\frac{1}{n_1} \cos \theta_1 (\hat{\mathbf{B}}_{1s} - \hat{\mathbf{B}}'_{1s}) = \frac{1}{n_2} \cos \theta_2 \hat{\mathbf{B}}_{2s}. \qquad (18\text{-}30)$$

Also, Eq. (18–21) simplifies to

$$\hat{\mathbf{B}}_{1s} + \hat{\mathbf{B}}'_{1s} = \hat{\mathbf{B}}_{2s}. \qquad (18\text{-}31)$$

We can write the solution of Eqs. (18–30) and (18–31) as

$$\hat{\mathbf{B}}'_{1s} = r_{12p} \hat{\mathbf{B}}_{1s}, \qquad \mathbf{B}_{2s} = \frac{n_2}{n_1} t_{12p} \hat{\mathbf{B}}_{1s}, \qquad (18\text{-}32)$$

where

$$r_{12p} = \frac{n_2 \cos \theta_1 - n_1 \cos \theta_2}{n_2 \cos \theta_1 + n_1 \cos \theta_2}, \qquad (18\text{-}33)$$

$$t_{12p} = \frac{2n_1 \cos \theta_1}{n_2 \cos \theta_1 + n_1 \cos \theta_2}. \qquad (18\text{-}34)$$

We write Eq. (18–32) in this way so that when we calculate the magnitudes of the E-vectors from Eq. (18–23) we get

$$\hat{E}'_{1p} = r_{12p} \hat{E}_{1p}, \qquad \hat{E}_{2p} = t_{12p} \hat{E}_{1p}. \qquad (18\text{-}35)$$

With these Fresnel coefficients we now have a complete solution of the boundary-value problem, since an incident wave of arbitrary polarization can be resolved into p- and s-components. It should be emphasized that the terms p- and s-polarization always refer to the direction of the E-vector.* For normal incidence $\theta_1 = 0$, and according to Snell's law $\theta_2 = 0$ as well. Thus Eqs. (18–33)–(18–35) reduce to Eqs. (18–6) and (18–7) for normal incidence. Equation (18–29) becomes identical to Eq. (18–34), but from Eq. (18–28) we find that $r_{12s} = -r_{12p}$ for normal incidence. For normal incidence the plane of incidence is undefined, and

---

* In older optics books, there occurs the term "plane of polarization," defined as the plane containing the B-vector and the propagation vector. We will avoid this term altogether.

so the physical result has to be independent of polarization. The difference arises only because $\mathbf{E}_{1p}$ and $\mathbf{E}'_{1p}$ point in opposite directions in Fig. 18–2 when θ_1 goes to zero, whereas $\mathbf{E}_{1s}$ and $\mathbf{E}'_{1s}$ point in the same direction. For s-polarization the Fresnel coefficients are relations between the E-vectors Eq. (18–27); for p-polarization they are relations only between their magnitudes Eq. (18–35), since the E-vectors all point in different directions with oblique incidence. Using Snell's law we could write

$$\cos \theta_2 = \sqrt{1 - (n_1/n_2)^2 \sin^2 \theta_1},$$

and thereby express the Fresnel coefficients entirely in terms of the material parameters n_1, n_2 and the given incident angle θ_1.

Relations between the intensities can again be obtained from the Fresnel coefficients, treating each polarization direction separately. We define the reflectance and transmittance as the component of the respective time-averaged Poynting vectors that is normal to the boundary, relative to the normal component of the incident Poynting vector,*

$$R_s = \frac{\mathbf{n} \cdot \bar{\mathbf{S}}'_{1s}}{\mathbf{n} \cdot \bar{\mathbf{S}}_{1s}}, \qquad T_s = \frac{\mathbf{n} \cdot \bar{\mathbf{S}}_{2s}}{\mathbf{n} \cdot \bar{\mathbf{S}}_{1s}}, \tag{18–36}$$

$$R_p = \frac{\mathbf{n} \cdot \bar{\mathbf{S}}'_{1p}}{\mathbf{n} \cdot \bar{\mathbf{S}}_{1p}}, \qquad T_p = \frac{\mathbf{n} \cdot \bar{\mathbf{S}}_{2p}}{\mathbf{n} \cdot \bar{\mathbf{S}}_{1p}}. \tag{18–37}$$

In terms of the Fresnel coefficients, we have†

$$R_s = r_{12s}^2, \qquad T_s = \frac{n_2 \cos \theta_2}{n_1 \cos \theta_1} t_{12s}^2, \tag{18–38}$$

$$R_p = r_{12p}^2, \qquad T_p = \frac{n_2 \cos \theta_2}{n_1 \cos \theta_1} t_{12p}^2. \tag{18–39}$$

The identities

$$R_s + T_s = 1, \qquad R_p + T_p = 1, \tag{18–40}$$

hold for oblique incidence on a nonconductor.

---

* In medium 1, $\mathbf{S} = (\mathbf{E}_1 + \mathbf{E}'_1) \times (\mathbf{H}_1 + \mathbf{H}'_1)$. It is possible to show that

$$\mathbf{n} \cdot \overline{(\mathbf{E}_1 \times \mathbf{H}'_1 + \mathbf{E}'_1 \times \mathbf{H}_1)} = 0$$

in a nonconducting medium (the only case we will consider), so that $\mathbf{n} \cdot \bar{\mathbf{S}} = \mathbf{n} \cdot (\bar{\mathbf{S}}_1 + \bar{\mathbf{S}}'_1)$ and it is meaningful to separate the Poynting vectors.

† Alternative definitions of the reflectance and transmittance are occasionally made, as ratios of the magnitudes of the Poynting vectors instead of normal components. This makes no difference in R, but removes the factor $(\cos \theta_2 / \cos \theta_1)$ from T.

For certain purposes, it is more convenient to have the Fresnel coefficients in the following form:

$$r_{12s} = \frac{\sin (\theta_2 - \theta_1)}{\sin (\theta_2 + \theta_1)}, \tag{18-41}$$

$$t_{12s} = \frac{2 \cos \theta_1 \sin \theta_2}{\sin (\theta_2 + \theta_1)}, \tag{18-42}$$

$$r_{12p} = \frac{\tan (\theta_1 - \theta_2)}{\tan (\theta_1 + \theta_2)}, \tag{18-43}$$

$$t_{12p} = \frac{2 \cos \theta_1 \sin \theta_2}{\sin (\theta_1 + \theta_2) \cos (\theta_1 - \theta_2)}. \tag{18-44}$$

These are easily seen to be equivalent to the above forms by using trigonometric identities and Snell's law.

18–3 BREWSTER'S ANGLE. CRITICAL ANGLE

We next consider the dependence of R and T on the angle of incidence for the case of two nonconducting media, using the Fresnel coefficients derived in the preceding section. In every case, $T = 1 - R$, so we will discuss only R. We have already examined the case of normal incidence $\theta_1 = 0$: The polarization does not matter, and R is greater the more the ratio n_2/n_1 differs from unity. For grazing incidence $\theta_1 = \pi/2$, $\cos \theta_1 = 0$, and $R_s = 1 = R_p$, as can be seen most easily from Eqs. (18–28) and (18–33). Near grazing incidence the reflectance is large; this high reflectance is the reason a calm lake is like a mirror. For intermediate angles of incidence, there are two particularly interesting angles.

Can there be a case of zero reflectance? Equations (18–41) and (18–43) show that there can. If $\theta_1 = \theta_2$, then $\tan (\theta_1 - \theta_2) = 0 = \sin (\theta_2 - \theta_1)$ and there is no reflected wave. Unfortunately, this can occur only if $n_1 = n_2$, that is, if the two media are optically indistinguishable. If, on the other hand, $\theta_1 + \theta_2 = \pi/2$, then $\tan (\theta_1 + \theta_2)$ is infinite and the amplitude of the p-polarized reflected wave is again zero. In this case the media are optically distinguishable. Since the s-polarization, $\mathbf{E}$ perpendicular to the plane of incidence, is partially reflected, unpolarized light incident at an angle satisfying $\theta_1 + \theta_2 = \pi/2$ will be polarized by reflection. Snell's law,

$$n_1 \sin \theta_1 = n_2 \sin \theta_2,$$

provides a means for determining the value of θ_1. Using $\theta_2 = \pi/2 - \theta_1$ in Snell's law gives

$$n_1 \sin \theta_B = n_2 \sin \left(\frac{\pi}{2} - \theta_B\right) = n_2 \cos \theta_B,$$

or

$$\tan \theta_B = \frac{n_2}{n_1}. \tag{18-45}$$

The quantity θ_B is known as *Brewster's angle*; the relationship between it and the indices of refraction as given in Eq. (18–45) is known as Brewster's law. Polarization at the Brewster angle is a practical means of producing polarized radiation, although not the most common one. In Fig. 18–3 the values of R_s and R_p are plotted for all values of θ_1, with $n_1 = 1$, $n_2 = 1.5$, as for an air–glass interface. The Brewster angle is $\theta_B = 56°$ for this case. The generally lower reflectance for p-polarized light accounts for the usefulness of Polaroid sunglasses. Since most outdoor reflecting surfaces are horizontal, the plane of incidence for most reflected glare reaching the eyes is vertical. On the assumption that one's head is usually erect, the polarizing lenses are oriented to pass light with E-vector in the vertical plane and eliminate the other more strongly reflected s-component.

There is another case besides grazing incidence in which $R_s = R_p = 1$. From Eqs. (18–28) and (18–33) we see that perfect reflection occurs for $\theta_2 = \pi/2$, as well as for $\theta_1 = \pi/2$. The incident angle for which $\theta_2 = \pi/2$ is called the *critical angle*, $\theta_1 = \theta_c$. From Snell's law

$$\sin \theta_c = \frac{n_2}{n_1}, \tag{18-46}$$

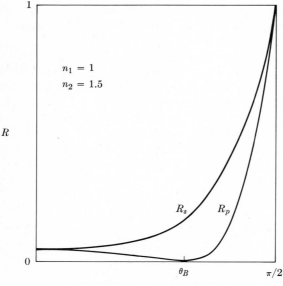

Figure 18–3 Reflectance for s- and p-polarization at an air–glass interface. Brewster's angle is $\theta_B = 56°$.

The critical angle is a real angle only if $n_2 < n_1$, but for any pair of transparent materials this will hold for incidence on one side or the other of their interface. From Eqs. (18–45) and (18–46)

$$\tan \theta_B = \sin \theta_c.$$

Since $\tan \theta$ is not restricted in value, there is always a real Brewster's angle; furthermore, since $\tan \theta > \sin \theta$, $\theta_B < \theta_c$. The case of a glass–air interface with $n_1 = 1.5$, $n_2 = 1$ is plotted in Fig. 18–4. For incidence from the glass side, $\theta_B = 34°$ and $\theta_c = 42°$. For incident angles greater than the critical angle, $\theta_1 > \theta_c$, Snell's law gives

$$\sin \theta_2 = \frac{n_1}{n_2} \sin \theta_1 > \frac{n_1}{n_2} \sin \theta_c.$$

Since $\sin \theta_c = n_2/n_1$, this requires

$$\sin \theta_2 > 1. \tag{18–47}$$

There is no such real angle, but this complication is not serious and is resolved in the next section. The result is that $R_s = R_p = 1$ for all $\theta_1 \geq \theta_c$. This perfect reflection is called *total internal reflection*. It is easily observed by looking into a glass prism or an aquarium or by looking up when swimming under water. It has a very important practical application in the light pipe, a fine glass fiber through which a light beam is transmitted, as in a waveguide for microwaves. (See Problem 18–4.)

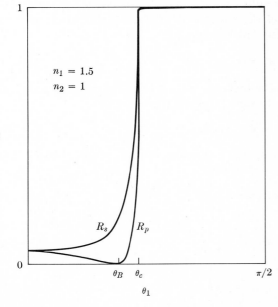

Figure 18–4 Reflectance for s- and p-polarization at a glass–air interface. Brewster's angle $\theta_B = 34°$ and the critical angle is $\theta_c = 42°$.

Our examples in this section have involved frequencies in the visible-light range, and transparent materials for which $n = \sqrt{K}$. For nonpolar materials the same relations hold at all lower frequencies (but not in the ultraviolet and higher frequency ranges). They do not hold at all lower frequencies for optically transparent polar materials made of polar molecules (e.g., water) or ions (e.g., rocksalt), since K is frequency dependent.

18–4 COMPLEX FRESNEL COEFFICIENTS.
REFLECTION FROM A CONDUCTING PLANE

The complication that arose in the last section for angles of incidence greater than the critical angle, namely $\sin \theta_2 > 1$, leads us to consider complex Fresnel coefficients. Since $\cos \theta = \sqrt{1 - \sin^2 \theta}$, a real value of $\sin \theta$ greater than 1 implies a pure imaginary value of $\cos \theta$, so that $\cos \hat{\theta}_2$ in the Fresnel coefficients is imaginary and they are complex. They would also be complex if medium 2 were conducting, since in that case $\hat{n}_2$ is complex. Snell's law

$$n_1 \sin \theta_1 = \hat{n}_2 \sin \hat{\theta}_2$$

shows that then $\sin \hat{\theta}_2$ has to be complex as well. We must, therefore, examine whether our derivation in Section 18–2 holds for complex angles and refractive indices. There is no way to draw Fig. 18–2 with a complex angle $\hat{\theta}_2$, and so we will have to be careful about the geometrical meaning of our conclusions. The derivation, however, did not appeal to the geometry of the figure. It was completely algebraic, and since all the algebraic vector relations hold as well for complex quantities as for real ones, the results are formally correct. We will be concerned only with cases in which one of the two media is transparent, which we now assume to be medium 1. Thus Eq. (18–16) becomes

$$\boldsymbol{\kappa}_1 \times \mathbf{n} = \hat{\boldsymbol{\kappa}}_2 \times \mathbf{n}, \tag{18–48}$$

so that the plane of incidence has a real unit normal $\mathbf{j}$ and the complex propagation vector $\hat{\boldsymbol{\kappa}}_2$ has no component in the $\mathbf{j}$ direction:

$$\hat{\boldsymbol{\kappa}}_2 \cdot \mathbf{j} = 0. \tag{18–49}$$

This is a restrictive assumption, but one which is valid in practical cases. The complex angle $\hat{\theta}_2$ is defined algebraically by

$$\hat{\boldsymbol{\kappa}}_2 \cdot \mathbf{n} = \hat{\kappa}_2 \cos \hat{\theta}_2. \tag{18–50}$$

Then Snell's law becomes

$$n_1 \sin \theta_1 = \hat{n}_2 \sin \hat{\theta}_2, \tag{18–51}$$

where

$$\sin \hat{\theta}_2 = \sqrt{1 - \cos^2 \hat{\theta}_2}. \tag{18–52}$$

All of the algebraic manipulations with the boundary conditions on the E- and B-fields are valid, and so the complex Fresnel coefficients are still given by Eqs. (18–28), (18–29), (18–33), and (18–34), with complex $\hat{n}_2$ and $\cos \hat{\theta}_2$. If they are expressed in polar form,

$$\hat{r}_{12s} = |\hat{r}_{12s}|e^{i\alpha_s}, \qquad \hat{r}_{12p} = |\hat{r}_{12p}|e^{i\alpha_p}, \ldots,$$

and used in Eqs. (18–27) and (18–35),

$$\hat{E}'_{1s} = |\hat{r}_{12s}|e^{i\alpha_s}\hat{E}_{1s}, \qquad \hat{E}'_{1p} = |\hat{r}_{12p}|e^{i\alpha_p}\hat{E}_{1p}, \ldots, \qquad (18\text{–}53)$$

it is clear that the reflected and transmitted E-fields are phase-shifted with respect to the incident E-field. The real reflectances for intensity defined by Eqs. (18–36) and (18–37) become

$$R_s = |\hat{r}_{12s}|^2, \qquad R_p = |\hat{r}_{12p}|^2, \qquad (18\text{–}54)$$

since the phases have no effect on the Poynting vectors given by Eq. (18–8). Considerable care is needed to obtain the correct transmittances from the complex quantities, but we shall have no use for these since in most cases they are not measurable in the conducting medium. In place of $R + T = 1$, the identities

$$\hat{r}_{12} = -\hat{r}_{21}, \qquad (18\text{–}55)$$

$$\hat{r}_{12}^2 + \hat{t}_{12}\hat{t}_{21} = 1 \qquad (18\text{–}56)$$

are useful when one medium is conducting. These hold for both s- and p-polarization.

For normal incidence from air on a conducting medium, with $n_1 = 1$, $\hat{n}_2 = n + ik$, the reflectance is

$$R_n = \frac{(n-1)^2 + k^2}{(n+1)^2 + k^2}. \qquad (18\text{–}57)$$

Since all the transmitted energy is eventually absorbed in a semi-infinite conducting medium, we define the *absorptance* as

$$A = 1 - R. \qquad (18\text{–}58)$$

For normal incidence

$$A_n = \frac{4n}{(n+1)^2 + k^2}. \qquad (18\text{–}59)$$

The absorptance is small (high reflectance) if $n \ll 1$, or $n \gg 1$, or $k \gg 1$. When $n \cong k \gg 1$ (case 3 in Section 17–4 with $K_i = g/\epsilon_0 \omega \gg 1$),

$$A_n \cong \frac{2}{k} \ll 1. \qquad (18\text{–}60)$$

In this case $k \cong \sqrt{K_i/2} = \sqrt{g/2\epsilon_0 \omega}$, so that

$$A_n \cong 2\sqrt{2\epsilon_0 \omega/g}. \qquad (18\text{–}61)$$

This is called the *Hagen-Rubens relation*; it should hold for moderately good conductors in the microwave region and below, and for metals into the infrared, with g the d-c conductivity. With the same values used in calculating the skin depth from Eq. (17–56), we find that for silver at $f = 10^{10}$ s^{-1} (3 cm wavelength)

$$A_n = 2\sqrt{2(8.854 \times 10^{-12})(2\pi \times 10^{10})/3 \times 10^7} = 3.9 \times 10^{-4},$$

$$R_n = 0.9996.$$

For sea water at $f = 6 \times 10^4$ s^{-1},

$$A_n = 25 \times 10^{-4},$$

$$R_n = 0.9975.$$

This high reflectance also manifests the problem in communicating with submarines. The absorptance is small in these cases because the skin depth is relatively small: From Eq. (17–48) and (18–60)

$$A_n = 4\pi \frac{\delta}{\lambda_1},$$

where λ_1 is the wavelength in air. In the frequency range of visible light, for silver $n \cong 0.05$, $k \cong 3$. For these values Eq. (18–60) is not valid, but Eq. (18–59) gives $R_n \cong 0.98$. More typical metallic values are perhaps those for nickel, $n \cong 2$, $k \cong 3$; these give $R_n \cong 0.56$. Such metals still look quite shiny, since the eye approximates a logarithmic detector. The corresponding values for silver and nickel are plotted as a function of the angle of incidence in Fig. 18–5. There is no Brewster's angle where R_p vanishes, but still R_p has a minimum and is always less than R_s. Some polarization by reflection still occurs. Equations (18–53) show that $\hat{E}'_{1s}$ and $\hat{E}'_{1p}$ have a relative phase difference $\alpha_s - \alpha_p$. Thus even if the incident wave is linearly polarized, the reflected wave may become elliptically polarized at oblique incidence.

The transmitted wave is important in problems such as those considered in the next section, even if it cannot be observed directly. Its amplitude and phase are given by $\hat{t}_{12s}$ and $\hat{t}_{12p}$, and its propagation vector by the $\hat{\kappa}_2$ that satisfies Eqs. (18–48)–(18–52). The latter will define the planes of constant phase and the phase velocity, as well as the planes of constant amplitude and the attenuation constant. These results are obtained by comparing the two equivalent expressions for $\hat{\kappa}_2$:

$$\hat{\kappa} = \kappa_r + i\kappa_i, \tag{18–62}$$

$$\hat{\kappa} = \hat{\kappa} \sin \hat{\theta} \mathbf{i} + \hat{\kappa} \cos \hat{\theta} \mathbf{k}. \tag{18–63}$$

(In this discussion we will drop the subscript 2.) The second form is justified by the restriction Eq. (18–49). The same restriction means that $\hat{\kappa} \times \mathbf{n}$ is real, and from Eq. (18–62)

$$\kappa_r \times \mathbf{n} = \kappa_1 \times \mathbf{n}, \tag{18–64}$$

$$\kappa_i \times \mathbf{n} = 0. \tag{18–65}$$

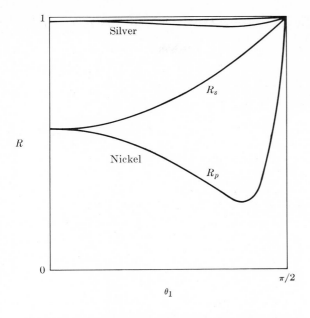

Figure 18–5 Reflectance for s- and p-polarization at an air–metal interface. Representative values for visible light are $n \cong 0.05$, $k \cong 3$ (silver); and $n \cong 2$, $k \cong 3$ (nickel).

Equation (18–65) shows that κ_i is parallel to $\mathbf{n} = \mathbf{k}$, and Eq. (18–64) is

$$\kappa_r \sin \Theta = \kappa_1 \sin \theta_1, \tag{18–66}$$

where Θ is the *real* angle between κ_r and $\mathbf{n}$. (See Fig. 18–6.) This could be called the real angle of refraction, in that it is the angle between the planes of constant phase and the boundary surface. The planes of constant amplitude, on the other hand, are parallel to the boundary surface, so that the wave is most rapidly attenuated directly into the conductor. We can rewrite Eq. (18–62)

$$\hat{\kappa} = \kappa_r \sin \Theta \, \mathbf{i} + \kappa_r \cos \Theta \, \mathbf{k} + i\kappa_i \mathbf{k}$$

$$= \kappa_1 \sin \theta_1 \mathbf{i} + (\kappa_r \cos \Theta + i\kappa_i)\mathbf{k}.$$

Comparing the components of this with those of Eq. (18–63), we see that

$$\kappa_1 \sin \theta_1 = \hat{\kappa} \sin \hat{\theta}, \tag{18–67}$$

$$\kappa_r \cos \Theta + i\kappa_i = \hat{\kappa} \cos \hat{\theta}. \tag{18–68}$$

The first is simply Snell's law again, but the second, together with Eq. (18–66), gives the relation of κ_r, κ_i, and Θ to n, k, and θ_1 which we seek. Let us make the abbreviation

$$\hat{\kappa} \cos \hat{\theta} = \frac{\omega}{c} (p + iq),$$

so that

$$\hat{n} \cos \hat{\theta} = p + iq. \tag{18–69}$$

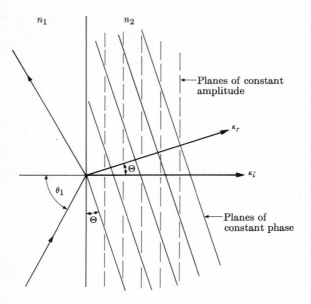

n_1 n_2

←— Planes of constant amplitude

κ_r

Θ

κ_i

θ_1

Θ

←— Planes of constant phase

Figure 18–6 Refraction into a conductor. The planes of constant phase are perpendicular to the propagation direction and make an angle Θ with the boundary, but the planes of constant amplitude are parallel to the boundary.

Then from Eq. (18–68) $\kappa_r \cos \Theta = (\omega/c)p$, and with Eq. (18–66)

$$\kappa_r = \frac{\omega}{c} \sqrt{p^2 + n_1^2 \sin^2 \theta_1}; \tag{18–70}$$

$$\kappa_i = \frac{\omega}{c} q. \tag{18–71}$$

It remains only to find the values of p and q, which follow immediately by squaring Eq. (18–69):

$$p^2 - q^2 + 2ipq = \hat{n}^2(1 - \sin^2 \hat{\theta}) = (n + ik)^2 - n_1^2 \sin^2 \theta_1$$

$$= n^2 - k^2 - n_1^2 \sin^2 \theta_1 + 2ink,$$

where we have used Snell's law $\hat{n} \sin \hat{\theta} = n_1 \sin \theta_1$ again. In terms of the dielectric constant $\hat{K} = \hat{n}^2$,

$$p^2 - q^2 + 2ipq = K_r - K_1 \sin^2 \theta_1 + iK_i.$$

Equating real and imaginary parts we get

$$K_r - K_1 \sin^2 \theta_1 = p^2 - q^2,$$

$$K_i = 2pq,$$

equations that are almost the same as Eq. (17–51) for n and k. The solutions

$$p = \sqrt{\tfrac{1}{2}[(K_r - K_1 \sin^2 \theta_1) + \sqrt{(K_r - K_1 \sin^2 \theta_1)^2 + K_i^2}]},$$

$$q = \sqrt{\tfrac{1}{2}[-(K_r - K_1 \sin^2 \theta_1) + \sqrt{(K_r - K_1 \sin^2 \theta_1)^2 + K_i^2}]}, \tag{18–72}$$

are the same as Eq. (17–52) for n and k except that K_r is replaced by $(K_r - K_1 \sin^2 \theta_1)$. Thus p and q play the role of generalized versions of n and k that depend on the angle of incidence θ_1. When $\theta_1 = 0$, $p = n$ and $q = k$. The attenuation length is given directly by q according to Eq. (18–71). Equation (18–70) can be written

$$\kappa_r = N \frac{\omega}{c}, \tag{18–73}$$

which defines a real refractive index $N(\theta_1)$,

$$N = \sqrt{p^2 + n_1^2 \sin^2 \theta_1}, \tag{18–74}$$

and gives the phase velocity as c/N. From Eq. (18–66), N satisfies a real version of Snell's law,

$$N \sin \Theta = n_1 \sin \theta_1; \tag{18–75}$$

$$N \cos \Theta = p. \tag{18–76}$$

Either of these can be used to find the real angle Θ. Although N is not the real part of $\hat{n}$, nor Θ the real part of $\hat{\theta}$, $N \cos \Theta$ is the real part of $\hat{n} \cos \hat{\theta}$. With this we have a complete solution for the complex case. The results are a fairly simple generalization of the real case, and the application to any particular problem is straightforward. Equations (18–72) are so complicated, however, that not much more can be said in general about propagation into a conducting medium at oblique incidence, except in extreme cases. For example, when K_i is very large, the case that leads to the Hagen-Rubens formula,

$$p \cong n \cong q \cong k \gg 1.$$

From Eq. (18–74)

$$N \gg 1$$

and from Eq. (18–75)

$$\Theta \cong 0.$$

The propagation direction is almost directly into the medium for any angle of incidence, although the attenuation is very strong; the velocity and wavelength are much reduced. The skin depth that was defined for normal incidence holds approximately for any angle of incidence.

Returning to the case of total internal reflection, we can apply the results to complete the solution of that problem also. Here $n_2 = \sqrt{K_{2r}}$, $K_{2i} = 0$; but $\cos \hat{\theta}_2$ is imaginary when $\theta_1 > \theta_c$. (We now restore the subscript 2.)

$$\cos \hat{\theta}_2 = \sqrt{1 - \sin^2 \theta_2} = \sqrt{1 - (n_1/n_2)^2 \sin^2 \theta_1},$$

$$\cos \hat{\theta}_2 = i\sqrt{(\sin \theta_1/\sin \theta_c)^2 - 1}, \tag{18–77}$$

since

$$\sin \theta_c = n_2/n_1.$$

Combining Eqs. (18–77) and (18–69), we find

$$n_2 \cos \hat{\theta}_2 = in_2\sqrt{(\sin \theta_1/\sin \theta_c)^2 - 1} = p + iq,$$

so that

$$p = 0, \qquad q = n_2\sqrt{(\sin \theta_1/\sin \theta_c)^2 - 1}. \tag{18–78}$$

(The same result is given by Eq. (18–72).) With Eq. (18–78), the Fresnel coefficient for s-polarized reflection Eq. (18–28) becomes

$$\hat{r}_{12s} = \frac{n_1 \cos \theta_1 - iq}{n_1 \cos \theta_1 + iq}.$$

The numerator is the complex conjugate of the denominator for any $\theta_1 > \theta_c$, so $\hat{r}_{12s}$ has the form

$$\hat{r}_{12s} = \frac{z^*}{z}, \qquad \hat{r}^*_{12s} = \frac{z}{z^*}.$$

Therefore

$$R_s = |\hat{r}_{12s}|^2 = \hat{r}_{12s}\hat{r}^*_{12s} = 1.$$

Likewise it is clear from Eq. (18–33) that $R_p = 1$ for all $\theta_1 > \theta_c$. We did not calculate T in the complex case, but energy conservation requires $T = 0$ when $R = 1$. On the other hand, the Fresnel transmission coefficients $\hat{t}_{12}$ are not zero; there are nonvanishing E- and B-fields in medium 2. This seeming paradox is most easily resolved by finding κ_2. With $p = 0$ in Eq. (18–74),

$$N = n_1 \sin \theta_1 = n_2 (\sin \theta_1/\sin \theta_c). \tag{18–79}$$

The real refractive index N of medium 2 varies between n_2 and n_1 as θ_1 increases from θ_c to $\pi/2$,

$$n_2 \le N \le n_1. \tag{18–80}$$

From Eq. (18–76) with $p = 0$, $N \ne 0$,

$$\cos \Theta = 0, \tag{18–81}$$

for any angle of incidence $\theta_1 \ge \theta_c$. That is, κ_{2r} is always parallel to the boundary, and as a consequence there is no energy flow perpendicular to the boundary in medium 2. We already know that κ_{2i} is perpendicular to the boundary. There is attenuation of the wave amplitude in this direction according to Eq. (18–71), since $q \ne 0$. The attenuation length is $\delta = 1/\kappa_{2i}$,

$$\delta = \frac{c}{\omega q} = \frac{c}{n_2 \omega} \frac{1}{\sqrt{(\sin \theta_1/\sin \theta_c)^2 - 1}}, \tag{18–82}$$

from Eqs. (18–71) and (18–78). Replacing n_2 with N from Eq. (18–79) and introducing the wavelength $\lambda_2/2\pi = c/N\omega$, we get

$$\delta = \frac{\lambda_2}{2\pi\sqrt{1 - (\sin \theta_c/\sin \theta_1)^2}}.$$

The attenuation length approaches infinity as θ_1 approaches θ_c (but then of course our assumption of an idealized semi-infinite geometry becomes unrealistic). As θ_1 approaches $\pi/2$, δ approaches $\lambda_2/2\pi \cos \theta_c$, which for glass–air is $\lambda_2/4.68$. Thus the behavior for $\theta_1 > \theta_c$ is a reasonable extension of the behavior for $\theta_1 \leq \theta_c$: As θ_1 increases up to θ_c, R increases and θ_2 increases; at θ_c, $R = 1$ and $\theta_2 = \pi/2$. As θ_1 increases beyond θ_c, R remains 1 and the real angle of refraction Θ remains $\pi/2$, but the infinite penetration of medium 2 is gradually reduced to a fraction of a wavelength. At the same time the phase velocity and the wavelength in medium 2 decrease from that characteristic of medium 2 to that of medium 1. Two other interesting features are that a linearly polarized incident wave becomes elliptically polarized on reflection, because of the complex (but different) $\hat{r}_{12s}$ and $\hat{r}_{12p}$; and that the wave in medium 2 is not transverse—$\hat{\mathbf{E}}_{2p}$ has a longitudinal component.

18–5 REFLECTION AND TRANSMISSION BY A THIN LAYER. INTERFERENCE

As a more complicated and more realistic boundary-value problem, we next consider *two* plane parallel infinite surfaces of discontinuity. This represents a slab of material bounded on either side by semi-infinite media, which may have different properties from each other. To the left of the plane $z = 0$, we assume medium 1; to the right of the plane $z = d$, medium 3; and in between, medium 2. A straightforward application of the boundary conditions at each of the two planes of discontinuity, patterned after the calculation of Section 18–2, leads to results for the E- and B-fields in each of the three regions. (See Problem 18–11.) An alternative approach, which leads to the same answer, is based on the results already obtained in Section 18–2, and this approach is more informative in certain respects. The idea is to consider an incident wave in medium 1, which is partly reflected and partly transmitted at the first interface; the transmitted wave is partly reflected and partly transmitted at the second interface; this reflected wave arrives back at the first interface, where it is partly reflected and partly transmitted; and so on. Since the Fresnel coefficients we already derived give the fractions reflected and transmitted at each interface, we have only to add up all the different contributions to the net wave reflected back into medium 1 and the net wave transmitted into medium 3. Although this procedure sounds infinite, it is in fact fairly simple.

The only new problem encountered before the waves are added up is that the different amplitudes must be added with their proper phase differences.* Each time the wave makes another pass through the layer, the phase shifts because of the change in $\boldsymbol{\kappa}_2 \cdot \mathbf{r}$ in the exponent. The situation is shown in Fig. 18–7. Two incident rays perpendicular to the plane wave front in medium 1 strike the front

---

* We assume that the layer is actually thin enough and smooth enough that the coherent phase differences among all the multiple waves are meaningful. One of the advantages of the present approach is that if this assumption is not valid and the phase differences are more or less random, the summation procedure to be developed is still applicable, provided it is applied to the wave intensities instead of the amplitudes.

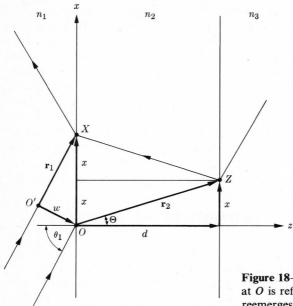

Figure 18–7 The ray that enters medium 2 at O is reflected from the back surface and reemerges to combine with the ray that is reflected at X.

surface of medium 2. One of them is partly reflected at X; the other is partly refracted at O, partly reflected off the back surface at Z, and partly refracted at X to reemerge into medium 1 and combine with the first ray. Since the phase is the same at the two points on the wave front O and O', we must calculate the phase difference between the two paths $O'X$ and OZX. This calculation is no more difficult when medium 2 is a conductor, and so we let the angle of refraction be Θ, the real angle found in the last section; for a nonconductor $\Theta = \theta_2$. The phase difference is

$$\hat{\beta} = 2\hat{\kappa}_2 \cdot \mathbf{r}_2 - \kappa_1 \cdot \mathbf{r}_1.$$

We resolve $\mathbf{r}_2$ into components

$$\mathbf{r}_2 = x\mathbf{i} + d\mathbf{k},$$

and $\mathbf{r}_1$ into

$$\mathbf{r}_1 = 2x\mathbf{i} - w\mathbf{p}_1,$$

where

$$\mathbf{p}_1 = \mathbf{s} \times \mathbf{u}_1 = \mathbf{j} \times \mathbf{u}_1$$

is perpendicular to $\kappa_1 = \kappa_1\mathbf{u}_1$. Then

$$\hat{\beta} = 2x(\hat{\kappa}_2 \cdot \mathbf{i} - \kappa_1 \cdot \mathbf{i}) + 2d\hat{\kappa}_2 \cdot \mathbf{k}.$$

Now

$$\hat{\kappa}_2 \cdot \mathbf{i} - \kappa_1 \cdot \mathbf{i} = \hat{\kappa}_2 \sin \hat{\theta}_2 - \kappa_1 \sin \theta_1 = 0$$

according to Snell's law, and $\hat{\kappa}_2 \cdot \mathbf{k} = \hat{\kappa}_2 \cos \hat{\theta}_2$, so that

$$\hat{\beta} = 2d\hat{\kappa}_2 \cos \hat{\theta}_2 = 2d \frac{\omega}{c} \hat{n}_2 \cos \hat{\theta}_2. \tag{18–83}$$

From Eq. (18–69)

$$\hat{\beta} = 2d \frac{\omega}{c} (p + iq). \tag{18–84}$$

In a nonconducting medium $p = n_2 \cos \theta_2$ and $q = 0$ for all angles of incidence. In a conducting medium at normal incidence, $p = n$ and $q = k$. The real part of $\hat{\beta}$ gives the real phase shift and the imaginary part of $\hat{\beta}$ gives the attenuation due to two traversals of the slab.

To add up all the contributions to the net amplitude reflection coefficient $\hat{r}$, we use the Fresnel coefficients for each boundary together with the phase shift $\hat{\beta}$. The Fresnel coefficients are different for s- and p-polarization, but we will drop the subscripts s and p for the present, remembering that the two polarizations have to be treated separately. From Fig. 18–8 we see that

$$\hat{r} = \hat{r}_{12} + \hat{t}_{12}\hat{r}_{23}\hat{t}_{21}e^{i\beta} + \hat{t}_{12}\hat{r}_{23}\hat{r}_{21}\hat{r}_{23}\hat{t}_{21}e^{2i\beta} + \cdots$$

$$= \hat{r}_{12} + \hat{t}_{12}\hat{r}_{23}\hat{t}_{21}e^{i\beta}[1 + \hat{r}_{21}\hat{r}_{23}e^{i\beta} + (\hat{r}_{21}\hat{r}_{23}e^{i\beta})^2 + \cdots].$$

Since

$$1 + z + z^2 + \cdots = \frac{1}{1 - z},$$

$$\hat{r} = \hat{r}_{12} + \frac{\hat{t}_{12}\hat{t}_{21}\hat{r}_{23}e^{i\beta}}{1 - \hat{r}_{21}\hat{r}_{23}e^{i\beta}}$$

$$= \frac{\hat{r}_{12} + \hat{r}_{23}(\hat{t}_{12}\hat{t}_{21} - \hat{r}_{12}\hat{r}_{21})e^{i\beta}}{1 - \hat{r}_{21}\hat{r}_{23}e^{i\beta}}.$$

Using the identities Eqs. (18–55) and (18–56), we get

$$\hat{r} = \frac{\hat{r}_{12} + \hat{r}_{23}e^{i\beta}}{1 + \hat{r}_{12}\hat{r}_{23}e^{i\beta}}. \tag{18–85}$$

A similar calculation gives the net amplitude transmitted into medium 3:

$$\hat{t} = \frac{\hat{t}_{12}\hat{t}_{23}e^{(1/2)i\beta}}{1 + \hat{r}_{12}\hat{r}_{23}e^{i\beta}}. \tag{18–86}$$

Note that the numerators give the effect of the front and back surfaces each acting alone, and the denominator accounts for all of the actual multiple reflections.

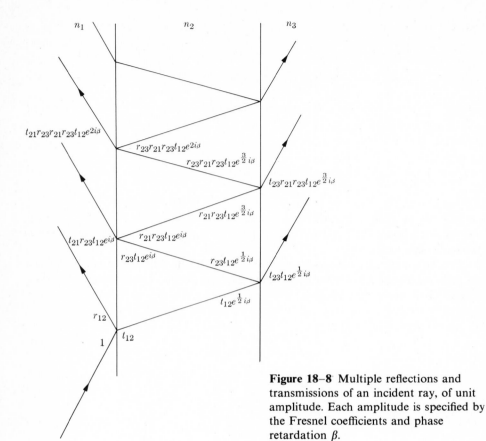

Figure 18–8 Multiple reflections and transmissions of an incident ray, of unit amplitude. Each amplitude is specified by the Fresnel coefficients and phase retardation β.

Since medium 1 and 3 have been assumed to be nonconducting, we can calculate the net intensity reflectance and transmittance:

$$R = \hat{r}\hat{r}^*, \qquad T = \frac{n_3 \cos \theta_3}{n_1 \cos \theta_1} \widetilde{t}\widetilde{t}^*. \qquad (18\text{-}87)$$

These are different for s- and p-polarizations. For a nonconducting slab

$$R + T = 1,$$

but for a conductor

$$R + T + A = 1, \qquad (18\text{-}88)$$

since energy can be absorbed in the conductor by Joule heating.

For a conductor, Eqs. (18–87) become extremely complicated when they are expressed in terms of n and k, even for normal incidence. They are nevertheless important, because measurement of R and T for thin films of metal provides one method to determine experimentally the optical constants. Computer calculations are necessary to solve the equations for n and k in terms of experimental values of R and T. The transmittance T is proportional to $\widetilde{t}\widetilde{t}^{*}$, and $\widetilde{t}\widetilde{t}^{*}$ is proportional to

$$e^{(1/2)i\beta}e^{-(1/2)i\beta*} = e^{(1/2)i(\beta-\beta*)}$$

$$= e^{-2d(\omega/c)q}.$$

At normal incidence, $q = k$, so that T contains the factor

$$e^{-2d/\delta},$$

where $\delta = c/k\omega$ is the skin depth. If medium 1 is air, $\omega/c = 2\pi/\lambda_1$, so that

$$e^{-2d/\delta} = e^{-4\pi kd/\lambda_1}.$$

For metals ($k \approx 2$) and visible light ($\lambda_1 \approx 5000$ Å), d must be less than about 1000 Å for appreciable transmission of light. When this exponential factor is small, the denominator in Eqs. (18–85) and (18–86) is approximately 1.

For nonconductors $q = 0$ (except for total internal reflection), so there is no attenuation due to this factor, but the equations still predict some interesting effects. For β and the Fresnel coefficients all real,

$$R = \frac{r_{12}^2 + r_{23}^2 + 2r_{12}r_{23}\cos\beta}{1 + r_{12}^2 r_{23}^2 + 2r_{12}r_{23}\cos\beta}. \tag{18–89}$$

At normal incidence,

$$r_{12} = \frac{n_1 - n_2}{n_1 + n_2}, \qquad r_{23} = \frac{n_2 - n_3}{n_2 + n_3}, \qquad \beta = 2d\frac{\omega}{c}n_2.$$

Suppose medium 1 is air with $n_1 = 1$, 3 is glass with $n_3 = 1.5$, and 2 is a thin layer of material with $n_2 = 1.3$. Then

$$R = \frac{0.0221 + 0.0186\cos\beta}{1.0001 + 0.0186\cos\beta},$$

$$\beta = 4\pi n_2(d/\lambda_1) = 16.3(d/\lambda_1).$$

Thus R varies between 0.040 and 0.004, with the maxima* at integral multiples of $\beta = 2\pi$, or d/λ_1 integral multiples of 0.39, as shown in Fig. 18–9. The most interesting feature of this result is that R can be less than $r_{12}^2 = 0.017$, the reflectance from the front face alone; this kind of effect can happen only because of destructive interference. The variation of R is between $r_{13}^2 = 0.040$, the value without any covering layer, to something less than $r_{23}^2 = 0.005$, the value from the back face alone. In fact the minimum value of R can be made zero if a material can be found such that $n_2 = \sqrt{n_1 n_3}$. (Problem 18–10.) The effect is exploited to produce nonreflecting lenses. Camera lenses are often coated to have nearly zero

---

* Note that the positions of the extrema of Eq. (18–89) are just the same as in an elementary calculation that takes only two reflections into account.

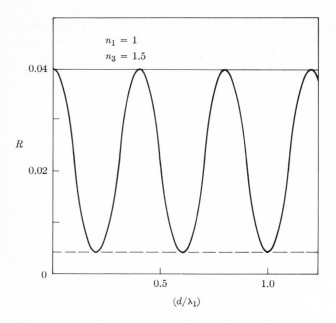

$n_1 = 1$

$n_3 = 1.5$

Figure 18–9 Interference effect on the reflectance from an air–glass interface coated with a thickness d of material with $n_2 = 1.3$.

reflectance near the middle of the visible spectrum; the condition for a minimum does not hold, however, at the red and blue ends of the spectrum, and so the lens has a purple cast when seen by its reflected light. The colors also depend on the angle of viewing, since for oblique incidence $\beta = 4\pi n_2 \cos \theta_2 (d/\lambda_1)$. If n_2 is greater than n_1 and n_3, R varies between a minimum of r_{13}^2 and a maximum that is greater than r_{12}^2 and r_{23}^2. The wavelength at which a minimum or maximum occurs depends on the thickness d of the layer; if this varies from point to point on the film, so will the predominantly reflected wavelengths. Such variation accounts for the colors seen in soap bubbles and oil films floating on water.

Another interesting effect occurs in nonconducting layers if n_2 is less than n_1. This is the case that leads to total internal reflection when there is only one boundary. One might first guess that if all the energy is reflected the presence of a second boundary could make no difference. This guess is wrong, however; the field always penetrates medium 2, a mean depth of δ. A second boundary spoils the perfect reflectance and there is a transmitted wave with $T = 1 - R$. This effect is called *frustrated total reflection*. Although n_1, n_2, and n_3 are real, $\hat{\theta}_2$ is complex when $\theta_1 > \theta_c$, making the Fresnel coefficients complex. We found for s-polarization

$$\hat{r}_{12} = \frac{n_1 \cos \theta_1 - iq}{n_1 \cos \theta_1 + iq}, \tag{18–90}$$

where

$$q = n_2 \sqrt{(\sin \theta_1 / \sin \theta_c)^2 - 1} \tag{18–91}$$
$$= \sqrt{(n_1 \sin \theta_1)^2 - n_2^2}$$

from Eq. (18–78) (for either polarization), with $\sin \theta_c = n_2/n_1$. Writing $\hat{r}_{12}$ in polar form, $\hat{r}_{12} = |\hat{r}_{12}| e^{i\alpha}$, we found that $|\hat{r}_{12}| = 1$. Thus

$$\hat{r}_{12} = e^{i\alpha},$$

where

$$\tan \alpha = \frac{\text{Im} \, (\hat{r}_{12})}{\text{Re} \, (\hat{r}_{12})}$$

$$= \frac{-2n_1 \cos \theta_1 q}{(n_1 \cos \theta_1)^2 - q^2}. \tag{18–92}$$

By the same argument, $|\hat{r}_{23}| = 1$. For simplicity let us assume that medium 3 is the same as medium 1 (for example, a second glass prism with an air layer between). Then $\hat{r}_{23} = \hat{r}_{21}$, and since $\hat{r}_{21} = -\hat{r}_{12}$ identically,

$$\hat{r}_{23} = -\hat{r}_{12} = -e^{i\alpha}. \tag{18–93}$$

From Eq. (18–84),

$$\hat{\beta} = 2d \frac{\omega}{c} (p + iq),$$

and in Eq. (18–78) we found that $p = 0$, so that $\hat{\beta} = i2dq(\omega/c)$ is purely imaginary. We therefore write

$$e^{i\hat{\beta}} = e^{-\gamma},$$

$$\gamma = 2dq(\omega/c) = 2 \frac{d}{\delta}, \tag{18–94}$$

since $q(\omega/c) = 1/\delta$ from Eq. (18–82). Now substituting these into Eq. (18–85), we get

$$\hat{r} = \frac{e^{i\alpha}(1 - e^{-\gamma})}{1 - e^{2i\alpha}e^{-\gamma}}.$$

Finally

$$R = \hat{r}\hat{r}^* = \frac{(1 - e^{-\gamma})^2}{1 + e^{-2\gamma} - 2e^{-\gamma} \cos 2\alpha},$$

$$T = 1 - R = \frac{2(1 - \cos 2\alpha)e^{-\gamma}}{1 + e^{-2\gamma} - 2e^{-\gamma} \cos 2\alpha}. \tag{18–95}$$

Note that for given n_1 and n_2, α depends only on the angle of incidence θ_1; it varies from 0 to π as θ_1 increases from θ_c to $\pi/2$. The exponent γ depends on the thickness d (as well as on θ_1 through δ). If γ is not too small,

$$T \cong 2(1 - \cos 2\alpha)e^{-2d/\delta}. \tag{18–96}$$

Note also that δ is proportional to λ, so frustrated total reflection can be observed with microwaves on a larger scale.

18–6 PROPAGATION BETWEEN PARALLEL CONDUCTING PLATES

Guided waves are another problem that can be treated by considering the interference between an incident and a reflected wave, or alternatively by starting with a new boundary-value problem that involves simultaneously satisfying the conditions at multiple boundaries. We again begin with the first approach. Now we are interested in the wave propagating in a dielectric medium, say air, which is bounded by conducting surfaces. Waveguides for microwaves are an application of this problem. As a simplification we idealize the conductivity of the metal to be infinite. Infinite g means infinite K_i for the metal, which means infinite $\hat{n}_2$ in Eqs. (18–28) and (18–33). Thus $\hat{r}_{12s} = -1$, $\hat{r}_{12p} = +1$, for reflection from a perfectly conducting plane at any angle of incidence. Actually we found $R_n = 0.9996$ for silver at 3 cm wavelength, so the approximation can be expected to be useful. In addition the dielectric medium is assumed to be vacuum.

As a preliminary to the study of waveguides, we now consider the propagation of electromagnetic waves in the region between two parallel, perfectly conducting plates. The region in which wave propagation is to be treated is that shown in Fig. 18–10. Since the x- and z-directions are physically indistinguishable, no generality is lost by considering only waves with wave vectors in the yz-plane—in particular, those making an angle θ with the y-axis in the plane of incidence. Such waves will impinge on the perfectly conducting surface at $y = a$ and will be reflected as waves whose propagation vectors make the angle θ with the minus y-axis. When these waves are reflected a second time, by the surface at $y = 0$, they become waves of the first type again. Thus it is seen that the propagation between two parallel conducting planes can be described in terms of the exponential factors

$$e^{i[\kappa(y\cos\theta + z\sin\theta) - \omega t]}$$

and (18–97)

$$e^{i[\kappa(-y\cos\theta + z\sin\theta) - \omega t]}.$$

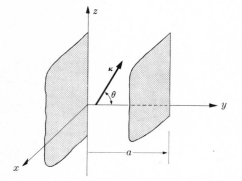

Figure 18–10 Wave propagation between two parallel, perfectly conducting planes.

For such waves there are two possible polarizations, which may be described by saying for the s-polarized one that **E** is parallel to the x-axis, and for the p-polarized one that **H** is parallel to the x-axis. These are known respectively as TE, transverse electric, and TM, transverse magnetic, waves in waveguide terminology. Only TE waves will be considered here. The treatment of TM waves will be left to an exercise.

The electric field in the region between the two conducting planes in the TE case is given by

$$\mathbf{E} = \mathbf{i}\{E_1 e^{i[\kappa(y\cos\theta + z\sin\theta) - \omega t]} + E_1' e^{i[\kappa(-y\cos\theta + z\sin\theta) - \omega t]}\}. \tag{18–98}$$

This electric field must vanish at $y = 0$, since E_t vanishes at the boundary of a perfect conductor. This condition is clearly satisfied for all z and all t if $E_1 = -E_1' = E$, as given by $r_s = -1$. Then **E** is given by

$$\mathbf{E} = iE(e^{i\kappa y\cos\theta} - e^{-i\kappa y\cos\theta})e^{i(\kappa z\sin\theta - \omega t)}. \tag{18–99}$$

In addition **E** must vanish at $y = a$ for all z and all t. This requirement imposes the condition

$$\kappa a \cos\theta = n\pi. \tag{18–100}$$

Thus for a given frequency ω, $\kappa = \omega/c$ and the angle that the waves make with the y-axis is fixed by Eq. (18–100). With this angle fixed, the apparent velocity in the z-direction is $v_p = c/\sin\theta$, which is always greater than the velocity of light in free space. This apparent contradiction of the special theory of relativity will be discussed in more detail later.

It is convenient to express the variation of the electric field in the y- and z-directions in terms of wavelengths. These wavelengths are

$$\lambda_g = \frac{2\pi}{\kappa\sin\theta} = \frac{\lambda_0}{\sin\theta} \quad \left(\lambda_0 = \frac{2\pi}{\kappa} = \frac{2\pi c}{\omega}\right) \tag{18–101}$$

for the z-direction, and

$$\lambda_c = \frac{2\pi}{\kappa\cos\theta} = \frac{\lambda_0}{\cos\theta} \tag{18–102}$$

for the y-direction. In terms of these wavelengths, the electric field, Eq. (18–98), is*

$$\mathbf{E} = \mathbf{i}E_0 \sin\frac{2\pi y}{\lambda_c} e^{i[(2\pi z/\lambda_g) - \omega t]}, \tag{18–103}$$

while Eq. (18–100) takes the form

$$\frac{a}{\lambda_c} = \frac{n}{2}. \tag{18–104}$$

---

* E_0 has been written for $2iE$.

From Eqs. (18–101) and (18–102) it follows immediately that

$$\frac{1}{\lambda_g^2} + \frac{1}{\lambda_c^2} = \frac{1}{\lambda_0^2}. \tag{18–105}$$

If the value $\lambda_c = 2a$, corresponding to $n = 1$ in Eq. (18–104), is considered, then as λ_0 increases, that is, as ω decreases, a point is reached where $1/\lambda_g^2$ must be negative in order to satisfy Eq. (18–105). In this case the coefficient of z in Eq. (18–103) is imaginary, and the exponential, instead of oscillating in z, becomes a decreasing exponential. To say this another way: if $\lambda_0 > 2a$ the electromagnetic wave will be exponentially damped in z, instead of propagating. If n is taken as 2, then $\lambda_c = 2a/2 = a$ and the longest wavelength propagated is a. The reason for the subscript c is now clear; it means "cutoff." The cutoff wavelength is the longest wavelength that can be propagated for a given mode (n value).

The velocity v_p, which was found earlier, always exceeds the velocity of light and, in fact, becomes infinite when the wavelength in free space equals λ_c, that is, when $\theta = 0$. This velocity is the phase velocity, by which is meant the velocity of a point of constant phase on the wave. Without dwelling on the relativistic aspects of the question, this represents an apparent contradiction of the postulate that no signal can be propagated with a velocity greater than the velocity of light. The resolution of this apparent difficulty is that energy is propagated down the guide with a smaller velocity than the velocity of light, namely, with the so-called group velocity. Signals are transmitted with the group velocity; they are not transmitted with the phase velocity.

To determine the velocity of energy propagation, we shall calculate the energy density. This energy density times the *group velocity* gives the energy flux, or Poynting vector. Thus by dividing the Poynting vector by the energy density, the velocity of energy propagation can be obtained. This result is a generalization of Eq. (17–35).

The magnetic induction in the guide is readily obtained from

$$\mathbf{V} \times \mathbf{E} = -\frac{\partial \mathbf{B}}{\partial t}. \tag{18–106}$$

By using Eq. (18–103) for $\mathbf{E}$, and assuming that $\mathbf{B}(\mathbf{r}, t) = \mathbf{B}(\mathbf{r})e^{-i\omega t}$, we quickly find

$$\mathbf{B}(\mathbf{r}, t) = \mathbf{j}E_0 \frac{2\pi}{\omega\lambda_g} \sin\frac{2\pi y}{\lambda_c} e^{i[(2\pi z/\lambda_g) - \omega t]} + i\mathbf{k}E_0 \frac{2\pi}{\omega\lambda_g} \cos\frac{2\pi y}{\lambda_c} e^{i[(2\pi z/\lambda_g) - \omega t]}. \tag{18–107}$$

The energy density is

$$u = \tfrac{1}{2}(\mathbf{E} \cdot \mathbf{D} + \mathbf{B} \cdot \mathbf{H}), \tag{18–108}$$

while the Poynting vector is

$$\mathbf{S} = \mathbf{E} \times \mathbf{H}. \tag{18–109}$$

Complex notation has been used for **E** and **B**, with the tacit assumption that the real part of each expression is to be taken. In calculating u and S then, the real parts should be taken and multiplied together. However, since the quantities to be used in calculating the group velocity are the time averages of Eqs. (18–108) and (18–109), Eq. (17–37) may be used to circumvent the taking of real parts.

The time average energy density is

$$\bar{u} = \tfrac{1}{4} \text{Re} \left[\mathbf{E}^* \cdot \mathbf{D} + \mathbf{B}^* \cdot \mathbf{H} \right] = \tfrac{1}{4} \text{Re} \left[\epsilon_0 E_0^* E_0 \sin^2 \left(\frac{2\pi y}{\lambda_c} \right) \right.$$

$$+ \frac{1}{\mu_0} E_0^* E_0 \left(\frac{2\pi}{\omega \lambda_g} \right)^2 \sin^2 \left(\frac{2\pi y}{\lambda_c} \right)$$

$$\left. + \frac{1}{\mu_0} E_0^* E_0 \left(\frac{2\pi}{\omega \lambda_c} \right)^2 \cos^2 \left(\frac{2\pi y}{\lambda_c} \right) \right]. \qquad (18\text{–}110)$$

Integrating in the y-direction, across the guide, effectively replaces each $\sin^2 (2\pi y/\lambda_c)$ and $\cos^2 (2\pi y/\lambda_c)$ by $a/2$. Thus

$$\int_0^a \bar{u} \, dy = \tfrac{1}{4} E_0^* E_0 \frac{a}{2} \left[\epsilon_0 + \frac{1}{\mu_0} \frac{4\pi^2}{\omega^2} \left(\frac{1}{\lambda_g^2} + \frac{1}{\lambda_c^2} \right) \right]$$

$$= \tfrac{1}{4} E_0^* E_0 \, \epsilon_0 \, a. \qquad (18\text{–}111)$$

The time average of the z-component of the Poynting vector is

$$\bar{S}_z = \tfrac{1}{2} \text{Re} \, E_x^* H_y$$

$$= \tfrac{1}{2} \text{Re} \left[E_0^* \sin \left(\frac{2\pi y}{\lambda_c} \right) \frac{1}{\mu_0} E_0 \frac{2\pi}{\omega \lambda_g} \sin \left(\frac{2\pi y}{\lambda_c} \right) \right] \qquad (18\text{–}112)$$

$$= \tfrac{1}{2} E_0^* E_0 \frac{2\pi}{\mu_0 \omega \lambda_g} \sin^2 \left(\frac{2\pi y}{\lambda_c} \right).$$

Integrating this expression from $y = 0$ to $y = a$ yields the total average power (per unit length in the x-direction) traveling down the guide:

$$\int_0^a \bar{S}_z \, dy = \tfrac{1}{4} E_0^* E_0 \frac{2\pi}{\mu_0 \omega \lambda_g} a. \qquad (18\text{–}113)$$

The velocity of energy propagation is the quotient of Eq. (18–113) divided by (18–111). Thus

$$v_g = \frac{2\pi}{\epsilon_0 \mu_0 \omega \lambda_g} = \frac{2\pi c^2}{\omega \lambda_g} = c \frac{\lambda_0}{\lambda_g}. \qquad (18\text{–}114)$$

From Eq. (18–101), we note that λ_g is greater than λ_0, and hence $\omega \lambda_g/2\pi$ is greater than c, which makes it clear that v_g is less than c.

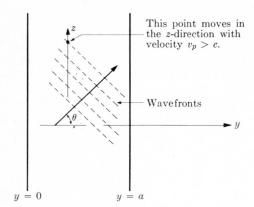

This point moves in the z-direction with velocity $v_p > c$.

Wavefronts

$y = 0$ $y = a$

Figure 18-11 Detailed motion of the wavefronts during wave propagation between conducting planes.

Our understanding of the difference between the group velocity, v_g, and the phase velocity, v_p, can be enhanced by noting that from Eq. (18–101) $\lambda_g = \lambda_0 / \sin \theta$. Using this result in Eq. (18–114), we find

$$v_g = c \sin \theta, \qquad (18\text{--}115)$$

and we have already seen that

$$v_p = \frac{c}{\sin \theta}. \qquad (18\text{--}116)$$

It is readily apparent that

$$v_g v_p = c^2, \qquad (18\text{--}117)$$

which is generally true for propagation in a waveguide. (Note that Eq. (18–117) does not necessarily apply to other kinds of wave propagation, in particular it does not apply to plane waves in unbounded nondispersive media where the phase and group velocities are identical.) Recalling that θ is the angle between the direction of propagation of one of the component waves and the y-axis makes it a simple matter to draw Fig. 18–11, which shows a section in the yz-plane of the region between the conducting planes. The intersection of a wavefront with the z-axis moves with the velocity $v_p = c/\sin \theta$; however, the component of c along the z-axis is $c \sin \theta = v_g$.

Many of the results obtained for the simple parallel plate waveguide persist for more complex cases. In particular, the common rectangular waveguide has very similar properties. In the next section some general aspects of other waveguides will be considered, with particular reference to rectangular guides.

18-7 WAVEGUIDES

In Section 16–4 it was shown that $\mathbf{E}$ and $\mathbf{H}$ both satisfy the wave equation in free space, that is,

$$\nabla^2 \mathbf{E} - \epsilon_0 \mu_0 \frac{\partial^2 \mathbf{E}}{\partial t^2} = 0, \qquad \nabla^2 \mathbf{H} - \epsilon_0 \mu_0 \frac{\partial^2 \mathbf{H}}{\partial t^2} = 0. \qquad (18\text{--}118)$$

For monochromatic waves, that is, waves of the form $\mathbf{E}(\mathbf{r}, t) = \mathbf{E}(\mathbf{r})e^{-i\omega t}$, these equations become

$$\nabla^2 \mathbf{E} + \frac{\omega^2}{c^2}\,\mathbf{E} = 0, \qquad \nabla^2 \mathbf{H} + \frac{\omega^2}{c^2}\,\mathbf{H} = 0. \tag{18–119}$$

In addition to these wave equations, Maxwell's equations must be satisfied. For the transverse electric (TE) case propagating in the z-direction, $E_z = 0$; furthermore, waves propagating in the z-direction have the remaining five field quantities proportional to $e^{i2\pi z/\lambda_g}$. Maxwell's curl equations in this case are

$\nabla \times \mathbf{E} - i\mu_0 \omega \mathbf{H} = 0$:

$$\frac{\partial E_y}{\partial x} - \frac{\partial E_x}{\partial y} - i\mu_0 \omega H_z = 0, \tag{a}$$

$$E_x = +\frac{\mu_0 \omega \lambda_g}{2\pi}\,H_y, \qquad \text{(b)} \tag{18–120}$$

$$E_y = -\frac{\mu_0 \omega \lambda_g}{2\pi}\,H_x. \tag{c}$$

$\nabla \times \mathbf{H} + i\epsilon_0 \omega \mathbf{E} = 0$:

$$\frac{\partial H_z}{\partial y} - \frac{2\pi i}{\lambda_g}\,H_y + i\epsilon_0 \omega E_x = 0, \tag{a}$$

$$\frac{2\pi i}{\lambda_g}\,H_x - \frac{\partial H_z}{\partial x} + i\epsilon_0 \omega E_y = 0, \qquad \text{(b)} \tag{18–121}$$

$$\frac{\partial H_y}{\partial x} - \frac{\partial H_x}{\partial y} = 0. \tag{c}$$

It is clear that (a) of Eq. (18–121) and (b) of Eq. (18–120) imply

$$\frac{\partial H_z}{\partial y} = \left(\frac{2\pi i}{\lambda_g} - i\,\frac{\epsilon_0 \mu_0 \omega^2 \lambda_g}{2\pi}\right)H_y, \tag{18–122}$$

and therefore that H_y can be found if H_z is known. Similarly, from (c) of Eq. (18–120) and (b) of Eq. (18–121), H_x can be found from H_z. Finally, E_x and E_y are simply related to H_y and H_x by (b) and (c) of Eq. (18–120). Thus if H_z is found, all the other field quantities may be found by differentiation. H_z itself must satisfy Eq. (18–118); therefore, taking cognizance of the $e^{i2\pi z/\lambda_g}$ z-dependence, we write

$$\frac{\partial^2 H_z}{\partial x^2} + \frac{\partial^2 H_z}{\partial y^2} + \left(\frac{\omega^2}{c^2} - \frac{4\pi^2}{\lambda_g^2}\right)H_z = 0. \tag{18–123}$$

Figure 18–12 Wave propagation inside a conducting cylinder.

It remains only to determine the boundary conditions to be imposed on the solutions of Eq. (18–123).

If a general cylindrical guide with perfectly conducting walls, such as that shown in Fig. 18–12, is under consideration, then the appropriate boundary conditions are that the tangential component of **E** and the normal component of **B** should vanish on S. The tangential component of **H** and the normal component of **D** are arbitrary. Imposing these conditions gives rise to a relationship connecting λ_g, ω, and the dimensions of the guide, exactly as Eq. (18–105) does for the parallel plane case.

To better understand the procedure, consider the rectangular waveguide shown in Fig. 18–13. Equation (18–123) can be separated by the usual method of separation of variables. The general solution consists of a sum of terms of the form

$$H_z(x, y, z) = (A \cos \kappa_x x \cos \kappa_y y + B \cos \kappa_x x \sin \kappa_y y$$
$$+ C \sin \kappa_x x \cos \kappa_y y + D \sin \kappa_x x \sin \kappa_y y)e^{2\pi i z/\lambda_g}, \quad (18\text{–}124)$$

with

$$-(\kappa_x^2 + \kappa_y^2) + [\omega^2/c^2 - (4\pi^2/\lambda_g^2)] = 0. \quad (18\text{–}125)$$

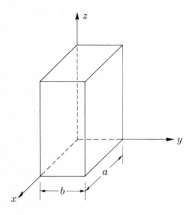

Figure 18–13 A rectangular waveguide.

From this H_z, we obtain E_x:

$$E_x = -\frac{\mu_0 \omega \lambda_g}{2\pi} \left(\frac{2\pi i}{\lambda_g} - i \frac{\epsilon_0 \mu_0 \omega^2 \lambda_g}{2\pi} \right)^{-1} \frac{\partial H_z}{\partial y}. \qquad (18\text{--}126)$$

The partial differentiation changes every $\cos \kappa_y y$ to a $\sin \kappa_y y$, and conversely. However, since E_x must vanish at $y = 0$ and at $y = b$, only terms involving $\sin \kappa_y y$ may survive in E_x, and these terms must have $\kappa_y = n\pi/b$. Thus only $\cos \kappa_y y$ terms survive in Eq. (18–124). A similar argument shows that only $\cos \kappa_x x$ terms may survive, and that these must have $\kappa_x = m\pi/a$. The allowed solutions for H_z, that is, those which give vanishing tangential components of $\mathbf{E}$ at the boundary, have the form

$$H_z = A \cos \frac{m\pi x}{a} \cos \frac{n\pi y}{b} e^{2\pi i z/\lambda_g}. \qquad (18\text{--}127)$$

Each possible pair of values of m and n is referred to as a mode. The notation TE_{mn} is used for modes of the form (18–127); TE means transverse electric, n and m count the number of half-waves in the narrower (n) and wider (m) dimensions.

Returning now to Eq. (18–125) and using $\kappa_x = m\pi/a$ and $\kappa_y = n\pi/b$, we obtain

$$\left(\frac{2\pi}{\lambda_g} \right)^2 = \left(\frac{2\pi}{\lambda_0} \right)^2 - \left(\frac{n\pi}{b} \right)^2 - \left(\frac{m\pi}{a} \right)^2, \qquad (18\text{--}128)$$

which clearly indicates that for fixed λ_0 the guide wavelength, and consequently the guide velocity $v_g = c\lambda_0 / \lambda_g$, depend on the mode. We also see that there are maximum wavelengths for the propagation of various modes. Clearly, if λ_0 is sufficiently large $(2\pi/\lambda_0)^2$ will be smaller than $(n\pi/b)^2 + (m\pi/a)^2$. In this case, the right side of Eq. (18–128) becomes negative and consequently the value of λ_g is imaginary. This leads to attenuation rather than propagation.

Rectangular waveguides are extensively used for the transmission of microwave power. It is usual to choose a waveguide size such that only the TE_{10} mode at the desired frequency propagates in the guide. A common size of waveguide is 0.4 in. $\times$ 0.9 in., inside dimensions. The maximum wavelength that will propagate in the TE_{10} mode is found by putting $m = 1$, $n = 0$, $a = 0.9$ in. $= 2.28$ cm and, $b = 0.4$ in. $= 1.01$ cm into Eq. (18–128). The result: $\lambda_{0,\max} = 4.57$ cm is obtained by putting $\lambda_g = \infty$; wavelengths longer than this will not propagate, but shorter wavelengths will. The mode with the next shorter cutoff wavelength is TE_{11} or TE_{20}, depending on the dimensions of the guide. If $b < a/\sqrt{3}$, the TE_{20} cutoff wavelength is greater than the TE_{11}. Calculation of the TE_{20} wavelength is very simple; it is just one-half the TE_{10} cutoff wavelength, or 2.28 cm. Imperfections in manufactured waveguides and high losses close to the TE_{10} cutoff wavelength make it necessary to restrict the TE_{10} band of commercial waveguides to the practical limits of 2.42 to 4.35 cm.

18–8 CAVITY RESONATORS

Another type of device closely related to waveguides and of considerable practical importance is the cavity resonator. Cavity resonators display the properties typical of resonant circuits in that they can store energy in oscillating electric and magnetic fields; furthermore, practical cavity resonators dissipate a fraction of the stored energy in each cycle of oscillation. In this latter respect, however, cavity resonators are usually superior to conventional *L-C* circuits by a factor of about twenty, that is, the fraction of the stored energy dissipated per cycle in a cavity resonator is about 1/20 the fraction dissipated per cycle in an *L-C* circuit. An additional advantage is that cavity resonators of practical size have resonant frequencies which range upward from a few hundred megahertz—just the region where it is almost impossible to construct ordinary *L-C* circuits.

The simplest cavity resonator is a rectangular parallelepiped with perfectly conducting walls. For such a cavity, the appropriate boundary conditions are the vanishing of the tangential component of $\mathbf{E}$ and the normal component of $\mathbf{B}$ at the boundary. The tangential component of $\mathbf{H}$ and the normal component of $\mathbf{D}$ are arbitrary. The electric and magnetic fields must satisfy the wave equations (18–118); thus E_x must satisfy

$$\frac{\partial^2 E_x}{\partial x^2} + \frac{\partial^2 E_x}{\partial y^2} + \frac{\partial^2 E_x}{\partial z^2} + \frac{\omega^2}{c^2} E_x = 0. \tag{18–129}$$

If the cavity consists of the region bounded by the six planes $x = 0$, $x = a$; $y = 0$, $y = b$; $z = 0$, $z = d$, then E_x must have the form

$$E_x = E_1 f_1(x) \sin \kappa_y y \sin \kappa_z z \, e^{-i\omega t}, \tag{18–130}$$

with $\kappa_y = m\pi/b$ and $\kappa_z = n\pi/d$, in order that E_x vanish at $y = 0$, at $z = 0$, at $y = b$, and at $z = d$. Furthermore, E_x alone cannot be a solution unless $f_1(x)$ is a constant, since $\nabla \cdot \mathbf{E}$ must vanish to satisfy one of Maxwell's equations. For E_y and E_z the situation is similar, and the solutions take the forms

$$E_y = E_2 \sin \kappa_x x \, f_2(y) \sin \kappa_z z \, e^{-i\omega t},$$

$$E_z = E_3 \sin \kappa_x x \sin \kappa_y y \, f_3(z) \, e^{-i\omega t}, \tag{18–131}$$

with κ_y and κ_z as in Eq. (18–130), and $\kappa_x = l\pi/a$. If the divergence of $\mathbf{E}$ is to vanish, then the equation

$$\left(E_1 \frac{df_1}{dx} \sin \kappa_y y \sin \kappa_z z + E_2 \sin \kappa_x x \frac{df_2}{dy} \sin \kappa_z z \right.$$

$$\left. + E_3 \sin \kappa_x x \sin \kappa_y y \frac{df_3}{dz} \right) e^{-i\omega t} = 0 \tag{18–132}$$

must be satisfied. This is accomplished if $f_1 = \cos \kappa_x x$, $f_2 = \cos \kappa_y y$, $f_3 = \cos \kappa_z z$, and

$$\kappa_x E_1 + \kappa_y E_2 + \kappa_z E_3 = 0, \tag{18–133}$$

which is just the condition that $\boldsymbol{\kappa}$ be perpendicular to $\mathbf{E}$. Returning to the wave equation, it is apparent that the resonant frequencies of the cavity are given by

$$\kappa_x^2 + \kappa_y^2 + \kappa_z^2 - \frac{\omega^2}{c^2} = 0, \qquad (18\text{–}134)$$

or

$$\frac{l^2}{a^2} + \frac{m^2}{b^2} + \frac{n^2}{d^2} - \frac{4f^2}{c^2} = 0. \qquad (18\text{–}135)$$

A typical cavity constructed from a waveguide 0.4 in. × 0.9 in. in size is characterized by $l = 1$, $m = 0$, $n = 2$ (a so-called TE_{102} cavity). The resonant frequency of such a cavity is clearly determined by the z-dimension, d. Many other aspects of the rectangular cavity resonator problem can be treated in detail; some of these are left as exercises.

Other forms of cavity resonators may be constructed; however, only the right circular cylinder and the rectangular parallelepiped are easily fabricated and amenable to an exact mathematical treatment. The treatment of the right circular cylinder involves functions that are more complicated than the sines and cosines, specifically the Bessel functions. Satisfying the boundary conditions requires finding the zeros of these functions in the same way the zeros of the sines entered the rectangular problem. Rather than enter into the elaborate discussion which results, we refer the interested reader to *Technique of Microwave Measurements* by C. G. Montgomery (New York: McGraw-Hill, 1947), p. 297 ff., where a brief but very useful treatment of the cylindrical cavity resonator is given.

18–9 SUMMARY

Practical problems of wave propagation usually involve boundaries between different media, where the complex dielectric constant $\hat{K}$ changes discontinuously. The conditions at a single plane boundary are expressed by Snell's law and the Fresnel coefficients, which depend on $\hat{n} = \sqrt{\hat{K}}$ and the angle of incidence. Problems involving multiple plane boundaries can be solved by superposition of the solutions for a single boundary, and show interference effects. Nonconducting media are a special case of conducting media in which the imaginary part of $\hat{K}$ and $\hat{n}$ vanishes.

1. Boundary conditions on the amplitudes cannot be satisfied unless the wave frequency is the same on either side of the boundary and the phase matches everywhere on the boundary. Hence the incident, reflected, and transmitted wave propagation vectors are all coplanar with the normal to the boundary, and the angle of reflection equals the angle of incidence, $\theta_1' = \theta_1$. The dispersion relation gives Snell's law

$$n_1 \sin \theta_1 = n_2 \sin \theta_2.$$

2. Continuity of the tangential components of the E- and H-fields is expressed by the Fresnel coefficients, the ratios of the reflected and transmitted E-field amplitudes to the incident amplitude. These differ for s- and p-polarization (E-vector perpendicular and parallel to the plane of incidence, respectively). For reflection

$$r_{12s} = \frac{n_1 \cos \theta_1 - n_2 \cos \theta_2}{n_1 \cos \theta_1 + n_2 \cos \theta_2},$$

$$r_{12p} = \frac{n_2 \cos \theta_1 - n_1 \cos \theta_2}{n_2 \cos \theta_1 + n_1 \cos \theta_2}.$$

3. The reflected and transmitted intensities are calculated in terms of the Fresnel coefficients from the normal component of the Poynting vectors. The reflectance is

$$R = r_{12} r_{12}^*,$$

For nonconducting media the transmittance is

$$T = 1 - R,$$

and for medium 2 conducting the absorptance is

$$A = 1 - R,$$

as a result of energy conservation. These equations are tractable for nonconductors or for normal incidence. They predict polarization by reflection at Brewster's angle, total internal reflection at the critical angle, and the Hagen-Rubens formula for r-f reflectance of a conductor.

4. For oblique incidence on a conductor, the propagation and attenuation are described by $\kappa_r = N\omega/c$, $\kappa_i = q\omega/c$, with the planes of constant phase at an angle Θ with the boundary and the planes of constant amplitude parallel to the boundary.

$$N = \sqrt{p^2 + n_1^2 \sin^2 \theta_1},$$

$$\sin \Theta = n_1 \sin \theta_1 / N.$$

The quantities p and q are generalizations of n and k which are related to $(K_r - K_1 \sin^2 \theta_1)$ and K_i just as n and k are for $\theta_1 = 0$. They also explain total reflection for angles greater than critical.

5. For two plane parallel boundaries, superposition gives the relative reflected amplitude as

$$r = \frac{r_{12} + r_{23} e^{i\beta}}{1 + r_{12} r_{23} e^{i\beta}},$$

where

$$\beta = 2d \frac{\omega}{c} n_2 \cos \theta_2 = 2d \frac{\omega}{c} (p + iq)$$

is the phase shift (and attenuation) on two traversals of the layer. The net reflectance is

$$R = rr^*.$$

$$1 - R = T + A,$$

where $A = 0$ if the layer is nonconducting. Through β, R shows interference effects which depend on (d/λ_1). With conduction or frustrated total reflection,

$$T \sim e^{-2d/\delta},$$

if $d \gtrsim \delta$, where δ is the attenuation length.

6. Guided waves propagate without attenuation (assuming perfectly reflecting boundaries) if the vacuum wavelength $\lambda_0 = 2\pi c/\omega$ is shorter than a cutoff wavelength λ_c, which is dependent on the dimensions of the waveguide.

$$\frac{1}{\lambda_g^2} = \frac{1}{\lambda_0^2} - \frac{1}{\lambda_c^2},$$

where λ_g is the wavelength of the guided wave. The group velocity (of energy propagation) is v_g, where

$$v_g v_p = c^2.$$

For a rectangular guide in the TE_{mn} mode,

$$\frac{1}{\lambda_c^2} = \left(\frac{m}{2a}\right)^2 + \left(\frac{n}{2b}\right)^2,$$

where $a > b$. Usually one restricts λ_0 so that only TE_{10} can propagate.

7. For plane waves that are attenuated or guided (and also for spherical waves), the amplitude is not constant over a surface of constant phase. Such waves are called *inhomogeneous*. The Poynting vector is more complicated than for a homogeneous plane wave propagating in a dielectric medium.

PROBLEMS

18–1 Calculate the Fresnel reflection coefficient for an s-polarized wave incident from air onto a dielectric at the Brewster angle, $\theta_1 = \theta_B$. Find the reflectance if $n = 1.5$.

18–2 A p-polarized wave is incident from air onto a dielectric surface near grazing incidence, $\theta_1 = \frac{1}{2}\pi - \delta$. Find the slope of the $R_p(\theta_1)$ curve as δ approaches zero, in terms of the dielectric constant K.

18–3 A p-polarized wave is incident from a transparent medium of dielectric constant K onto an air boundary at an angle slightly less than the critical angle, $\theta_1 = \theta_c - \delta$. Approximate R_p as a function of δ as δ approaches zero, and show that the slope of the $R_p(\theta_1)$ curve is infinite at θ_c.

18–4 Suppose an optical fiber has a refractive index $n = 1.55$. Compute the largest angle between the fiber axis and a light ray that will propagate along the fiber, if the fiber is surrounded by air. Compute the angle, if the fiber is surrounded by a cladding material with index 1.53.

18–5 A light wave with p-polarization in air is reflected from a metal surface. Calculate R_p under the assumption that $\cos \theta_2 \cong 1$, as is often the case. Find the value of θ_1 for which R_p is a minimum. Evaluate this θ_1 and the corresponding R_p if $n = 1$, $k = 6$ (appropriate for aluminum).

18–6 A plane wave is normally incident from air onto the plane boundary of a metal. Assume that the frequency is in the range where $n \cong k \gg 1$. From the Fresnel transmission coefficient find $|E_2|^2$ just inside the metal surface. Calculate the energy dissipation per unit volume near the surface, and evaluate it if the incident amplitude is $E_1 = 10$ V/cm and the frequency is $f = 10^{10}$ Hz.

18–7 A wave in air is obliquely incident on a conducting surface at angle θ_1, in the frequency range where the Hagen-Rubens relation holds. Show that Eq. (18–60) is replaced by

$$A_s = \frac{2 \cos \theta_1}{k}, \qquad A_p = \frac{2}{k \cos \theta_1}.$$

18–8 A wave in air is reflected at normal incidence from a conducting surface. From $\hat{r}_{12s}$, show that the phase shift of the E-vector is

$$\alpha_s = \tan^{-1} \frac{2k}{n^2 + k^2 - 1}.$$

Verify that this result goes to $\alpha_s = \pi$ for infinite conductivity.

18–9 Suppose a radio wave of $\omega = 10^7$ s^{-1} is reflected from the earth's surface at normal incidence. Compute the phase shift on reflection from the result of Problem 18–8, assuming that $K = 9$, $g = 10^{-4}$ $(\Omega m)^{-1}$ for this terrain.

18–10 A dielectric medium of refractive index n_3 is overcoated with a layer of index n_2, and a wave is incident from the dielectric medium n_1. Show that $r_{12} = r_{23}$ for $n_2 = \sqrt{n_1 n_3}$, and thus $R = 0$ for $\cos \beta = -1$, at normal incidence.

18–11 A beam of monochromatic light (frequency ω) in air is incident normally on a dielectric film of refractive index n. The thickness of the film is d. Calculate the reflection and transmission coefficients as a function of d and n, by satisfying the boundary conditions at both faces. Let there be waves traveling toward the right and left inside the film, E_2 and E_2', in addition to the incident E_1, reflected E_1', and transmitted E_3 waves.

18–12 Consider the matrix equation

$$\begin{pmatrix} E_m \\ E_m' \end{pmatrix} = (C_m) \begin{pmatrix} E_{m+1} \\ E_{m+1}' \end{pmatrix},$$

where

$$(C_m) = \frac{1}{t_{m,m+1}} \begin{pmatrix} e^{i(1/2)\beta m} & r_{m,m+1} e^{i(1/2)\beta m} \\ r_{m,m+1} e^{-i(1/2)\beta m} & e^{-i(1/2)\beta m} \end{pmatrix}.$$

a) Verify that the results for a single interface between two dielectrics are given by

$$\begin{pmatrix} E_1 \\ E_1' \end{pmatrix} = (C_1) \begin{pmatrix} E_2 \\ 0 \end{pmatrix},$$

with $\beta_1 = 0$.

b) Verify that the results for two interfaces are given by

$$\begin{pmatrix} E_1 \\ E_1' \end{pmatrix} = (C_1)(C_2) \begin{pmatrix} E_3 \\ 0 \end{pmatrix},$$

with $\beta_2 = 2d_2(\omega/c)n_2 \cos \theta_2$. This approach is convenient to generalize to a system of multiple layers separated by plane boundaries.

18–13 A metal surface is overcoated by a dielectric layer. Calculate the resulting reflectance R, assuming that the metal is a perfect conductor ($g = \infty$).

18–14 Radiation is normally incident on an unsupported metal film in air. Assume that the film is thick enough that multiple reflections can be neglected. Calculate the transmittance T in terms of n and k of the metal.

18–15 Consider a dielectric slab that is too thick for the multiply reflected beams to interfere coherently. Add up all the intensities to find the net reflectance in terms of the reflectances R_{12} and R_{23} of the individual boundaries. Specialize the result to the case of identical media on each side of the slab.

18–16 Find the surface charge density and the current per unit width on the surface of a perfect conductor on which plane electromagnetic waves are incident, when the electric vector is (1) perpendicular to the plane of incidence, and (2) parallel to the plane of incidence.

18–17 Determine **E** and **B** for TM waves propagating in the yz-plane between two parallel, perfectly conducting plates, at $y = 0$ and at $y = a$.

18–18 Consider a TM wave in a rectangular waveguide ($H_z = 0$) propagating in the z-direction with wavelength λ_g. Show that

$$E_z = A \sin \frac{m\pi x}{a} \sin \frac{n\pi y}{b} e^{2\pi i z/\lambda_g}$$

satisfies the wave equation (18–123) and the boundary conditions. What is the cutoff frequency of the TM_{11} mode? Why is there no TM_{10} mode?

18–19 Determine the limiting values of the width a of a waveguide of square cross section that will transmit a wave of length λ in the TE_{10} mode but not in the TE_{11} or TM_{11} modes.

18–20 Write down the **E** and **H** fields for the TE_{101} mode of a cubic cavity of side a. Sketch the nature of the field distributions throughout the cube.

CHAPTER 19 Optical Dispersion in Materials

How an electromagnetic wave propagates in a linear material medium is determined entirely by the optical constants n and k. These depend only on the dielectric constant K and the conductivity g of the material. So far the values of these latter parameters have been taken as given, but notice has been made that the values may (in fact always do) depend on the frequency of the wave, varying widely in the range from d-c to x-rays. In order to use judgment in applying the results of electromagnetic radiation theory, one must know something of the principles behind this variation, called *dispersion*. We will now present a microscopic model of materials that predicts their behavior. It is an extension to a-c of the models treated briefly in Chapters 5 and 7 for d-c. Known as the *Drude-Lorentz theory*, it is based on treating the charged particles that constitute the material as classical harmonic oscillators or as free particles.

19-1 DRUDE-LORENTZ HARMONIC OSCILLATOR MODEL

All ordinary matter is composed of negative electrons and positive nuclei. If, for the purpose at hand, some of the electrons (more or fewer than Z, the nuclear charge) can be considered as tightly bound to the nucleus and moving with it, the composite entity is a charged ion. The electrons or ions will be treated as harmonic oscillators—that is, particles bound to an equilibrium position by a linear restoring force. For generality we make it a damped harmonic oscillator, including a linear damping force proportional to the velocity. When an electromagnetic wave is present, the oscillator is driven by the electric field of the wave.* The response of the medium is obtained by adding up the motions of the particles; since the assumed forces are linear, the K and g that result from the model will be constant (i.e., independent of E, although they will depend on frequency). Applied to electrons, the model describes the bound electrons in atoms, but free electrons

---

* The Lorentz force is $\mathbf{F} = q(\mathbf{E} + \mathbf{v} \times \mathbf{B})$, but for a wave $B = (n/c)E$. The magnetic force is smaller by $n(v/c)$, and will be neglected. In any case the magnetic force does no work on the particle, since it is perpendicular to $\mathbf{v}$.

can be included as a special case simply by putting the restoring-force constant of the oscillator equal to zero.

The classical equation of motion* for the one-dimensional forced, damped oscillator is

$$m \frac{d^2x}{dt^2} + G \frac{dx}{dt} + Cx = eE_m, \tag{19-1}$$

or

$$\frac{d^2x}{dt^2} + \gamma \frac{dx}{dt} + \omega_0^2 x = \frac{eE_m}{m}, \tag{19-2}$$

where e and m are the charge and mass of the particle, and E_m is the "molecular field" discussed in Chapter 5. The damping constant $\gamma = G/m$ has the dimensions of frequency. The natural frequency of the undamped oscillator is ω_0 and is related to the force constant C by $m\omega_0^2 = C$. In Chapter 5 the force constant was expressed for bound outer-shell electrons in terms of the "radius" of the atom R_0. In the static case x is independent of t, so that Eq. (19-2) becomes identical to Eq. (5-12), with

$$\omega_0^2 = \frac{e^2}{4\pi\epsilon_0 mR_0^3}. \tag{19-3}$$

For free electrons, we put $\omega_0 = 0$ in Eq. (19-2), which becomes identical to Eq. (7-31), with

$$\gamma = \frac{1}{\tau} \tag{19-4}$$

where τ is the mean time between collisions. Appropriate values of ω_0 and γ for other cases will be discussed later. To take some account of mutual interactions among the particles, the driving field E_m is assumed to depend on the E-field of the wave E according to

$$E_m = E + \frac{v}{\epsilon_0} P, \tag{19-5}$$

where P is the polarization of the medium. It was shown in Eq. (5-7) that for an isotropic, nonpolar dielectric, $v = \frac{1}{3}$. For a metal, $v = 0$. It is not our main purpose here to deal with the difficult problem of the proper local field correction, and so we will just leave it as v and try to eliminate it from consideration in the following. We assume that E_m and P, like E, depend on position and time sinusoidally, or in complex form

$$E_m = \hat{E}_m e^{-i(\omega t - \kappa \cdot \mathbf{r})}.$$

---

* Quantum mechanics should of course be used for electrons. The classical solution is useful because it happens to be the same as the quantum one, with a suitable reinterpretation of the natural frequency ω_0.

In our applications the wavelength $\lambda = 2\pi/\kappa$ will be much greater than the size of the region in which the particle moves. For example, in Eq. (19–3) R_0 is 1 or 2 Å, and for visible light $\lambda \cong 5000$ Å. Thus the spatial variation of E_m is negligible over the positions of the particle; in other words, we can assume that

$$\kappa = 0, \tag{19–6}$$

and the field is uniform,

$$E_m = \hat{E}_m e^{-i\omega t}. \tag{19–7}$$

With Eq. (19–7) the steady-state solutions of Eq. (19–2) are obtained by the well-known procedure of substituting

$$x(t) = \hat{x}e^{-i\omega t} \tag{19–8}$$

and determining the unknown amplitude $\hat{x}$ so that the equation is satisfied for the given frequency ω. The result is

$$\hat{x} = \frac{eE_m/m}{\omega_0^2 - \omega^2 - i\gamma\omega}. \tag{19–9}$$

The amplitude of the particle motion is proportional to the driving field E_m and, as a function of driving frequency, is especially large for $\omega = \omega_0$ (resonance). In the absence of damping, the resonance amplitude would be infinite, and so a realistic model requires some damping.* In the mechanical oscillator problem (and also in the LRC-circuit problem, which is mathematically the same), one usually expresses the complex result in polar form in order to exhibit the real amplitude and phase, but here we leave it as is for now.

The connection between the mechanical displacement x of the microscopic charged particles that make up the material and the macroscopic electrical response of the medium is given by calculating the polarization density P. The dipole moment due to the displaced charge e is ex (assuming that the neutralizing charge $-e$ is remaining at rest). Then

$$P = Nex, \tag{19–10}$$

where N is the number of charges per unit volume. Now the applied field E (of the wave) and the response P are assumed to be proportional,

$$P = \chi E, \tag{19–11}$$

where

$$\epsilon = \epsilon_0 + \chi,$$

$$K = 1 + \chi/\epsilon_0. \tag{19–12}$$

---

* Also, a steady-state solution would never be reached without damping, since the transient solution would never die out.

From Eqs. (19-9)–(19-11),

$$\chi E = \frac{Ne^2/m}{\omega_0^2 - \omega^2 - i\gamma\omega} E_m,$$

and from Eqs. (19-5) and (19-11)

$$E_m = (1 + v\chi/\epsilon_0)E.$$

Thus

$$\frac{\chi}{1 + v\chi/\epsilon_0} = \frac{Ne^2/m}{\omega_0^2 - \omega^2 - i\gamma\omega}.$$

From (19-12)

$$\chi = \epsilon_0(K - 1),$$

so that the dimensionless dielectric constant is given by

$$\frac{K - 1}{1 + v(K - 1)} = \frac{Ne^2/\epsilon_0 m}{\omega_0^2 - \omega^2 - i\gamma\omega}.$$

It is convenient to make the abbreviation

$$\omega_p^2 = \frac{Ne^2}{\epsilon_0 m}, \tag{19-13}$$

in terms of which

$$\frac{K - 1}{1 + v(K - 1)} = \frac{\omega_p^2}{\omega_0^2 - \omega^2 - i\gamma\omega}. \tag{19-14}$$

This gives the sought-for relation between the macroscopic dielectric constant K and the microscopic properties of the charged particles that constitute the medium. Two important features of this result should be noted: First K is complex and secondly it is frequency dependent. Thus the simplest model we could set up automatically implies a conducting, dispersive medium.

Before discussing the consequences of the model, we should suggest a simple generalization, which extends its applicability to most real materials. Seldom do all the charged particles of a material have the same properties. For example, they may divide into electrons and ions, or into electrons in different inner and outer orbits of an atom, and so on. If there are N_i particles of charge e_i, mass m_i, natural resonant frequency ω_{0i}, and damping frequency γ_i, then following the same derivation as above leads to

$$\frac{K - 1}{1 + v(K - 1)} = \sum_i \frac{\omega_{pi}^2}{\omega_{0i}^2 - \omega^2 - i\gamma_i\omega}. \tag{19-15}$$

If all of the particles have the same charge and mass, (e.g., all are electrons in different orbits), this can be written

$$\frac{K-1}{1+v(K-1)} = \omega_p^2 \sum_i \frac{f_i}{\omega_{0i}^2 - \omega^2 - i\gamma_i\omega}, \tag{19-16}$$

where $f_i = N_i/N$ is the fraction of oscillators of type i. Since $\sum N_i = N$,

$$\sum f_i = 1. \tag{19-17}$$

(In the quantum mechanical interpretation of this result, f_i is called the *oscillator strength* and Eq. (19–17) is called the *f-sum rule*.) We shall limit our discussion to those cases that do not depend on subtleties of the local field correction v. If the Lorentz local field $v = \frac{1}{3}$ applies, Eq. (19–16) becomes

$$\frac{K-1}{K+2} = \tfrac{1}{3}\omega_p^2 \sum \frac{f_i}{\omega_{0i}^2 - \omega^2 - i\gamma_i\omega}. \tag{19-18}$$

Since ω_p^2 is proportional to N, this is a generalization of the Clausius-Mossotti equation. It is easy to solve Eqs. (19–15) or (19–18) for K, and for some purposes the result is useful, but it does not clarify the frequency dependence of K, which is the main subject of this chapter. However, if $v = 0$, Eq. (19–15) simplifies to

$$K - 1 = \sum \frac{\omega_{pi}^2}{\omega_{0i}^2 - \omega^2 - i\gamma_i\omega}. \tag{19-19}$$

This also holds, for any v, at frequencies where K is not too different from 1. In addition, at frequencies where one of the resonance peaks of Eq. (19–15) dominates all the others, we can eliminate v from consideration:

$$(K-1)(\omega_0^2 - \omega^2 - i\gamma\omega) = [1 + v(K-1)]\omega_p^2,$$

so that

$$K - 1 = \frac{\omega_p^2}{(\omega_0^2 - v\omega_p^2) - \omega^2 - i\gamma\omega}. \tag{19-20}$$

This is the same as a single resonance peak with zero local field correction and effective resonance frequency $\sqrt{\omega_0^2 - v\omega_p^2}$. Therefore, we will be able to limit our detailed discussion of the frequency dependence of $\hat{K}$ to the simple resonance peak

$$\hat{K} - 1 = \frac{\omega_p^2}{\omega_0^2 - \omega^2 - i\gamma\omega}, \tag{19-21}$$

and still be able to apply the results to many practical problems. Written in terms of real and imaginary parts, Eq. (19–21) is

$$K_r = 1 + \frac{\omega_p^2(\omega_0^2 - \omega^2)}{(\omega_0^2 - \omega^2)^2 + (\gamma\omega)^2}, \tag{19-22}$$

$$K_i = \frac{\omega_p^2\gamma\omega}{(\omega_0^2 - \omega^2)^2 + (\gamma\omega)^2}. \tag{19-23}$$

These quantities could then be substituted into Eqs. (17–52),

$$n = \sqrt{\tfrac{1}{2}[K_r + \sqrt{K_r^2 + K_i^2}]},$$
$$k = \sqrt{\tfrac{1}{2}[-K_r + \sqrt{K_r^2 + K_i^2}]}, \tag{19-24}$$

in order to have explicit expressions for the dispersion of n and k as a function of frequency. The results would be too complicated, however, to reveal any general features of the dispersion. In the following sections we consider various cases of practical interest where simplifying assumptions can be made.

Before proceeding to these examples, we should also comment on the other significant feature of the result Eq. (19–21)—that $\hat{K}$ is complex. Even though the model was set up for *bound* charges, the resulting complex $\hat{K}$ is characteristic of a *conducting* medium. There is a nonvanishing $g = K_i \epsilon_0 \omega$, without the intentional introduction of a conduction current density J. In addition, with $\omega_0 = 0$ for *free* charges, there is still a $K_r = K$ characteristic of a *dielectric* medium. The model automatically incorporates both real and imaginary parts of $\hat{K}$, corresponding to both the displacement current $\partial D/\partial t$ and the conduction current J in Maxwell's curl H equation. In our model we could have calculated

$$J = Ne\,\frac{dx}{dt} = -i\omega Nex$$

instead of $P = Nex$, and from this obtained an expression for the conductivity g. Clearly we would find that

$$J = \frac{dP}{dt} = -i\omega P,$$

and

$$\hat{g} = -i\omega\hat{\chi}. \tag{19-25}$$

We have to calculate P or J but not both, since they are equivalent expressions of the fact that the particle displacement has a component in phase with the E-field and also a component $90°$ out of phase, or that the particle velocity has an out-of-phase component and also an in-phase component. For static fields, bound charges have displacement proportional to the field (in phase) and free charges have velocity proportional (in phase); but at high frequencies both bound and free charges can each have in-phase and out-of-phase components of displacement and velocity. All of the following discussion could instead be formulated in terms of the complex conductivity according to Eq. (19–25), and in certain contexts it is more usual to do so. The relations to the complex dielectric constant are

$$\hat{g} = -i\epsilon_0 \omega(\hat{K} - 1), \tag{19-26}$$

$$g_r = \epsilon_0 \omega K_i, \qquad g_i = -\epsilon_0 \omega(K_r - 1).$$

As expected, $g_r = g$ is the actual real conductivity, and g_i is related to the real dielectric constant.

19–2 RESONANCE ABSORPTION BY BOUND CHARGES

For our first example we will consider the application of Eqs. (19–22)–(19–24) to bound charges, that is, to materials that are nonconductors for d-c. To estimate the expected size of the parameters in Eqs. (19–22) and (19–23) for valence electrons, let us put $R_0 = 2$ Å in Eq. (19–3):

$$\omega_0 = \sqrt{\frac{(1.6 \times 10^{-19})^2}{4\pi(8.854 \times 10^{-12})(0.91 \times 10^{-30})(2 \times 10^{-10})^3}} = 5.6 \times 10^{15} \text{ s}^{-1}.$$

This electron resonance frequency corresponds to a wavelength in air of 3350 Å (335 nm), which is in the ultraviolet region just beyond the visible spectrum. Larger or smaller R_0 would give a longer or shorter resonance wavelength. Combining Eq. (19–3) with Eq. (19–13) which defines ω_p, we find

$$\frac{\omega_p^2}{\omega_0^2} = 4\pi R_0^3 N = 3NV_a, \tag{19–27}$$

where V_a is the volume of our "atom." Thus if the atoms are closely packed, as when they are condensed in a liquid or solid,

$$\frac{\omega_p^2}{\omega_0^2} \approx 1. \tag{19–28}$$

When they are less densely aggregated, as in a gas (or a solution), this ratio is correspondingly less, since N is smaller but usually ω_0 is not strongly affected by the state of aggregation. The ratio is also much less than 1 for the more tightly bound inner-shell electrons, which have smaller orbits. Reasonable values of the damping frequency γ are more difficult to estimate. To get further insight into its meaning, let us return to the equation of motion (19–2) and consider the case of the free oscillator with $E_m = 0$. The solution of this equation is also well known:

$$x = x_0 e^{-\gamma t/2} e^{-i\omega_0' t}, \tag{19–29}$$

where

$$\omega_0' = \sqrt{\omega_0^2 - (\gamma/2)^2}$$

is nearly the same as ω_0 if γ is small. The amplitude of the oscillation decays as $e^{-\gamma t/2}$, and the energy of oscillation, which is proportional to the square of the amplitude, decays as $e^{-\gamma t}$. Thus we can write

$$\gamma = \frac{1}{\tau}, \tag{19–30}$$

where τ is the mean decay time for the oscillator energy. The significance of τ in this equation is not the same as in Eq. (19–4) for free electrons, but it is analogous: in both cases τ is the mean time of energy loss—energy of free harmonic oscillation or energy of free flight, respectively. The oscillating particle necessarily radiates electromagnetic energy at the expense of its energy of oscillation; in the next chapter it is shown that the rate of radiation corresponds to a decay rate

$$\frac{\gamma}{\omega_0} = \frac{4\pi}{3} \frac{R_e}{\lambda_0}, \tag{19-31}$$

where

$$R_e = \frac{e^2}{4\pi\epsilon_0 mc^2} = 2.81 \times 10^{-15} \text{ m} \tag{19-32}$$

and $\lambda_0 = 2\pi c/\omega_0$ is the vacuum wavelength corresponding to ω_0; R_e is the so-called classical radius of the electron. (See Problem 6–6.*) Even for x-rays ($\lambda_0 \approx 1$ Å $= 10^{-10}$ m) the damping due to this mechanism is extremely small relative to the resonance frequency. Usually there are additional modes of decay due to interactions with other particles (collisions), which make the damping frequency enormously greater than the single-particle rate due to radiation. Still, however, we assume that

$$\frac{\gamma}{\omega_0} \ll 1$$

for the applications of this section.

Since γ is assumed to be small, a first approximation with $\gamma = 0$ is informative, even though unphysical. Equation (19–22) with $\gamma = 0$ is

$$K_r - 1 = \frac{\omega_p^2}{\omega_0^2 - \omega^2}, \tag{19-33}$$

as shown in Fig. 19–1. Equation (19–23) with $\gamma = 0$ gives $K_i = 0$ at all frequencies except $\omega = \omega_0$, where it is undefined. But Eq. (19–23) in general gives

$$K_i = \frac{\omega_p^2}{\omega_0} \frac{1}{\gamma} \text{ at } \omega = \omega_0, \tag{19-34}$$

so that K_i at the resonance peak is proportional to $1/\gamma$ and goes to infinity in the limit as γ goes to zero. This limit can be expressed as a function of frequency by the Dirac delta function, which was used earlier to represent a point charge distribution as a function of position.

$$K_i = \frac{\omega_p^2}{\omega_0} \frac{\pi}{2} \delta(\omega - \omega_0). \tag{19-35}$$

The coefficient $\pi/2$ is chosen so that the integral over ω,

$$\int_0^\infty K_i \, d\omega = \frac{\omega_p^2}{\omega_0} \frac{\pi}{2},$$

---

* The charge distribution in an electron is not physically observable, so Eq. (19–32) is a more convenient definition than the one for a uniform charge distribution.

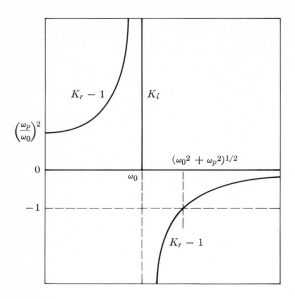

Figure 19–1 Dielectric constants as a function of frequency for an extremely narrow, strong absorption line at ω_0.

is the correct limit for $\gamma \to 0$, as will be shown later. Since K_i is proportional to g, which specifies the energy dissipation per unit volume in the material, this approximation to $K_i(\omega)$ represents the case of an infinitely narrow absorption line at $\omega = \omega_0$, as is also indicated in Fig. 19–1. The most interesting feature of the figure is that K_r has values significantly different from 1 (for vacuum) at frequencies very far away from ω_0, the only frequency where K_i differs from 0 (for vacuum). The effect of energy absorption in the optical region is manifested in the dielectric constant K (but not in the conductivity g) at all frequencies, even d-c. For $\omega = 0$

$$K_r(0) = K_0 = 1 + \frac{\omega_p^2}{\omega_0^2}, \tag{19–36}$$

which may differ appreciably from 1 according to Eq. (19–28). The value $K_0 = 5.5$ for diamond, for example, is understandable, even though diamond is a nonconductor at frequencies below the ultraviolet. Conversely, a nonconductor can have K different from 1 only if it is conducting (absorbing) in some other frequency region; a medium in which $g(\omega) = 0$ for all frequencies necessarily has $K(\omega) = 1$ also, i.e., it is identical to vacuum, as we will see below. On the other hand, at frequencies far above the absorption frequency, K goes to 1 in all cases.

The optical constants n and k are obtained for this case by using the approximations (19–33) and (19–35) in (19–24). Except at $\omega = \omega_0$, $K_i = 0$, and so the approximations of (17–53) and (17–54) apply. From Eq. (19–33), $K_r = 0$ at $\omega = \sqrt{\omega_0^2 + \omega_p^2}$; above this frequency and below ω_0, K_r is positive. In these regions

$$n = \sqrt{K_r}, \qquad k = 0. \tag{19–37}$$

In the intermediate region K_r is negative, so that

$$n = 0, \qquad k = \sqrt{-K_r}. \tag{19–38}$$

These functions are shown in Fig. 19–2. At frequencies below ω_0 and above $\sqrt{\omega_0^2 + \omega_p^2}$, the material is transparent ($k = 0$). Below the resonance, n is greater than 1 and increases with increasing frequency (shorter wavelength). This is the "normal dispersion" behavior of a glass prism in the visible region, which shows a greater refraction for blue than for red, and it is typical of all transparent materials. Above the resonance absorption, n increases with frequency, but it is less than 1, also characteristic of all materials in the x-ray region. Immediately above the resonance a wave is attenuated ($k > 0$), but not, however, because of absorption (except exactly at ω_0 where $K_i \neq 0$). The wave is attenuated because of perfect reflection at the surface where it enters the medium:

$$R = \frac{(n-1)^2 + k^2}{(n+1)^2 + k^2},$$

which for $n = 0$ gives $R = 1$. Although glass has a reflectance of only $R = 0.04$ in the visible, in the ultraviolet it should be highly reflecting.

To examine the detailed behavior in the immediate vicinity of ω_0, suppose that $\gamma > 0$, but still $\gamma \ll \omega_0$. Near ω_0 (i.e., in range $\omega_0 \pm$ several times γ), the approximation

$$\omega_0^2 - \omega^2 = (\omega_0 + \omega)(\omega_0 - \omega) \cong 2\omega_0(\omega_0 - \omega)$$

is valid, and the dielectric constants Eqs. (19–22) and (19–23) simplify to

$$K_r - 1 = \frac{\omega_p^2}{\omega_0} \frac{\frac{1}{2}(\omega_0 - \omega)}{(\omega_0 - \omega)^2 + (\gamma/2)^2}, \qquad (19\text{–}39)$$

$$K_i = \frac{\omega_p^2}{\omega_0} \frac{\gamma/4}{(\omega_0 - \omega)^2 + (\gamma/2)^2}. \qquad (19\text{–}40)$$

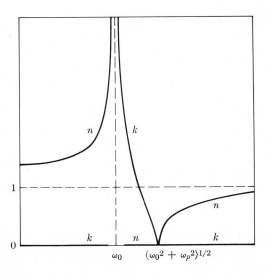

Figure 19–2 Optical constants as a function of frequency, derived from the dielectric constants of Fig. 19–1.

The function K_i has the so-called *Lorentzian* line shape shown in Fig. 19–3, which also shows $K_r - 1$. These functions have especially simple properties: K_i is even with respect to $(\omega_0 - \omega)$, and $K_r - 1$ is odd. It is easy to show that the width of K_i at half-maximum is γ and the extremes of $K_r - 1$ occur at these same points. The integral of K_i over all frequencies can be carried out with the result $(\omega_p^2/\omega_0)(\pi/2)$, thus justifying the coefficient in Eq. (19–35). Since the width of the peak is γ and the height is proportional to $1/\gamma$, the area under the curve is independent of γ. The maximum value of K_i at resonance,

$$M = \frac{\omega_p^2}{\omega_0 \gamma} = \frac{\omega_p^2}{\omega_0^2} \frac{\omega_0}{\gamma},$$

can be quite large. For a gas at atmospheric pressure, $(\omega_p/\omega_0)^2 \approx 10^{-3}$, but $(\omega_0/\gamma) \approx 10^5$, so that M may be greater than 100. This value gives a conductivity $g = \epsilon_0 \omega K_i$ at resonance that is comparable with the d-c conductivity of a metal.

The shape of the n and k curves associated with the Lorentzian dielectric constants depends very much on the size of M. If M is large, $K_i \gg 1$ at ω_0; since $K_r = 1$ at ω_0, the approximation

$$n \cong k \cong \sqrt{K_i/2}$$

holds at ω_0. The shape of the curves, shown in Fig. 19–4, is rather different from the Lorentzian shape. It is a more realistic picture of the behavior near ω_0 than Fig. 19–2. In the other extreme when $M \ll 1$, $K_i \ll 1$ everywhere. Since $|K_r - 1| < K_i$ near the maximum, $|K_r - 1| \ll 1$, or $K_r \cong 1$. In this case

$$n \cong \sqrt{K_r}, \qquad k = K_i/2n.$$

Since K_r differs from 1 by a small amount, the square root can be expanded to give

$$n - 1 \cong \tfrac{1}{2}(K_r - 1), \qquad k \cong \tfrac{1}{2}K_i. \tag{19–41}$$

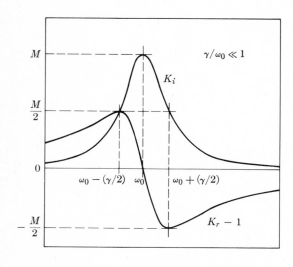

Figure 19–3 Dielectric constants for a narrow line at ω_0. (The origin of frequency is displaced far to the left.) For a weak line with $M \ll 1$, $n - 1$ and k are just one half of $K_r - 1$ and K_i, respectively.

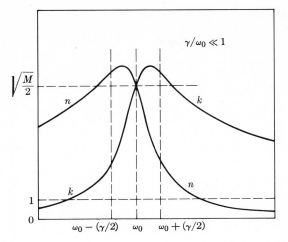

Figure 19–4 Optical constants for a narrow, strong line, derived from the dielectric constants of Fig. 19–3 with $M = 100$. (The origin of frequency is displaced far to the left.)

The shape of the $n - 1$ and k curves is just the same Lorentzian shape as the $K_r - 1$ and K_i curves, respectively. This approximation may hold, for example, in a rather dilute solution or in a gas at a fairly low pressure.

Far from the resonance line, when $|\omega_0 - \omega| \gg \gamma$, γ can simply be neglected in the denominators of Eqs. (19–22) and (19–23), which then simplify to

$$K_r - 1 = \frac{\omega_p^2}{\omega_0^2 - \omega^2},$$

$$K_i = \frac{\omega_p^2 \gamma \omega}{(\omega_0^2 - \omega^2)^2}. \qquad (19\text{–}42)$$

In this approximation $K_i \ll |K_r|$, and far on the low frequency side $K_r > 0$. Equations (19–41) apply, so

$$n - 1 \cong \frac{1}{2} \left(\frac{\omega_p}{\omega_0}\right)^2 \left[1 - \left(\frac{\omega}{\omega_0}\right)^2\right]^{-1}$$

$$\cong \frac{1}{2} \left(\frac{\omega_p}{\omega_0}\right)^2 \left[1 + \left(\frac{\omega}{\omega_0}\right)^2\right].$$

In terms of wavelength,

$$n - 1 \cong \frac{1}{2} \left(\frac{\lambda_0}{\lambda_p}\right)^2 \left[1 + \left(\frac{\lambda_0}{\lambda}\right)^2\right]. \qquad (19\text{–}43)$$

This is known as the *Cauchy relation*, and it is a useful formula for fitting the refractive index of a transparent material.

If γ is not so small, none of the foregoing simple relations holds quantitatively, but the qualitative behavior of $\hat{K}$ and $\hat{n}$ is still similar. The case of $\omega_p/\omega_0 = 1$, $\gamma/\omega_0 = \frac{1}{2}$, is plotted in Fig. 19–5. The imaginary part always has a peak with its

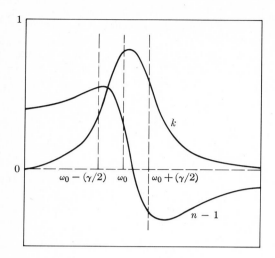

Figure 19–5 Optical constants for a wide, moderately strong absorption band at ω_0.

maximum near ω_0. In the region of this peak, the real part always has a region of negative slope, called the *anomalous dispersion* region. As the frequency increases from zero, there are successive regions of transparency, absorption, high reflectance, and transparency, just as in the simplest delta-function approximation. In actual materials there are always many electronic absorption peaks in the ultraviolet, perhaps extending into the visible, and in solids they may be wide and strongly overlapping.

If the vibrating charged particles are heavy ions instead of electrons, the resonance frequency ω_0 is a few hundred times smaller: The linear restoring-force constant is about the same, since it also derives from the Coulomb force, but the ion mass is 4 or 5 orders of magnitude greater than the electron mass. The frequency ω_p is also proportionately smaller, since it too is inversely proportional to the square root of the mass. For example, in an ionic crystal

$$\frac{\omega_p}{\omega_0} \cong 1, \qquad \frac{\gamma}{\omega_0} \cong 0.1.$$

The absorption peak is similar to the electronic one, but it is located in the infrared instead of the visible or the ultraviolet. The corresponding K_r or n makes no contribution at higher frequency, but it does at lower frequency. Thus for rocksalt the static dielectric constant is about 6, in comparison to about 2 $(= 1.5^2)$ in the visible. The latter is due to the electronic absorption in the ultraviolet; the former includes the effect of the ionic absorption in the infrared as well.

19–3 THE DRUDE FREE-ELECTRON THEORY

In some important states of matter, notably metals and plasmas, the electrons belonging to the outer atomic orbits are not localized (bound), but are *free* to contribute to d-c conduction. At high frequency their behavior is modified by

Table 19–1 Some Typical Particle Densities N and Plasma Frequencies ω_p for Electron Plasmas

	Density (m^{-3})	Plasma frequency (s^{-1})
Metal	10^{28}	10^{16}
Semiconductor (doped)	10^{24}	10^{14}
Semiconductor (pure)	10^{20}	10^{12}
Fusion experiment	10^{20}	10^{12}
Ionosphere	10^{11}	10^{7}
Interplanetary space	10^{7}	10^{5}

inertial effects; but it is explained by our same microscopic model, if the assumed restoring force for these electrons is put equal to zero. Then, with $\omega_0 = 0$, Eq. (19–21) simplifies to

$$\hat{K} - 1 = -\frac{\omega_p^2}{\omega(\omega + i\gamma)}. \qquad (19\text{–}44)$$

The real and imaginary parts are

$$K_r - 1 = -\frac{\omega_p^2}{\omega^2 + \gamma^2}, \qquad (19\text{–}45)$$

$$K_i = \frac{\omega_p^2 \gamma}{\omega(\omega^2 + \gamma^2)}. \qquad (19\text{–}46)$$

In the free-particle context,

$$\omega_p = \sqrt{\frac{Ne^2}{\epsilon_0 m}} \qquad (19\text{–}13)$$

is called the *plasma frequency*,* and it is identical to that defined in Chapter 14. Now the frequency ranges in which interesting features of the dispersion occur are determined by ω_p. Some typical values are listed in Table 19–1. They span from the radio frequency to the ultraviolet regions. As noted previously, the damping constant γ is the reciprocal collision time,

$$\gamma = \frac{1}{\tau}. \qquad (19\text{–}4)$$

The time τ is also the time constant for the decay of a current with no driving field, as can be seen by putting $\omega_0 = 0$ and $E_m = 0$ in Eq. (19–2): $dx/dt = v$,

$$\frac{dv}{dt} + \gamma v = 0,$$

$$v = v_0 e^{-\gamma t} = v_0 e^{-t/\tau}. \qquad (19\text{–}47)$$

---

* The plasma frequency is usually seen expressed in Gaussian units, in which $\omega_p = \sqrt{4\pi Ne^2/m}$.

For metals at room temperature, $\gamma \approx 10^{14}$ s^{-1}, so that

$$\frac{\gamma}{\omega_p} \ll 1.$$

This relation usually holds for semiconductor electronic materials and gaseous plasmas as well. Equation (19–44) assumes a simpler form when it is expressed in terms of the complex conductivity $\hat{g}$ using Eq. (19–26) and $\gamma = 1/\tau$:

$$\hat{g} = \frac{g_0}{1 - i\omega\tau}, \qquad (19\text{–}48)$$

where

$$g_0 = \epsilon_0 \omega_p^2 \tau = \frac{Ne^2\tau}{m}$$

is the d-c conductivity. Nevertheless we will treat $\hat{K}$, since the aim is to find the optical constants n and k from it.

When γ is small, we may again look first at the case $\gamma = 0$:

$$K_r = 1 - \frac{\omega_p^2}{\omega^2},$$

$$\qquad (19\text{–}49)$$

$$K_i = \frac{\omega_p^2}{\omega} \frac{\pi}{2} \delta(\omega).$$

Figures 19–1 and 19–2 still show the behavior, if the origin of ω is taken at $\omega_0 = 0$. The transparent region below ω_0 is eliminated, and the frequency where $K_r = 0$ is ω_p. Between $\omega = 0$ and $\omega = \omega_p$ the material is perfectly reflecting ($n = 0$), and above ω_p it is transparent ($k = 0$). The transition between high reflectance and transparency at ω_p accounts for the well-known fact that the ionosphere reflects radio waves in the AM broadcast band ($f = \omega/2\pi \lesssim 1.5 \times 10^6$ s^{-1}), but is transparent to FM and TV ($f \approx 10^8$ s^{-1}). In another frequency range, sodium metal is highly reflecting in the visible, but is transparent for ultraviolet wavelengths shorter than $\lambda_p = 2100$ Å (210 nm) corresponding to its plasma frequency. Just at the plasma frequency, $n = 0$ and $k = 0$; the charges all move back and forth in phase (infinite wavelength) with no attenuation. This motion is the free plasma oscillation discussed in Chapter 14. It is an example of the longitudinal wave* that can occur when $\hat{K} = 0$. Although the no-damping approximation accounts for interesting effects near the plasma frequency, it is an oversimplification for lower frequencies, where it predicts infinite conductivity and infinite dielectric constant for d-c. In particular it does not yield the Hagen-Rubens formula nor its associated skin depth.

---

* Since the wavelength is infinite and the propagation velocity zero, it is meaningless to say it is longitudinal or transverse in this case. We have calculated $\hat{K}(\kappa, \omega)$ only for $\kappa = 0$, however—cf. (19–6). A more sophisticated calculation including pressure effects shows the existence of longitudinal plasma waves with finite wavelength, $\kappa \neq 0$.

To improve on the delta-function approximation, it is necessary to go back to Eqs. (19–45) and (19–46). We assume that the damping is small, $\gamma \ll \omega_p$, so there are three frequency ranges to consider.

1. $\omega \ll \gamma$:

$$K_r \cong \frac{-\omega_p^2}{\gamma^2},$$ (19–50)

$$K_i \cong \frac{\omega_p^2}{\omega\gamma}.$$ (19–51)

The real dielectric constant K_r is a (large) negative constant; K_i goes to infinity as ω goes to zero, but

$$g = \epsilon_0 \omega K_i = \frac{\epsilon_0 \omega_p^2}{\gamma} = \frac{Ne^2\tau}{m} = g_0$$

is constant in this range and equals the d-c conductivity g_0. Since $K_i/|K_r| = \gamma/\omega \gg 1$, Eq. (17–55) gives

$$n \cong k \cong \sqrt{K_i/2} = \frac{\omega_p}{\sqrt{2\omega\gamma}} \gg 1.$$ (19–52)

This is the case that gives the Hagen-Rubens formula for the low frequency absorptance,

$$A \cong \frac{2}{k} = 2\sqrt{\frac{2\epsilon_0\omega}{g_0}},$$

and the skin depth

$$\delta = \frac{c}{k\omega} \cong \sqrt{\frac{2}{\mu_0\omega g_0}}.$$

2. $\gamma \ll \omega \ll \omega_p$:

$$K_r \cong -\frac{\omega_p^2}{\omega^2},$$ (19–53)

$$K_i \cong \frac{\omega_p^2\gamma}{\omega^3}.$$ (19–54)

Since $K_i/|K_r| = \gamma/\omega \ll 1$,

$$k \cong \sqrt{-K_r} = \frac{\omega_p}{\omega},$$

$$n = \frac{K_i}{2k} \cong \frac{\gamma\omega_p}{2\omega^2}.$$ (19–55)

In this region $k/n = 2\omega/\gamma \gg 1$, so that the absorptance Eq. (18–59) is

$$A \cong \frac{4n}{k^2} \cong \frac{2\gamma}{\omega_p} = \frac{2}{\omega_p \tau}$$

(sometimes called the Mott-Zener formula), and the skin depth is

$$\delta = \frac{c}{k\omega} = \frac{c}{\omega_p} = \lambda_p/2\pi,$$

where λ_p is the vacuum wavelength corresponding to the plasma frequency. For metals these results apply to the infrared region. Since $\omega_p \tau$ is large, $A = 1 - R$ is only a few percent; the skin depth is very small and independent of frequency.

3. $\omega_p \ll \omega$:

$$K_r \cong 1, \tag{19–56}$$

$$K_i \cong \frac{\omega_p^2 \gamma}{\omega^3}. \tag{19–57}$$

$$n \cong \sqrt{K_r} \cong 1,$$

$$k = \frac{K_i}{2n} \cong \frac{\omega_p^2 \gamma}{2\omega^3}. \tag{19–58}$$

Here the material is nearly transparent. In metals, however, except for the alkali metals, the onset of this transparency is obscured by the resonance absorption of bound electrons belonging to inner orbits. This additional absorption, discussed in the last section, also increases the values of K_i and n for most metals in the visible region, reducing the reflectance and accounting for the characteristic colors of copper and gold. The free-electron results for n and k are plotted on a log-log scale in Fig. 19–6. The parameter values chosen are $\omega_p = 9 \times 10^{15}$ s^{-1}, $\gamma = 3.6 \times 10^{13}$ s^{-1}; these are appropriate for sodium metal at room temperature, for which the free-electron theory gives good agreement with experiment. The three straight-line regions of the figure are plots of Eqs. (19–52), (19–55), and (19–58). The intermediate points at $\omega = \gamma$ and $\omega = \omega_p$ are easily calculated from Eqs. (19–45) and (19–46), since $K_i = -K_r$ and $K_i \gg K_r$, respectively. The region well above γ agrees with the delta-function approximation ($\gamma = 0$), with some rounding of the kinks at ω_p. Examples with $\gamma \gtrsim \omega_p$ are less common; their treatment is left as an exercise.

All the results of this section also apply when the charged particles are heavy ions instead of electrons. Since the medium has been assumed to be electrically neutral, positive ions must always be present with an average number density N equal to that of the electrons. In metals the positive ions are not freely mobile, of course, but in gaseous plasmas they are, and their motion is often important. The plasma frequency ω_p is inversely proportional to the square root of the particle mass m, and so the ion plasma frequency is about two orders of magnitude smaller than the electron plasma frequency given in Table 19–1, for example.

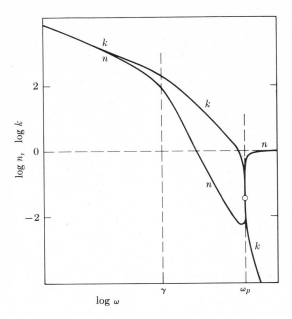

Figure 19–6 Log-log plot of the optical constants of free charges versus frequency, with $\gamma/\omega_p = 0.004$. Power-law approximations are valid over most of the frequency range.

*19–4 DIELECTRIC RELAXATION. ELECTROLYTIC CONDUCTION

Electrical resonance absorption does not usually occur in materials at frequencies lower than the heavy-ion peaks in the infrared (though it does in man-made structures, of course). There is, however, another kind of absorption mechanism, known as *dielectric loss*, which can occur at lower frequencies (but not at higher). It is frequently important as a loss mechanism at microwave frequencies and below, and its concomitant dispersion of the real dielectric constant explains, for example, the difference between the static dielectric constant of water, 81, and the optical value of about 1.8 ($= 1.33^2$). This effect is an extension to a-c of the second type of static polarization discussed in Chapter 5—namely, the orientational polarization of permanent dipoles. This case can also be based on the model described by Eq. (19–1), though with some strain on the physical interpretation of the quantities involved. It arises when the damping and restoring forces are important, but inertial effects (acceleration) can be neglected. In this case the equation of motion becomes

$$G \frac{dx}{dt} + Cx = eE_m, \tag{19–59}$$

or

$$\gamma \frac{dx}{dt} + \omega_0^2 x = eE_m/m.$$

The appropriate solution is obtained from Eq. (19–21) by neglecting ω^2,

$$\hat{K} - 1 = \frac{\omega_p^2}{\omega_0^2 - i\gamma\omega}.$$

The parameters in this solution all contain m; however, m does not really enter because the inertial term has been neglected in obtaining the solution. It is better to rewrite the solution as

$$\hat{K} - 1 = \frac{Ne^2/\epsilon_0}{C - iG\omega},$$

where m is eliminated entirely. From Eq. (19–59) with $E_m = 0$, the return to equilibrium is described by

$$x = x_0 e^{-Ct/G}.$$

Therefore, we write

$$\frac{C}{G} = \frac{1}{\tau}, \tag{19–60}$$

where τ in this context is called the *relaxation time*.* In terms of it

$$\hat{K} - 1 = \frac{K_0 - 1}{1 - i\omega\tau}, \tag{19–61}$$

where

$$K_0 - 1 = \frac{Ne^2}{\epsilon_0 C}. \tag{19–62}$$

Clearly K_0 is the static value of K, for $\omega = 0$; it is the same as the static value for resonance absorption, except that C has a different meaning. This can be obtained from the calculation in Chapter 5:

$$K_0 - 1 = \frac{\chi}{\epsilon_0} = \frac{N}{\epsilon_0} \frac{p_0^2}{3k_B T},$$

where k_B is Boltzmann's constant. If we let the permanent dipole moment be

$$p_0 = ea,$$

where a is the effective separation of charges $\pm e$, then

$$K_0 - 1 = \frac{Ne^2 a^2}{3\epsilon_0 k_B T}. \tag{19–63}$$

---

* Note that in this process $1/\tau \neq \gamma$. We are using τ consistently to mean the time constant for return to equilibrium, regardless of its relation to γ or the determining physical quantities.

Numerically this can be, at the maximum, of the order of 10^2 at room temperature. Comparing Eq. (19–63) with Eq. (19–62) shows that

$$C = \frac{3k_B T}{a^2}.$$

(19–64)

The restoring force is therefore due to the thermal energy $k_B T$ rather than to a mechanical elastic potential energy.*

What is new in this discussion is the relaxation time τ, which determines the a-c behavior. Its value can be estimated most easily for certain *solid* polar materials in which the dipoles consist of mobile pairs of positive and negative ions of charge e separated by a distance a. The ions have certain fixed equilibrium positions in the solid, but they can make jumps to other neighboring allowed positions. By means of such jumps the dipole can rotate into alignment with an applied field, or return to a random orientation when the field is removed. The time τ is therefore the mean time for a jump. This is given by

$$\frac{1}{\tau} = \frac{1}{\tau_0} e^{-\Delta U/k_B T} \quad \text{(solid)}$$

(19–65)

The factor $1/\tau_0$ is approximately the vibrational frequency of the ion around its equilibrium position, and $\exp(-\Delta U/k_B T)$ is the statistical Boltzmann factor. A jump is attempted with each vibration, but only a fraction succeeds, depending on the energy ΔU required in order to squeeze through the barrier to the neighboring equilibrium position. The frequency $1/\tau_0 = \omega_0$ is the ionic vibrational frequency considered previously, $\omega_0 \approx 10^{13} \text{ s}^{-1}$. The barrier energy ΔU must be appreciable in order for this model to hold at all, since the very existence of equilibrium positions between which jumps occur depends on the barriers. Thus

$$\frac{1}{\tau} \lesssim 10^{11} \text{ s}^{-1},$$

and $1/\tau$ will be very much smaller than this at temperatures much below the melting temperature. For liquids the calculation of τ is more difficult. The *Debye relaxation time* is given by

$$\frac{1}{\tau} = \frac{k_B T}{4\pi\eta R_0^3}, \quad \text{(liquid)}$$

(19–66)

where η is the viscosity (not electrical resistivity) and R_0 is the radius of the molecule. This gives reasonable agreement with experimental values of η and predicts a frequency in the microwave region for most polar liquids at room temperature, since the dipoles in a liquid are more freely rotating than in a solid.

---

* The restoring force is a generalized "thermodynamic force," not a mechanical one. What relaxes after an applied force is removed (with the relaxation time τ) is not the energy but rather the entropy.

The viscosity is temperature dependent, but still Eq. (19–66) is less strongly temperature dependent than Eq. (19–65).

The frequency dependence of Eq. (19–61) is seen by separating it into real and imaginary parts, giving the *Debye equations*

$$K_r - 1 = \frac{K_0 - 1}{1 + (\omega\tau)^2},$$ (19–67)

$$K_i = \frac{(K_0 - 1)\omega\tau}{1 + (\omega\tau)^2}.$$ (19–68)

These are plotted in Fig. 19–7, which shows an absorption peak in K_i with maximum at $\omega = 1/\tau$ and height $\frac{1}{2}(K_0 - 1)$, and K_r falling from K_0 to 1 with the halfway point at $\omega = 1/\tau$. The dispersion of K_r has a shape characteristically different from that for resonance absorption. The shape of the curves is quite similar to that for a strongly overdamped oscillation. For frequencies much below or much above $1/\tau$, $K_i \ll K_r$ $(K_r > 0)$, so that $n = \sqrt{K_r}$, $k = K_i/2n$. As an example, these results can be applied to fresh water in which the d-c conductivity can be neglected; the reflectance is given by Eq. (18–57). For electrical measurements on laboratory-sized samples, the results are usually expressed in terms of the *loss angle* θ:

$$\tan\theta = \frac{K_i}{K_r} = \frac{(K_0 - 1)\omega\tau}{K_0 + (\omega\tau)^2}.$$ (19–69)

The physical reason for this kind of frequency dependence is that above $1/\tau$ the relaxation cannot follow the applied field.

Electrolytic conduction is a phenomenon that, although it involves d-c conduction by "free" charges, proceeds by a mechanism that is closely related to that of dielectric relaxation. The charge carriers are mobile ions that move (in solids)

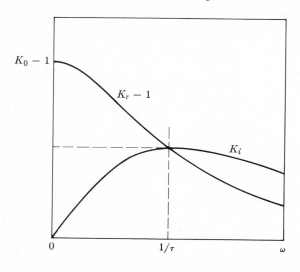

Figure 19–7 Dielectric constants as a function of frequency for Debye relaxation.

by jumping or (in liquids) by a mechanism related to viscosity. Again, inertial effects are negligible and the ion mass is not directly involved. The difference from dielectric relaxation is that positive-negative pairs of ions are not inseparable, but the positive or negative ions are free to migrate individually, contributing to a transport current. The d-c conductivity is again more easily calculated for a solid electrolyte; the result is

$$g_0 = \frac{Ne^2a^2}{k_B T \tau}, \quad \text{(solid)} \tag{19-70}$$

where τ is a mean jump time, perhaps different from that for dipolar reorientation but still given by an equation like Eq. (19–65), and a is the distance of a jump. In a liquid electrolyte the result involves the viscosity η. The *Einstein-Stokes* formula,

$$g_0 = \frac{Ne^2}{6\pi\eta R_0}, \quad \text{(liquid)} \tag{19-71}$$

is applicable to dilute solutions. For liquid electrolytes g_0 can be of the order of 10^2 $(\Omega \text{ m})^{-1}$, and also for a special class of solid materials called *superionic conductors*. In most solid ionic materials g_0 is less than 1 $(\Omega \text{ m})^{-1}$, usually far less at room temperature.

The a-c conductivity is obtained most directly from Eq. (19–48),

$$\hat{g} = \frac{g_0}{1 - i\omega\tau}, \tag{19-48}$$

where τ was shown in Eq. (19–47) to be the relaxation time for the conduction (or the startup time for steady-state conduction), which in this case is the jump time in a solid electrolyte. Equation (19–48) must necessarily be the result of a linear theory, regardless of the microscopic interpretation of the parameters. Since Eq. (19–48) has exactly the same functional form as Eq. (19–61), the real and imaginary parts of the conductivity have the same shape as the curves shown in Fig. 19–7. To find the optical constants, Eq. (19–48) can be converted, by means of Eq. (19–26), to

$$\hat{K} - 1 = i\,\frac{\hat{g}}{\epsilon_0\omega} = \frac{-g_0/\epsilon_0}{\omega(\omega\tau + i)}, \tag{19-72}$$

the real and imaginary parts of which are

$$K_r - 1 = \frac{-\tau g_0/\epsilon_0}{1 + (\omega\tau)^2}, \tag{19-73}$$

$$K_i = \frac{g_0/\epsilon_0}{\omega[1 + (\omega\tau)^2]}. \tag{19-74}$$

The contribution of Eq. (19–73) to the static dielectric constant is

$$K_0 - 1 = -\frac{\tau g_0}{\epsilon_0},$$

which for a solid electrolyte is

$$K_0 - 1 = -\frac{Ne^2a^2}{\epsilon_0 k_B T}.$$

Although this has the same form as Eq. (19–63) for dielectric relaxation (but *opposite* sign), the values of N may differ widely in the two cases. In pure water every molecule is involved in dielectric relaxation, but only a few parts per million contribute to conduction. Electrolytic conduction and dielectric relaxation often occur together in the same material. The two contributions to the loss angle can be separated by the different frequency dependences of K_i in simple cases. In aqueous solutions, however, the picture is complicated by the local field correction (not the Lorentz one) and by microscopic interactions between the ions and the dipoles.

19–5 THE KRAMERS-KRONIG RELATIONS

In all the examples so far examined, an absorption peak in K_i had associated with it a dispersion of K_r, with a more or less characteristic shape. A sum of several peaks would have a corresponding composite dispersion curve. We now show that there is a general relation between K_r and K_i such that, whatever the mechanism of absorption, if $K_i(\omega)$ is known at all frequencies then $K_r(\omega)$ is uniquely determined at each frequency by a certain definite integral over K_i. In particular, if $K_i(\omega) = 0$, then $K_r(\omega) = 1$ at all frequencies. This relation has considerable theoretical and practical usefulness, since if $K_i(\omega)$ can somehow be calculated or measured then it is unnecessary to calculate or measure K_r separately; it can be determined from the Kramers-Kronig relation. The general relation can be found by assuming that any arbitrary absorption as a function of ω can be represented as the sum of a continuous distribution of very narrow harmonic oscillator absorption lines. The integral of Eq. (19–35) or (19–40) over frequency is

$$\int_0^\infty K_i \, d\omega = \frac{\omega_p^2}{\omega_0}\frac{\pi}{2}, \tag{19–75}$$

which we call $K_i(\omega_0) \, \Delta\omega_0$, the total contribution of the oscillators of natural frequency ω_0 in the small range $\Delta\omega_0$ to the overall K_i. If ω_p^2 is expressed in terms of $K_i(\omega_0) \, \Delta\omega_0$,

$$\omega_p^2 = \frac{2}{\pi}\,\omega_0 K_i(\omega_0) \, \Delta\omega_0.$$

When this is put into Eq. (19–33) for K_r, the resulting

$$\Delta(K_r - 1) = \frac{2}{\pi}\frac{\omega_0}{\omega_0^2 - \omega^2}\,K_i(\omega_0) \, \Delta\omega_0$$

represents the contribution to K_r at a given ω due to the oscillators in the range $\Delta\omega_0$ around ω_0. The sum of all the contributions from different ω_0 to K_r at this ω

is obtained by changing $\Delta\omega_0$ to a differential $d\omega_0$ (and incidentally designating the variable of integration as ω' instead of ω_0):

$$K_r(\omega) - 1 = \frac{2}{\pi} \int_0^\infty \frac{\omega' K_i(\omega')\, d\omega'}{\omega'^2 - \omega^2}. \tag{19–76}$$

This is the *Kramers-Kronig relation*, which was first used in 1926 to find the refractive index for x-rays from the measured absorption. More recently it has been widely applied in other spectral regions.

As the simplest indication of the power of the relation, inserting the delta-function K_i Eq. (19–35) into Eq. (19–76) immediately gives the associated K_r of Eq. (19–33). Except in other rather artificial examples, however, the integration cannot be carried out analytically and must be performed numerically. One problem in the integration is the singularity at $\omega' = \omega$. This is not serious—the negative contributions for $\omega' < \omega$ cancel the positive ones for $\omega' > \omega$—but the problem should be recognized. As a second example, which will illustrate this and other problems encountered in the application of the Kramers-Kronig integral, consider the low-frequency approximation for free particles* Eq. (19–51),

$$K_i = \frac{\omega_p^2}{\gamma\omega}.$$

This gives an easy integral in Eq. (19–76),

$$K_r - 1 = \frac{2}{\pi} \frac{\omega_p^2}{\gamma} \int_0^\infty \frac{d\omega'}{\omega'^2 - \omega^2}, \tag{19–77}$$

and we also know the result to be expected, from Eq. (19–50). To handle the singularity, write†

$$\int_0^\infty = \lim_{\delta\to 0} \left[\int_0^{\omega-\delta} + \int_{\omega+\delta}^\infty \right].$$

Then

$$\int_0^{\omega-\delta} \frac{d\omega'}{\omega'^2 - \omega^2} = -\frac{1}{2\omega} \ln \left| \frac{\omega + \omega'}{\omega - \omega'} \right| \bigg|_0^{\omega-\delta} = -\frac{1}{2\omega} \ln \frac{2\omega - \delta}{\delta}. \tag{19–78}$$

$$\int_{\omega+\delta}^\infty \frac{d\omega'}{\omega'^2 - \omega^2} = -\frac{1}{2\omega} \ln \left| \frac{\omega + \omega'}{\omega - \omega'} \right| \bigg|_{\omega+\delta}^\infty = \frac{1}{2\omega} \ln \frac{2\omega + \delta}{\delta}. \tag{19–79}$$

Adding these integrals gives

$$\int_0^\infty \frac{d\omega'}{\omega'^2 - \omega^2} = \lim_{\delta\to 0} \frac{1}{2\omega} \ln \frac{2\omega + \delta}{2\omega - \delta} = 0. \tag{19–80}$$

---

* Strictly speaking, the theory is not applicable to $\hat{K}$ of a free-electron conductor, because of the singularity in $K_i(\omega)$ at $\omega = 0$. Equation (19–76) still holds, however; see Problem 19–16.

† This is called the *Cauchy principal part* of the integral.

Thus the result of Eq. (19–77) is

$$K_r = 1. \tag{19–81}$$

This result contains some instructive features. First, the singularity in the integrand does not make the integral infinite. (In fact, it is zero, but that is a peculiarity of this special case.) It is clear that the singularity will cause no trouble so long as K_i is continuous at ω. Secondly the result Eq. (19–81) for K_r is a constant, but the constant disagrees with the constant $-(\omega_p/\gamma)^2$ required by Eq. (19–50). This discrepancy emphasizes that $K_i(\omega)$ must be known over the entire frequency range from 0 to ∞ in order to get the correct result from Eq. (19–76). In this example Eq. (19–51) is correct at frequencies well below γ, but wrong at higher frequencies. The true value of K_r at low frequency comes from this higher frequency behavior of K_i. The integral always gives the correct value of K_r when ω goes to infinity, namely, $K_r = 1$, as can be seen directly in Eq. (19–76). Thus if K_r is constant it must give the constant 1. In using the integral with experimental data, it is necessary to make some reasonable extrapolation above and below the frequency range in which the measurements were made, even to find K_r only within this range. A more obvious cause of error would be the omission of an unknown absorption peak above the range of measurement.

The integral in Eq. (19–76) can be transformed in various ways. Since the integral in Eq. (19–80) is zero, any constant multiple of it can be subtracted from Eq. (19–76) without affecting the result. Subtracting $(2/\pi)\omega K_i(\omega)$ times it gives

$$K_r(\omega) - 1 = \frac{2}{\pi} \int_0^\infty \frac{[\omega' K_i(\omega') - \omega K_i(\omega)]\, d\omega'}{\omega'^2 - \omega^2}.$$

This form is useful for numerical integrations because it has a nonsingular integrand, the numerator vanishing at the same point as the denominator. An integration by parts in Eq. (19–76) gives

$$K_r(\omega) - 1 = \frac{1}{\pi} \int_0^\infty \frac{dK_i(\omega')}{d\omega'} \ln \frac{1}{|\omega'^2 - \omega^2|}\, d\omega'.$$

This gives a qualitative idea of the behavior of K_r to be expected. The integrand is heavily weighted for frequencies near ω by the second factor, so that the magnitude of K_r is strongly affected by the *slope* of K_i at nearby frequencies. This kind of relation can be seen in all the particular examples treated in previous sections.

Although the Kramers-Kronig relation has been derived from the harmonic oscillator model of the medium, the only aspect of the model it actually depends on is its *linearity*. Since the harmonic oscillator is the prototype of linear systems, it is not surprising that the general result follows from it, but a model-independent derivation brings out some other interesting features. A rigorous derivation using this approach is based on the theory of complex integration* and will not be

---

* See, for example, L. Landau and E. Lifshitz, *Electrodynamics of Continuous Media* (Reading, Mass.: Addison-Wesley, 1960), Sections 58–62.

attempted, but the basic ideas are fairly simple. Because of their generality, similar results can also be obtained for other complex response functions, such as the Fresnel reflection coefficient, complex a-c impedance (for which the theory was first fully developed*), and even systems in nuclear and elementary-particle physics. Such relations are also called *dispersion relations*. In addition to linearity, the other basic assumption is that the response of the system is *causal*. That is, the response will not anticipate the applied force but will occur only *after* the force is applied.

To express the assumed linearity and causality, we write the polarization P in terms of the applied field E as

$$P(t) = \int_0^\infty dt' f(t') E(t - t'), \qquad (19\text{--}82)$$

where $f(t')$ is real. That is, P at the present time (t) is proportional to contributions from E now and in the past $(t - t'$, with $t' \geq 0)$ but not in the future $(t' < 0)$. Now let P and E be expressed as a superposition of plane waves (i.e., take Fourier transforms), in order to introduce the susceptibility:

$$E(t) = \int_{-\infty}^\infty d\omega \hat{E}(\omega) \exp(-i\omega t),$$

$$P(t) = \int_{-\infty}^\infty d\omega \hat{\chi}(\omega) \hat{E}(\omega) \exp(-i\omega t).$$

Inserting the first of these equations into Eq. (19–82) and comparing with the second, we see that

$$\hat{\chi}(\omega) = \int_0^\infty dt' f(t') \exp(i\omega t'),$$

or

$$\chi_r(\omega) = \int_0^\infty dt' f(t') \cos \omega t',$$

$$\chi_i(\omega) = \int_0^\infty dt' f(t') \sin \omega t'.$$

Thus χ_r and χ_i are not independent because they are each uniquely related to $f(t')$. To get their relation to each other, one must solve one of these equations for $f(t')$, as can be done using the Fourier theorem, e.g.,

$$f(t') = (2/\pi) \int_0^\infty d\omega' \chi_i(\omega') \sin \omega' t'. \qquad (19\text{--}83)$$

Substituting this into the integral for χ_r, one can carry out the integration over t',

$$\chi_r(\omega) = (2/\pi) \int_0^\infty d\omega' \omega' \chi_i(\omega')/(\omega'^2 - \omega^2). \qquad (19\text{--}84)$$

---

* See the book by Bode cited in Chapter 13.

Similarly,

$$\chi_i(\omega) = -(2\omega/\pi) \int_0^\infty d\omega' \chi_r(\omega')/(\omega'^2 - \omega^2). \qquad (19\text{-}85)$$

The first of these equations is identical to Eq. (19-76) and the second is complementary to it.

We may use Eqs. (19-82) and (19-83) to calculate an elementary example of the response of a dispersive medium to a *non*sinusoidal applied field. Once again we consider the case of a medium with a single very narrow absorption peak. Let

$$\chi_i(\omega') = (\pi \omega_0 \chi_0/2)\, \delta(\omega' - \omega_0),$$

so that $\chi_r(0) = \chi_0$. Then from Eq. (19-83) the response function of this medium is $f(t') = \chi_0 \omega_0 \sin \omega_0 t'$. Now assume that a step-function electric field is applied (locally),

$$E(t) = 0, \qquad t < 0,$$
$$= E_0. \qquad t > 0.$$

With this E, Eq. (19-82) becomes

$$P(t) = E_0 \int_0^t dt' f(t'),$$

and with the calculated $f(t')$

$$P(t) = \chi_0 E_0(1 - \cos \omega_0 t).$$

After the sudden application of the electric field E_0, the medium "rings" at the absorption (resonance) frequency ω_0. If the damping were nonzero, the oscillations of P would eventually damp out to $\chi_0 E_0$ as for a d-c field. We see that there is no simple proportionality between $P(t)$ and $E(t)$, and their (time dependent) ratio is not a material property, emphasizing again that the material constants are ratios between Fourier transforms.

19–6 SUMMARY

The complex dielectric constant of a material is calculated as a function of the frequency of the electric field by treating the electrons and ions either as classical damped harmonic oscillators or free particles. The result is

$$\hat{K} - 1 = \frac{\omega_p^2}{\omega_0^2 - \omega^2 - i\gamma\omega},$$

where ω_0 is the natural frequency, γ the damping frequency, and

$$\omega_p = \sqrt{\frac{Ne^2}{\epsilon_0 m}}$$

(the plasma frequency for free particles). On the basis of this the typical kinds of frequency dependence of the real dielectric constant and conductivity can be catalogued, depending on whether the inertial, damping, or restoring forces can be neglected. If local-field effects can be ignored, the dielectric responses of different groups of particles are additive. The frequency dependence of the optical constants n and k depends on that of the dielectric constants and also on the relative magnitudes of the real and imaginary parts. The real and imaginary parts are not independent of each other, but are related by the Kramers-Kronig relations.

1. Resonance absorption occurs when inertial and restoring forces dominate. For small damping the dielectric constant has the simple Lorentzian shape. If K_i is small, the optical constants have the same shape. In any case "anomalous dispersion" occurs in the region of the absorption peak. The peaks are in the visible or ultraviolet for electrons and in the infrared for ions.

2. The Drude free-electron theory results by putting the restoring force (ω_0) equal to zero. The low-frequency skin depth and absorptance (Hagen-Rubens formula) with the d-c conductivity result when $\omega \ll \gamma$. For $\omega \gg \gamma$, ω_p, the free particles make a very small contribution to conduction or absorption. If $\gamma \ll \omega_p$, there is an intermediate frequency region (in the infrared for metals) of low absorptance (high reflectance),

$$A = 2/\omega_p \tau,$$

where τ is the collision frequency, and small skin depth,

$$\delta = \lambda_p/2\pi,$$

where λ_p is the plasma wavelength. The plasma frequency is in the ultraviolet for metals, and at much lower frequencies in other common electron plasmas.

3. Inertial effects are negligible in electrolytic conduction and dielectric relaxation. The former has the same frequency dependence as free-particle conduction, although the mobility and decay-time mechanisms are different. The latter has an "over damped" frequency dependence, with no "normal dispersion" region. The mechanism is that of the orientational polarizability of permanent dipoles.

4. The Kramers-Kronig relation

$$K_r(\omega) - 1 = \frac{2}{\pi} \int_0^\infty \frac{\omega' K_i(\omega')\, d\omega'}{\omega'^2 - \omega^2}$$

between the real and imaginary parts of the dielectric response function depends only on linearity of the medium (and on causality). It or variations of it have useful applications in all linear systems.

PROBLEMS

19-1 The density and refractive index of liquid benzene at 20°C are 0.879 g/cm^3 and 1.50 (for $\lambda = 589$ nm), respectively. From the Clausius-Mossotti equation, compute the refractive index of benzene vapor at 20°C, where the vapor pressure is 0.1 atmosphere; also at the boiling point, 80°C.

19–2 Prove that the width of the Lorentzian curve Eq. (19–40) at half maximum is γ and that the area under the curve is

$$\frac{\pi}{2}\frac{\omega_p^2}{\omega_0}.$$

19–3 The refractive index of diamond at $\lambda = 5893$ Å is 2.417; take the static dielectric constant to be 5.50. Fit these data to the simple model which has a single δ-function absorption at λ_0, in order to determine λ_0.

19–4 Use the Cauchy formula to estimate the refractive index of hydrogen gas at standard conditions for wavelengths of 4000 and 7000 Å. Assume that $\lambda_0 = 1216$ Å (the Lyman-α line).

19–5 Experimental absorption peaks sometimes have a shape that is closer to a Gaussian curve than a Lorentzian. Plot a Gaussian and a Lorentzian with the same peak height and width-at-half-maximum on the same graph, so as to show the difference between them.

19–6 The dielectric function for bound ions oscillating with negligible damping can be written

$$K(\omega) = K_\infty + \frac{\omega_p^2}{\omega_T^2 - \omega^2},$$

where ω_T is the resonance frequency for the long wavelength transverse vibrations of the ions, and K_∞ approximates the contribution of electronic motions which resonate at frequencies much higher than ω_T. Note that $K(\omega) \to \infty$ for $\omega = \omega_T$. If $K(\omega) = 0$ for $\omega = \omega_L$, so that long wavelength longitudinal oscillations of frequency ω_L can occur, show that

$$\frac{\omega_L^2}{\omega_T^2} = \frac{K_0}{K_\infty},$$

where $K_0 = K(0)$ is the d-c dielectric constant. This is called the *Lyddane-Sachs-Teller relation*.

19–7 Suppose a dilute solution consists of N atomic oscillators per unit volume dissolved in a transparent medium of refractive index n_∞. Assuming that $K_i \ll 1$ at the resonance frequency and $n = n_\infty$, find k and the absorption coefficient α. Neglect local field effects. Show that

$$N = \frac{\epsilon_0\,mc}{e^2}\,n_\infty\,\gamma\alpha.$$

Such a relation (known as Smakula's equation or Chako's equation) is often used to find N from the measured optical absorption height α and width γ.

19–8 Consider a medium containing free particles with collision time τ and d-c conductivity g_0. Calculate the real and imaginary parts of the conductivity for the frequency $\omega = 1/\tau$. What is the real dielectric constant K?

19–9 For a free-electron plasma with $\gamma/\omega_p = 10^{-2}$, calculate approximate values of n and k at $\omega = \omega_p$.

19–10 Suppose that in the dielectric function of a free-electron plasma with negligible damping, the contribution from bound inner-orbit electrons which resonate at a higher frequency can be approximated by K_∞:

$$K(\omega) = K_\infty - \frac{\omega_p^2}{\omega^2}.$$

Find the frequency of longitudinal oscillations. In silver the value calculated from the free (valence) electron density is $\omega_p = 13.8 \times 10^{15}$ s^{-1}, while longitudinal plasma osciliations are observed at $\omega = 5.8 \times 10^{15}$ s^{-1}. What is K_∞?

19–11 In a free-electron plasma, the longitudinal plasma waves occur at the frequency $\omega = Ne^2/\epsilon_0 m$, where m is the electron mass. If the positive ions of mass M are also freely mobile, show that the longitudinal waves occur at $\omega = Ne^2/\epsilon_0 \mu$, where $\mu = mM/(M + m)$ is the reduced mass, assuming that the contributions to K are additive.

19–12 Discuss the dielectric behavior of free particles for the case $\gamma \gg \omega_p$. That is, find approximate expressions for $\hat{K}$ and $\hat{n}$ in the various frequency regions. Does the Hagen-Rubens relation hold, and if so in what frequency range?

19–13 The dielectric loss angle of a polar dielectric has a maximum as a function of frequency. Calculate the frequency at which the maximum occurs, and find K_r and K_i at this frequency if $K_0 \gg 1$.

19–14 Use the Kramers-Kronig relations to prove the following formulas (called *sum rules*) for the real conductivity g and susceptibility χ:

$$\int_0^\infty \chi(\omega')\, d\omega' = \frac{\pi}{2} \lim_{\omega \to \infty} g(\omega) = 0,$$

$$\int_0^\infty g(\omega')\, d\omega' = -\frac{\pi}{2} \lim_{\omega \to \infty} \omega^2 \chi(\omega) = \frac{\pi Ne^2}{2m}.$$

19–15 Suppose a medium is characterized by a single absorption peak $\chi_i(\omega') = \frac{1}{2}\pi\omega_0 \chi_0\, \delta(\omega' - \omega_0)$. Find the response $P(t)$ to a pulsed E-field,

$$E(t) = \frac{E_0}{\omega_0}\, \delta(t).$$

19–16 By considering $J(t)$ instead of $P(t)$, show that the complex conductivity obeys the dispersion relations

$$g_r(\omega) = \frac{2}{\pi} \int_0^\infty \frac{\omega' g_i(\omega')\, d\omega'}{\omega'^2 - \omega^2}, \qquad g_i(\omega) = -\frac{2}{\pi} \omega \int_0^\infty \frac{g_r(\omega')\, d\omega'}{\omega'^2 - \omega^2}.$$

These hold for a conductor as well as an insulator, since $\hat{g}(\omega)$ has no singularity at $\omega = 0$.
 a) Using the relation $\hat{g} = -i\omega\hat{\chi}$, compare these dispersion relations with Eqs. (19–84) and (19–85).
 b) Derive the "sum rules"

$$\int_0^\infty \chi(\omega')\, d\omega' = -\frac{\pi}{2} g(0), \qquad \int_0^\infty g(\omega')\, d\omega' = \frac{\pi}{2} \frac{Ne^2}{m}.$$

Note that they are consistent with those derived in Problem 19–14 for an insulator $(g(0) = 0)$.

CHAPTER 20 Radiation Emission

A convenient starting point for the study of electromagnetic wave generation is with the vector potentials satisfying the inhomogeneous wave equation with sources. In this chapter we consider several idealized radiation sources as well as more complicated systems. Approximations are made that limit the validity of the solutions to fields produced by slowly moving (nonrelativistic) charges, that is, those with velocity v small compared to the velocity of light, $v \ll c$. Nevertheless the results are applicable both to the emission of radio waves from antennas and of light from atoms. We begin with the simplest antenna example, and then develop a more general procedure.

20–1 RADIATION FROM AN OSCILLATING DIPOLE

A simple example of radiation from a prescribed, time-dependent charge-current distribution is provided by calculation of the radiation from an oscillating electric dipole. The dipole will be assumed to consist of spheres located at $z = \pm l/2$ connected by a wire of negligible capacitance, as shown in Fig. 20–1. The charge on the upper sphere is q, and that on the lower sphere is $-q$. Conservation of charge requires that the current in the connecting wire be given by

$$I = +\dot{q}, \tag{20–1}$$

where I is positive in the plus z-direction. It must be noted that the condition of negligible capacitance of the wire and its concomitant uniform current can be satisfied only if the length l of the dipole is small compared with the wavelength of the radiation (see the discussion at the beginning of Chapter 13).

The vector potential due to the current distribution specified by Eq. (20–1) is, in vacuum, from Eq. (16–84),

$$A_z(\mathbf{r}, t) = \frac{\mu_0}{4\pi} \int_{-l/2}^{l/2} \frac{I(z', t - |\mathbf{r} - z'\mathbf{k}|/c)\, dz'}{|\mathbf{r} - z'\mathbf{k}|}. \tag{20–2}$$

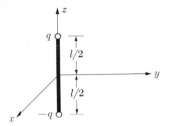

Figure 20–1 An oscillating electric dipole.

This rather cumbersome expression can be quickly simplified if we examine the quantity $|\mathbf{r} - \mathbf{k}z'|$. It is clear that

$$|\mathbf{r} - \mathbf{k}z'| = (r^2 - 2z'\mathbf{k} \cdot \mathbf{r} + z'^2)^{1/2}. \qquad (20\text{–}3)$$

If l is small compared with r, that is, if we consider the field only at large distances from the dipole, the right side of Eq. (20–2) can be expanded in the form

$$|\mathbf{r} - \mathbf{k}z'| = r - z' \cos \theta, \qquad (20\text{–}4)$$

where θ is the angle between $\mathbf{r}$ and the z-axis. The quantity in (20–4) is involved twice in the expression for $\mathbf{A}$. In the denominator, $z' \cos \theta$ can simply be neglected if r is large enough. In the retardation term, however, $z' \cos \theta$ can be neglected only if $z' \cos \theta/c$ is negligible compared with the time during which the current changes significantly, e.g., compared with the period for harmonically varying currents. Since $z' \cos \theta \lesssim l/2$, this means that $z' \cos \theta/c$ can be neglected in the retardation term only if

$$\frac{l}{2} \ll cT = \lambda. \qquad (20\text{–}5)$$

Thus, if the dipole is small compared with one wavelength, and the observation point is far, compared with l, from the dipole, then $\mathbf{A}$ is given by

$$A_z(\mathbf{r}, t) = \frac{\mu_0}{4\pi} \frac{1}{r} \, l \, I\left(t - \frac{r}{c}\right). \qquad (20\text{–}6)$$

The scalar potential φ can be found either by applying the Lorentz condition or by using the appropriate expression for the retarded potential. Both methods give the same final result; however, because the electric potential due to a dipole is the difference between two large terms, great care must be used in approximating the retarded potential. Since this difficulty is circumvented in the Lorentz condition calculation, the scalar potential will be obtained by solving

$$\nabla \cdot \mathbf{A} + \frac{1}{c} \frac{\partial \varphi}{\partial t} = 0, \qquad (20\text{–}7)$$

with **A** as given by Eq. (20–6). Thus

$$\frac{\partial \varphi}{\partial t} = -\frac{l}{4\pi\epsilon_0} \frac{\partial}{\partial z} \frac{1}{r} I\left(t - \frac{r}{c}\right)$$

$$= \frac{l}{4\pi\epsilon_0} \left[\frac{z}{r^3} I\left(t - \frac{r}{c}\right) + \frac{z}{r^2 c} I'\left(t - \frac{r}{c}\right)\right], \qquad (20\text{–}8)$$

where I' represents the derivative of I with respect to its argument. This equation is readily integrated by noting that $I = +q'$ and thus that

$$\varphi(\mathbf{r}, t) = \frac{l}{4\pi\epsilon_0} \frac{z}{r^2} \left[\frac{q(t - r/c)}{r} + \frac{I(t - r/c)}{c}\right]. \qquad (20\text{–}9)$$

Having obtained the scalar and vector potentials, we now need only to differentiate them to obtain the electromagnetic field. Before doing so, it is convenient to specialize the charge-current distribution to one which varies harmonically with the time. The particular choice

$$q\left(t - \frac{r}{c}\right) = q_0 \cos \omega \left(t - \frac{r}{c}\right),$$

$$I = I_0 \sin \omega \left(t - \frac{r}{c}\right) = -\omega q_0 \sin \omega \left(t - \frac{r}{c}\right) \qquad (20\text{–}10)$$

will be made. Resolving **A** into spherical components, we obtain

$$A_r = \frac{\mu_0}{4\pi} \frac{I_0 l}{r} \cos \theta \sin \omega \left(t - \frac{r}{c}\right),$$

$$A_\theta = -\frac{\mu_0}{4\pi} \frac{I_0 l}{r} \sin \theta \sin \omega \left(t - \frac{r}{c}\right), \qquad (20\text{–}11)$$

$$A_\phi = 0,$$

and it becomes obvious that only the ϕ-component of **B** is different from zero. This component is

$$B_\phi = \frac{1}{r} \frac{\partial}{\partial r} (r A_\theta) - \frac{1}{r} \frac{\partial A_r}{\partial \theta}$$

$$= \frac{\mu_0}{4\pi} \frac{I_0 l}{r} \sin \theta \left[\frac{\omega}{c} \cos \omega \left(t - \frac{r}{c}\right) + \frac{1}{r} \sin \omega \left(t - \frac{r}{c}\right)\right]. \qquad (20\text{–}12)$$

The computation of the electric field is somewhat more complex, since not only **A** but also φ is involved. The result of performing the differentiations is

$$E_r = -\frac{\partial \varphi}{\partial r} - \frac{\partial A_r}{\partial t} = \frac{2 l I_0 \cos \theta}{4\pi\epsilon_0} \left[\frac{\sin \omega(t - r/c)}{r^2 c} - \frac{\cos \omega(t - r/c)}{\omega r^3}\right],$$

$$E_\theta = -\frac{1}{r}\frac{\partial\varphi}{\partial\theta} - \frac{\partial A_\theta}{\partial t}$$

$$= -\frac{lI_0 \sin\theta}{4\pi\epsilon_0}\left[\left(\frac{1}{\omega r^3} - \frac{\omega}{rc^2}\right)\cos\omega\left(t - \frac{r}{c}\right) - \frac{1}{r^2 c}\sin\omega\left(t - \frac{r}{c}\right)\right],$$

$$E_\phi = -\frac{1}{r\sin\theta}\frac{\partial\varphi}{\partial\phi} - \frac{\partial A_\phi}{\partial t} = 0. \tag{20-13}$$

It is interesting to compute the rate at which the dipole radiates energy. This is done by integrating the normal component of the Poynting vector over a sphere of radius R. Thus

$$\oint \mathbf{S}\cdot\mathbf{n}\,da = \frac{1}{\mu_0}R^2\int_0^\pi E_\theta B_\phi 2\pi\sin\theta\,d\theta. \tag{20-14}$$

Equations (20–12) and (20–13) make it possible to evaluate completely the integral that appears in Eq. (20–14); however, it is perhaps more instructive to evaluate only the portion which does not vanish as $R \to \infty$. This is done by selecting the term proportional to $1/r$ in E_θ and B_ϕ. The result is

$$\oint \mathbf{S}\cdot\mathbf{n}\,da = \frac{(I_0 l)^2}{6\pi\epsilon_0}\frac{\omega^2}{c^3}\cos^2\omega\left(t - \frac{r}{c}\right). \tag{20-15}$$

This is the instantaneous radiated power; the average radiated power (since the average of $\cos^2$ is one-half) is

$$\bar{P} = \frac{l^2\omega^2}{6\pi\epsilon_0 c^3}\frac{I_0^2}{2}. \tag{20-16}$$

A more conventional form of Eq. (20–16) is obtained by introducing $\lambda = 2\pi c/\omega$ and $c = 1/\sqrt{\epsilon_0\mu_0}$. The result is

$$\bar{P} = \frac{2\pi}{3}\sqrt{\frac{\mu_0}{\epsilon_0}}\left(\frac{l}{\lambda}\right)^2\frac{I_0^2}{2}. \tag{20-17}$$

A resistance R carrying a current $I_0 \cos\omega t$ dissipates energy at an average rate $\bar{P} = RI_0^2/2$. Comparing this with Eq. (20–17), we see that it is sensible to define the *radiation resistance* of a dipole by

$$R_r = \frac{2\pi}{3}\sqrt{\frac{\mu_0}{\epsilon_0}}\left(\frac{l}{\lambda}\right)^2, \tag{20-18}$$

or

$$R_r = 789\left(\frac{l}{\lambda}\right)^2 \text{ ohms,} \quad \text{(free space)}.$$

For a material medium μ_0 and ϵ_0 are replaced by μ and ϵ, and $\lambda = 2\pi/\omega\sqrt{\epsilon\mu}$. One might be tempted to use Eq. (20–18) to describe the radiation from a radio antenna. Unfortunately, several defects prevent obtaining good results in this way.

The principal defects are (1) the effect of the proximity of the earth is neglected, (2) ordinarily antennas are not capacitively loaded at the ends, and (3) antennas are very seldom short compared with the wavelength they radiate. Removal of the last two defects will be discussed in the next section; however, discussion of the perturbing effect of the earth is beyond the scope of this text.

20–2   RADIATION FROM A HALF-WAVE ANTENNA

The restriction to lengths small compared with one wavelength can be removed in some cases by relatively simple means. In particular, a wire that is just one-half wavelength in length can be broken into infinitesimal elements, to each of which the method of the preceding section can be applied. Let the wire lie along the z-axis from $-\lambda/4$ to $+\lambda/4$ and carry a current

$$I(z', t) = I_0 \sin \omega t \cos \left(\frac{2\pi z'}{\lambda} \right). \tag{20–19}$$

This vanishes at the ends of the wire. The nonuniformity of the current requires a varying charge density, which is largest at the ends of the wire. An element dz' at z' contributes, in vacuum,

$$dE_\theta = I_0 \frac{\sin \theta}{4\pi\epsilon_0 Rc^2} \omega \cos \omega \left(t - \frac{R}{c} \right) \cos \left(\frac{2\pi z'}{\lambda} \right) dz' \tag{20–20}$$

to E_θ. Here R is the distance from dz' to the point of observation, and terms of order $1/R^2$ have been neglected. In the same way,

$$dB_\phi = \frac{\mu_0}{4\pi} \frac{I_0 \omega}{Rc} \sin \theta \cos \omega \left(t - \frac{R}{c} \right) \cos \left(\frac{2\pi z'}{\lambda} \right) dz'. \tag{20–21}$$

The problem in calculating E_θ and B_ϕ is reduced to evaluating

$$K = \int_{-\pi/2}^{\pi/2} \frac{1}{R} \cos \omega \left(t - \frac{R}{c} \right) \cos u \, du, \tag{20–22}$$

where $u = 2\pi z'/\lambda$. As before, $R = r - z' \cos \theta$, and hence by choosing r sufficiently large $z' \cos \theta$ can be made negligible. In the argument of the cosine, however, more care is required, and K is written as

$$K = \frac{1}{r} \int_{-\pi/2}^{\pi/2} \cos \left[\omega \left(t - \frac{r}{c} \right) + u \cos \theta \right] \cos u \, du.$$

The cosine can be expanded to give

$$K = \frac{1}{r} \cos \omega \left(t - \frac{r}{c} \right) \int_{-\pi/2}^{\pi/2} \cos (u \cos \theta) \cos u \, du$$

$$- \frac{1}{r} \sin \omega \left(t - \frac{r}{c} \right) \int_{-\pi/2}^{\pi/2} \sin (u \cos \theta) \cos u \, du.$$

The second integral vanishes, and the first can be evaluated by expressing the cosines as exponentials or by using standard tables. The result is

$$K = \frac{2}{r} \cos \omega \left(t - \frac{r}{c} \right) \frac{\cos \left[(\pi/2) \cos \theta \right]}{\sin^2 \theta}. \tag{20-23}$$

Having evaluated K, we find that

$$E_\theta = \frac{I_0}{2\pi\epsilon_0 rc} \cos \omega \left(t - \frac{r}{c} \right) \frac{\cos \left[(\pi/2) \cos \theta \right]}{\sin \theta}$$

$$B_\phi = \frac{\mu_0 I_0}{2\pi r} \cos \omega \left(t - \frac{r}{c} \right) \frac{\cos \left[(\pi/2) \cos \theta \right]}{\sin \theta}. \tag{20-24}$$

The integrated average Poynting vector is

$$\bar{P} = \frac{1}{4\pi} \sqrt{\frac{\mu_0}{\epsilon_0}} I_0^2 \int_0^\pi \frac{\cos^2 \left[(\pi/2) \cos \theta \right]}{\sin^2 \theta} \sin \theta \, d\theta. \tag{20-25}$$

The remaining integral can be evaluated only as an infinite series, but we simply note that for a half-wave antenna the result is

$$\bar{P} = 73.1 \text{ ohms } \frac{I_0^2}{2}, \quad \text{(free space)}. \tag{20-26}$$

This method can be applied to more complicated problems; however, the technical details become rather formidable.

20–3 RADIATION FROM A GROUP OF MOVING CHARGES

In this section we shall derive an expression for the power radiated by a group of moving charges, or equivalently, by a charge-current distribution. The motion of the charges is *arbitrary* except for the following restrictions: during the time it takes for radiation to propagate from the neighborhood of the charges to the observation point, we can imagine that all charges and currents in the distribution are contained in a volume V_1 whose dimensions are small compared to the source-observer distance (see Fig. 20–2). Furthermore, the dimensions of V_1 are small compared to the dominant wavelengths of emitted radiation. The above restrictions also imply that, compared to the speed of light, the charges are moving slowly. It is assumed that the charges are moving in vacuum.

As a first step toward a solution to the problem, we must calculate the electromagnetic potentials. These are just the retarded potentials, which were discussed in Section 16–6. The origin of coordinates O is taken inside the volume V_1 and the position of a charge element is denoted by $\mathbf{r}'$ (see Fig. 20–2). The field point P is at distance $\mathbf{r}$ from the origin. For convenience, the auxiliary distance $\mathbf{R}$, which denotes the position of the field point relative to a charge element, is introduced. Clearly,

$$\mathbf{r}' + \mathbf{R} = \mathbf{r}. \tag{20-27}$$

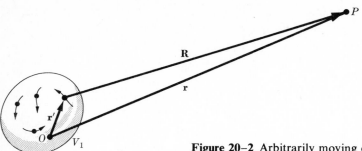

Figure 20–2 Arbitrarily moving charges contained in a volume V_1. The fields are to be calculated at P.

Since $r \gg r'$

$$R = |\mathbf{r} - \mathbf{r}'| \approx r - \frac{\mathbf{r}' \cdot \mathbf{r}}{r}. \tag{20–28}$$

The retarded scalar potential φ at the field point P now takes the form

$$\varphi(\mathbf{r}, t) = \frac{1}{4\pi\epsilon_0} \int_{V_1} \frac{\rho(\mathbf{r}', t - R/c)\, dv'}{R}$$

$$= \frac{1}{4\pi\epsilon_0} \int_{V_1} \frac{\rho(\mathbf{r}', t - r/c + \mathbf{r}' \cdot \mathbf{r}/cr)\, dv'}{r - (\mathbf{r}' \cdot \mathbf{r})/r}. \tag{20–29}$$

Using the Binomial Theorem,

$$\left(r - \frac{\mathbf{r}' \cdot \mathbf{r}}{r}\right)^{-1} = r^{-1} + r^{-2}\frac{\mathbf{r}' \cdot \mathbf{r}}{r} + \cdots, \tag{20–30}$$

and Taylor's series expansion,

$$\rho\left(\mathbf{r}', t - \frac{r}{c} + \frac{\mathbf{r}' \cdot \mathbf{r}}{cr}\right) = \rho\left(\mathbf{r}', t - \frac{r}{c}\right) + \frac{\mathbf{r}' \cdot \mathbf{r}}{cr} \frac{\partial \rho}{\partial t}\bigg|_{\mathbf{r}', t - r/c} + \cdots, \tag{20–31}$$

we obtain

$$\varphi(\mathbf{r}, t) = \frac{1}{4\pi\epsilon_0 r} \int_{V_1} \rho\left(\mathbf{r}', t - \frac{r}{c}\right) dv' + \frac{1}{4\pi\epsilon_0 r^3} \mathbf{r} \cdot \int_{V_1} \mathbf{r}'\rho\left(\mathbf{r}', t - \frac{r}{c}\right) dv'$$

$$+ \frac{1}{4\pi\epsilon_0 r^2 c} \mathbf{r} \cdot \frac{d}{dt} \int_{V_1} \mathbf{r}'\rho\left(\mathbf{r}', t - \frac{r}{c}\right) dv' + \text{higher-order terms}.$$

$$\tag{20–32}$$

The first integral in Eq. (20–32) is the total charge Q of the distribution. This is a constant, independent of time. The second (and also the third) integral is the electric dipole moment $\mathbf{p}$ of the distribution, evaluated at the time $t - r/c$. The higher-order terms fall off with a larger power of r'/r and depend on a higher multipole moment of the distribution. Because of the restrictions imposed at the beginning of this section, these terms do not contribute appreciably (see below) to the distant electromagnetic field of the charge distribution. Thus,

$$\varphi(\mathbf{r}, t) = \frac{1}{4\pi\epsilon_0} \left[\frac{Q}{r} + \frac{\mathbf{r} \cdot \mathbf{p}(t - r/c)}{r^3} + \frac{\mathbf{r} \cdot \dot{\mathbf{p}}(t - r/c)}{cr^2} \right], \qquad (20\text{--}33)$$

with $\dot{\mathbf{p}} \equiv d\mathbf{p}/dt$. As a result of the Taylor series expansion only one retarded time appears in the terms explicitly retained.

The retarded vector potential $\mathbf{A}$ at the field point is given by

$$\mathbf{A}(\mathbf{r}, t) = \frac{\mu_0}{4\pi} \int_{V_1} \frac{\mathbf{J}(\mathbf{r}', t - r/c + \mathbf{r}' \cdot \mathbf{r}/cr)\, dv'}{r - (\mathbf{r}' \cdot \mathbf{r})/r}$$

$$= \frac{\mu_0}{4\pi r} \int_{V_1} \mathbf{J}\left(\mathbf{r}', t - \frac{r}{c}\right) dv' + \text{higher-order terms}. \qquad (20\text{--}34)$$

The higher-order terms do not have to be written out explicitly since again they depend on a higher multipole moment of the distribution. In other words, Eq. (20–34) is already consistent with Eq. (20–32). From the results of Problem 7–2 one may write the last equation as

$$\mathbf{A}(\mathbf{r}, t) = \frac{\mu_0}{4\pi r} \dot{\mathbf{p}}\left(t - \frac{r}{c}\right). \qquad (20\text{--}35)$$

The scalar potential (20–33) could also have been obtained from (20–35) and the Lorentz condition.

The electric and magnetic fields may be obtained from the usual relationships

$$\mathbf{E} = -\frac{\partial \mathbf{A}}{\partial t} - \nabla\varphi, \qquad \mathbf{B} = \nabla \times \mathbf{A}.$$

We shall restrict our interest here to the radiation zone fields, i.e., to contributions to $\mathbf{E}$ and $\mathbf{B}$ which fall off as r^{-1}, since these contributions are sufficient to determine the power radiated by the charge distribution. The calculation of $\partial\mathbf{A}/\partial t$ is straightforward; to obtain $\nabla\varphi$ we note that since $\dot{\mathbf{p}}$ is a function of $t - r/c$,

$$\frac{\partial}{\partial r} \dot{\mathbf{p}} \equiv -\frac{1}{c} \ddot{\mathbf{p}}. \qquad (20\text{--}36)$$

Thus:

$$\mathbf{E}(\mathbf{r}, t) = \frac{-\mu_0}{4\pi r} \ddot{\mathbf{p}}\left(t - \frac{r}{c}\right) + \frac{1}{4\pi\epsilon_0} \frac{\mathbf{r} \cdot \ddot{\mathbf{p}}(t - r/c)}{c^2 r^3} \mathbf{r}$$

$$+ \text{ terms which fall off more rapidly than } (1/r). \qquad (20\text{--}37)$$

To calculate $\mathbf{B}(\mathbf{r}, t)$ we must take the curl of Eq. (20–35):

$$\nabla \times \left[\frac{1}{r} \dot{\mathbf{p}}\left(t - \frac{r}{c}\right)\right] = \nabla \frac{1}{r} \times \dot{\mathbf{p}} + \frac{1}{r} \nabla \times \dot{\mathbf{p}}$$

$$= -\frac{1}{r^2} \frac{\mathbf{r}}{r} \times \dot{\mathbf{p}} + \frac{1}{r} \frac{\mathbf{r}}{r} \times \frac{\partial \dot{\mathbf{p}}}{\partial r}$$

$$= -\frac{1}{r^2} \frac{\mathbf{r}}{r} \times \dot{\mathbf{p}} - \frac{1}{cr} \frac{\mathbf{r}}{r} \times \ddot{\mathbf{p}}.$$

The last step comes about through the use of Eq. (20–36). The first term in the curl may be disregarded since it falls off more rapidly than r^{-1};

$$\nabla \times \mathbf{A}(\mathbf{r}, t) = \frac{-\mu_0}{4\pi cr^2} \mathbf{r} \times \ddot{\mathbf{p}}\left(t - \frac{r}{c}\right). \tag{20–38}$$

The radiation zone fields therefore are given by

$$\mathbf{B}(\mathbf{r}, t) = -\frac{\mu_0}{4\pi cr^2} \mathbf{r} \times \ddot{\mathbf{p}}, \tag{20–39}$$

$$\mathbf{E}(\mathbf{r}, t) = \frac{1}{4\pi\epsilon_0 c^2} \left[\frac{(\mathbf{r} \cdot \ddot{\mathbf{p}})\mathbf{r} - r^2\ddot{\mathbf{p}}}{r^3}\right] \tag{20–40}$$

$$= \frac{1}{4\pi\epsilon_0 c^2 r^3} \mathbf{r} \times (\mathbf{r} \times \ddot{\mathbf{p}})$$

$$= -\frac{c}{r} \mathbf{r} \times \mathbf{B}(\mathbf{r}, t),$$

where $\ddot{\mathbf{p}}$ is evaluated at the retarded time.

It is evident that $\mathbf{E}$ and $\mathbf{B}$ are perpendicular to each other, and are each perpendicular to $\mathbf{r}$. This result could have been anticipated from Eq. (17–22), since $\mathbf{r}/r = \boldsymbol{\kappa}/\kappa = \mathbf{u}$. Thus it is sufficient to calculate only $\mathbf{A}$ when only the radiation field is sought ($r \gg \lambda$). The Poynting vector $\mathbf{S} = (1/\mu_0)(\mathbf{E} \times \mathbf{B})$ has the direction of $\mathbf{r}$, and is given by

$$\mathbf{S} = \frac{c}{\mu_0 r} \mathbf{B} \times (\mathbf{r} \times \mathbf{B}) = \frac{c}{\mu_0 r} \mathbf{r}B^2 \tag{20–41}$$

or

$$\mathbf{S} = \frac{1}{16\pi^2\epsilon_0 c^3 r^5} \mathbf{r}(\mathbf{r} \times \ddot{\mathbf{p}})^2.$$

If the z-axis is taken in the direction of $\ddot{\mathbf{p}}$,

$$\mathbf{S} = \frac{\ddot{\mathbf{p}}^2 \sin^2 \theta}{16\pi^2\epsilon_0 c^3 r^2} \frac{\mathbf{r}}{r}. \tag{20–42}$$

The maximum power is radiated at $90°$ to $\ddot{\mathbf{p}}$. The total radiated power is obtained by integrating the Poynting vector over a closed surface surrounding the charge distribution. A convenient choice for such a surface is a sphere, centered in the charge distribution, with sufficiently large radius so that all parts of its surface are in the radiation zone. Then

$$P_R = -\frac{dW}{dt} = \oint_s \mathbf{S} \cdot \mathbf{n}\, da$$

$$= \frac{\ddot{p}^2}{16\pi^2 \epsilon_0 c^3} \int \frac{\sin^2 \theta}{r^3}\, \mathbf{r} \cdot \frac{\mathbf{r}}{r}\, r^2 \sin \theta\, d\theta\, d\phi,$$

from which one readily obtains the important result

$$P_R = -\frac{dW}{dt} = \frac{1}{4\pi\epsilon_0}\frac{2}{3}\frac{\ddot{\mathbf{p}}^2}{c^3} \tag{20-43}$$

for the power radiated by a group of charges moving slowly compared to the speed of light.

Equation (20–43) is an expression for the power radiated by arbitrarily moving charges in terms of their electric dipole moment $\mathbf{p}$. The previously derived expression for the radiation from an oscillating dipole, Eq. (20–16), is a special example of (20–43); in that case $p = (lI_0/\omega) \cos \omega(t - r/c)$. Now it could happen that due to a special symmetry of the system, the electric dipole moment vanishes or is independent of time. In this case, the power radiated is not necessarily zero, but more terms would have to be retained in the expansions of φ and $\mathbf{A}$ (Eqs. 20–32 and 20–34) to make the calculation. In fact, we would find that the power radiated depends on some higher multipole moment of the system in this case. The various multipole radiations get progressively less intense as the multipole order is increased; e.g., quadrupole radiation is roughly a factor $(a/\lambda)^2$ smaller than dipole radiation, where a is the dimension of the system and λ the wavelength of the emitted radiation. Thus, if $\ddot{p}$ does not vanish for the system under consideration, Eq. (20–43) gives the major contribution to the radiated power.

Equations (20–39), (20–40), and (20–43) can also be applied to the radiation from a single accelerated charge q. The dipole moment of the charge is $q\mathbf{r}'$, where $\mathbf{r}'$ is measured from an arbitrary origin. Then

$$\dot{\mathbf{p}} = q\dot{\mathbf{r}}' = q\mathbf{v},$$

where $\mathbf{v}$ is the velocity of the charge. This is independent of the origin. Finally

$$\ddot{\mathbf{p}} = q\dot{\mathbf{v}},$$

where $\dot{\mathbf{v}}$ is the acceleration of the charge. Substituting this last result into Eq. (20–43), we obtain

$$P_R = -\frac{dW}{dt} = \frac{q^2}{4\pi\epsilon_0}\frac{2}{3}\frac{\dot{v}^2}{c^3} \tag{20-44}$$

for the power radiated by a slowly moving accelerated charge.

*20–4 NEAR AND INTERMEDIATE ZONE FIELDS

Only terms in the fields proportional to $1/r$ contribute to the radiated energy, but higher powers of $1/r$ dominate in the region close to the radiating charges. These fields are of interest for comparison to the static case, and also in practical problems in the immediate vicinity of antennas. In this section we shall calculate the entire E- and B-fields of a point electric dipole. To do so, we start with the dipole potentials Eqs. (20–33) and (20–35), and we retain the terms in the differentiations that were discarded in the preceding section. The contribution to $\mathbf{E}$ from $-\partial\mathbf{A}/\partial t$ is part of the radiation field and has already been calculated. Applying the vector identity (1–1–6) to Eq. (20–33) with $Q = 0$, we find

$$-\nabla\varphi = \frac{-1}{4\pi\epsilon_0} \left[\frac{\mathbf{r}}{cr^2} \cdot \nabla\dot{\mathbf{p}} + \frac{\mathbf{r}}{cr^2} \times (\nabla \times \dot{\mathbf{p}}) + \frac{\mathbf{r}}{r^3} \cdot \nabla\mathbf{p} \right.$$

$$\left. + \frac{\mathbf{r}}{r^3} \times (\nabla \times \mathbf{p}) + \dot{\mathbf{p}} \cdot \nabla \frac{\mathbf{r}}{cr^2} + \mathbf{p} \cdot \nabla \frac{\mathbf{r}}{r^3} \right],$$

with the terms arranged in order of increasing powers of $1/r$. The first two terms, when added to $-\partial\mathbf{A}/\partial t$, give the *radiation field* of Eq. (20–40),

$$\mathbf{E}_1 = \frac{1}{4\pi\epsilon_0} \frac{1}{c^2 r} \left[\frac{\mathbf{r}}{r} \cdot \ddot{\mathbf{p}} \frac{\mathbf{r}}{r} - \ddot{\mathbf{p}} \right] = \frac{1}{4\pi\epsilon_0} \frac{1}{c^2 r} \frac{\mathbf{r}}{r} \times \left(\frac{\mathbf{r}}{r} \times \ddot{\mathbf{p}} \right), \qquad (20\text{–}40)$$

which varies as $1/r$ and depends on $\ddot{\mathbf{p}}$. The last term, when expanded, is just the *static dipole field* of Eq. (2–36),

$$\mathbf{E}_3 = \frac{1}{4\pi\epsilon_0} \frac{1}{r^3} \left[3 \frac{\mathbf{r}}{r} \cdot \mathbf{p} \frac{\mathbf{r}}{r} - \mathbf{p} \right] = \frac{1}{4\pi\epsilon_0} \frac{1}{r^3} \left[\frac{\mathbf{r}}{r} \times \left(\frac{\mathbf{r}}{r} \times \mathbf{p} \right) + 2 \frac{\mathbf{r}}{r} \cdot \mathbf{p} \frac{\mathbf{r}}{r} \right], \qquad (2\text{–}36)$$

which varies as $1/r^3$ and depends on $\mathbf{p}$. The dipole $\mathbf{p}$ is varying with time in this case, of course, but the spatial dependence of $\mathbf{E}_3$ is instantaneously that of a static dipole field. The third, fourth, and fifth terms (which are similar in form to the first, second, and last terms) lead to the transition field or the *induction field*,

$$\mathbf{E}_2 = \frac{1}{4\pi\epsilon_0} \frac{1}{cr^2} \left[3 \frac{\mathbf{r}}{r} \cdot \dot{\mathbf{p}} \frac{\mathbf{r}}{r} - \dot{\mathbf{p}} \right] = \frac{1}{4\pi\epsilon_0} \frac{1}{cr^2} \left[\frac{\mathbf{r}}{r} \times \left(\frac{\mathbf{r}}{r} \times \dot{\mathbf{p}} \right) + 2 \frac{\mathbf{r}}{r} \cdot \dot{\mathbf{p}} \frac{\mathbf{r}}{r} \right], \qquad (20\text{–}45)$$

which varies as $1/r^2$ and depends on $\dot{\mathbf{p}}$. The total E-field is then

$$\mathbf{E} = \mathbf{E}_1 + \mathbf{E}_2 + \mathbf{E}_3;$$

$\mathbf{p}$ and its derivatives are to be evaluated at the retarded time, $t' = t - r/c$. The B-field is obtained by taking the curl of Eq. (20–35). If we retain the term that was neglected in the derivation of Eq. (20–39), we see from Eqs. (20–40) and (20–45) that

$$\mathbf{B} = \frac{\mathbf{r}}{cr} \times (\mathbf{E}_1 + \mathbf{E}_2).$$

Thus $\mathbf{B}$ contains a radiation and an induction field, but there is no "static" B-field.

From the second of the expressions given above for $E_{1,2,3}$, it is clear that all three fields have a transverse component, which is perpendicular to $\mathbf{r}$ and in the plane defined by $\mathbf{r}$ and $\mathbf{p}$. The radiation field is purely transverse, but E_2 and E_3 have in addition a longitudinal component in the direction of $\mathbf{r}$. If we choose spherical coordinates with the polar axis in the direction of $\mathbf{p}$, then the transverse components are E_θ and are proportional to $\sin \theta$; the longitudinal components are E_r and are proportional to $\cos \theta$. The B-field has only a ϕ-component (TM).

Let us specialize to the case of a dipole in the $\mathbf{k}$-direction with a sinusoidally oscillating magnitude: $\mathbf{p}(t) = p\mathbf{k}e^{-i\omega t}$. Then

$$\mathbf{p}(t') = p\mathbf{k}e^{-i(\omega t - \kappa r)},$$

where κ is the propagation constant of the wave in the radiation zone,

$$\kappa = \omega/c.$$

Inserting this $\mathbf{p}(t')$ into Eqs. (20–40), (2–36), and (20–45), we find for the components of $\mathbf{E} = \mathbf{E}_1 + \mathbf{E}_2 + \mathbf{E}_3$

$$E_\theta = \frac{p\kappa^3}{4\pi\epsilon_0} \sin \theta \left[-\frac{1}{\kappa r} - i \frac{1}{(\kappa r)^2} + \frac{1}{(\kappa r)^3} \right] e^{-i(\omega t - \kappa r)},$$

$$E_r = \frac{p\kappa^3}{4\pi\epsilon_0} 2 \cos \theta \left[-i \frac{1}{(\kappa r)^2} + \frac{1}{(\kappa r)^3} \right] e^{-i(\omega t - \kappa r)}.$$

The radiation field or the static field dominates when $\kappa r \gg 1$ or $\kappa r \ll 1$, respectively; that is, when r is large or small compared to the wavelength of the emitted radiation. The transition between the "static" dipole field and the transverse radiation field is shown schematically by the field lines of Fig. 20–3. In the radiation zone the field is an outward propagating spherical wave.

20–5 RADIATION DAMPING. THOMSON CROSS SECTION

The radiated power calculated in Section 20–3 is lost by the system of charges, and in a steady-state situation it must be made up by another source. For an antenna, this source is the transmitter, and the loss is expressed by the radiation resistance. For an electron in a material medium through which a wave is propagating, the power source is the E-field of the wave and the loss is expressed by the damping frequency. We now want to relate the damping frequency used in the last chapter to the rate of energy loss due to radiation by a charged particle, as given in Eq. (20–44),

$$P = \frac{e^2}{4\pi\epsilon_0} \frac{2}{3} \frac{\dot{v}^2}{c^3}. \tag{20–44}$$

The power loss due to a force F is $P = -Fv$. Equating this to the radiation loss of Eq. (20–44), we have for the damping force

$$F = -\frac{2}{3} \frac{e^2}{4\pi\epsilon_0 c^3} \frac{\dot{v}^2}{v}. \tag{20–46}$$

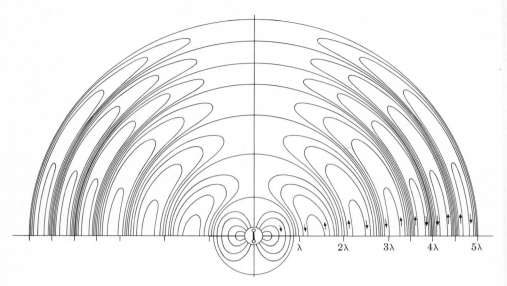

Figure 20–3 Electric field lines produced by an oscillating electric dipole.

The assumed linear damping force in Eq. (19–1) was $F = -Gv$, so that the damping frequency $\gamma = G/m$ is

$$\gamma = \frac{2}{3} \frac{e^2}{4\pi\epsilon_0 mc^3} \frac{\dot{v}^2}{v^2}.$$

For an oscillating charge, $v = v_0 \sin \omega t$, and

$$\gamma = \frac{2}{3} \frac{e^2\omega^2}{4\pi\epsilon_0 mc^3} \frac{\cos^2 \omega t}{\sin^2 \omega t}.$$

The trouble with this result is that γ is not a constant, but varies from zero to infinity over a cycle of the harmonic motion. Nevertheless the average effect of the damping is very small in one cycle, and since $\bar{P} = m\gamma\overline{v^2} \propto \overline{\dot{v}^2}$ and

$$\overline{\cos^2 \omega t} = \overline{\sin^2 \omega t},$$

the effective damping frequency is taken to be*

$$\gamma = \frac{2}{3} \frac{e^2\omega^2}{4\pi\epsilon_0 mc^3}. \tag{20–47}$$

---

* This last step of the argument is admittedly somewhat crude. The problem is not due to a failure of classical mechanics, since a quantum mechanical calculation gives the same result. The problem of a correct incorporation of the radiation reaction force Eq. (20–46) into dynamical theory has been discussed many times since Lorentz's presentation of Eq. (20–47) in 1909. It is a fundamental problem, but we cannot go into it here.

In terms of the "classical radius of the electron," $R_e = e^2/4\pi\epsilon_0 mc^2 = 2.81 \times 10^{-15}$ m,

$$\frac{\gamma}{\omega} = \frac{4\pi}{3} \frac{R_e}{\lambda}. \tag{20-48}$$

Since $\gamma = \Delta\omega$ is the width of a narrow absorption peak, and since $\Delta\omega/\omega = \Delta\lambda/\lambda$,

$$\Delta\lambda = \frac{4\pi}{3} R_e = 1.16 \times 10^{-4} \text{ Å}.$$

This *natural line width* is narrower than that ordinarily seen in absorption spectra, even in gases at low pressure, because there are other damping mechanisms that are usually greater than the radiation damping. This is, however, a lower limit on the possible damping.

Another immediate consequence of Eq. (20–44) that is more commonly observable is the Thomson cross section for the scattering of x-rays. X-ray frequencies (photon energies) are large compared to the resonance frequencies (binding energies) of many of the electrons in matter. These can be treated as free electrons accelerated by the E-field of the x-rays,

$$m\dot{v} = eE, \tag{20-49}$$

so that

$$P = \frac{1}{4\pi\epsilon_0} \frac{2}{3} \frac{e^4}{m^2c^3} E^2$$

is the total power radiated by one electron. The *Thomson scattering cross section* σ_T is defined as P divided by the incident Poynting vector (power per unit area)

$$S_0 = \frac{1}{\mu_0 c} E^2; \tag{20-50}$$

thus

$$\sigma_T = \frac{2}{3} \frac{1}{4\pi\epsilon_0^2} \frac{e^4}{m^2 c^4} = \frac{8\pi}{3} R_e^2, \tag{20-51}$$

where R_e is the classical radius of the electron. The smallness of this cross section compared to the size of an atom is the reason x-rays are penetrating. The angular dependence of the scattered radiation is given by the *differential cross section*, defined from Eq. (20–42):

$$d\sigma_T = \frac{dP}{S_0},$$

where

$$dP = S\, da = Sr^2\, d\Omega$$

is the power scattered (i.e., re-radiated) into the element of solid angle

$$d\Omega = \sin\theta \, d\theta \, d\phi.$$

With S from Eq. (20–42), the differential cross section is

$$\frac{d\sigma_T}{d\Omega} = R_e^2 \sin^2\theta. \tag{20–52}$$

Here θ is the angle between the observation direction and the E-vector of the incident wave (perpendicular to its propagation direction). If the incident x-rays are unpolarized (the usual case), a more useful expression is the average of this over all polarization directions. The result (problem 20–10), is

$$\frac{d\sigma_T}{d\Omega} = R_e^2 \frac{1 + \cos^2\beta}{2}, \tag{20–53}$$

where β is the angle between the observation direction and the incident propagation direction. This cross section is measurable in the *Compton effect*, which is the incoherent* scattering of x-rays at frequencies too high for resonance absorption. This classical formula ceases to hold at very high frequencies; the photon energy ($\hbar\omega$) must be much less than the electron rest-energy mc^2, or $\lambda \gg 0.02$ Å.

The Thomson cross section Eq. (20–51) also follows from the theory of Chapter 19, if we investigate the attenuation of the incident Poynting vector instead of the magnitude of the scattered one. From Eq. (20–50),

$$S_0 = \frac{1}{\mu_0 c} E^2 = \frac{1}{\mu_0 c} E_0^2 e^{-2z/\delta},$$

where

$$\delta = \frac{c}{k\omega}$$

is the "skin depth." Here $\delta/2$ plays the role of a mean-free path for the incident photons. Since the mean-free path equals $1/N\sigma$, where N is the number of electrons per unit volume,

$$\sigma_T = \frac{2}{N\delta} = \frac{2k\omega}{Nc}. \tag{20–54}$$

Now when $n \approx 1$, $k \ll 1$, as for x-rays,

$$k = \tfrac{1}{2}K_i = \frac{\omega_p^2 \gamma}{2\omega^3}.$$

---

* Incoherent scattering is opposed to the coherent scattering (with the same individual cross sections) which gives x-ray diffraction in crystals, under special conditions on wavelength and angle of incidence (Bragg's law).

This is the high frequency result for free electrons, and it also holds for bound electrons when $\omega \gg \omega_0$. Using $\omega_p^2 = Ne^2/\epsilon_0 m$, $\gamma = (4\pi/3)(R_e/\lambda)\omega$ from Eq. (20–48), and $\lambda = 2\pi c/\omega$, we have

$$k = \frac{Ne^2}{2\epsilon_0 m}\frac{4\pi}{3}\frac{R_e\omega^2}{2\pi c\omega^3} = \frac{4\pi}{3}\frac{Ne^2}{4\pi\epsilon_0 mc\omega}R_e,$$

and from Eq. (20–54),

$$\sigma_T = \frac{8\pi}{3}R_e^2. \tag{20–51}$$

The damping frequency γ which gives the Thomson cross section is the radiation damping frequency. At such high frequencies ($\omega \gtrsim 10^{19}$ s^{-1}) it dominates the collision frequency $1/\tau$. This approach, of looking at the propagation of the incident radiation, does not of course give the angle dependence Eq. (20–52) of the scattered radiation.

20–6 SUMMARY

The fields produced by time dependent charge and current distributions are calculated from the integral solutions for $\mathbf{A}$ and φ. (It would be sufficient to treat only $\mathbf{A}$, since φ is related to it by the Lorentz condition.) To find solutions valid at large distances $r \gg a$, where a is the size of the region in which the source currents are located, the integrand is expanded by the multipole expansion. The *electric dipole* term treated here is the most important, unless it happens to vanish in which case higher order terms would have to be considered. The integral is further simplified for a *point dipole* $a \ll \lambda$, since all the source current elements are approximately in phase. (This condition implies that $v \ll c$.) The result is

$$\mathbf{A}(\mathbf{r}, t) = \frac{\mu_0}{4\pi r}\dot{\mathbf{p}}(t - r/c)$$

where

$$\dot{\mathbf{p}}(t) = \int \mathbf{J}(\mathbf{r}', t)dv'.$$

From this, with a prescribed $\mathbf{J}$, one finds $\mathbf{B}$, $\mathbf{E}$, and $\mathbf{S}$. Finding them is particularly easy in the radiation field at $r \gg \lambda$. Then

$$\mathbf{E} = -c\,\frac{\mathbf{r}}{r} \times \mathbf{B}$$

and

$$\mathbf{S} = \frac{c}{\mu_0}B^2\,\frac{\mathbf{r}}{r},$$

just as for a plane wave. (It is unnecessary to find φ.)

1. The Poynting vector in the radiation field of a point dipole is

$$\mathbf{S} = \frac{\ddot{\mathbf{p}}^2 \sin^2 \theta}{16\pi^2 \epsilon_0 c^3 r^2} \frac{\mathbf{r}}{r}.$$

Integrating this over the entire sphere gives the total power radiated as

$$P = \frac{1}{4\pi\epsilon_0} \frac{2}{3} \frac{\ddot{\mathbf{p}}^2}{c^3}.$$

2. The radiation from a short linear (electric dipole) antenna is given by (1) with $p(t) = (lI_0)/\omega) \cos \omega t$. In terms of the radiation resistance R_r,

$$\bar{P} = \tfrac{1}{2} R_r I_0^2,$$

$$R_r = 789 \left(\frac{l}{\lambda}\right)^2 \text{ ohms} \qquad (l \ll \lambda)$$

in free space.

3. The radiation from a longer antenna can be found by integrating the results for a short antenna, assuming $v \ll c$, as is always true for a radio antenna. For a half-wave antenna,

$$R_r = 73.1 \text{ ohms.}$$

4. For a slowly moving point charge, $\ddot{p} = q\dot{v}$,

$$P = \frac{1}{4\pi\epsilon_0} \frac{2}{3} \frac{q^2 \dot{v}^2}{c^3}.$$

5. For an oscillating electron, this leads to the radiation damping frequency

$$\gamma = \frac{4\pi}{3} \frac{R_e}{\lambda} \omega$$

and the Thomson cross section for a free electron

$$\sigma_T = \frac{8\pi}{3} R_e^2,$$

where

$$R_e = \frac{1}{4\pi\epsilon_0} \frac{e^2}{mc^2} = 2.81 \times 10^{-15} \text{ m}$$

is the "classical radius of the electron."

PROBLEMS

20–1 Suppose a spherically symmetric charge distribution is oscillating purely in the radial direction, so that it remains spherically symmetric at every instant. Prove that no radiation is emitted.

20-2 (a) Determine, as a function of the angles θ and ϕ, the average power density radiated into vacuum by an oscillating dipole. (b) Calculate the total power radiated by a dipole of length 3 m at a frequency of 500 kHz if the current in the dipole is 2 A (effective value). (c) What is the radiation resistance of the dipole oscillator in part (b)?

20-3 A circular loop of wire carrying the current $I = I_0 \cos \omega t$ constitutes an *oscillating magnetic dipole*. Determine the radiation fields **E** and **B** for this oscillator, and the total power radiated.

20-4 As sources of electromagnetic radiation, determine the relative efficiency of an electric dipole of length 2 m compared with a magnetic dipole of the same diameter at a frequency of 1 MHz.

20-5 (a) Compute the maximum current to a half-wave antenna which radiates 1 kW. (b) What is the E-field corresponding to the average power density at a distance of 10 km from the antenna. Ignore effects of the earth.

20-6 Verify that **A** and φ in Eqs. (20-35) and (20-33) satisfy the Lorentz condition.

20-7 A measure of the directivity of an antenna is the power per steradian in a direction in which this is a maximum, divided by $1/(4\pi)$ times the total power radiated. Calculate the directivity for an oscillating electric dipole.

20-8 Suppose an electric dipole **p** rotates with a constant angular velocity ω about an axis perpendicular to the dipole moment. Find the radiation field and the Poynting vector. (*Hint*: Treat the rotating dipole as the superposition of two sinusoidally varying dipoles at right angles to each other.)

20-9 The classical model of the hydrogen atom has the electron revolving in a circular orbit of radius r and kinetic energy

$$E_k = \frac{1}{2} \frac{e^2}{4\pi\epsilon_0 r}.$$

(a) Calculate the fractional energy radiated per revolution, PT/E_k, where T is the orbital period. (b) Quantum mechanics prescribes that in the nth level

$$\frac{v}{c} = \frac{1}{n} \frac{1}{137}.$$

Evaluate PT/E_k for $n = 2$.

20-10 An *unpolarized* x-ray beam of intensity I_0 is incident on matter containing free electrons. Considering one electron only and using the expressions from Section 20-3, show that the intensity of the scattered beam is given by

$$I_s = I_0 \frac{(1 + \cos^2 \beta)}{2} \frac{R_e^2}{r^2},$$

where β is the angle between OP and the original x-ray beam. Point O is the position of the electron, and P is the point where the scattered beam is to be measured.

20-11 X-rays of wavelength 0.2 Å are attenuated in aluminum primarily by Compton scattering. Calculate the absorption coefficient $\alpha = 2/\delta$ from the Thomson cross section. There are 6.06×10^{28} Al atoms/m^3.

CHAPTER 21 Electrodynamics

The field produced by a fast moving point charge can be calculated from the retarded potentials. There are, however, certain difficulties associated with this calculation, which are related to the retardation and reflect the fact that the present charge distribution (in space) must be extrapolated back to the appropriate retarded time. This procedure would be essentially trivial except that different portions of the charge distribution require different retarded times. Although one might expect this effect to disappear for point charges, it actually does not. The appropriate scalar and vector potentials for a moving point charge are the Lienard-Wiechert potentials, which we will now derive.

21–1 THE LIENARD-WIECHERT POTENTIALS

The Lienard-Wiechert potentials are, as noted above, the scalar and vector potentials produced by a moving point charge. One might think that $q/4\pi\epsilon_0 R$, with R the appropriate retarded distance, would give the scalar potential due to a moving point charge. This, however, is not the case, as can be shown in several ways. One of the most instructive procedures is to consider a moving volume carrying with it a fixed charged distribution, for example a uniformly charged spherical volume, moving through space along a prescribed trajectory. The field due to a point charge is the properly taken limit of the field due to such a distribution.

The scalar potential due to a moving charge distribution, at point ξ and time t, is given by the retarded potential*

$$\varphi(\xi, t) = \frac{1}{4\pi\epsilon_0} \int \frac{\rho(\mathbf{r}', t')}{|\xi - \mathbf{r}'|} \, dv'. \qquad (21\text{–}1)$$

The crux of the difficulty is now apparent; namely, t' is not fixed and hence the volume of integration, i.e., the volume in which ρ is different from zero, cannot be readily specified. To obviate the difficulty, some fixed time t_1 may be chosen and the integration over $\mathbf{r}'$ changed to an integration over $\mathbf{r}_1$. The most convenient

_____

* Throughout this chapter only free space is considered.

470

choice for t_1 is the retarded time for some point in the interior of the charge distribution. If at time t_1 the charged volume is moving with a velocity $\mathbf{v}(t_1)$, then the important relationships are

$$\rho(\mathbf{r}', t') = \rho(\mathbf{r}_1, t_1), \tag{21-2}$$

$$\mathbf{r}_1 = \mathbf{r}' - \mathbf{v}(t')(t' - t_1) - \tfrac{1}{2}\dot{\mathbf{v}}(t')(t' - t_1)^2 + \cdots, \tag{21-3}$$

where $\dot{\mathbf{v}}$ is the time derivative of $\mathbf{v}$. It is important to understand that t' in Eq. (21–3) is not constant, but depends on $\mathbf{r}'$. The remaining problem is that of relating dv' to dv_1, which is, of course, accomplished through the Jacobian determinant. The relationship is

$$dv_1 = \frac{\partial(x_1, y_1, z_1)}{\partial(x', y', z')} \, dv', \tag{21-4}$$

where the Jacobian, $\partial(x_1, y_1, z_1)/\partial(x', y', z')$, is given by

$$\frac{\partial(x_1, y_1, z_1)}{\partial(x', y', z')} = \begin{vmatrix} \dfrac{\partial x_1}{\partial x'} & \dfrac{\partial x_1}{\partial y'} & \dfrac{\partial x_1}{\partial z'} \\[2mm] \dfrac{\partial y_1}{\partial x'} & \dfrac{\partial y_1}{\partial y'} & \dfrac{\partial y_1}{\partial z'} \\[2mm] \dfrac{\partial z_1}{\partial x'} & \dfrac{\partial z_1}{\partial y'} & \dfrac{\partial z_1}{\partial z'} \end{vmatrix}. \tag{21-5}$$

The derivatives are

$$\frac{\partial x_1}{\partial x'} = 1 - v'_x \frac{\partial t'}{\partial x'} - \dot{v}'_x(t' - t_1) \frac{\partial t'}{\partial x'} + \cdots,$$

and

$$\frac{\partial x_1}{\partial y'} = -v'_x \frac{\partial t'}{\partial y'} - \dot{v}'_x(t' - t_1) \frac{\partial t'}{\partial y'} + \cdots, \tag{21-6}$$

where v'_x is $v_x(t')$, the x-component of the velocity at the retarded time t'. The retarded time t' is related to the retarded position simply by

$$t' = t - \frac{|\mathbf{r}' - \boldsymbol{\xi}|}{c} \,; \tag{21-7}$$

hence

$$\frac{\partial t'}{\partial x'} = -\frac{n'_x}{c}, \tag{21-8}$$

where $\mathbf{n}'$ is a unit vector in the direction $\mathbf{r}' - \boldsymbol{\xi}$. A straightforward, though tedious, expansion of the Jacobian, using Eqs. (21–6) and (21–8), gives

$$\frac{\partial(x_1, y_1, z_1)}{\partial(x', y', z')} = 1 + \frac{\mathbf{v} \cdot \mathbf{n}'}{c} + \frac{\dot{\mathbf{v}} \cdot \mathbf{n}'(t' - t_1)}{c} + \cdots, \tag{21-9}$$

where the higher terms involve the second and higher derivatives of $\mathbf{v}'$.

Equation (21–9) can be used in Eq. (21–1) to obtain the scalar potential. However, since the principal interest is in small charged volumes (point charges), it is appropriate to note that if

$$\frac{\dot{\mathbf{v}}' \cdot \mathbf{n}'}{c} (t' - t_1) \cong \frac{\dot{v}d}{c^2} \ll 1,$$

where d measures the size of the charge distribution, then this term can certainly be neglected in the limiting point charge case. Similar criteria exist for the terms involving higher derivatives; however, we need not consider these. Finally, then,

$$\varphi(\boldsymbol{\xi}, t) = \frac{1}{4\pi\epsilon_0} \int \frac{\rho(\mathbf{r}_1, t_1)}{|\boldsymbol{\xi} - \mathbf{r}'|} \frac{dv_1}{1 + \mathbf{v}' \cdot \mathbf{n}'/c}. \tag{21-10}$$

Again, if $d \ll |\boldsymbol{\xi} - \mathbf{r}'|$, then $|\boldsymbol{\xi} - \mathbf{r}'|$ can be replaced by R_{t_1}, the distance from the interior point (chosen earlier) to the observation point at time t_1. Thus*

$$\varphi(\boldsymbol{\xi}, t) = \frac{1}{4\pi\epsilon_0} \frac{1}{R_{t_1}(1 + \mathbf{v}' \cdot \mathbf{n}'/c)} \int \rho(\mathbf{r}_1, t_1) \, dv_1 \tag{21-11}$$

or, since the integral is now over a well-defined volume,

$$\varphi(\boldsymbol{\xi}, t) = \frac{1}{4\pi\epsilon_0} \frac{q}{R_{t_1}(1 + \mathbf{v}' \cdot \mathbf{n}'/c)}, \tag{21-12}$$

which is the scalar Lienard-Wiechert potential. The vector potential is found to be

$$\mathbf{A}(\boldsymbol{\xi}, t) = \frac{\mu_0}{4\pi} \frac{q\mathbf{v}'}{R_{t_1}(1 + \mathbf{v}' \cdot \mathbf{n}'/c)}. \tag{21-13}$$

These expressions are often written as

$$\varphi(\boldsymbol{\xi}, t) = \frac{q}{4\pi\epsilon_0} \left[\frac{1}{R(1 + \mathbf{v} \cdot \mathbf{n}/c)} \right]_{\text{ret}},$$

and

$$\mathbf{A}(\boldsymbol{\xi}, t) = \frac{\mu_0 q}{4\pi} \left[\frac{\mathbf{v}}{R(1 + \mathbf{v} \cdot \mathbf{n}/c)} \right]_{\text{ret}}, \tag{21-14}$$

which simply means that the quantities in brackets must be evaluated at our t_1.

21–2 THE FIELD OF A UNIFORMLY MOVING POINT CHARGE

The most direct application of the Lienard-Wiechert potentials is to the calculation of the field of a point charge moving in a straight line with constant velocity. The geometry of such a situation is shown in Fig. 21–1. The field at point P is to be calculated at time t, at which time the charge is at x. The retarded position x' and the retarded time t' are determined by

$$R'^2 = c^2(t - t')^2 = (x_0 - x')^2 + b^2. \tag{21-15}$$

---

* Note that $\mathbf{v}' \cdot \mathbf{n}' = \mathbf{v}(t_1) \cdot \mathbf{n}(t_1)$ to the approximation involved in Eq. (21–11).

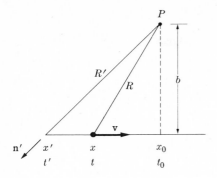

The scalar potential is given by

$$\varphi(P,\, t) = \frac{q}{4\pi\epsilon_0} \frac{1}{R'[1 + (\mathbf{v} \cdot \mathbf{n}'/c)]}. \tag{21-16}$$

From the diagram it is clear that

$$R' \frac{\mathbf{v} \cdot \mathbf{n}'}{c} = -R' \frac{v}{c} \frac{x_0 - x'}{R'} = -\frac{v(x_0 - x')}{c}. \tag{21-17}$$

Even after Eq. (21–17) is substituted in Eq. (21–16), a multitude of variables appear in the expression for φ. In calculating the electric field by taking the gradient of φ, etc., these variables would have to be differentiated very carefully and would cause the calculation to be quite cumbersome. Rather than follow this procedure, it is preferable to eliminate the undesirable variables in φ and obtain an expression that involves only the coordinates of P, the present time t, and parameters that describe the path of the charged particle.

Since the charge moves from x' to x_0 in time $t_0 - t'$, it is clear that

$$c^2(t - t')^2 = v^2(t_0 - t')^2 + b^2. \tag{21-18}$$

If this equation is solved for t' the result is

$$t' = \frac{c^2 t - v^2 t_0 \pm \sqrt{v^2 c^2 (t_0 - t)^2 + b^2(c^2 - v^2)}}{c^2 - v^2}. \tag{21-19}$$

The minus sign must be used in this equation to ensure that t' is retarded with respect to t. To verify this, it is only necessary to observe that at $t = t_0 = 0$, $t' = \pm\sqrt{b^2(c^2 - v^2)}/(c^2 - v^2)$ and hence only the negative sign gives an earlier time. Having found t', we find $x_0 - x'$ from

$$x_0 - x' = v(t_0 - t')$$

$$= v \left(\frac{t_0(c^2 - v^2) - c^2 t + v^2 t_0 + \sqrt{v^2 c^2 (t_0 - t)^2 + b^2(c^2 - v^2)}}{c^2 - v^2} \right),$$

$$\tag{21-20}$$

while R' is shown to be

$$R' = c\left(\frac{t(c^2 - v^2) - c^2t + v^2t_0 + \sqrt{v^2c^2(t_0 - t)^2 + b^2(c^2 - v^2)}}{c^2 - v^2}\right). \tag{21-21}$$

Equations (21–20) and (21–21) may be used to evaluate the denominator which appears in Eq. (21–16). This denominator becomes

$$R^* = R' - \frac{v(x_0 - x')}{c} \tag{21-22}$$

through the use of Eq. (21–17), and subsequently, it becomes

$$R^* = (c^2 - v^2)^{-1}[v^2c(t_0 - t) + c\sqrt{v^2c^2(t_0 - t)^2 + b^2(c^2 - v^2)}$$

$$- v^2c(t_0 - t) - \frac{v^2}{c}\sqrt{v^2c^2(t_0 - t)^2 + b^2(c^2 - v^2)}]$$

$$= \sqrt{v^2(t_0 - t)^2 + b^2(1 - v^2/c^2)} \tag{21-23}$$

through Eqs. (21–20) and (21–21). The scalar potential is

$$\varphi(P, t) = \frac{q}{4\pi\epsilon_0} \frac{1}{\sqrt{v^2(t_0 - t)^2 + b^2(1 - v^2/c^2)}}, \tag{21-24}$$

while the vector potential is

$$\mathbf{A}(P, t) = \frac{\mu_0 q}{4\pi} \frac{\mathbf{v}}{\sqrt{v^2(t_0 - t)^2 + b^2(1 - v^2/c^2)}}. \tag{21-25}$$

It is important to realize that Eqs. (21–24) and (21–25) contain only the position and time of the observation point, and the parameters $(\mathbf{v}, t_0)$ that describe the path of the charged particle.

To make this statement more concrete and to put the potentials in a form more suitable for calculating the fields, the coordinate system must be fixed more carefully. Since the charge moves along the x-axis, and since this is an axis of symmetry for the problem, it is only necessary to specify the origin on the x-axis. This is conveniently accomplished by taking $x = 0$ to be the position of the charge at $t = 0$. Then $x = vt$ and, in particular, $x_0 = vt_0$.

If the point P is specified by the cartesian coordinates ξ, η, ζ, then

$$\xi = x_0 = vt_0 \quad \text{and} \quad \eta^2 + \zeta^2 = b^2. \tag{21-26}$$

Using these results in Eq. (21–25) and letting $\boldsymbol{\xi} = (\xi, \eta, \zeta)$, we obtain

$$\varphi(\boldsymbol{\xi}, t) = \frac{q}{4\pi\epsilon_0} \frac{1}{\sqrt{(\xi - vt)^2 + (\eta^2 + \zeta^2)(1 - v^2/c^2)}}$$

and

$$\tag{21-27}$$

$$\mathbf{A}(\boldsymbol{\xi}, t) = \frac{\mu_0 q}{4\pi} \frac{\mathbf{v}}{\sqrt{(\xi - vt)^2 + (\eta^2 + \zeta^2)(1 - v^2/c^2)}}.$$

It must be remembered that these equations apply only if **v** is along the x-axis; other directions require modification of the formulas.

The important thing about equations (21–27) is that they are in a form ideally suited for the calculation of the fields. Thus

$$\mathbf{E}(\xi, t) = -\frac{\partial \mathbf{A}}{\partial t} - \nabla_\xi \varphi$$

$$= -\frac{\mu_0 q}{4\pi} \mathbf{v} \, \frac{v(\xi - vt)}{R^{*3}}$$

$$+ \frac{q}{4\pi\epsilon_0} \frac{1}{R^{*3}} [(\xi - vt)\mathbf{i} + \eta(1 - v^2/c^2)\mathbf{j} + \zeta(1 - v^2/c^2)\mathbf{k}]. \qquad (21\text{–}28)$$

Noting that $\mathbf{v} = v\mathbf{i}$, $\epsilon_0 \mu_0 = 1/c^2$, and $\xi - vt = x_0 - x$ makes it possible to rewrite Eq. (21–28) in the form

$$\mathbf{E}(\xi, t) = \frac{q}{4\pi\epsilon_0} \frac{\mathbf{R}}{R^{*3}} (1 - v^2/c^2), \qquad (21\text{–}29)$$

where **R** is a vector from the position of the charge at time t to the point P and

$$R^* = R' - \frac{\mathbf{R}' \cdot \mathbf{v}}{c}.$$

The magnetic induction can be found by evaluating $\mathbf{B} = \nabla \times \mathbf{A}$; however, a much simpler procedure is to note that

$$\mathbf{A} = \mu_0 \epsilon_0 \mathbf{v}\varphi \qquad (21\text{–}30)$$

and hence that

$$\mathbf{B} = \mu_0 \epsilon_0 \nabla \times (\mathbf{v}\varphi) = -\mu_0 \epsilon_0 \mathbf{v} \times \nabla\varphi. \qquad (21\text{–}31)$$

Since **v** is along the x-axis, only the y- and z-components of $\nabla\varphi$ are important in the cross product. These components are just the negatives of the y- and z-components of **E**. In this way we find

$$\mathbf{B} = \frac{1}{c^2} \mathbf{v} \times \mathbf{E}, \qquad (21\text{–}32)$$

which completes the computation of the fields.

It is interesting to note that although the source of the field is the retarded position, lines of **E** are directed away from the instantaneous position of the charge. The lines of **B** are circles with centers on the charge path. The E-field is not spherically symmetric as it is in the static case, but is stronger in the direction perpendicular to the velocity. (See Problem 21–1.)

Having obtained the field vectors, we are in a position to calculate other electromagnetic quantities; however, rather than pursue these possibilities, we refer the reader to more advanced texts* that deal at length with such problems.

21-3 THE FIELD OF AN ACCELERATED POINT CHARGE

If an accelerated point charge is to be considered, certain simplifications that appear in the constant-velocity case are no longer possible. The major difficulty here is a direct result of the fact that the Lienard-Wiechert potentials can no longer be expressed in terms of the present position of the charge; instead, the retarded position and time appear explicitly. The potentials

$$\varphi = \frac{q}{4\pi\epsilon_0} \left[\frac{1}{R(1 + \mathbf{v} \cdot \mathbf{n}/c)} \right]_{\text{ret}}$$

and

$$\mathbf{A} = \frac{q}{4\pi\epsilon_0} \left[\frac{\mathbf{v}/c^2}{R(1 + \mathbf{v} \cdot \mathbf{n}/c)} \right]_{\text{ret}}$$

(21–33)

are still correct; however, in differentiating them to obtain the fields it must be noted that derivatives with respect to the position of the field point must be taken at constant observation time, and derivatives with respect to the observation time at fixed field points. Since the retarded time appears explicitly in the potentials, care must be used to obtain the correct derivatives.

To clarify the differentiation problem, we note that the potentials are functions of the field point ξ, the observation time t, the retarded position $\mathbf{r}'$ of the charge, and the retarded time t'. The trajectory of the particle is specified by giving $\mathbf{r}'$ as a function of t', so that the dependence on $\mathbf{r}'$ can be removed. Furthermore, the retardation condition

$$(\xi - x')^2 + (\eta - y')^2 + (\zeta - z')^2 = c^2(t - t')^2$$

(21–34)

provides a single relationship among the remaining variables. Thus it is clear that although the potentials depend superficially on eight variables, only four of these are really independent. In computing the fields $\mathbf{E}$ and $\mathbf{B}$ it is necessary to differentiate the potentials with respect to each of ξ, η, ζ, and t, holding the other three fixed; for example, $\mathbf{A}$ must be differentiated with respect to t holding ξ, η, and ζ constant. Since it is t' that appears explicitly in the potentials, the calculation of these derivatives causes some difficulty.

To keep track of the variables that are being held constant during various differentiations, the following notation will be adopted: A partial derivative in which all other variables, dependent or independent, are held constant will be designated by the usual partial derivative symbol. If not all other variables are

---

* For example: Panofsky and Phillips, *Classical Electricity and Magnetism*, Second Edition (Reading, Mass.: Addison-Wesley, 1962).

held constant, then those that are will be indicated by subscripts. Thus the derivative of $\mathbf{A}$ that is needed in computing $\mathbf{E}$ is $(\partial \mathbf{A}/\partial t)_\xi$, while those of φ are $(\partial \varphi/\partial \xi)_{\eta,\zeta,t}$, etc. To transform $(\partial \mathbf{A}/\partial t)_\xi$ into a derivative with respect to t', we write

$$\left(\frac{\partial \mathbf{A}}{\partial t}\right)_\xi = \left(\frac{\partial \mathbf{A}}{\partial t}\right) + \left(\frac{\partial \mathbf{A}}{\partial t'}\right)\left(\frac{\partial t'}{\partial t}\right)_\xi \qquad (21\text{-}35)$$

and

$$\left(\frac{\partial \mathbf{A}}{\partial t'}\right)_\xi = \left(\frac{\partial \mathbf{A}}{\partial t'}\right) + \left(\frac{\partial \mathbf{A}}{\partial t}\right)\left(\frac{\partial t}{\partial t'}\right)_\xi. \qquad (21\text{-}36)$$

The retardation condition, Eq. (21-34), together with the equation specifying the trajectory, $\mathbf{x}' = \mathbf{x}'(t')$, is equivalent to an equation of the form

$$f(\boldsymbol{\xi}, t, t') = 0.$$

This relationship implies that $(\partial t/\partial t')_\xi = 1/(\partial t'/\partial t)_\xi$, which, when combined with Eqs. (21-35) and (21-36), gives

$$\left(\frac{\partial \mathbf{A}}{\partial t}\right)_\xi = \left(\frac{\partial \mathbf{A}}{\partial t'}\right)_\xi \left(\frac{\partial t'}{\partial t}\right)_\xi. \qquad (21\text{-}37)$$

In calculating the time derivatives of the potentials, this equation is just what is required to get the electric and magnetic fields. The other derivatives are all of the form $(\partial \varphi/\partial \xi)_t$. Such derivatives are readily evaluated by noting that

$$\left(\frac{\partial \varphi}{\partial \xi}\right)_t = \left(\frac{\partial \varphi}{\partial \xi}\right)_{t,t'} + \left(\frac{\partial \varphi}{\partial t'}\right)_{\xi,t}\left(\frac{\partial t'}{\partial \xi}\right)_t, \qquad (21\text{-}38)$$

in which all of the subscripts have been included to avoid any possibility of confusion.

From Eqs. (21-37) and (21-38) it is clear that the calculation of $\mathbf{E}$ requires that the derivatives $(\partial t'/\partial t)_\xi$ and $(\partial t'/\partial \xi)_t$ must be evaluated. Each of these may be easily evaluated by differentiating the square root of Eq. (21-34),

$$[(\xi - x')^2 + (\eta - y')^2 + (\zeta - z')^2]^{1/2} = c(t - t'), \qquad (21\text{-}39)$$

in the appropriate way. If the derivative with respect to t (holding ξ constant) is taken, the equation

$$-\frac{1}{R'}\mathbf{R}' \cdot \left(\frac{\partial \mathbf{r}'}{\partial t}\right)_\xi = c\left[1 - \left(\frac{\partial t'}{\partial t}\right)_\xi\right] \qquad (21\text{-}40)$$

results. In this equation, $\mathbf{r}' = \mathbf{i}x' + \mathbf{j}y' + \mathbf{k}z'$ and $\mathbf{R}' = \boldsymbol{\xi} - \mathbf{r}'$. Since $\mathbf{r}'$ depends explicitly only on t', the derivative on the left is easily changed, to give

$$-\frac{1}{R'}\mathbf{R}' \cdot \mathbf{v}'\left(\frac{\partial t'}{\partial t}\right)_\xi = c\left[1 - \left(\frac{\partial t'}{\partial t}\right)_\xi\right], \qquad (21\text{-}41)$$

where $\mathbf{v}' = \partial\mathbf{r}'/\partial t'$ is the velocity of the charge at the retarded time t'. Solving this for $(\partial t'/\partial t)_\xi$ leads to

$$\left(\frac{\partial t'}{\partial t}\right)_\xi = \frac{R'}{R' - \mathbf{R}' \cdot \mathbf{v}'/c} = \frac{R'}{R^*}. \tag{21-42}$$

A similar calculation in which Eq. (21–39) is differentiated with respect to ξ at constant (η, ζ, t) gives

$$\left(\frac{\partial t'}{\partial \xi}\right)_t = -\frac{(\xi - x')}{(R' - \mathbf{R}' \cdot \mathbf{v}'/c)c}. \tag{21-43}$$

Computing the other two components, and writing the result as a vector equation, we obtain

$$(\boldsymbol{\nabla}_\xi t')_t = -\frac{\mathbf{R}'/c}{R' - \mathbf{R}' \cdot \mathbf{v}'/c} = -\frac{\mathbf{R}'}{R^* c}. \tag{21-44}$$

With these derivatives at hand, the electric field due to an accelerated point charge is readily computed from the Lienard-Wiechert potentials. Thus

$$\mathbf{E}(\xi, t) = -(\boldsymbol{\nabla}_\xi \varphi)_t - \left(\frac{\partial \mathbf{A}}{\partial t}\right)_\xi$$

$$= -(\boldsymbol{\nabla}_\xi \varphi)_{tt'} - \left(\frac{\partial \varphi}{\partial t'}\right)_{\xi t}(\boldsymbol{\nabla}_\xi t')_t - \left(\frac{\partial \mathbf{A}}{\partial t'}\right)_\xi\left(\frac{\partial t'}{\partial t}\right)_\xi. \tag{21-45}$$

The derivatives of the potentials that appear in this equation are easily found to be

$$(\boldsymbol{\nabla}_\xi \varphi)_{tt'} = -\frac{q}{4\pi\epsilon_0}\frac{\mathbf{R}'/R' - \mathbf{v}'/c}{(R' - \mathbf{R}' \cdot \mathbf{v}'/c)^2}, \tag{21-46}$$

$$\left(\frac{\partial \varphi}{\partial t'}\right)_{\xi t} = \frac{q}{4\pi\epsilon_0}\left[\frac{\mathbf{R}' \cdot \mathbf{v}'}{R'} - \frac{v'^2}{c} + \frac{\mathbf{R}' \cdot \dot{\mathbf{v}}'}{c}\right]\frac{1}{R^{*2}}, \tag{21-47}$$

$$\left(\frac{\partial \mathbf{A}}{\partial t'}\right)_\xi = \frac{q}{4\pi\epsilon_0}\left[\frac{\dot{\mathbf{v}}'}{R^* c^2} + \frac{\mathbf{v}'}{c^2}\frac{1}{R^{*2}}\left(\frac{\mathbf{R}' \cdot \mathbf{v}'}{R'} - \frac{v'^2}{c} + \frac{\mathbf{R}' \cdot \dot{\mathbf{v}}'}{c}\right)\right]\frac{1}{R^{*2}}. \tag{21-48}$$

Using these results in Eq. (21–45), we find

$$\mathbf{E}(\xi, t) = \frac{q}{4\pi\epsilon_0}\left[\frac{1}{R^{*3}}\left(\mathbf{R}' - \frac{R'\mathbf{v}'}{c}\right)\left(1 - \frac{v'^2}{c^2}\right)\right.$$

$$\left. + \left(\mathbf{R}' - \frac{R'\mathbf{v}'}{c}\right)\frac{\dot{\mathbf{v}}' \cdot \mathbf{R}'}{R^{*3}c^2} - \frac{\dot{\mathbf{v}}'R'}{R^{*2}c^2}\right]. \tag{21-49}$$

A similar calculation gives

$$\mathbf{B}(\xi, t) = \frac{q}{4\pi\epsilon_0 c^2}\left\{\frac{\mathbf{v}' \times \mathbf{R}'}{R^{*3}}\left(1 - \frac{v'^2}{c^2}\right)\right.$$

$$\left. + \frac{1}{R^{*3}c}\frac{\mathbf{R}'}{R'} \times \left[\mathbf{R}' \times \left(\left[\mathbf{R}' - \frac{R'\mathbf{v}'}{c}\right] \times \dot{\mathbf{v}}'\right)\right]\right\}. \tag{21-50}$$

These results may be used to explain many such important phenomena as radiation damping and the classical Bremsstrahlung. Most of these calculations are readily available in various texts on electrodynamics and except for one example will be omitted here in the interest of brevity.

21–4 RADIATION FIELDS FOR SMALL VELOCITIES

The result of the calculation of the fields from the potentials for an arbitrarily moving point charge, which was carried out in the last section, may be rewritten as

$$\mathbf{E}(\xi, t) = \frac{q}{4\pi\epsilon_0} \frac{1}{R^{*3}} \left\{ \left(\mathbf{R}' - \frac{R'\mathbf{v}'}{c}\right)\left(1 - \frac{v'^2}{c^2}\right) \right. \tag{21–51}$$

$$\left. + \frac{1}{c^2} \mathbf{R}' \times \left[\left(\mathbf{R}' - \frac{R'\mathbf{v}'}{c}\right) \times \dot{\mathbf{v}}'\right]\right\},$$

$$\mathbf{B}(\xi, t) = \frac{\mathbf{R}' \times \mathbf{E}}{R'c}. \tag{21–52}$$

From Eq. (21–52) we note that the B-field of a point charge in vacuum is always perpendicular to the E-field at the same point and time, and also perpendicular to the line connecting the field point with the *retarded* position of the particle, $\mathbf{R}'$. The fields contain two terms, the second one proportional to the acceleration $\dot{\mathbf{v}}'$ and the first one independent of it. Both terms depend on v'/c. For uniform motion ($\dot{\mathbf{v}}' = 0$) the first term in Eq. (21–51) gives the previous result (21–29), since $R'/c = t - t'$. Even for nonuniform motion at high velocity, the first term does not contribute to the radiation from the charge, because its magnitude falls off with distance like $1/R'^2$. We have seen previously that the fields must fall off like $1/R'$ in order to contribute to the Poynting vector at large distances; the second, acceleration-dependent term does decrease like $1/R'$. We see that this part of the E-field is likewise perpendicular to $\mathbf{R}'$: in the radiation field $\mathbf{E}$, $\mathbf{B}$, and $\mathbf{R}'$ are mutually perpendicular, and $\mathbf{S} = (1/\mu_0 c)E^2(\mathbf{R}'/R')$.

For simplicity, we consider here only the case of a slowly moving charge. If the velocity of the charge is small compared with the velocity of light, that is, if $v'/c \ll 1$, then the approximations

$$\mathbf{R}' - \frac{R'\mathbf{v}'}{c} \approx \mathbf{R}'$$

and

$$R^* = R' - \frac{\mathbf{R}' \cdot \mathbf{v}'}{c} \approx R'$$

may be made in Eqs. (21–51) and (21–52). If, in addition, only the radiation field, i.e., the part of the field proportional to $1/R'$, is considered, then Eqs. (21–51) and

(21–52) become

$$E(\xi, t) = \frac{q}{4\pi\epsilon_0} \frac{\mathbf{R}' \times (\mathbf{R}' \times \dot{\mathbf{v}}')}{R'^3 c^2} \tag{21-53}$$

and

$$\mathbf{B}(\xi, t) = \frac{q}{4\pi\epsilon_0} \frac{\mathbf{R}' \times [\mathbf{R}' \times (\mathbf{R}' \times \dot{\mathbf{v}}')]}{R'^4 c^3} = \frac{q}{4\pi\epsilon_0 c^2} \frac{\dot{\mathbf{v}}' \times \mathbf{R}'}{R'^2 c}. \tag{21-54}$$

From these field vectors the Poynting vector is found to be

$$\mathbf{S} = \mathbf{E} \times \mathbf{H} = \frac{q^2}{16\pi^2 \epsilon_0^2 \mu_0 c^2} \frac{1}{R'^5 c^3} [\mathbf{R}' \times (\mathbf{R}' \times \dot{\mathbf{v}}')] \times [\dot{\mathbf{v}}' \times \mathbf{R}'], \tag{21-55}$$

which, through the use of vector identities, reduces to

$$\mathbf{S} = \frac{q^2}{16\pi^2 \epsilon_0 c^3} \frac{\mathbf{R}'(\mathbf{R}' \times \dot{\mathbf{v}}')^2}{R'^5}. \tag{21-56}$$

The total radiated power is obtained by integrating this Poynting vector over a closed surface surrounding the charge. A convenient choice for such a surface is a sphere centered at the retarded position of the charge. If, furthermore, the z-axis is chosen in the direction of $\dot{\mathbf{v}}'$, then

$$P_R = -\frac{dW}{dt} = \int_S \mathbf{S} \cdot \mathbf{n} \, da$$

$$= \frac{q^2}{16\pi^2 \epsilon_0 c^3} \int \frac{R'^2 \dot{v}'^2 \sin^2 \theta}{R'^5} \mathbf{R}' \cdot \frac{\mathbf{R}'}{R'} R'^2 \sin \theta \, d\theta \, d\phi, \tag{21-57}$$

from which one readily obtains

$$P_R = -\frac{dW}{dt} = \frac{q^2}{4\pi\epsilon_0} \frac{2}{3} \frac{\dot{v}'^2}{c^3} \tag{21-58}$$

for the power radiated from a slowly moving, accelerated charge, in agreement with our earlier result (20–44).

This completes our brief survey of radiation from moving charges. The basic ideas have been presented and some elementary applications have been given in detail. For the details of other calculations, reference should be made to various published works, and particularly to *Classical Electricity and Magnetism*, Second Edition, by W. K. H. Panofsky and M. Phillips (Reading, Mass.: Addison-Wesley, 1962); *The Classical Theory of Fields*, Second Edition, by L. D. Landau and E. M. Lifshitz (Reading, Mass.: Addison-Wesley, 1962); and *Classical Electrodynamics*, Second Edition, by J. D. Jackson (New York: Wiley, 1975).

21–5 SUMMARY

The scalar and vector potentials of a fast moving point charge are obtained from the integral solutions for the retarded potentials. The results at the point P and time t are

$$\varphi(P, t) = \frac{q}{4\pi\epsilon_0} \frac{1}{R^*},$$

$$\mathbf{A}(P, t) = \frac{\mathbf{v'}}{c^2} \varphi(P, t),$$

where

$$R^* = R' - \mathbf{R'} \cdot \mathbf{v'}/c,$$

$\mathbf{R'}$ is the vector to the field point P from the position of the particle of charge q at the *retarded* time $t' = t - R'/c$, and $\mathbf{v'}$ is the particle velocity at the *retarded* time. The fields are obtained from the potentials by differentiating them correctly:

$$\mathbf{E}(P, t) = \frac{q}{4\pi\epsilon_0} \frac{1}{R^{*3}} \left\{ (\mathbf{R'} - R'\mathbf{v'}/c)(1 - v'^2/c^2) + \frac{1}{c^2} \mathbf{R'} \times \left[\left(\mathbf{R'} - R' \frac{\mathbf{v'}}{c} \right) \times \dot{\mathbf{v}}' \right] \right\},$$

$$\mathbf{B}(P, t) = \mathbf{R'} \times \mathbf{E}(P, t)/R'c.$$

1. The E-field of a uniformly moving ($\dot{v} = 0$) point charge is symmetric about a plane perpendicular to $\mathbf{v}$ through the *present* position of the charge at t; it is stronger in the direction perpendicular to $\mathbf{v}$ than in the direction along $\mathbf{v}$. It is always pointing away from the *present* position of the charge.

2. Only the acceleration terms contribute to radiation. In the radiation field $\mathbf{R'}$, $\mathbf{E}$, and $\mathbf{B}$ are mutually perpendicular. For $v'/c \ll 1$, the result is the same as the one derived in Chapter 20.

PROBLEMS

21–1 (a) For a charge in uniform motion with velocity $\mathbf{v}$, show that

$$R^* = R \sqrt{1 - \left(\frac{v}{c} \sin \theta \right)^2},$$

where $\mathbf{R}$ is the vector from the present position of the charge to the field point and $\mathbf{R} \cdot \mathbf{v} = Rv \cos \theta$. (b) Plot a polar graph of $|\mathbf{E}|$ as a function of θ for a fixed R, assuming $v/c = 0.8$.

21–2 Use the result of Problem 21–1(a) to find the B-field of a charge dq moving uniformly with velocity v. Letting $v \, dq = I \, dx$, where I is the current in a long straight wire, integrate $d\mathbf{B}$ over the length of the wire. Note that the result agrees with that obtained from Ampere's law, even for $v/c \approx 1$.

21–3 Suppose the acceleration of a fast particle is in the same direction as its velocity. Show that the radiation is zero along the direction of motion.

CHAPTER 22 The Special Theory of Relativity

As discussed in Chapters 2 and 8, the interaction between groups of charges (or currents) is usually described by separating the phenomenon into two parts: (1) the establishment of an electromagnetic field by the source, and (2) the interaction of a second group of charges (and/or currents) with the field. The field itself may be sampled by an observer, using test charges and currents. This decomposition of the interaction is not unique; in fact, the detailed nature of the electromagnetic field depends on the state of motion of the observer.

For example, consider two observers, A and B. Observer A is at rest relative to a group of fixed charges and sees only an electric field associated with them. Observer B is in motion relative to A; Observer B, therefore, sees a group of moving charges and hence a magnetic as well as an electric field.

Are both observers entitled to use Maxwell's equations to describe their physical observations? And, if so, how does an observer transform the electric and magnetic fields in one frame of reference to give the field components in a second reference frame moving relative to the first? These are questions that we shall attempt to answer in the present chapter.

22–1 PHYSICS BEFORE 1900

Maxwell's basic ideas about the electromagnetic field were first published in 1862. In the next forty years the mathematical structure of the laws of electricity and magnetism was gradually developed (particularly by H. A. Lorentz), and many consequences of the theory were observed experimentally. Yet there was a number of problems that bothered theoretical physicists, particularly those concerned with the mathematical structure of physical laws.

All previous experience with wave motion indicated that a medium was required for the propagation of waves. In Chapter 16 we explained that Maxwell's equations in free space are compatible with, and in fact lead to, a wave equation in which the waves are propagated with velocity $c = 1/\sqrt{\epsilon_0 \mu_0}$. Hence, it was natural to assume that some sort of ethereal medium pervaded all space (including vacuum) for the propagation of electromagnetic waves. Maxwell, himself, felt a need for this medium and called it the *ether*. But the existence of a medium poses a problem because it introduces a preferred reference frame, namely, the one in which the medium is at rest.

It was known that Newton's laws of motion were unaffected by a *Galilean transformation*, i.e., by a coordinate transformation between two frames of reference (coordinate systems) in relative motion. For example, let Σ be a coordinate system at rest, and let Σ' be a coordinate system moving in the x-direction with uniform velocity $\mathbf{u}$ (see Fig. 22–1). The relation between the coordinates and times in the two systems is given by (a Galilean transformation)

$$x' = x - ut, \qquad y' = y, \qquad z' = z, \qquad t' = t. \qquad (22\text{–}1)$$

Newton's fundamental laws of motion are of the same form in Σ and Σ'. In fact, it is impossible to determine the absolute velocity of any reference frame by mechanical experiments.

What happens to Maxwell's equations under a Galilean transformation? We are not in a position to answer this question because we do not yet know how the fields transform, but we can sidestep the question by looking instead at the scalar wave equation (which is homogeneous and contains only one field component). In free space,

$$\frac{\partial^2 \varphi}{\partial x^2} + \frac{\partial^2 \varphi}{\partial y^2} + \frac{\partial^2 \varphi}{\partial z^2} = \frac{1}{c^2} \frac{\partial^2 \varphi}{\partial t^2}, \qquad (22\text{–}2)$$

where φ stands for one of the field components. If we substitute

$$\frac{\partial}{\partial t} = \frac{\partial x'}{\partial t} \frac{\partial}{\partial x'} + \frac{\partial t'}{\partial t} \frac{\partial}{\partial t'} + \cdots$$

into Eqs. (22–2), using Eq. (22–1) to evaluate $\partial x'/\partial t$, etc., we find that the transformed wave equation is no longer of the same form as Eq. (22–2). This is just a mathematical statement of the fact, which we know to be true for mechanical waves, that the wave motion is propagated with a fixed velocity relative to the

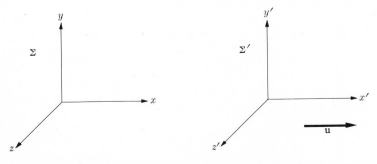

Figure 22–1 Two coordinate systems moving relative to one another (in the x-direction) with uniform velocity $\mathbf{u}$.

stationary medium. But in a reference frame moving with respect to the medium, the wave propagation appears more complicated.

What, then, is the situation with regard to the laws of electricity and magnetism? A number of possibilities was proposed before the subject was clarified by Lorentz and Poincaré, and particularly by Einstein in 1905. Briefly stated, these possibilities are

1. Maxwell's equations are inadequate to explain electromagnetic phenomena.

2. There is a preferred reference frame, that of the stationary ether. Maxwell's equations require modification in other reference frames.

3. Maxwell's equations have the same form in all reference frames that are moving with uniform velocity relative to one another. The Galilean transformation is not adequate to connect different reference frames when electromagnetic fields are involved.

As we know, the third choice presented here is the correct one, and it is, in fact, a partial statement of the *principle of relativity*. The predictions of Maxwell's equations have been verified experimentally, at least in our terrestrial environment, and attempts to measure an absolute ether frame have not been successful. There is no *single* experiment that killed the "ether hypothesis" and forced us to accept relativity, but the combined results of a large number of experiments turned out to be compatible with no other possibility. Three fundamental experiments are*

1. The aberration of starlight (the small shift in apparent position of distant stars in the direction of the earth's orbital motion).

2. Measurement of the velocity of light in moving fluids (Fizeau, 1859).

3. The Michelson-Morley experiment (1887).

The Michelson-Morley experiment attempted to measure the velocity of the earth, relative to an absolute frame (in which light waves are propagated with velocity c). The results of the experiment indicated either that there is no preferred frame or that the earth is always in the preferred reference frame. This experiment in itself would appear to negate the absolute ether-frame hypothesis, since the earth is continually changing its velocity in its motion around the sun. However, it would be possible for the earth to remain in the preferred frame if it dragged the ether along with it; i.e., a massive celestial object such as the earth could perhaps drag the ether along with it in its motion.

On the other hand, the first two experiments cited are not compatible with an "ether drag." From measurements made during the course of one year it is seen

---

* The reader who would like to pursue the history of this subject in more detail is referred to R. S. Shankland, "Michelson-Morley Experiment," *Am. J. Phys.*, vol. 32, p. 16 (1964); A. Einstein et al., *The Principle of Relativity* (New York: Dodd, Mead, 1923); E. T. Whittaker, *History of the Theories of Aether and Electricity*, Vol. II (New York: Philosophical Library, 1951).

that the apparent position of a star traces out a small elliptical path on the celestial sphere, the angular shift in position being of the order of v/c, where v is the earth's orbital speed. This aberration of starlight would presumably be absent if the ether were dragged along with the earth. The experiments with moving fluids can be made compatible with an ether-drag hypothesis only if one assumes (rather artificially) that objects less massive than the earth are only partially successful in dragging the ether along with them.

22–2 THE LORENTZ TRANSFORMATION AND EINSTEIN'S POSTULATES OF SPECIAL RELATIVITY

In 1904 H. A. Lorentz discovered a curious and remarkable transformation that leaves the form of Maxwell's equations unaltered, provided the field components are suitably changed. Let us again consider two coordinate systems Σ and Σ' that are moving relative to one another in the x-direction with uniform velocity $\mathbf{u}$ (see Fig. 22–1). Instead of the Galilean transformation, we now assume (a Lorentz transformation)

$$x' = \frac{1}{\sqrt{1 - u^2/c^2}} (x - ut),$$

$$y' = y,$$

$$z' = z,$$

$$t' = \frac{1}{\sqrt{1 - u^2/c^2}} \left(t - \frac{u}{c^2} x \right). \tag{22-3}$$

Again, we shall sidestep the question of how the electric and magnetic fields transform and look at the wave equation (22–2). Let us substitute

$$\frac{\partial}{\partial x} = \frac{\partial x'}{\partial x} \frac{\partial}{\partial x'} + \frac{\partial t'}{\partial x} \frac{\partial}{\partial t'},$$

$$\frac{\partial}{\partial y} = \frac{\partial y'}{\partial y} \frac{\partial}{\partial y'},$$

$$\frac{\partial}{\partial z} = \frac{\partial z'}{\partial z} \frac{\partial}{\partial z'},$$

$$\frac{\partial}{\partial t} = \frac{\partial x'}{\partial t} \frac{\partial}{\partial x'} + \frac{\partial t'}{\partial t} \frac{\partial}{\partial t'}$$

into the wave equation, using Eqs. (22–3) to evaluate the partial derivatives $(\partial x'/\partial x)$, etc. For example,

$$\frac{\partial x'}{\partial x} = \frac{1}{\sqrt{1 - u^2/c^2}}, \qquad \frac{\partial x'}{\partial t} = -\frac{u}{\sqrt{1 - u^2/c^2}}.$$

If we make the indicated substitutions, combine terms, and cancel common factors on each side of the equation, we obtain

$$\frac{\partial^2\varphi}{\partial x'^2} + \frac{\partial^2\varphi}{\partial y'^2} + \frac{\partial^2\varphi}{\partial z'^2} = \frac{1}{c^2}\frac{\partial^2\varphi}{\partial t'^2}. \tag{22-4}$$

Since this equation is homogeneous in φ, it is reasonable to expect that we can replace φ by φ' (its value in the primed coordinate system) in Eq. (22-4) without disturbing the equality. Thus the form of the wave equation is invariant to a Lorentz transformation.

Although the Lorentz transformation provides a basis for the development of special relativity, the far-reaching consequences of relativity were not discovered by Lorentz. He still believed in the ether hypothesis at this time, and tried very hard to fit his newly discovered transformation into the ether picture of electromagnetics. The development of special relativity as we now know it was left to H. Poincaré and A. Einstein.

As early as 1899, and again in 1900 and in 1904, Poincaré suggested that the experimental result of Michelson and Morley (i.e., their failure to observe an absolute ether frame) was a manifestation of a general principle: that absolute motion cannot be detected by laboratory experiments of any kind, and this implied that the laws of nature must be the same for two observers in uniform motion relative to each other. He called this the *Principle of Relativity*. Poincaré also concluded that a new type of dynamics would have to be developed, which would be characterized, among other things, by the rule that no velocity can exceed the velocity of light. In 1905 Einstein published his *Electrodynamics of Moving Bodies*, in which he developed the *special theory of relativity* from two basic postulates: (1) the relativity principle and (2) the constancy of the speed of light. Einstein derived the way in which various physical quantities had to transform in going from one reference frame to another and showed how Newton's laws of mechanics had to be modified.

The Einstein postulates are

1) *The laws of nature are the same in all coordinate systems moving with uniform motion relative to one another.*
2) *The velocity of light in empty space is the same in all reference systems and is independent of the motion of the emitting body.*

Again we consider two coordinate systems Σ and Σ', which are moving relative to one another in the x-direction with uniform velocity $\mathbf{u}$ (see Fig. 22-1). At times $t = 0$ and $t' = 0$, the origins of the two systems coincide, and at that instant a light source at the common origin emits a light pulse. An observer in the unprimed system, with appropriate detectors at various distances from the origin, sees the light signal propagate out as a spherical wave front. The coordinates of a point (x, y, z) on the wave front satisfy

$$x^2 + y^2 + z^2 - c^2t^2 = 0, \tag{22-5}$$

whereas for a point (x_1, y_1, z_1) in advance of the wave front (at the same time t),

$$x_1^2 + y_1^2 + z_1^2 - c^2 t^2 > 0, \qquad (22\text{–}6)$$

and for a point (x_2, y_2, z_2) behind,

$$x_2^2 + y_2^2 + z_2^2 - c^2 t^2 < 0. \qquad (22\text{–}7)$$

An observer in the primed coordinate system also sees the light signal propagate outward. And according to Einstein's two postulates, the observer sees a spherical wave front propagated with velocity c. Thus Eqs. (22–5), (22–6), and (22–7) also hold in the primed coordinates. Since the primed and unprimed coordinates are presumably related by a linear transformation, we are led to the result*

$$x^2 + y^2 + z^2 - c^2 t^2 = x'^2 + y'^2 + z'^2 - c^2 t'^2 \qquad (22\text{–}8)$$

with (x, y, z, t) an arbitrary *space-time point* and (x', y', z', t') its transform in the primed system. Equivalently we might have written

$$(\Delta x)^2 + (\Delta y)^2 + (\Delta z)^2 - c^2(\Delta t)^2 = (\Delta x')^2 + (\Delta y')^2 + (\Delta z')^2 - c^2(\Delta t')^2$$

$$(22\text{–}9)$$

for the relationship between an arbitrary *space-time interval* in Σ and the corresponding interval in Σ'.

Having found a quantity that is invariant to a change of reference frame, we now look for a transformation that will leave the "invariant quantity" unaffected. The *Lorentz transformation* is just such a transformation; the quantity

$$x^2 + y^2 + z^2 - c^2 t^2$$

is unaffected by the Lorentz transformation, Eq. (22–3), as can be verified by direct substitution. Thus the application of Einstein's two postulates leads directly to the Lorentz transformation.

If the Lorentz transformation is the proper one for transforming coordinates from one reference frame to another, then the more intuitive Galilean transformation, Eq. (22–1), clearly cannot be correct. The Galilean transformation is never precisely correct, but it does form a valid approximation in the limit when all velocities are small compared to the speed of light. Newtonian mechanics too must be modified since the correct laws of motion must transform properly under a Lorentz transformation, not under a Galilean transformation.

In the following sections the relativistic transformation will be discussed in more detail and the transformation laws for other physical quantities will be obtained. Before proceeding, however, we pause to discuss three simple con-

---

* We have not ruled out the possibility that

$$x^2 + y^2 + z^2 - c^2 t^2 = K(u)(x'^2 + y'^2 + z'^2 - c^2 t'^2)$$

with $K(u)$ a constant of proportionality depending on u. Such a scale change can be ruled out by considering the inverse transformation, however, so that $K(u) \equiv 1$.

sequences of the Lorentz transformation: (1) modification of the concept of simultaneity, (2) the Lorentz contraction, and (3) the time dilation.

Two events occur *simultaneously* if they happen at the same time. Since the events may occur at widely separated spatial positions, such a statement implies that we have a way of synchronizing clocks so that each event may be timed separately. Now let us suppose that two events at positions x_1 and x_2 in the Σ-frame occur simultaneously, i.e., the times t_1 and t_2 at which the two events occur are the same. But according to the Lorentz transformation (22–3) the times in the Σ'-system are not the same:

$$t_1' - t_2' = \frac{(u/c^2)}{\sqrt{1 - (u/c)^2}} [x_2 - x_1]. \tag{22-10}$$

Thus we must modify our intuitive concept of simultaneity: if two events occur simultaneously in one reference frame, they are not necessarily simultaneous in another. Having accepted the Lorentz transformation, we necessarily have to give up the concept of a "universal time."

A simple example will perhaps make this point more plausible. Suppose observer A is traveling in a space ship with velocity u relative to an earthbound observer B. Observer A wants to do an experiment involving the simultaneous detection of a light signal at two different positions and so places one detector in the front of the space ship, the other near the rear, and carefully measures the distance between the two detectors. Observer A then places a light source midway between the two detectors. Since the light signal propagates in spherical waves from the source, A's two detectors do in fact detect the signal simultaneously. But what about observer B? In the earthbound frame, B again sees the light signal propagate out from the source in spherical waves, but the detector at the front end of the space ship is moving away from the expanding wave front whereas the detector in the rear is moving toward it. Thus the detection is not simultaneous in B's system.

The apparent contraction of a moving object in the direction of its motion is called the *Lorentz contraction*. In a length measurement, the length of the object to be measured is compared with a standard scale. This offers no special problem if the object and scale are at rest relative to one another. Suppose, however, that an observer in the Σ-frame wanted to measure the length of a moving object (an object at rest in the Σ'-frame). Because the object is in motion relative to the observer and scale, it is important to compare the two ends of the object with the scale at the same time; i.e., if position x_1 is determined at time t_1, and x_2 at time t_2, then t_1 must equal t_2 in order for the length measurement to be meaningful. But according to the Lorentz transformation, Eq. (22–3),

$$x_1' - x_2' = \frac{1}{\sqrt{1 - \beta^2}} (x_1 - x_2) \tag{22-11}$$

with $\beta \equiv u/c$. Now $l' = x_1' - x_2'$ may be regarded as the "true" length of the object (its length measured by an observer at rest relative to it). Its apparent length (the length seen by an observer in the Σ-system)

$$l = l'\sqrt{1 - \beta^2} \tag{22-12}$$

appears contracted. It is easy to verify that the transverse dimensions, those in the y- and z-directions, are unaffected by the motion.

The *time dilation*, the apparent slowing down of temporal events associated with a moving object, can best be obtained from Eq. (22-9). This equation may be written

$$(dt)^2 \left[c^2 - \left(\frac{dx}{dt}\right)^2 - \left(\frac{dy}{dt}\right)^2 - \left(\frac{dz}{dt}\right)^2 \right] = (dt')^2 \left[c^2 - \left(\frac{dx'}{dt'}\right)^2 - \left(\frac{dy'}{dt'}\right)^2 - \left(\frac{dz'}{dt'}\right)^2 \right].$$
$$\tag{22-13}$$

Let Σ' be the *rest system* of the object, i.e., the system in which the object is at rest. Then (velocity)$'_x \equiv (dx'/dt') = 0$, etc., and

$$u^2 = \left[\left(\frac{dx}{dt}\right)^2 + \left(\frac{dy}{dt}\right)^2 + \left(\frac{dz}{dt}\right)^2 \right].$$

Equation (22-13) becomes

$$\Delta t = \frac{\Delta t'}{\sqrt{1 - \beta^2}}. \tag{22-14}$$

Thus the apparent time interval Δt (the interval measured by an observer in the Σ system) appears to be lengthened over the intrinsic time interval $\Delta t'$. Another way of interpreting Eq. (22-14) is to say that clocks appear to slow down when they are in motion relative to an observer.

22-3 GEOMETRY OF SPACE-TIME

The Lorentz transformation of the preceding section is a linear transformation connecting the space coordinates and time in one frame of reference to the corresponding quantities in a second frame, which is in uniform motion relative to the first one. It thus appears that we might be able to construct a four-dimensional geometry in which the space coordinates and time show up on an equal footing, with the Lorentz transformation acting as some kind of geometric operation in this four-dimensional space. Now we have noted that a certain quadratic function of the coordinates and time, namely,

$$x^2 + y^2 + z^2 - c^2t^2,$$

is an invariant, i.e., it has the same value in all reference frames. This recalls to mind that the length of a vector, specifically the length l of the position vector, with

$$l^2 = x^2 + y^2 + z^2,$$

is invariant to rotations of the coordinate axes in ordinary (three-dimensional) space. (See Appendix I.)

In trying to extend this formalism to four dimensions and treat $x^2 + y^2 + z^2 - c^2t^2$ as the square of a "length" in space-time, we run into the obvious problem that the fourth component, ct, enters the expression with a minus sign. This means that space-time is basically a non-Euclidean four-dimensional space. We can get around many of the difficulties that this implies by defining the four "coordinates" as

$$x_1 = x, \qquad x_2 = y, \qquad x_3 = z, \qquad x_4 = ict, \qquad (22\text{–}15)$$

with i the unit imaginary number. This four-dimensional space (which was introduced by H. Minkowski) is not, strictly speaking, Euclidean because it involves an imaginary coordinate. Many of its properties may, however, be derived by treating it as a Euclidean space. This approach will be used here. The quantity

$$\sum_\mu x_\mu^2 = x_1^2 + x_2^2 + x_3^2 + x_4^2$$

is invariant to certain transformations. These transformations (which include, of course, the Lorentz transformations) have many of the properties of rotations or orthogonal transformations, but because they may have imaginary components, they are properly called *complex orthogonal transformations*. In what follows, however, this distinction does not have important consequences and the Lorentz transformation in Minkowski space will be treated as an orthogonal transformation.*

The quantity defined by (x_1, x_2, x_3, x_4) is a four-dimensional vector. We shall have occasion to define other four-component vectors (i.e., quantities whose components transform as (x_1, x_2, x_3, x_4) under a Lorentz transformation). Four-

---

* Because of the common use of abstract spaces in contemporary physics, it is appropriate to point out the source of the difficulty in using "Euclidean" to describe Minkowski space and "orthogonal" to describe the Lorentz transformations. This is that both *Euclidean* and *orthogonal* represent ideas that were developed to deal with real variables. If one tries to generalize by admitting complex coordinates, then the most fruitful generalization of the length of a vector is

$$\left\{ \sum_i x_i^* x_i \right\}^{1/2},$$

with x_i^* the complex conjugate of x_i. Transformations which leave this length invariant are unitary transformations, characterized in the notation of Appendix I by $\sum_i a_{ij} a_{ik}^* = \delta_{jk}$. The Lorentz transformation does not fall into this classification and consequently requires an entirely separate but parallel development for its complete elucidation. There is one crucial difference between Minkowski space and either unitary or orthogonal space: For the last two spaces the length of any component of a vector is less than or equal to the length of the vector itself, while the "length" of a component of a four-vector in Minkowski space is not so restricted. Similarly, in unitary and orthogonal transformations, the magnitudes of all coefficients are less than or equal to unity, but this is not true for Lorentz transformations. These points are important, but further discussion here would lead us too far afield.

dimensional vectors are called *four vectors* or *world vectors* to distinguish them from ordinary three-dimensional vectors. A quantity that is unchanged by a Lorentz transformation is called a *world scalar*. A point in space-time is called a *world point*, and the trajectory of a particle in space-time is called a *world line*.

22–4 THE LORENTZ TRANSFORMATION AS AN ORTHOGONAL TRANSFORMATION

The results of the formalism of orthogonal transformations applied to ordinary three-dimensional vectors (which is developed in Appendix I) may be directly applied to four-dimensional space-time by adding the fourth component, $x_4 = ict$. All summations now go from 1 to 4. It is customary to use Greek indices to describe four-dimensional quantities and reserve Latin indices for three-dimensional entities. Thus F_i represents the ith component of an ordinary three-dimensional vector, but $T_{\mu\nu}$ represents the μ,νth component of a four-dimensional tensor.

The Lorentz transformation (22–3) for transforming the unprimed to the primed system of Fig. 22–1 may be written as

$$x'_1 = \frac{1}{\sqrt{1 - \beta^2}} x_1 + 0 \cdot x_2 + 0 \cdot x_3 + \frac{i\beta}{\sqrt{1 - \beta^2}} x_4,$$

$$x'_2 = 0 \cdot x_1 + x_2 + 0 \cdot x_3 + 0 \cdot x_4,$$

$$x'_3 = 0 \cdot x_1 + 0 \cdot x_2 + x_3 + 0 \cdot x_4,$$

$$x'_4 = -\frac{i\beta}{\sqrt{1 - \beta^2}} x_1 + 0 \cdot x_2 + 0 \cdot x_3 + \frac{1}{\sqrt{1 - \beta^2}} x_4, \qquad (22\text{–}16)$$

with $\beta \equiv u/c$. The matrix of this transformation is

$$\mathbf{A} = \begin{bmatrix} \dfrac{1}{\sqrt{1 - \beta^2}} & 0 & 0 & i\dfrac{\beta}{\sqrt{1 - \beta^2}} \\ 0 & 1 & 0 & 0 \\ 0 & 0 & 1 & 0 \\ -i\dfrac{\beta}{\sqrt{1 - \beta^2}} & 0 & 0 & \dfrac{1}{\sqrt{1 - \beta^2}} \end{bmatrix} \qquad (22\text{–}17)$$

We can easily verify that Eq. (22–17) is an orthogonal transformation,* i.e., that its components satisfy (I–6) in Appendix I.

Matrix **A** is particularly simple (it has only six nonzero entries) in this case because the Lorentz transformation relates two systems that are in relative motion

---

* We use the term "orthogonal" instead of the more accurate term "complex orthogonal"; cf. the discussion of Section 22–3.

along one of the coordinates axes (namely, the x-axis). Thus, x and t are transformed into x' and t' but the y- and z-directions are unaffected. In the general case when the direction of relative motion is not along a coordinate axis, the transformation is more complicated, but the matrix components still satisfy the orthogonality relationships. Since the coordinate directions can generally be chosen to fit the needs of a particular problem, we shall restrict ourselves in this book to Lorentz transformations of the type (22–17), or equivalently to transformations linking the coordinate systems in Fig. 22–1.

The Lorentz transformation (22–16) can be interpreted as a rotation in the $x_1 x_4$-plane. If this is the case, then the rotation angle θ is determined from

$$x_1' = x_1 \cos \theta + x_4 \sin \theta$$

or

$$\tan \theta = i\beta = i(u/c). \tag{22–18}$$

Thus the rotation angle is not a real angle.* Mathematically, the Lorentz transformation acts like a rotation in our orthogonal four-dimensional space, but a rotation through an imaginary angle.

The inverse Lorentz transformation, i.e., the transformation taking us from the primed system to the unprimed system, is given by the transpose matrix to (22–17):

$$\tilde{\mathbf{A}} = \begin{bmatrix} \dfrac{1}{\sqrt{1-\beta^2}} & 0 & 0 & -i\,\dfrac{\beta}{\sqrt{1-\beta^2}} \\ 0 & 1 & 0 & 0 \\ 0 & 0 & 1 & 0 \\ i\,\dfrac{\beta}{\sqrt{1-\beta^2}} & 0 & 0 & \dfrac{1}{\sqrt{1-\beta^2}} \end{bmatrix} \tag{22–19}$$

22–5 COVARIANT FORM OF THE ELECTROMAGNETIC EQUATIONS

The fundamental equations of electromagnetic theory, the Maxwell equations, which we discussed in Chapter 16, are written in terms of time and space derivatives of the $\mathbf{E}$- and $\mathbf{B}$-fields. In the usual three-dimensional system, time enters the equations as a scalar but the three space derivatives enter in certain symmetric combinations (divergence or curl). We may display the symmetry more directly by writing the divergence equation (e.g., Gauss's Law) as

$$\sum_i \frac{\partial E_i}{\partial x_i} = \frac{1}{\epsilon_0}\,\rho, \tag{22–20}$$

---

* This reflects the unboundedness of the transformation alluded to earlier.

and the curl equation (e.g., Ampère's Law) as

$$\frac{\partial B_j}{\partial x_i} - \frac{\partial B_i}{\partial x_j} = \mu_0 J_k + \mu_0 \epsilon_0 \frac{\partial E_k}{\partial t}. \tag{22-21}$$

This last equation actually represents three equations (the three components of the vector curl equation) with i, j, k standing for a cyclic permutation of x, y, z.

In the preceding sections, however, we have noted that the Lorentz transformation mixes up the space coordinates and the time and may be regarded as a rotation in $x_1 x_2 x_3 x_4$-space. Thus, x_1, x_2, x_3, and x_4 should enter the Maxwell equations in a symmetrical way. We should, in fact, be able to write the Maxwell equations in terms of four-dimensional curls and divergences. A formulation of the equations of electricity and magnetism that treats space coordinates and time on an equal footing is called a *covariant* formulation. We must proceed with a certain amount of caution, however, because a vector quantity in three dimensions does not necessarily become part of a four-dimensional vector.

We start with the equation of continuity:

$$\mathbf{V} \cdot \mathbf{J} + \frac{\partial \rho}{\partial t} = 0. \tag{22-22}$$

Since J_x, J_y, and J_z are not independent of the charge density ρ, these four quantities form a natural four-vector. In fact, if we define the *four-vector current density* $\mathfrak{I}_\nu$ by its components $(\mathfrak{I}_1, \mathfrak{I}_2, \mathfrak{I}_3, \mathfrak{I}_4 = ic\rho)$, we may write the equation of continuity in covariant form:

$$\sum_\nu \frac{\partial \mathfrak{I}_\nu}{\partial x_\nu} = 0, \tag{22-23}$$

where the summation is understood to run from $\nu = 1$ to $\nu = 4$; or equivalently

$$\square \cdot \mathfrak{I} = 0, \tag{22-23a}$$

with $\square \cdot$ standing for a four-dimensional divergence.

Now the vector potential $\mathbf{A}$ and the scalar potential φ satisfy inhomogeneous wave equations:

$$\nabla^2 \mathbf{A} - \frac{1}{c^2} \frac{\partial^2 \mathbf{A}}{\partial t^2} = -\mu_0 \mathbf{J},$$

$$\nabla^2 \varphi - \frac{1}{c^2} \frac{\partial^2 \varphi}{\partial t^2} = -\frac{1}{\epsilon_0} \rho. \tag{22-24}$$

Since $\mathbf{J}$ and ρ are the components of a four-vector, Eq. (22-24) must represent the four components of a four-vector equation and $\mathbf{A}$ and φ must also combine to form a four-vector. If we define the *four-potential* or *world potential* $\mathfrak{A}_\lambda$ by its components, $\mathfrak{A}_1 = A_1$, $\mathfrak{A}_2 = A_2$, $\mathfrak{A}_3 = A_3$, $\mathfrak{A}_4 = i\varphi/c$, then Eqs. (22-24) may be written as

$$\sum_\nu \frac{\partial^2 \mathfrak{A}_\lambda}{\partial x_\nu^2} = -\mu_0 \mathfrak{I}_\lambda. \tag{22-25}$$

They may also be expressed as

$$\square^2 \mathfrak{A} = -\mu_0 \mathfrak{J}, \tag{22-25a}$$

where $\square^2 \equiv \nabla^2 - (1/c^2)\partial^2/\partial t^2$ is the four-dimensional Laplacian operator, or the *d'Alembertian* operator. The Lorentz condition, Eq. (16–63), takes the form

$$\sum_v \frac{\partial \mathfrak{A}_v}{\partial x_v} = 0,$$

or

$$\square \cdot \mathfrak{A} = 0. \tag{22-26}$$

We are now in a position to examine the electromagnetic field components. These may be obtained from the customary three-dimensional equations

$$\mathbf{B} = \nabla \times \mathbf{A}, \tag{16-57}$$

$$\mathbf{E} = -\nabla\varphi - \frac{\partial \mathbf{A}}{\partial t}. \tag{16-60}$$

But $\mathbf{A}$ and $i\varphi/c$ form a four-vector, so the last equation may be written (in component form) as

$$i\frac{1}{c} E_1 = \frac{\partial \mathfrak{A}_1}{\partial x_4} - \frac{\partial \mathfrak{A}_4}{\partial x_1}, \text{ etc.} \tag{22-27}$$

Thus, $\mathbf{B}$ and $i\mathbf{E}/c$ together make up the four-dimensional curl of $\mathfrak{A}$. The curl operation applied to a vector actually produces an antisymmetric tensor.* This is evident from the form of Eq. (22–27) since it is clear that a two-index quantity must be generated. We define the electromagnetic field tensor $\mathbf{F}$ by the expression

$$F_{\mu v} = \frac{\partial \mathfrak{A}_v}{\partial x_\mu} - \frac{\partial \mathfrak{A}_\mu}{\partial x_v}. \tag{22-28}$$

Here

$$F_{11} = F_{22} = F_{33} = F_{44} = 0,$$

$$F_{14} = -F_{41} = -iE_1/c,$$

$$F_{24} = -F_{42} = -iE_2/c,$$

$$F_{34} = -F_{43} = -iE_3/c,$$

$$F_{12} = -F_{21} = B_3,$$

$$F_{23} = -F_{32} = B_1,$$

$$F_{31} = -F_{13} = B_2.$$

---

* In three dimensions an antisymmetric tensor has three independent components, T_{12}, T_{23}, T_{31}, and these transform under space rotations like the components of a vector. It was therefore satisfactory to treat the curl of a vector as a vector. For an antisymmetric tensor, note that $T_{11} = 0$, $T_{21} = -T_{12}$, etc.

In four dimensions an antisymmetric tensor has six independent components, and the tensorial character of the quantity cannot be simplified.

In matrix form,

$$
\mathbf{F} = \begin{bmatrix}
0 & B_3 & -B_2 & \dfrac{-iE_1}{c} \\[1.2em]
-B_3 & 0 & B_1 & \dfrac{-iE_2}{c} \\[1.2em]
B_2 & -B_1 & 0 & \dfrac{-iE_3}{c} \\[1.2em]
\dfrac{iE_1}{c} & \dfrac{iE_2}{c} & \dfrac{iE_3}{c} & 0
\end{bmatrix}.
\tag{22-29}
$$

Suppose we now take the divergence of the field tensor. Because of its form, Eq. (22–28), we obtain

$$
\sum_\nu \frac{\partial F_{\mu\nu}}{\partial x_\nu} = \frac{\partial}{\partial x_\mu} \sum_\nu \frac{\partial \mathfrak{A}_\nu}{\partial x_\nu} - \sum_\nu \frac{\partial^2 \mathfrak{A}_\mu}{\partial x_\nu^2}.
\tag{22-30}
$$

In view of Eqs. (22–25) and (22–26) this becomes

$$
\sum_\nu \frac{\partial F_{\mu\nu}}{\partial x_\nu} = \mu_0 \mathfrak{J}_\mu
\tag{22-31}
$$

or

$$
\square \cdot \mathbf{F} = \mu_0 \mathfrak{J}.
\tag{22-31a}
$$

This is a four-vector equation which represents a covariant formulation of two of the Maxwell equations, namely, $\mathbf{V} \cdot \mathbf{E} = \rho/\epsilon_0$ and $\mathbf{V} \times \mathbf{B} = \mu_0 \mathbf{J} + (1/c^2)\, \partial \mathbf{E}/\partial t$. Furthermore we have the identity

$$
\frac{\partial F_{\mu\nu}}{\partial x_\lambda} + \frac{\partial F_{\nu\lambda}}{\partial x_\mu} + \frac{\partial F_{\lambda\mu}}{\partial x_\nu} = 0,
\tag{22-32}
$$

where μ, ν, and λ are all different and stand for any three of the subscripts 1, 2, 3, and 4. Equation (22–32) follows immediately from the form of $F_{\mu\nu}$ (Eq. 22–28). We can easily verify that Eq. (22–32) represents the other two Maxwell equations.

The majority of advanced books on relativistic electromagnetic theory achieve a certain compactness of notation by introducing what is known as the *summation convention*. In this formalism all summation signs are suppressed, but summation is implied by means of a repeated index. Thus, for example, the equation of continuity becomes

$$
\frac{\partial \mathfrak{J}_\nu}{\partial x_\nu} = 0,
$$

and the wave equation for the potential becomes

$$
\frac{\partial^2 \mathfrak{A}_\mu}{\partial x_\nu\, \partial x_\nu} = -\mu_0 \mathfrak{J}_\mu.
$$

Other equations may be written by analogy. We do not make explicit use of the summation convention in this book, but have introduced it here primarily as an aid to further reading.

22–6 TRANSFORMATION LAW FOR THE ELECTROMAGNETIC FIELD

Since the electromagnetic field is a tensor quantity in the four-dimensional formulation, its components will transform like those of a second rank tensor under a Lorentz transformation:

$$F'_{\mu\nu} = \sum_\alpha \sum_\beta a_{\mu\alpha} a_{\nu\beta} F_{\alpha\beta}. \tag{22-33}$$

This is just Eq. (I–16) of Appendix I, rewritten to include the fact that $(\tilde{a})_{\beta\nu} = a_{\nu\beta}$.

Again we take the primed system to be one moving with velocity u in the x-direction, relative to the unprimed system. The Lorentz transformation is given by Eq. (22–17). Thus

$$B'_x = F'_{23} = \sum_\alpha \sum_\beta a_{2\alpha} a_{3\beta} F_{\alpha\beta}$$

$$= F_{23} = B_x, \tag{22-34}$$

and

$$B'_y = F'_{31} = \sum_\alpha \sum_\beta a_{3\alpha} a_{1\beta} F_{\alpha\beta}$$

$$= \frac{1}{\sqrt{1-\beta^2}} F_{31} + i \frac{\beta}{\sqrt{1-\beta^2}} F_{34}$$

$$= \frac{1}{\sqrt{1-\beta^2}} [B_y + (\beta/c)E_z]. \tag{22-35}$$

Similarly, we find that

$$B'_z = \frac{1}{\sqrt{1-\beta^2}} [B_z - (\beta/c)E_y]. \tag{22-36}$$

As far as the electric field is concerned,

$$E'_x = icF'_{14} = ic \sum_\alpha \sum_\beta a_{1\alpha} a_{4\beta} F_{\alpha\beta}$$

$$= ic \left[-\frac{i\beta}{1-\beta^2} F_{11} + \frac{1}{1-\beta^2} F_{14} + \frac{\beta^2}{1-\beta^2} F_{41} + \frac{i\beta}{1-\beta^2} F_{44} \right]$$

$$= \frac{ic}{1-\beta^2} \left[-i\frac{E_x}{c} + \beta^2 i \frac{E_x}{c} \right] = E_x. \tag{22-37}$$

Finally, we verify that

$$E'_y = \frac{1}{\sqrt{1-\beta^2}} [E_y - c\beta B_z], \qquad (22\text{--}38)$$

$$E'_z = \frac{1}{\sqrt{1-\beta^2}} [E_z + c\beta B_y]. \qquad (22\text{--}39)$$

Thus the components of $\mathbf{E}$ and $\mathbf{B}$ in the direction of motion are unaffected, but the transverse components are modified.

The above results may be summarized by the following three-dimensional equations:

$$\mathbf{E}'_{\parallel} = \mathbf{E}_{\parallel}, \qquad \mathbf{E}'_{\perp} = \frac{1}{\sqrt{1-\beta^2}} [\mathbf{E}_{\perp} + \mathbf{u} \times \mathbf{B}];$$

$$\mathbf{B}'_{\parallel} = \mathbf{B}_{\parallel}, \qquad \mathbf{B}'_{\perp} = \frac{1}{\sqrt{1-\beta^2}} \left[\mathbf{B}_{\perp} - \frac{1}{c^2} \mathbf{u} \times \mathbf{E}\right], \qquad (22\text{--}40)$$

where $\parallel$ and $\perp$ mean components parallel to and perpendicular to the velocity $\mathbf{u}$ of the Lorentz transformation.

The inverse transformation is obviously given by

$$\mathbf{E}_{\parallel} = \mathbf{E}'_{\parallel}, \qquad \mathbf{E}_{\perp} = \frac{1}{\sqrt{1-\beta^2}} [\mathbf{E}'_{\perp} - \mathbf{u} \times \mathbf{B}'];$$

$$\mathbf{B}_{\parallel} = \mathbf{B}'_{\parallel}, \qquad \mathbf{B}_{\perp} = \frac{1}{\sqrt{1-\beta^2}} \left[\mathbf{B}'_{\perp} + \frac{1}{c^2} \mathbf{u} \times \mathbf{E}'\right]. \qquad (22\text{--}41)$$

This completes our discussion of the transformation law for the electromagnetic field components. These results will be used in the following section.

22–7 THE FIELD OF A UNIFORMLY MOVING POINT CHARGE

To demonstrate the utility of the Lorentz transformation, we shall calculate the electric and magnetic fields of a point charge in uniform motion. Let us assume that the point charge q is moving with the velocity u along the x-axis. The geometry is shown in Fig. 22–2. We construct a second coordinate system (the primed system) that is moving along with the charge; for convenience we make the origin of this system, O', coincide with the charge itself.

In the primed system the charge is at rest. Therefore, at the field point P,

$$\mathbf{B}' = 0,$$

$$\mathbf{E}' = \frac{q\mathbf{r}'}{4\pi\epsilon_0(r')^3}. \qquad (22\text{--}42)$$

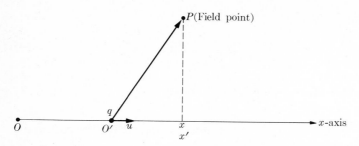

Figure 22–2 Coordinate system for determining the **E** and **B** fields of a charge in uniform motion. The heavy line qP is the source-field vector (**r′** in the primed system; **R** in the laboratory system). Here x is the coordinate in the laboratory system and $x′$ is the coordinate in the primed system.

The fields in the laboratory system may be obtained by using Eq. (22–41). Thus

$$E_x = E_{\parallel} = E'_x = \frac{qx'}{4\pi\epsilon_0(r')^3}, \qquad \mathbf{E}_{\perp} = \gamma \mathbf{E}'_{\perp} = \frac{\gamma q \mathbf{r}'_{\perp}}{4\pi\epsilon_0(r')^3}, \qquad (22\text{–}43)$$

where $\gamma \equiv 1/\sqrt{1 - \beta^2}$. Now from Eq. (22–3),

$$x' = \gamma(x - ut), \qquad y' = y, \qquad z' = z,$$

where t is the time elapsed (in the laboratory system) from the moment at which the two origins coincide. The vector **r′** is therefore given by the components:

$$\mathbf{r}' = \{\gamma(x - ut), y, z\}. \qquad (22\text{–}44)$$

It is convenient to define a quantity **R*** by

$$\gamma \mathbf{R}^* = \{\gamma(x - ut), y, z\}. \qquad (22\text{–}45)$$

Thus Eq. (22–43) becomes

$$E_x = \frac{q}{4\pi\epsilon_0} \frac{\gamma(x - ut)}{\gamma^3(R^*)^3},$$

$$E_y = \frac{q}{4\pi\epsilon_0} \frac{\gamma y}{\gamma^3(R^*)^3}, \qquad (22\text{–}46)$$

$$E_z = \frac{q}{4\pi\epsilon_0} \frac{\gamma z}{\gamma^3(R^*)^3}$$

or

$$\mathbf{E} = \frac{q}{4\pi\epsilon_0} \frac{\mathbf{R}}{(R^*)^3}(1 - \beta^2), \qquad (22\text{–}46a)$$

where **R** is defined by

$$\mathbf{R} = \{x - ut, \ y, \ z\}. \tag{22-47}$$

The electric field is directed radially away from the instantaneous position of the point charge, but in contrast to the static case it is no longer spherically symmetric. In fact, for a fast moving charge, the field is strongly concentrated in the plane at right angles to its motion.

The magnetic field is given by

$$B_x = B_{\parallel} = 0,$$

$$\mathbf{B}_{\perp} = \gamma \, \frac{1}{c^2} \, \mathbf{u} \times \mathbf{E}' = \gamma \, \frac{1}{c^2} \, \mathbf{u} \times \mathbf{E}'_{\perp} \tag{22-48}$$

$$= \frac{1}{c^2} \, \mathbf{u} \times \mathbf{E}_{\perp}$$

or

$$\mathbf{B} = \frac{1}{c^2} \, \mathbf{u} \times \mathbf{E}. \tag{22-48a}$$

The magnetic field lines are circles with centers on the charge path.

22–8 SUMMARY

It is an experimental conclusion that electromagnetic radiation propagates with the constant velocity c in vacuum, in *every* uniformly moving coordinate system. Thus the wave equation must be unchanged by a transformation to a uniformly moving coordinate system. This end is accomplished by the Lorentz transformation of coordinates, which leaves the quadratic form $(x^2 + y^2 + z^2 - c^2 t^2)$ invariant. The transformation is conveniently expressed in terms of the geometry of a four-dimensional, complex (nonEuclidean) Minkowski space, in which ict is the fourth coordinate. The Lorentz transformation (and also an ordinary spatial rotation) is represented by a complex orthogonal matrix, which operates on a four-vector, or world-vector. The Lorentz transformation to a system moving with velocity **u** in the x-direction is

$$\mathbf{A} = \begin{pmatrix} \gamma & 0 & 0 & i\beta\gamma \\ 0 & 1 & 0 & 0 \\ 0 & 0 & 1 & 0 \\ -i\beta\gamma & 0 & 0 & \gamma \end{pmatrix},$$

where $\beta = u/c$, $\gamma = 1/\sqrt{1 - \beta^2}$. (This is a rotation through an imaginary angle.) The Maxwell equations are covariant—i.e., they have the same form if the E- and B-field components are properly transformed by the Lorentz transformation.

1. The principal electromagnetic four-vectors, the field tensor, and the relations among them are as follows:

Four-vectors:

$$\text{Space-time } \mathbf{x} = (x, y, z, ict),$$
$$\text{Current-charge } \mathfrak{J} = (J_x, J_y, J_z, ic\rho),$$
$$\text{Potential } \mathfrak{A} = (A_x, A_y, A_z, i\varphi/c).$$

Field tensor:

$$F_{\mu\nu} = \frac{\partial \mathfrak{A}_\nu}{\partial x_\mu} - \frac{\partial \mathfrak{A}_\mu}{\partial x_\nu}, \qquad \mathbf{F} = \square \times \mathfrak{A}.$$

Maxwell's equations:

$$\sum_\nu \frac{\partial F_{\mu\nu}}{\partial x_\nu} = \mu_0 \mathfrak{J}_\mu, \qquad \square \cdot \mathbf{F} = \mu_0 \mathfrak{J};$$

$$\frac{\partial F_{\mu\nu}}{\partial x_\lambda} + \frac{\partial F_{\nu\lambda}}{\partial x_\mu} + \frac{\partial F_{\lambda\mu}}{\partial x_\nu} = 0.$$

Wave equation for potential:

$$\sum_\nu \frac{\partial^2 \mathfrak{A}_\mu}{\partial x_\nu^2} = -\mu_0 \mathfrak{J}_\mu, \qquad \square^2 \mathfrak{A} = -\mu_0 \mathfrak{J}.$$

Lorentz condition:

$$\sum_\nu \frac{\partial \mathfrak{A}_\nu}{\partial x_\nu} = 0, \qquad \square \cdot \mathfrak{A} = 0,$$

Equation of continuity:

$$\sum_\nu \frac{\partial \mathfrak{J}_\nu}{\partial x_\nu} = 0, \qquad \square \cdot \mathfrak{J} = 0.$$

2. The field tensor transforms according to

$$\mathbf{F'} = \mathbf{AFA}^{-1} = \mathbf{AF\tilde{A}}.$$

In three dimensions,

$$\mathbf{E'}_{\parallel} = \mathbf{E}_{\parallel}, \qquad \mathbf{E'}_{\perp} = \gamma(\mathbf{E}_{\perp} + \mathbf{u} \times \mathbf{B});$$

$$\mathbf{B'}_{\parallel} = \mathbf{B}_{\parallel}, \qquad \mathbf{B'}_{\perp} = \gamma\left(\mathbf{B}_{\perp} - \frac{1}{c^2}\,\mathbf{u} \times \mathbf{E}\right).$$

3. The field of a uniformly moving point charge is easily obtained by transforming from the rest system of the charge.

PROBLEMS

22–1 Transform the wave equation to the primed system of coordinates by using the Galilean transformation, Eq. (22–1). Show that

$$\varphi = F\{x - (c - u)t\} + G\{x + (c + u)t\},$$

where F and G are arbitrary functions of their arguments, is a solution to the transformed equation.

22–2 By making two consecutive Lorentz transformations, first to the Σ'-system moving with velocity u relative to the Σ-system, and then to Σ'' moving with velocity u' relative to Σ', prove the relativistic addition theorem for velocities:

$$u'' = \frac{u + u'}{1 + uu'/c^2}.$$

22–3 Given uniform electric and magnetic fields $\mathbf{E}$ and $\mathbf{B}$, find a Lorentz transformation that will make $\mathbf{E}$ and $\mathbf{B}$ parallel to each other. (*Hint:* Choose the velocity $\mathbf{u}$ of the Σ'-system in a direction perpendicular to both $\mathbf{E}$ and $\mathbf{B}$, and determine the magnitude of $u/(1 + \beta^2)$ in terms of E^2, B^2, and $\mathbf{E} \times \mathbf{B}$.)

22–4 Equation (2–30) gives the electric field of a long straight wire containing λ units of charge per unit length. Make a Lorentz transformation to a system moving with velocity u in a direction parallel to the wire. Calculate the B-field in the new system and compare with the B-field of a current-carrying wire, Eq. (8–35). Is there an E-field in the moving system? What is the physical difference between a charged wire moving in the direction of its length and a current-carrying wire?

22–5 Show that the scalar product $\mathbf{E} \cdot \mathbf{B}$ is unchanged by a Lorentz transformation. Show the same for $E^2 - c^2 B^2$.

22–6 Determine the Poynting vector for the uniformly moving point charge of Section 22–7, and show that the total power radiated is zero.

Appendixes

APPENDIX I Coordinate Transformations, Vectors, and Tensors

In order to achieve a compactness in notation we shall use x_1, x_2, x_3 for the Cartesian coordinates x, y, z. A coordinate transformation is *linear* if the new coordinates can be expressed as a linear combination of the old coordinates. Thus

$$x'_1 = a_{11}x_1 + a_{12}x_2 + a_{13}x_3,$$
$$x'_2 = a_{21}x_1 + a_{22}x_2 + a_{23}x_3, \qquad\qquad \text{(I–1)}$$
$$x'_3 = a_{31}x_1 + a_{32}x_2 + a_{33}x_3$$

or

$$x'_i = \sum_j a_{ij}x_j \qquad\qquad \text{(I–1a)}$$

is a linear transformation. It is understood that the summation runs from $j = 1$ to $j = 3$. Furthermore i can take on any of the values 1, 2 or 3. The set of coefficients $\{a_{ij}\}$ describes the transformation.

Consider, for example, the coordinate transformation described by

$$x' = x \cos \theta + y \sin \theta,$$
$$y' = -x \sin \theta + y \cos \theta, \qquad\qquad \text{(I–2)}$$
$$z' = z.$$

This linear transformation describes a rotation about the z-axis through the angle θ, in which the x- and y-axes are transformed into the x'- and y'- axes, respectively. This is illustrated in Fig. I–1. The length of the position vector, $l = \sqrt{x^2 + y^2 + z^2}$, is left invariant by the transformation since it is evident from (I–2) that

$$x^2 + y^2 + z^2 = x'^2 + y'^2 + z'^2. \qquad\qquad \text{(I–3)}$$

Equation (I–2) is an example of an *orthogonal transformation* in three dimensions, an orthogonal transformation being one which is real and leaves the length of a

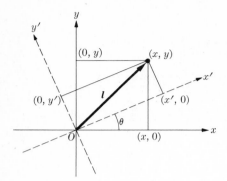

Figure I–1 Coordinate system rotation in two dimensions. The heavy line is the projection of vector l onto the xy-plane.

vector unchanged. Properties of orthogonal transformations will be discussed in more detail in the following paragraphs.

A transformation is orthogonal if it leaves the length of the displacement vector (or equivalently if it leaves $\sum x_i^2$) unchanged. Let us assume that (I–1a) is an orthogonal transformation; then

$$\sum_i (x_i')^2 = \sum_k x_k^2. \tag{I–4}$$

But

$$(x_i')^2 = \sum_j \sum_k a_{ij} a_{ik} x_j x_k$$

and

$$\sum_i (x_i')^2 = \sum_j \sum_k \sum_i a_{ij} a_{ik} x_j x_k. \tag{I–5}$$

Equations (I–4) and (I–5) agree only if the quantity

$$\sum_i a_{ij} a_{ik} = \begin{cases} 0, & j \neq k, \\ 1, & j = k. \end{cases} \tag{I–6}$$

This last equation may be written more compactly by introducing the Kronecker delta δ_{jk}, which was defined on p. 42:

$$\sum_i a_{ij} a_{ik} = \delta_{jk}. \tag{I–6a}$$

Equation (I–6a) *is the condition which must be imposed to make the transformation* $\{a_{ij}\}$ *orthogonal.* We easily verify that the rotation, (I–2), satisfies this criterion.

The transformation (I–1) may be written symbolically as

$$\mathbf{X'} = \mathbf{AX}, \tag{I–7}$$

where $\mathbf{X'}$ is the position vector with transformed components (x_1', x_2', x_3'), $\mathbf{X}$ is the position vector with original components (x_1, x_2, x_3), and $\mathbf{A}$ is regarded as a matrix operator. In fact **A** is the matrix of the coefficients $\{a_{ij}\}$:

$$\mathbf{A} = \begin{bmatrix} a_{11} & a_{12} & a_{13} \\ a_{21} & a_{22} & a_{23} \\ a_{31} & a_{32} & a_{33} \end{bmatrix}. \tag{I-8}$$

If we display the vectors **X'** and **X** as column matrices, we may write (I–7) as

$$\begin{bmatrix} x_1' \\ x_2' \\ x_3' \end{bmatrix} = \begin{bmatrix} a_{11} & a_{12} & a_{13} \\ a_{21} & a_{22} & a_{23} \\ a_{31} & a_{32} & a_{33} \end{bmatrix} \begin{bmatrix} x_1 \\ x_2 \\ x_3 \end{bmatrix}. \tag{I-7a}$$

The transformation (I–1) follows from this last equation through the usual laws of matrix multiplication. Thus the matrix **A** given by (I–8) and Eq. (I–1a) are equivalent ways of describing the coordinate transformation.

The inverse of a transformation must bring us back again to the original set of coordinates. Thus, if $\{b_{ij}\}$ is the inverse transformation to $\{a_{ij}\}$, then

$$x_j = \sum_i b_{ji} x_i'. \tag{I-9}$$

Combining this with (I–1a), we find

$$x_j = \sum_k \sum_i b_{ji} a_{ik} x_k,$$

which is an identity if

$$\sum_i b_{ji} a_{ik} = \delta_{jk}. \tag{I-10}$$

Equation (I–10) is the condition imposed in order for **B** to be the inverse transformation to **A**. If, in addition, **A** is an orthogonal transformation, then (I–6a) holds. Comparison of (I–6a) and (I–10) shows that if matrix **B** is constructed so that

$$b_{ji} = a_{ij}, \tag{I-11}$$

then **B** will indeed describe the inverse transformation. According to (I–11), matrix **B** is constructed from **A** by interchanging its rows and columns. The new matrix is called the transpose of **A** and is given the symbol $\tilde{\mathbf{A}}$. Thus *the inverse of an orthogonal transformation is the transpose of the original transformation.*

In Chapter 1 we defined a vector function as a quantity possessing both direction and magnitude at each point in space. An alternative definition is:

> *A vector is a quantity whose components transform under an orthogonal transformation as the components of the position vector.*

Thus, if **F** is any vector function, its transform **F'**, resulting from an orthogonal transformation **A**, is

$$\mathbf{F}' = \mathbf{A}\mathbf{F}. \tag{I-12}$$

Scalar functions of position, such as the length of a vector or the dot product of two vectors, are unchanged by an orthogonal transformation, or are *invariant*.

In addition to scalars and vectors, there are other more complicated quantities. One of these, whose transformation properties we need, is the *second-rank tensor* or simply *tensor*. A tensor is a quantity with two-index components; thus a component of tensor **T** is T_{ij}, where i and j can each take on the values 1, 2, and 3. The reader will be exposed to the quadrupole moment tensor Q_{ij} (p. 43), the dielectric tensor ϵ_{ij} for anisotropic media (bottom of p. 338), and the four-dimensional field tensor $F_{\mu\nu}$ (Chapter 22). A familiar example from mechanics is the moment of inertia tensor. A linear relationship between two vector quantities can be expressed in terms of a second-rank tensor. Thus the angular momentum **L** of a rigid body can be related to its angular velocity **ω** by means of the moment of inertia tensor **I**:

$$\mathbf{L} = \mathbf{I}\boldsymbol{\omega}$$

or

$$L_i = \sum_j I_{ij}\omega_j.$$

The tensor itself can be expressed in matrix form:

$$\mathbf{I} = \begin{bmatrix} I_{11} & I_{12} & I_{13} \\ I_{21} & I_{22} & I_{23} \\ I_{31} & I_{32} & I_{33} \end{bmatrix}.$$

Suppose two vectors **F** and **X** are linearly related through the tensor relationship

$$\mathbf{F} = \mathbf{T}\,\mathbf{X}. \tag{I–13}$$

Now if an orthogonal transformation **A** is performed, **X** will be converted into **X′** and **F** into **F′**. We should be able to express Eq. (I–13) in the transformed system as

$$\mathbf{F'} = \mathbf{T'X'}, \tag{I–14}$$

where **T′** is the transform of **T**. But

$$F'_i = \sum_j a_{ij}F_j = \sum_j \sum_k a_{ij}T_{jk}X_k$$

$$= \sum_j \sum_k \sum_m a_{ij}T_{jk}(\tilde{a})_{km}X'_m$$

$$= \sum_m \left[\sum_j \sum_k a_{ij}T_{jk}(\tilde{a})_{km} \right] X'_m \tag{I–15}$$

with $(\tilde{a})_{km} \equiv a_{mk}$. For Eqs. (I–14) and (I–15) to be compatible,

$$T'_{im} = \sum_j \sum_k a_{ij}T_{jk}(\tilde{a})_{km}. \tag{I–16}$$

This last equation expresses the transformation law of a second-rank tensor under an orthogonal transformation. Equation (I–16) also expresses the rule for multiplying three matrices together to find the i, m-component of the resulting matrix.

Thus (I–16) is written symbolically as

$$\mathbf{T}' = \mathbf{A}\mathbf{T}\tilde{\mathbf{A}} = \mathbf{A}\mathbf{T}\mathbf{A}^{-1}. \tag{I-16a}$$

Let us finally determine the transformation law for the vector differential operator $\mathbf{V}$. Consider the ith component of the gradient of a scalar invariant function φ; we think of φ as a function of the coordinates x_j, which are in turn functions of the new coordinate x_i'. Using the chain rule of partial differentiation, we find

$$\frac{\partial \varphi}{\partial x_i'} = \sum_j \frac{\partial \varphi}{\partial x_j} \frac{\partial x_j}{\partial x_i'}.$$

From Eq. (I–9),

$$\frac{\partial \varphi}{\partial x_i'} = \sum_j b_{ji} \frac{\partial \varphi}{\partial x_j}, \tag{I-17}$$

and from (I–11) for an orthogonal transformation

$$\frac{\partial}{\partial x_i'} = \sum_j a_{ij} \frac{\partial}{\partial x_j}. \tag{I-18}$$

Thus the components of $\mathbf{V}$ transform like those of the position vector, (I–1a), under an orthogonal transformation.*

---

* Under a more general transformation, quantities that transform by (I–17) have to be distinguished from those that transform by (I–1a). They are called *covariant vectors* and *contravariant vectors*, respectively, but we will not need this distinction.

APPENDIX II Systems of Units

In this book the charge-rationalized mks system of units has been used. This system has the virtue of including the practical electrical units of potential difference (volt), current (ampere), resistance (ohm), etc. As a result, the system rapidly gained favor with electrical engineers and is now the international standard. In other areas, notably atomic, solid-state, plasma, and nuclear physics, another system, known as the *Gaussian system*, has remained popular. Most other systems have faded from use, and hence only the Gaussian system will be discussed here at length.

In the mks system of units the appearance of the numbers ϵ_0 and μ_0 in the formulation of Coulomb's law and the Biot law, respectively, causes an apparent difficulty. The difficulty is simply that Coulomb's law,

$$\mathbf{F}_2 = \frac{1}{4\pi\epsilon_0} \frac{q_1 q_2 \mathbf{r}_{12}}{r_{12}^3}, \tag{II-1}$$

cannot be used to define the coulomb unless ϵ_0 is known. By the same token, it cannot be used to define ϵ_0 unless the coulomb is previously defined. A technical point is that since ϵ_0 is taken as an experimentally determined number, using (II-1) to define the coulomb would result in a coulomb that would change every time ϵ_0 was redetermined. Thus it is clear that (II-1) should be used to define ϵ_0, with the coulomb otherwise defined.

A corresponding difficulty does not arise in the magnetic case because one takes $\mu_0 = 4\pi \times 10^{-7}$ T · m/A, by definition. As a result, the expression

$$\frac{F}{l} = \frac{\mu_0}{2\pi} \frac{II'}{r} \tag{II-2}$$

for the force per unit length between two parallel, current-carrying wires can be used to define the ampere, viz.:

One ampere is that steady current which, when present in each of two long parallel conductors separated by a distance of one meter, results in a force per meter of length between them numerically equal to 2×10^{-7} N/m.

Of course, any other geometry could be used and would result in an equally unambiguous (and, in fact, numerically identical) definition of the ampere.

Having thus defined the ampere, the coulomb is defined as the charge transported by a steady current of one ampere flowing for one second. This in turn makes it feasible to use (II–1) to define ϵ_0. There is thus no real problem, but only an artificial one arising from a desire to treat the mathematically simpler case of electrostatics before discussing the magnetic interaction of currents.

It is sometimes thought that this problem does not arise if the cgs Gaussian system of units is used. This is true only in the sense that the coefficient in Coulomb's law is chosen to be 1 dyne cm²/esu², which places the burden of agreeing with experiment on the magnetic interactions. This means that the velocity of light appears in the expression for the force between current-carrying conductors. Since in the usual treatment the conveniently defined quantity occurs first, this problem is less painfully obvious in the Gaussian system than in the mks system.

The Gaussian system is a combination of two earlier systems: the electrostatic system, esu, and the electromagnetic system, emu. The electrostatic system resulted from writing Coulomb's law in the form

$$\mathbf{F}_2 = \frac{q_1 q_2 \mathbf{r}_{12}}{r_{12}^3},$$ (II–3)

and defining the esu of charge as that charge which when placed one centimeter from an exactly similar charge experiences a force of one dyne. It is obvious that the esu of charge is much smaller than the coulomb (in fact 1 coulomb = 3 × 10^9 esu). The electromagnetic system resulted from writing the Biot law, Eq. (8–25), without the factor $\mu_0/4\pi$ and defining the abampere as the current that, when present in a long straight wire, results in a force of 1 dyne/cm when the wire is placed 1 cm from a parallel conductor carrying the same current. From $|\mu_0/4\pi| = 10^{-7}$ and 1 newton = 10^5 dynes, it is found that 1 abampere = 10 ampere.

Either of the two starting points noted above could be used to initiate the development of a complete system of units. Historically, however, esu was used primarily for electrostatic problems and emu for electromagnetic problems. This being the case, it was natural that a hybrid system using esu for electrical quantities and emu for magnetic quantities should develop. The system that evolved in this way is known as the Gaussian system. The principal point of contact of esu and emu in the Gaussian system is in the current density, where

$$\mathbf{J}_{emu} = \frac{\mathbf{J}_{esu}}{c}.$$ (II–4)

In the Gaussian system we use $\mathbf{J}_{esu}$ and explicitly exhibit the velocity of light in the magnetic equations. Thus the Biot law has the form

$$d\mathbf{F}_2 = \frac{I_1 I_2}{c^2} \frac{d\mathbf{l}_2 \times (d\mathbf{l}_1 \times \mathbf{r}_{12})}{r_{12}^3}$$ (II–5)

with I in esu/s.

In Gaussian units, Maxwell's equations are

$$\nabla \times \mathbf{E} + \frac{1}{c}\frac{\partial \mathbf{B}}{\partial t} = 0,$$

$$\nabla \cdot \mathbf{D} = 4\pi\rho,$$

$$\nabla \times \mathbf{H} - \frac{1}{c}\frac{\partial \mathbf{D}}{\partial t} = \frac{4\pi \mathbf{J}}{c},$$

$$\nabla \cdot \mathbf{B} = 0. \tag{II-6}$$

The fields are derived from scalar and vector potentials by means of

$$\mathbf{B} = \nabla \times \mathbf{A} \quad \text{and} \quad \mathbf{E} = -\nabla\varphi - \frac{1}{c}\frac{\partial \mathbf{A}}{\partial t}, \tag{II-7}$$

and the Lorentz force is

$$\mathbf{F} = q\left(\mathbf{E} + \frac{\mathbf{v}}{c} \times \mathbf{B}\right). \tag{II-8}$$

$\mathbf{D}$ and $\mathbf{B}$ are related to $\mathbf{E}$ and $\mathbf{H}$ by

$$\mathbf{D} = \mathbf{E} + 4\pi\mathbf{P} \quad \text{and} \quad \mathbf{B} = \mathbf{H} + 4\pi\mathbf{M}, \tag{II-9}$$

where $\mathbf{P}$ is the electric dipole moment $(\mathbf{p} = q\mathbf{l})$ per unit volume, and $\mathbf{M}$ is the magnetic dipole moment $(\mathbf{m} = IA\mathbf{n}/c)$ per unit volume. These equations are substantially sufficient to define the Gaussian system of units. In addition, the energy density is

$$u = \frac{1}{8\pi}(\mathbf{E} \cdot \mathbf{D} + \mathbf{B} \cdot \mathbf{H})$$

and the Poynting vector is

$$\mathbf{S} = \frac{c}{4\pi}\mathbf{E} \times \mathbf{H}.$$

For convenience Table II-1 gives the numerical relationships of Gaussian units to mks units.

Some advantages of the Gaussian system, which account for its continued use in the physics literature, are that the factor $4\pi\epsilon_0$ does not occur in all the equations of atomic physics where the Coulomb potential plays a central role, and that the velocity v always enters in the dimensionless form v/c prescribed by the Lorentz transformation. In relativity it is natural for $\mathbf{E}$ and $\mathbf{B}$ to have the same units, since their components are simply different elements of the field tensor. There is no reason for the auxiliaries $\mathbf{D}$ and $\mathbf{H}$ to have other units. It is a convenience that in material media the permittivity and susceptibility are dimensionless and related by

$$\epsilon = 1 + 4\pi\chi_e$$

like the corresponding magnetic quantities,

$$\mu = 1 + 4\pi\chi_m.$$

Table II–1

Quantity	Symbol	Gaussian units*	mks units
Capacitance	C	9×10^{11} cm (esu/statvolt)	= 1 F
Charge	Q	3×10^9 esu	= 1 C
Conductivity	g	9×10^9 s^{-1}	= 1 $(\Omega \cdot m)^{-1}$
Current	I	3×10^9 esu/s $(= 10^{-1}$ abamp)	= 1 A
Electric displacement	D	$12\pi \times 10^5$ dyne/esu	= 1 C/m^2
Electric field	E	$\frac{1}{3} \times 10^{-4}$ dyne/esu	= 1 V/m
Energy	U	10^7 erg	= 1 J
Force	F	10^5 dyne	= 1 N
Inductance	L	$\frac{1}{9} \times 10^{-11}$ s^2/cm	= 1 H
Magnetic flux	Φ	10^8 maxwell	= 1 Wb
Magnetic induction	B	10^4 gauss	= 1 T
Magnetic intensity	H	$4\pi \times 10^{-3}$ oersted (gauss)	= 1 A/m
Potential	φ	1/300 erg/esu (statvolt)	= 1 V
Resistance	R	$\frac{1}{9} \times 10^{-11}$ s/cm (statvolt $\cdot$ s/esu)	= 1 Ω

* The factor 3 $(9 = 3^2)$ in the conversions comes from the velocity of light, $c \cong 3 \times 10^8$ m/s. For greater accuracy, use 2.9979 instead of 3.

APPENDIX **III** Vector Differential Operators

1. RECTANGULAR COORDINATES

$$\nabla\varphi = \mathbf{i}\,\frac{\partial\varphi}{\partial x} + \mathbf{j}\,\frac{\partial\varphi}{\partial y} + \mathbf{k}\,\frac{\partial\varphi}{\partial z},$$

$$\nabla\cdot\mathbf{F} = \frac{\partial F_x}{\partial x} + \frac{\partial F_y}{\partial y} + \frac{\partial F_z}{\partial z},$$

$$\nabla\times\mathbf{F} = \mathbf{i}\left(\frac{\partial F_z}{\partial y} - \frac{\partial F_y}{\partial z}\right) + \mathbf{j}\left(\frac{\partial F_x}{\partial z} - \frac{\partial F_z}{\partial x}\right) + \mathbf{k}\left(\frac{\partial F_y}{\partial x} - \frac{\partial F_x}{\partial y}\right).$$

2. CYLINDRICAL COORDINATES

$$\nabla\varphi = \mathbf{a}_r\,\frac{\partial\varphi}{\partial r} + \mathbf{a}_\theta\,\frac{1}{r}\,\frac{\partial\varphi}{\partial\theta} + \mathbf{k}\,\frac{\partial\varphi}{\partial z},$$

$$\nabla\cdot\mathbf{F} = \frac{1}{r}\,\frac{\partial}{\partial r}\,(rF_r) + \frac{1}{r}\,\frac{\partial F_\theta}{\partial\theta} + \frac{\partial F_z}{\partial z},$$

$$\nabla\times\mathbf{F} = \mathbf{a}_r\left(\frac{1}{r}\,\frac{\partial F_z}{\partial\theta} - \frac{\partial F_\theta}{\partial z}\right) + \mathbf{a}_\theta\left(\frac{\partial F_r}{\partial z} - \frac{\partial F_z}{\partial r}\right) + \mathbf{k}\,\frac{1}{r}\left(\frac{\partial}{\partial r}\,(rF_\theta) - \frac{\partial F_r}{\partial\theta}\right).$$

3. SPHERICAL COORDINATES

$$\nabla\varphi = \mathbf{a}_r\,\frac{\partial\varphi}{\partial r} + \mathbf{a}_\theta\,\frac{1}{r}\,\frac{\partial\varphi}{\partial\theta} + \mathbf{a}_\phi\,\frac{1}{r\sin\theta}\,\frac{\partial\varphi}{\partial\phi},$$

$$\nabla\cdot\mathbf{F} = \frac{1}{r^2}\,\frac{\partial}{\partial r}\,(r^2F_r) + \frac{1}{r\sin\theta}\,\frac{\partial}{\partial\theta}\,(F_\theta\sin\theta) + \frac{1}{r\sin\theta}\,\frac{\partial F_\phi}{\partial\phi},$$

$$\mathbf{\nabla} \times \mathbf{F} = \mathbf{a}_r \, \frac{1}{r \sin \theta} \left[\frac{\partial}{\partial \theta} \left(F_\phi \sin \theta \right) - \frac{\partial F_\theta}{\partial \phi} \right]$$

$$+ \mathbf{a}_\theta \, \frac{1}{r} \left[\frac{1}{\sin \theta} \frac{\partial F_r}{\partial \phi} - \frac{\partial (r F_\phi)}{\partial r} \right] + \mathbf{a}_\phi \, \frac{1}{r} \left[\frac{\partial (r F_\theta)}{\partial r} - \frac{\partial F_r}{\partial \theta} \right].$$

The Laplacian of a scalar quantity, $\nabla^2 \varphi$, is given on p. 52 for the three coordinate systems. Vector identities are given on p. 19.

APPENDIX IV Dirac Delta Function

In one dimension, the delta function is defined by

$$\delta(x) = 0 \text{ for } x \neq 0,$$

$$\int_{-\infty}^{+\infty} \delta(x)\, dx = 1. \tag{IV-1}$$

This can be understood in an elementary way as the limit of an ordinary continuous function. For example, the Gaussian function

$$\frac{1}{\sqrt{\pi}\,\epsilon}\, e^{-x^2/\epsilon^2}$$

has a peak at $x = 0$, which is higher and narrower the smaller is ϵ, and its integral from $-\infty$ to $+\infty$ is 1 for any value of ϵ. Thus we could define

$$\delta(x) = \lim_{\epsilon \to 0} \frac{1}{\sqrt{\pi}\,\epsilon}\, e^{-x^2/\epsilon^2}. \tag{IV-2}$$

Another useful representation is

$$\delta(x) = \lim_{\epsilon \to 0} \frac{\epsilon}{\pi} \frac{\sin^2 x/\epsilon}{x^2}. \tag{IV-3}$$

As a third example, we may use the Lorentzian function introduced in Chapter 19:

$$\delta(x) = \lim_{\epsilon \to 0} \frac{\epsilon}{\pi} \frac{1}{x^2 + \epsilon^2}. \tag{IV-4}$$

An important application of the delta function involves the Fourier theorem. If a function $f(x)$ is expressed by a Fourier integral,

$$f(x) = \int_{-\infty}^{+\infty} g(k) e^{ikx}\, dx,$$

the theorem gives the Fourier transform

$$g(k) = \frac{1}{2\pi} \int_{-\infty}^{+\infty} f(x) e^{-ikx} \, dx.$$

For $f(x) = \delta(x)$, we find $g(k) = 1/2\pi$ (see below), so that

$$\delta(x) = \frac{1}{2\pi} \int_{-\infty}^{+\infty} e^{ikx} \, dk. \qquad \text{(IV-5)}$$

Thus the Fourier transform of the delta function is a constant function, and conversely. Taking the real part of (IV-5) gives

$$\delta(x) = \frac{1}{2\pi} \int_{-\infty}^{+\infty} \cos kx \, dk. \qquad \text{(IV-6)}$$

Equations (IV-5) and (IV-6) may be considered as other representations of the delta function.

The mean-value theorem of calculus gives

$$\int_{-a}^{+a} F(x) f(x) \, dx \cong F(0) \int_{-a}^{+a} f(x) \, dx,$$

so that as $f(x) \to \delta(x)$

$$\int F(x) \, \delta(x) \, dx = F(0). \qquad \text{(IV-7)}$$

In all of the above equations one can replace the variable x by $(x - x_0)$; in particular,

$$\int F(x) \, \delta(x - x_0) \, dx = F(x_0). \qquad \text{(IV-8)}$$

In any equation involving the delta function, the domain of integration can be reduced to any interval containing the point where the argument of the delta function vanishes.

An extension to three dimensions can be achieved by writing

$$\delta(\mathbf{r}) \, dv = \delta(x) \, \delta(y) \, \delta(z) \, dx \, dy \, dz. \qquad \text{(IV-9)}$$

In Chapter 2 we found that

$$\delta(\mathbf{r}) = \nabla^2 \left(\frac{-1}{4\pi r} \right). \qquad \text{(IV-10)}$$

APPENDIX V Static Electrification

The oldest branch of our subject—electricity and magnetism—is that of static electrification, i.e., the generation of large potentials by bringing two dissimilar materials into close contact and then separating them. This interesting phenomenon is usually dismissed rather abruptly after a brief discussion of a few classic experiments which serve to introduce to the reader the concepts of charge and charge separation. A complete discussion of the phenomenon is beyond the scope of this book, since it would involve digressions into thermodynamics and the quantum theory of matter; nevertheless, a few remarks about the subject are in order.

When two dissimilar materials are brought into close contact, they try to establish various types of equilibrium. One of these is thermal equilibrium; thus heat flows from the hotter to the cooler material in an attempt to make their temperatures the same. Another type, and the one of direct importance to our discussion, involves an attempt to equalize the *electrochemical potentials* of the two substances. The electrochemical potential is a thermodynamic potential which governs the flow of charged particles (electrons) from one substance to another. Thus electrons flow from the material with the initially higher electrochemical potential to the material of lower potential until the potentials are equalized.

Electrochemical potential differences between two substances are typically of the order of a few tenths of an electron-volt; i.e., just after contact each electron moves from one substance to the other under "chemical forces" as if it were subjected to a voltage difference of a few tenths of a volt. (The electrochemical potential difference between two metals is equal to the difference in their *work functions*. Incidentally, two dissimilar metals may be electrified in the usual way, provided they are held with insulating handles so that they do not discharge during the process.) But the transfer of charge actually creates an electrostatic voltage difference. Thus when two dissimilar materials are brought together, sufficient charge is transferred from one to the other to establish a voltage difference of (typically) a few tenths of a volt.

A voltage difference of this order of magnitude produced by chemical forces is perhaps not surprising. But what then is the origin of the large voltages (10^4 to 10^5 volts) actually observed in static electrification experiments? These come about in the process of drawing the two materials apart, and the energy required to create these potentials is supplied by the experimenter or the machine that separates the materials.

Now, as we have seen, bringing the two dissimilar materials together causes a transfer of charge. The two materials may be regarded as the "plates" of a capacitor. Drawing the materials apart is analogous to drawing apart the plates of a capacitor at *constant charge Q*. From the basic definition of capacitance C,

$$Q = C(\Delta\varphi),$$

we see that the voltage difference $\Delta\varphi$ will increase as the capacitance of the system decreases. As an approximation, we might regard the electrified materials as forming the plates of a parallel-plate capacitor. This is perhaps not too bad an approximation so long as the separation between the plates is small. In this case (Chapter 6) the capacitance is inversely proportional to their separation. Thus in pulling the materials apart from an initial separation of the order of 10 angstrom units to a separation of one millimeter we might expect a magnification of the potential difference by a factor of 10^6.

The final voltage achieved in the process depends on the detailed geometry. This, for example, determines the capacitance of the system. Furthermore, the surface roughness determines how close the materials can be brought into contact (on the average), and therefore their initial separation; it also determines the uniformity of the charge spread over their surfaces. Concentrations of charge at sharp surface irregularities will give rise to large electric fields that may exceed the dielectric breakdown strength of air during the drawing-apart process. Thus a spark may develop, producing a partial discharge of the electrified bodies.

Answers to Odd-Numbered Problems

CHAPTER 1

1–1. $(\mathbf{A} - \mathbf{B}) \times (\mathbf{C} - \mathbf{D}) = 0$, $1:3$

1–3. $\mathbf{A} \cdot \mathbf{B} = 0$, $\mathbf{A} + \mathbf{B} = \mathbf{C}$

1–7. The angle between $\mathbf{R}$ and $\mathbf{R} - \mathbf{A}$, which is 90°, may be inscribed in a semicircle with $\mathbf{A}$ as diameter. As $\mathbf{R}$ varies, the various semicircles describe the surface of a sphere.

1–13. $\mathbf{n} = (ax\mathbf{i} + by\mathbf{j} + cz\mathbf{k})/\sqrt{a^2x^2 + b^2y^2 + c^2z^2}$

1–15. $\mathbf{V} \cdot \mathbf{F} = \dfrac{1}{r} \dfrac{\partial}{\partial r}(rF_r) + \dfrac{1}{r} \dfrac{\partial F_\theta}{\partial \theta} + \dfrac{\partial F_z}{\partial z}$

1–17. No.

CHAPTER 2

2–1. $\tan^3 \theta/(1 + \tan^2 \theta) = q^2/16\pi\epsilon_0 \, mg \, l^2$

2–3. $E = 2296$ V/m (along diagonal)

2–5. (a) $E = (\sigma/2\epsilon_0)(1 - z/\sqrt{z^2 + R^2})$

(b) $E = (\beta/2\epsilon_0)\left[\dfrac{L}{2}\left(\dfrac{L}{2} - \sqrt{\dfrac{L^2}{4} + R^2}\right) + R^2 \log\left(\dfrac{L}{2R}\sqrt{1 + \dfrac{L^2}{2R^2}}\right)\right]$

2–7. $x = \sqrt{2}\,a/(\sqrt{2} - 1)$, saddlepoint

2–9. $\varphi = (\rho/4\epsilon_0)\left[(z + \tfrac{1}{2}L)\{(z + \tfrac{1}{2}L)^2 + R^2\}^{1/2} - 2zL\right.$

$\qquad - (z - \tfrac{1}{2}L)\{(z - \tfrac{1}{2}L)^2 + R^2\}^{1/2}$

$\qquad \left. + R^2 \log\dfrac{\left|z + \tfrac{1}{2}L + \sqrt{(z + \tfrac{1}{2}L)^2 + R^2}\right|}{\left|z - \tfrac{1}{2}L + \sqrt{(z - \tfrac{1}{2}L)^2 + R^2}\right|}\right]$

2–11. (a) 300,000 V, (b) 4,730 km

2-13. 1.1×10^{-12} C/m^3, positive

2-15. (a) $\varphi = (A/\epsilon_0)(R - \frac{1}{2}r)$ for $r \leq R$

$\varphi = AR^2/2\epsilon_0 r$ for $r \geq R$

(b) $\varphi = (\rho_0/2\epsilon_0)(R^2 - \frac{1}{3}r^2)$ for $r \leq R$

$\varphi = R^3\rho_0/3\epsilon_0 r$ for $r \geq R$

2-17. $\nabla \times \mathbf{r}/r^a = 0$, $\nabla \cdot \mathbf{r}/r^a = (3 - a)/r^a$

$\rho = (3 - a)q/4\pi r^a$; $\varphi = q/4\pi\epsilon_0(a - 2)r^{a-2}$ $(a \neq 2)$

2-19. $\mathbf{E} = \dfrac{q}{4\pi\epsilon_0} \left(\dfrac{1}{r} + \dfrac{1}{\lambda} \right) e^{-r/\lambda} \mathbf{r}/r^2$

$\rho = q[\delta(\mathbf{r}) - 1/4\pi\lambda^2 r]e^{-r/\lambda}$

2-21. Treat dipole as two equal but oppositely charged point charges separated by a small distance.

2-23. $Q_{11} = Q_{22} = -2ql^2$; $Q_{33} = 4ql^2$; other components zero

2-25. (a) $\mathbf{p} = q\mathbf{l}_1 + q\mathbf{l}_2$, $p = 2ql \cos \theta/2$

(b) $q = 5.26 \times 10^{-20}$ C $= 0.328$ e

CHAPTER 3

3-1. Between: $\varphi = \dfrac{r_b\varphi_b - r_a\varphi_a + (\varphi_a - \varphi_b)r_a r_b/r}{r_b - r_a}$;

for $r > r_b$: $\varphi = \varphi_b r_b/r$

3-9. $\varphi(r, \theta) = \dfrac{p}{4\pi\epsilon_0} \dfrac{P_1 (\cos \theta)}{r^2} \left(1 - \dfrac{r^3}{a^3} \right)$, $r \leq a$

3-11. $\varphi = -(1 - a^3/r^3)E_0 r \cos \theta + Q/4\pi\epsilon_0 r$

3-13. $\sigma = -\epsilon_0 A/2r^{1/2}$ on upper surface

3-15. The mirror image of the charge distribution with ρ replaced by $-\rho$

3-17. $F = -\dfrac{q^2}{16\pi\epsilon_0 d^2} \left[\dfrac{1}{u^2} + \sum_{n=1}^{\infty} \left(\dfrac{1}{(n + u)^2} - \dfrac{1}{(n - u)^2} \right) \right]$

$\cong -\dfrac{q^2}{16\pi\epsilon_0 d^2} \left[\dfrac{1}{u^2} - \dfrac{1}{(1 - u)^2} + 1 - 2u \right]$, where $u = x/d$

3-21. $M = (x_0/a) + \sqrt{(x_0/a)^2 - 1}$, image at $x_0(M^2 - 1)/(M^2 + 1)$

3-23. $3p^2/32\pi\epsilon_0 d^4$, attraction

CHAPTER 4

4-1. $\rho_P = -2ax$; Q_P (on ends) $= A(aL^2 + b)$, $-bA$

4-3. $E_z = \dfrac{P}{2\epsilon_0} \left[\dfrac{\frac{1}{2}L - z}{\sqrt{(\frac{1}{2}L - z)^2 + R^2}} + \dfrac{\frac{1}{2}L + z}{\sqrt{(\frac{1}{2}L + z)^2 + R^2}} \right]$ outside rod

4–5. $E = (1/\epsilon_0)P \cos \gamma$

4–7. $\dfrac{\tan \theta_1}{\tan \theta_2} = \dfrac{K_1}{K_2}$

4–9. $q' = [(\epsilon_1 - \epsilon_2)/(\epsilon_1 + \epsilon_2)]q$; $q'' = 2\epsilon_2 q/(\epsilon_1 + \epsilon_2)$

4–11. $\varphi_1 = \dfrac{3}{K + 2}\dfrac{p}{4\pi\epsilon_0}\dfrac{\cos \theta}{r^2}$, $r \geq a$

$\varphi_2 = \dfrac{p}{4\pi K\epsilon_0}\left[\dfrac{\cos \theta}{r^2} + \dfrac{K - 1}{K + 2}\dfrac{2}{a^3} r \cos \theta\right]$, $r \leq a$

4–13. $E_1 = \dfrac{K_2}{K_1} E_2$, $P_1 = \dfrac{K_1 - 1}{K_1} K_2 \epsilon_0 E_2$, $\sigma_{P1} = \pm P_1$

4–15. $D = K\epsilon_0 \Delta\varphi/[K d - (K - 1)t]$

4–17. $E = Q/2\pi(\epsilon_1 + \epsilon_2)r^2$

4–19. Inside: $\mathbf{E} = -\mathbf{P}/3\epsilon_0$

outside: $E_r = \dfrac{2R^3 P}{3\epsilon_0 r^3} \cos \theta$

$E_\theta = \dfrac{R^3 P}{3\epsilon_0 r^3} \sin \theta$

CHAPTER 5

5–1. $\alpha = 9.7 \times 10^{-41}$ C $\cdot$ m^2/V; $R_0 = 0.96 \times 10^{-10}$ m

5–3. 2.6×10^{-16} m

5–5. 2.94×10^{-30} C $\cdot$ m

CHAPTER 6

6–1. 18.75 cm

6–3. $U = -\dfrac{2q^2}{4\pi\epsilon_0}\left[\dfrac{1}{l} - \dfrac{1}{2d + l} + \dfrac{1}{4d} + \dfrac{1}{4(d + l)}\right]$ (since $E = 0$ behind the plane)

6–5. $4\pi R^5 \rho_0^2/15\epsilon_0$

6–9. (a) $\Delta\varphi = 10.79$ V, $U = 1.618\ \mu$J

(b) $\Delta\varphi = 21.58$ kV, $U = 3.24$ mJ

6–11. $-(R/d)q$

6–13. $\epsilon_1 \epsilon_2/(\epsilon_1 d_2 + \epsilon_2 d_1)$

6–15. 23.8 V

6–17. $C = 4\pi\epsilon/\left(\dfrac{1}{r_1} - \dfrac{1}{r_2}\right)$

6–19. (a) $Kl(\Delta\varphi)_0/[l + (K - 1)x]$
(b) $(K - 1)Q^2 d/2\epsilon_0 w[l + (K - 1)x]^2$

6–21. $[4 \, mg/2\pi\epsilon_0(K - 1)]^{1/2}$

6–25. $F_x = \frac{1}{2}(1 - 1/K)\epsilon_0 E^2 A$

CHAPTER 7

7–1. (a) $v = 0.739 \times 10^{-3}$ m/s
(b) $\tau = 2.5 \times 10^{-14}$ s

7–3. $\varphi_{\text{int}} = \dfrac{\varphi_1 g_1(d - a) + \varphi_2 g_2 a}{g_2 a + g_1(d - a)},$

$\sigma = \dfrac{(g_1 \epsilon_2 - g_2 \epsilon_1)(\varphi_1 - \varphi_2)}{g_2 a + g_1(d - a)}$

7–5. (a) $\varphi_1 = -E_0 r \cos\theta - \frac{1}{2}E_0 a^3 \dfrac{\cos\theta}{r^2}$ outside,

$\varphi_2 = -\frac{3}{2}E_0 r \cos\theta$ inside. (b) $\sigma = -\frac{3}{2}KE_0 \cos\theta$

7–7. $I = 2\pi g \, \Delta\varphi/\ln (r_2/r_1)$

7–11. $I = \pi g s \, \Delta\varphi/\cosh^{-1} (b/2a)$

7–13. (a) 675 W/m², (b) 645 W/m²

7–15. 20 Ω

7–17. (a) $I = (\mathcal{V}_2 R_1 + \mathcal{V}_1 R_2)/(R_1 R + R_2 R + R_1 R_2)$
(b) $R_1 R_2/(R_1 + R_2)$

7–19. (a) $4R/5$ (b) $(11/20)R$

7–21. (a) $(R_4 R_5 - R_3 R_6) \mathcal{V}_1/[D + R_g(R_3 + R_4)(R_5 + R_6)]$,
where $D = R_3 R_4 R_5 + R_4 R_5 R_6 + R_5 R_6 R_3 + R_6 R_3 R_4$

7–23. One part in 4×10^6

CHAPTER 8

8–3. (a) 0.0048 cm
(b) 1.64×10^{-7} s

8–5. $f = \frac{1}{2}\mu_0 \dfrac{N}{L} I^2 = 6.28$ N/m

8–7. $B = \sqrt{3}\,\mu_0 I/\pi a$

8–9. $\mu_0 IN/4a$

8–11. $\mu_0 NI$

8–13. (a) $\dfrac{\partial B_r}{\partial z} = \dfrac{\partial B_z}{\partial r}$

8–15. $\nabla \times \nabla \times \mathbf{B} = \mu_0 \nabla \times \mathbf{J} = 0$

8–17. $B = \frac{1}{2}\mu_0 J s$

8–19. $B = \mu_0 NI/2\pi r, \ b/a = 4/3$

8–21. $A_z = (\mu_0 I/2\pi) \ln (r/b)$ between the conductors

8–23. (b) $\nabla \times (\mathbf{A} + \nabla\psi) = \nabla \times \mathbf{A}$

(c) $\nabla \cdot (\mathbf{A} + \nabla\psi) = f = \nabla^2\psi, \ \psi = -\frac{1}{4\pi} \int \frac{f}{R} \, dv$

8–25. φ^* is not single valued.

8–27. (d) $B_r = (\mu_0 I/2a) \left[\cos\theta - \frac{3r^2}{4a^2} (5\cos^3\theta - 3\cos\theta) + \cdots \right]$

$B_\theta = (\mu_0 I/2a) \left[-\sin\theta + \frac{3r^2}{4a^2} (5\cos^2\theta - 1)\sin\theta + \cdots \right]$

CHAPTER 9

9–1. $\mathbf{J}_M = \nabla \times \mathbf{M} = 0$

$\mathbf{j}_M = \mathbf{M} \times \mathbf{n}; j_M = M$ on cylindrical surfaces, $j_M = 0$ on sides

9–3. (b) $(4/3)\pi R^3\mathbf{M}_0$

9–5. $\sigma_M = M_0 x/[x^2 + (b^4/a^4)(y^2 + z^2)]^{1/2}$

$\rho_M = 0$

9–7. (b) $B_z = \frac{1}{2}\mu_0 M \left[\dfrac{\frac{1}{2}L - z}{\sqrt{(\frac{1}{2}L - z)^2 + R^2}} + \dfrac{\frac{1}{2}L + z}{\sqrt{(\frac{1}{2}L + z)^2 + R^2}} \right]$

9–9. (a) 0.25 T

(b) 0.95 T

(c) 1.52 T

9–11. $\mathbf{B}_i = 2(\mu/\mu_0)\mathbf{B}_0 /(1 + \mu/\mu_0)$

9–13. $J_M = 0, j_M = \dfrac{\chi I}{2\pi a}$ (inner surface), $j_M = \dfrac{-\chi I}{2\pi b}$ (outer surface)

$I_M = \chi I$ (inner), $I_M = -\chi I$ (outer)

9–15. (a) $\theta_1 = 0, I$ parallel; (b) $\theta_1 = 90°, I$ antiparallel

9–19. (a) Sintered oxide 0.4 T

35 percent Co steel 0.22 T

(b) Sintered oxide 0.53 T

35 percent Co steel 0.96 T

9–21. 0.64 T

CHAPTER 10

10–3. 3.69×10^{-4}

10–5. $\gamma = 976$

CHAPTER 11

11-5. $\mathscr{V} = \mathscr{V}_0 \cos \omega t$, $\mathscr{V}_0 = 2\pi f NAB = 12.56$ V

11-7. (a) From b to a
(b) From b to a
(c) From a to b

11-9. $\frac{1}{2}B^2 a^2 r^2 \omega g t$

11-11. $(\mu_0 L/2\pi) \ln (R_2/R_1)$

11-13. $\mu_0 \pi a^2 b^2 / 4r^3$

11-15. $M = (\mu_0 h/2\pi) \ln (1 + d/r)$

11-17. $(\mu_0/\pi) \ln (d/a)$

11-23. $E = Kr/2$ inside, $E = Ka^2/2r$ outside

11-25. Let $\nabla \times \mathbf{B} = \alpha \mathbf{B}$. $\mathbf{J} = \mathbf{J}_0 \, e^{-(\alpha^2/g\mu_0)t}$

CHAPTER 12

12-3. $U = \dfrac{1}{s + 2} [L_1 I_1^2 + 2M_{12} I_1 I_2 + L_2 I_2^2]$

12-5. $\dfrac{N^2 I^2 A \mu^2}{\mu_0 (l + d)^2}$

12-7. (a) $F = B_0^2 \chi_m A/2\mu_0 (1 + \chi_m)$
(b) 1.76×10^{-4} N

12-11. (a) and (b) $\frac{1}{2}\mu_0 \dfrac{N}{l} I^2$

12-13. $F = -\frac{1}{2}\mu_0 N^2 I^2 \left(\dfrac{b}{\sqrt{b^2 - a^2}} - 1 \right)$; collapse

12-15. $I = I_0 \dfrac{L}{L + M}$

12-17. Commercial iron: 0.018 W/cm^3
Tungsten steel: 0.395 W/cm^3

12-19. $-d(\mathbf{m} \cdot \mathbf{B})$

CHAPTER 13

13-1. (a) $I = 0.605$ A,
$dI/dt = 1.59$ A/s
(b) $I = 1.295$ A,
$dI/dt = 0.558$ A/s
(c) $I = 1.663$ A,
$dI/dt = 0.0062$ A/s

13-3. $Q = C\mathscr{V}_0[1 - e^{-t/RC}]$

13–5. $|Z| = \sqrt{\dfrac{R^2(1 - \omega^2 LC)^2 + (\omega L)^2}{1 + (\omega RC)^2}}$; $|Z| = R$ for $\omega = 0$;

$|Z| \cong L/RC$ for $\omega = 1/\sqrt{LC}$; $|Z| \cong \omega L$ for $\omega \to \infty$.

13–7. $Z = \dfrac{R\alpha - \omega C'R(\omega L - 1/\omega C) + i[\omega C'R^2 + (\omega L - 1/\omega C)\alpha]}{\alpha^2 + \omega^2 C'^2 R^2}$,

where $\alpha = 1 + \omega^2 C'L - C'/C$

13–9. (a) 3.2×10^{-2} deg
   (b) zero to 1.8 MHz
13–13. (a) $f = 1.78 \times 10^3$ Hz
   (b) $f = 1.78 \times 10^3$ Hz
   (c) $f = 0.796 \times 10^3$ Hz
13–15. (a) $L/C = 2R^2$
   (b) $L = \sqrt{2}\,R/\omega_c$; $C = 1/\sqrt{2}\,R\omega_c$
13–17. $0.0713 - 0.0034i$ mA
13–19. $1/\sqrt{3LC}$
13–21. $V_1 = -100 - 700i$ V
   $V_2 = 150 - 750i$ V

CHAPTER 14

14–5. $F_1 = (2/5)\pi a^3 \sigma B_0^2 v_0$

CHAPTER 15

15–1. $\mathbf{B} = B_0 \mathbf{i} - B_0(a/r)^2(\mathbf{a}_r \cos\theta + \mathbf{a}_\theta \sin\theta)$ outside
   $\mathbf{B} = \mathbf{H} = 0$ inside
   $\mathbf{j}_S = -2B_0\mu_0^{-1} \sin\theta\,\mathbf{k}$
15–3. $\mathbf{B} = B_0\mathbf{i} - b(a/r)^2(\mathbf{a}_r \cos\theta + \mathbf{a}_\theta \sin\theta)$ outside
   $\mathbf{B} = c[\mathbf{a}_r(\lambda/r)I_1(r/\lambda)\cos\theta - \mathbf{a}_\theta I_1'(r/\lambda)\sin\theta]$ inside, with I_1
   modified Bessel function of the first kind, and I_1' its first derivative

$$b = B_0 \frac{[I_1'(a/\lambda) - (\lambda/a)I_1(a/\lambda)]}{I_1'(a/\lambda) + (\lambda/a)I_1(a/\lambda)}$$

$$c = 2B_0/[I_1'(a/\lambda) + (\lambda/a)I_1(a/\lambda)]$$

15–5. For $r = a - \delta$ with $\delta \ll a$ and $\lambda \ll a$
   $u(r) = 3B_0(\lambda/a)e^{-\delta/\lambda}$
   $v(r) = -(3/2)B_0\,e^{-\delta/\lambda}$

CHAPTER 16

16–1. (a) $Q = C(\Delta\varphi)e^{-gt/\epsilon}$
   (b) $-(g/\epsilon)C(\Delta\varphi)e^{-gt/\epsilon}$
   (c) Zero

16-3. (a) $H = \frac{1}{2}a\, \partial D/\partial t = I/2\pi a$
(b) $S = \frac{1}{2}aE\, \partial D/\partial t$, inward
(c) $\pi a^2\, dE\, \partial D/\partial t$

16-7. $\mathbf{B} = -\mathbf{i}E_0\sqrt{\epsilon\mu}\,\sin\omega(\sqrt{\epsilon\mu}\,z - t) + \mathbf{j}E_0\sqrt{\epsilon\mu}\,\cos\omega(\sqrt{\epsilon\mu}\,z - t)$

$\mathbf{S} = \mathbf{k}\sqrt{\dfrac{\epsilon}{\mu}}\,E_0^2$

16-9. $\mathbf{A} = -\mathbf{i}(\lambda E_0/2\pi c)\,\cos\left[2\pi(z - ct)/\lambda\right]$

16-11. $\mathbf{V}\cdot\mathbf{A} = -\epsilon\mu\,\dfrac{\partial\varphi}{\partial t} - \mu g\varphi$

CHAPTER 17

17-5. $E = 700$ V/m; $B = 2.33 \times 10^{-6}$ T, rms

17-7. $\bar{S} = \dfrac{1}{c\mu_0}\,\overline{E^2} = \dfrac{1}{c\mu_0}\,\frac{1}{2}(E_1^2 + E_2^2 + 2E_1 E_2\,\cos\phi)$

17-11. $n = 1.099\sqrt{K_r}$, $k = 0.455\sqrt{K_r}$; $\delta/\lambda = 0.384$
17-13. (b) $g = 0.92 \times 10^{-6}$ $(\Omega\text{m})^{-1}$

CHAPTER 18

18-1. $r_{12s} = \dfrac{1 - n^2}{1 + n^2}$; for $n = 1.5$, $R_s = 14.8$ percent

18-3. $R_p \cong 1 - 4K\sqrt{\dfrac{2\delta}{\sqrt{K - 1}}}$

18-5. $R_p \cong \dfrac{(n\cos\theta_1 - 1)^2 + k^2\cos^2\theta_1}{(n\cos\theta_1 + 1)^2 + k^2\cos^2\theta_1}$

$\cos\theta_1 = 1/\sqrt{n^2 + k^2}$; $\theta_1 = 80.5°$, $R_p = 72$ percent
18-9. $K_i = 1.13$; $n \cong 3$, $k \cong 0.188$; $\alpha = 2.7°$
18-13. $\hat{r}_{23} = \pm 1$, $R = 1$

18-15. $R = \dfrac{R_{12} + R_{23} - 2R_{12}R_{23}}{1 - R_{12}R_{23}}$; for $R_{12} = R_{23}$, $R = \dfrac{2R_{12}}{1 + R_{12}}$

18-17. $B_x = B_1\cos(\kappa y\cos\theta)e^{i(\kappa z\sin\theta - \omega t)}$
$E_y = -cB_1\sin\theta\cos(\kappa y\cos\theta)e^{i(\kappa z\sin\theta - \omega t)}$
$E_z = icB_1\cos\theta\sin(\kappa y\cos\theta)e^{i(\kappa z\sin\theta - \omega t)}$

18-19. $\dfrac{\lambda}{2} < a < \dfrac{\lambda}{\sqrt{2}}$

CHAPTER 19

19–1. $n_v - 1 = 1.63 \times 10^{-4};\ 1.36 \times 10^{-3}$
19–3. $\lambda_0 = 1560$ Å
19–7. $k \cong K_i/2n_\infty,\ \alpha = 2k\omega_0/c$
19–9. $n \cong k \cong 0.071$
19–13. $\omega = \sqrt{K_0/\tau};\ K_r \cong 1,\ K_i \cong \sqrt{K_0}$
19–15. $P = \chi_0 E_0 \sin \omega_0 t$

CHAPTER 20

20–3. $$B_\theta = \frac{\mu_0 I_0}{4\pi} A \frac{\omega^2}{c^2 r} \sin \theta \cos \omega \left(t - \frac{r}{c}\right)$$

$$E_\phi = -\frac{I_0 A}{4\pi\epsilon_0} \frac{\omega^2}{c^3 r} \sin \theta \cos \omega \left(t - \frac{r}{c}\right),$$

where A is the area of the circular loop.

$$P = \frac{\mu_0}{6\pi} I_0^2 A^2 \frac{\omega^4}{c^3} \cos^2 \omega \left(t - \frac{r}{c}\right)$$

20–5. (a) 5.23 A; (b) 2.45×10^{-2} V/m
20–7. $\frac{3}{2}$
20–9. (a) $8\pi/3(v/c)^3$; (b) 4.07×10^{-7}
20–11. 52.1 m^{-1}

CHAPTER 22

22–3. $\dfrac{\mathbf{u}}{1 + u^2/c^2} = \dfrac{\mathbf{E} \times \mathbf{B}}{(E^2/c^2 + B^2)}$

Index